Springer Series in Advanced Manufacturing

Xun Xu · Andrew Y.C. Nee

Advanced Design and Manufacturing Based on STEP

Springer

Assoc.Prof. Xun Xu
University of Auckland
Dept. Mechanical Engineering
Auckland 1142
New Zealand
x.xu@auckland.ac.nz

Prof. Andrew Y.C. Nee
National University of Singapore
Dept. of Mechanical Engineering
Singapore 119260
Singapore
mpeneeyc@nus.edu.sg

ISSN 1860-5168
ISBN 978-1-84882-738-7 e-ISBN 978-1-84882-739-4
DOI 10.1007/978-1-84882-739-4
Springer Dordrecht Heidelberg London New York

British Library Cataloguing in Publication Data
A catalogue record for this book is available from the British Library
Library of Congress Control Number: 2009935947

Cover design: eStudioCalamar, Figueres/Berlin

Printed on acid-free paper

Springer is part of Springer Science+Business Media (www.springer.com)

Preface

Design and manufacturing is the essential element in any product development lifecycle. Industry vendors and users have been seeking a common language to be used for the entire product development lifecycle that can describe design, manufacturing and other data pertaining to the product. Many solutions were proposed, the most successful being the **St**adndard for **E**xchange of **P**roduct model (STEP). STEP provides a mechanism that is capable of describing product data, independent from any particular system. The nature of this description makes it suitable not only for neutral file exchange, but also as a basis for implementing, sharing and archiving product databases. ISO 10303-AP203 is the first and perhaps the most successful AP developed to exchange design data between different CAD systems. Going from geometric data (as in AP203) to features (as in AP224) represents an important step towards having the right type of data in a STEP-based CAD/CAM system. Of particular significance is the publication of STEP-NC, as an extension of STEP to NC, utilising feature-based concepts for CNC machining purposes.

The aim of this book is to provide a snapshot of the recent research outcomes and implementation cases in the field of design and manufacturing where STEP is used as the primary data representation protocol. The 20 chapters are contributed by authors from most of the top research teams in the world. These research teams are based in national research institutes, industries as well as universities.

As the title suggests, the first chapter gives an overview of STEP. It touches briefly on the history and objectives of STEP. This is followed by discussions on technical details of information modelling, data representation, the STEP integrated resources, application protocols (APs), and STEP-NC. The chapter closes with some comments on the future of STEP. The appendix at the end gives some additional sources of information about STEP.

The second chapter discusses the four STEP-related process planning languages. The first two of these are STEP standard process planning languages, i.e. all of AP 240 and part of AP 238. The third is part of ISO 14649. The fourth, called FBICS-ALPS, is a version of the ALPS language adapted for FBICS, a system that does automatic feature-based process planning. The machining features available in AP 238, ISO 14649, and AP 224 are summarised. The way in which FBICS does feature-based process planning is also presented and a system is described that translates FBICS plans into ISO 14649 plans.

Also targeting at process planning, Chapter 3 presents an approach for enabling automatic generation of STEP-NC based Workplans using graph theories. Workplan is mapped into a directed graph and then the shortest path and a Hamiltonian Path (HP) inside this directed graph as optimal sequenced solution are found. Thus, the

Workplan is structured and reordered accordingly. Finally, the corresponding NC machining codes are generated and distributed to the machine tool. The focus of this chapter is the investigation of heuristic algorithms in order to sequence STEP-NC machining operations.

Chapter 4 describes the newly developed STEPNC++ – an effective tool for feature-based CAM/CNC. STEPNC++ can realize direct translation of feature-based CAM files into feature-based conversational CNC part program files. The chapter argues that a much greater benefit can be found using STEP-NC to reduce the cost associated with the impedance mismatch between CAM and CNC resulting in the prohibitive loss of CAM process information and a CNC trapped into a motion-based machining realm. This is evidenced by a demonstration system that incorporates part representation using STEP-NC Part 21 files, reading and analysing feature-based elements of the STEP-NC, translation into a generic feature-based canonical representation, and generation of actual CNC programs relying on conversational programming "Canned Cycles".

With a focus on turning operations, Chapter 5 explores how ISO 14649 can be used to combine turning and milling operations to support interoperable CNC manufacturing of rotational asymmetric components at a single turning centre. The major contribution is the creation of a computational environment for a STEP-NC compliant system for turning operations, namely SCSTO. SCSTO is the experimental part of the research, supported by specification of information models and constructed using a structured methodology and object-oriented methods.

While the STEP-NC standard has shown its benefits for conventional metal cutting, Chapter 6 discusses applications of STEP-NC in a stone cutting process, which features outsourcing, collaboration and shop-floor integration in today's stone manufacturing arena. The chapter also discusses how STEP and STEP-NC standards may be extended to accommodate saw blade stone cutting technology.

An open, STEP-compliant CNC software and hardware platform is presented in Chapter 7. The chapter analyses the requirements of such a platform. The software structure of the platform contains a Decision Unit (consisting of intelligent control rules and algorithms), Function Description Data (consisting of NC functions) and System Engine. With this software architecture, a corresponding hardware platform is designed, where an Ethernet-based industrial fieldbus is used.

Staying with machining but with a goal to optimise the process, Chapter 8 presents a method to manage cutting forces based on the cross-sectional geometry of each tool-path over the course of the machining process. ISO 10303 AP 238 is used as the data structure to implement the tool-path geometry information into the process optimisation. The chapter explains the fundamentals of cutting force calculations, the tool path cross-sectional geometry in machining operations and its parameterisation in ISO 10303 AP 238.

In an effort to achieve a STEP-NC enabled NC programming environment, Chapter 9 proposes an implementation method. In the second half of the chapter, the STEP-NC Platform for Advanced and Intelligent Programming (SPAIM) is presented. This platform controls current industrial machine tools directly from STEP-NC files. It also includes new tool-path programming methods, such as pattern strategies for trochoidal milling and plunging tool-paths. The platform demonstrates the benefits of STEP-NC for industry and also forms a basis for future STEP-NC related research and validation.

Chapter 10 discusses the current status of STEP-compliant CNC systems, and also predicts the future directions of the research. The concept of the future digital factory is presented, which requires so-called "Smart and Ubiquitous NC machining workstations" that represent the brain and knowledge repository of the manufacturing system. Standardisation of the whole process is emphasised and the use of STEP and related standards to form an interoperable and adaptable solution across varied equipments and devices is suggested as a key enabling tool.

Extending the scope to process control, Chapter 11 presents a STEP-compliant process control solution called Renishaw Productivity+. The authors use STEP-NC standards to integrate machining and inspection processes for CNC machine tools. Based on the standards, a process control information model for CNC manufacturing is specified. The final part of the chapter describes a standardised process control system together with a computational prototype based on this system and its application with a simple piece.

In Chapter 12, a STEP-NC compliant methodology for representing technological manufacturing resources is presented. This modelling approach is based on mechanical elements that constitute machine tools and other manufacturing hardware together with their kinematic links. It is developed with a focus on supporting process planning decisions. Models for various types of machines are presented at the end of the chapter to highlight the flexibility of this approach in modelling manufacturing resources.

A similar piece of research work is presented in Chapter 13, where modelling and implementation of a digital semantic machining model is proposed. The model is based on STEP technology, which includes representation and quality inspection of shape data, feature model and feature recognition process model, and feature-based CNC machining process model.

In support of decentralised manufacturing, Chapter 14 presents a procedure for the design of decentralised STEP-NC compliant manufacturing solutions. Considering the problem of dispatching and managing a network of STEP-NC compliant machines, a three-tiered architecture is proposed to deal with remote machining requests. An ad hoc STEP-NC Network Management Protocol describes the exchange of messages between tiers.

Moving into product development supporting tools, Chapter 15 proposes a Generic Product Modelling Framework (GPMF) to overcome the problems of information exchange and sharing. This framework uses STEP as a foundation, and consists of four functional components: an EXPRESS Data Model, a STEP–based modelling environment, a "five-phase" modelling method, and three EDM data exchange and sharing methods. The case study demonstrates the capability of integrating information in product design, manufacturing and assembly.

STEP standards have also been extensively used in the domain of Product Data Management (PDM). Chapter 16 discusses the schemas that deal with PDM and are spread over several Application Protocols of STEP. These schemas are collectively known as STEP PDM Schema. The chapter also describes the fast growing Web services over the Internet that are also built upon the STEP standards.

Extending Product Data Management to Product Lifecycle Management (PLM), Chapter 17 explores the suitability of STEP to support PLM. It summarises the relevant standards and discusses their potentials to support PLM. The chapter focuses on engineering change management. In order to capture the change evaluation data, STEP-compliant extensions to the existing engineering change data model are proposed.

Chapter 18 discusses the prospective of using STEP to coordinate data exchange in an extended enterprise. EXPRESS data models, Business Processes and agent technology are combined and systems for coordinating data sharing along the networked enterprise are constructed. STEP-based repositories are built to store and retrieve information about them.

The issue of heterogeneous object models is discussed in Chapter 19. This is about transporting and exchanging the included geometry, topology as well as material distribution between CAD modellers, CAE tools and CAM facilities. The chapter focuses on an XML implementation for data exchanges. A prototype CAD module is developed to construct an XML-based heterogeneous material model, and the XML model is then exported to SolidWorks to test the validity of the proposed approach.

The last chapter presents a module-based platform for seamless interoperable CAD-CAM-CNC planning. With an extension of the STEP standards for CAM related data and functional interfaces, this framework is able to embed software modules that encapsulate specific functionalities. Its realisation relies on known techniques from service-oriented architecture. The developed platform can be used for simulation systems or process data acquisition, which can further improve NC planning processes.

This book also contains an Appendix that summarises some of the commonly used tools for working with STEP data. Many of these tools are Open Source or freeware. These tools can support STEP data generation, validation, conversion and interpretation. One of the major functions of these tools is the ability to compile EXPRESS schemas into other computer languages. There are tools that can produce the equivalent EXPRESS-G diagram given an EXPRESS schema. There are also tools for visualising STEP data.

Industrial use of STEP has shown evidence of significant cost savings, higher quality, and reduced time-to-market. Its important function will only gain more attention and see more implementation in the rapidly changing economy that is increasingly globalised, collaborative, distributed, interoperable and integrated. One thing is certain – STEP is more than an international standard for exchanging product data. It is about design reuse, data archiving, and solving challenging manufacturing and business problems.

Taking this opportunity, the editors would like to express their deep appreciation to all the authors for their significant contributions to this book. Their commitment, enthusiasm, and technical expertise made this book what it is. We are also grateful to the publisher for supporting this project. The authors are thankful to the International Organisation for Standardisation (ISO) for allowing the authors to quote some of the content of the STEP standards. The authors are also grateful to Dr. Matthieu Rauch for his assistance in compiling and formatting the book. It is our sincere hope that readers will find this book both informative and thought-provoking.

Auckland, New Zealand *Xun Xu*

Singapore *Andrew Y.C. Nee*

2009

Contents

List of Contributors

Kiyoshi Akama
Graduate School of Information Science and Technology
Hokkaido University
Kita-14 Nishi-9, Sapporo
Hokkaido
Japan

Klaus-Peter Arnold
Automation Technology
Brandenburg University of Technology
Siemens-Halske-Ring 14, Cottbus
Germany

Ulrich Berger
Automation Technology
Brandenburg University of Technology
Siemens-Halske-Ring 14, Cottbus
Germany

Christian Brecher
Werkzeugmaschinenlabor (WZL)
RWTH Aachen University
Steinbachstraße 19, 52074 Aachen
Germany

Amedeo Buonanno
Senseable City Laboratory
Massachusetts Institute of Technology
77 Massachusetts avenue
Cambridge 02139, MA
USA

Francesco Calabrese
Senseable City Laboratory
Massachusetts Institute of Technology
77 Massachusetts avenue
Cambridge 02139, MA
USA

Keith Case
Wolfson School of Mechanical and Manufacturing Engineering
Loughborough University
Leicestershire LE11 3TU
UK

Wanlin Chen
Department of Mechanical Engineering
School of Engineering
University of Auckland
20 Symonds Street, Auckland
New Zealand

Debasish Dutta
Department of Mechanical Science and Engineering
University of Illinois at Urbana-Champaign
204 Coble Hall, MC-322, 801 S. Wright St., Champaign, IL 61820
USA

Julio Garrido Campos
Automation and Systems Engineering Department
University of Vigo
E.T.S.I. Industriales, 36200 Vigo
Spain

Jean-Yves Hascoet
Modelisation Optimisation Process Production
Research Institute of Communications and Cybernetics of Nantes (IRCCyN)
UMR CNRS 6597, 1 rue de la Noe,
BP92101, 44321 Nantes Cedex 03
France

Tianliang Hu
School of Mechanical Engineering
Shandong University
27 Jingshi Road, Jinan 250061
People's Republic of China

Takeshi Kishinami
Kushiro National College of Technology
2-32-1, Otanoshike, Kushiro
Hokkaido
Japan

Tsukasa Kondo
Department of Mechanical Engineering
Hakodate National College of Technology
14-1, Tokura-Cho, Hakodate
Hokkaido
Japan

X.Y. Kou
Department of Mechanical Engineering
The University of Hong Kong
Pokfulam Road
Hong Kong
People's Republic of China

Thomas Kramer
Intelligent Systems Division, MS8230
National Institute of Standards and Technology
100 Bureau Drive, Gaithersburg
MD 20899
USA

Ralf Kretzschmann
Automation Technology
Brandenburg University of Technology
Siemens-Halske-Ring 14, Cottbus
Germany

Sanjeev Kumar
Innovative Design and Manufacturing Research Centre
Department of Mechanical Engineering
University of Bath
Claverton Down, Bath, BA2 7AY
United Kingdom

Raphael Laguionie
Modelisation Optimisation Process Production
Research Institute of Communications and Cybernetics of Nantes (IRCCyN)
UMR CNRS 6597, 1 rue de la Noe,
BP92101, 44321 Nantes Cedex 03
France

Riliang Liu
School of Mechanical Engineering
Shandong University
27 Jingshi Road, Jinan 250061
People's Republic of China

Wolfram Lohse
Werkzeugmaschinenlabor (WZL)
RWTH Aachen University
Steinbachstraße 19, 52074 Aachen
Germany

Karla López de la Cruz
Centro de Investigación en Tecnologías de Información y Sistemas
Instituto de Ciencias Básicas e Ingeniería
Universidad Autónoma del Estado de Hidalgo
Carretera Pachuca – Tulancingo Km. 4.5,
Pachuca, Hidalgo
México

Omar López-Ortega
Centro de Investigación en Tecnologías de Información y Sistemas
Instituto de Ciencias Básicas e Ingeniería
Universidad Autónoma del Estado de Hidalgo
Carretera Pachuca – Tulancingo Km. 4.5,
Pachuca, Hidalgo
México

Chandresh Mehta
Department of Mechanical Engineering
University of Michigan
2250 G.G. Brown, 2350 Hayward St
Ann Arbor, MI 48105
USA

John Michaloski
Intelligent Systems Division, MS8230
National Institute of Standards and Technology
100 Bureau Drive, Gaithersburg
MD 20899
USA

Satoshi Mitsui
Department of Information System Engineering
Asahikawa National College of Technology
2-2-1-6, Shunkoudai, Asahikawa
Hokkaido
Japan

Aydin Nassehi
Innovative Design and Manufacturing Research Centre
Department of Mechanical Engineering
University of Bath
Claverton Down, Bath, BA2 7AY
United Kingdom

Stephen T. Newman
Innovative Design and Manufacturing Research Centre
Department of Mechanical Engineering
University of Bath
Claverton Down, Bath, BA2 7AY
United Kingdom

Van Khai Nguyen
CADCAMation SA
103 route de Chancy, 1213 Onex
Geneva
Switzerland

David Odendahl
The Boeing Company
P.O. Box 3707, Seattle
Washington 98124-2207
USA

Masahiko Onosato
Graduate School of Information Science and Technology
Hokkaido University
Kita-14 Nishi-9, Sapporo
Hokkaido
Japan

Lalit Patil
Department of Mechanical Engineering
University of Michigan
2250 G.G. Brown, 2350 Hayward St
Ann Arbor, MI 48105
USA

Frederick Proctor
Intelligent Systems Division, MS8230
National Institute of Standards and Technology
100 Bureau Drive, Gaithersburg
MD 20899
USA

Matthieu Rauch
Modelisation Optimisation Process Production
Research Institute of Communications and Cybernetics of Nantes (IRCCyN)
UMR CNRS 6597, 1 rue de la Noe,
BP92101, 44321 Nantes Cedex 03
France

Vijay Srinivasan
IBM and Columbia University
New York
N.Y.
U.S.A.

John Stark
CADCAMation SA
103 route de Chancy, 1213 Onex
Geneva
Switzerland

S.T. Tan
Department of Mechanical Engineering
The University of Hong Kong
Pokfulam Road
Hong Kong
People's Republic of China

Fumiki Tanaka
Graduate School of Information Science and Technology
Hokkaido University
Kita-14 Nishi-9, Sapporo
Hokkaido
Japan

Sid Venkatesh
The Boeing Company
P.O. Box 3707, Seattle
Washington 98124-2207
USA

Parag Vichare
Innovative Design and Manufacturing
Research Centre
Department of Mechanical Engineering
University of Bath
Claverton Down, Bath, BA2 7AY
United Kingdom

Mirco Vitr
Werkzeugmaschinenlabor (WZL)
RWTH Aachen University
Steinbachstraße 19, 52074 Aachen
Germany

Shane Q. Xie
Department of Mechanical Engineering
School of Engineering
University of Auckland
20 Symonds Street, Auckland
New Zealand

Liangji Xu
The Boeing Company
P.O. Box 3707, Seattle
Washington 98124-2207
USA

Xun Xu
Department of Mechanical Engineering
School of Engineering
University of Auckland
20 Symonds Street, Auckland
New Zealand

Makoto Yamada
Department of Mechanical Engineering
Hakodate National College of
Technology
14-1, Tokura-Cho, Hakodate
Hokkaido
Japan

Lin Yang
School of Mechanical Engineering
Shandong University
27 Jingshi Road, Jinan 250061
People's Republic of China

Yusri Yusof
Department of Mechanical and
Manufacturing Engineering
University of Tun Hussein Onn Malaysia
(UTHM)
Parit Raja, 86400 Johor
Malaysia

Chengrui Zhang
School of Mechanical Engineering
Shandong University
27 Jingshi Road, Jinan 250061
People's Republic of China

1

STEP in a Nutshell

Thomas Kramer[1] and Xun Xu[2]

[1] Intelligent Systems Division, MS8230, National Institute of Standards and Technology
100 Bureau Drive, Gaithersburg, MD 20899, USA
Email: thomas.kramer@nist.gov

[2] Department of Mechanical Engineering, University of Auckland, 20 Symonds Street, Auckland, New Zealand
Email: x.xu@auckland.ac.nz

Abstract
This chapter gives an overview of STEP (Standard for the Exchange of Product model data), assuming that the reader may not be familiar with STEP. The chapter gives a short example of an industrial use of STEP. Then it touches briefly on the history and objectives of STEP, followed by the organization of the Parts of STEP. Next, it goes into technical details of (1) information modelling using EXPRESS and EXPRESS-G, (2) data representation using STEP Part 21 files and other methods, (3) the STEP integrated resources, (4) application protocols (APs), including application activity models (AAMs), application reference models (ARMs), units of functionality (UOFs), application interpreted models (AIMs), application interpreted constructs (AICs), application modules (AMs), and conformance classes, and (5) STEP-NC. The chapter closes with some comments on the future of STEP. The Appendix at the end gives some additional sources of information about STEP.

1.1 Introduction

STEP, the **S**tandard for the **E**xchange of **P**roduct Model Data, is a large and powerful set of ISO (International Organisation for Standardisation) standards, all under ISO 10303. The overall objective of STEP is to provide a mechanism that describes a complete and unambiguous product definition throughout the life cycle of a product. STEP provides both broadly useful data modelling methods and data models focused on specific industrial uses. The STEP standards contain several dozen separate documents.

Here is an example typical of the use of STEP today. An automobile engine designer working with a commercially available computer-aided design (CAD) system designs an engine block. The CAD system's native representation of the design is proprietary to the vendor of the system, but a STEP output module has been included within the CAD system that translates the proprietary representation into a representation using the STEP application protocol for configuration controlled design (AP 203) [1.22]. The AP 203 representation is saved in a STEP data file using Part 21 of STEP [1.10]. The engine block design is sent to a manufacturing plant by sending the STEP Part 21 file for the design. At the manufacturing plant, a manufacturing engineer using a CAD system from a different vendor tells the CAD system to read the STEP file. This is possible because the second CAD vendor has also implemented STEP AP 203. The system has a module that can read the STEP file and build a representation of the design in the second CAD system's native format. With the design now resident in the CAD system, the manufacturing engineer goes to work figuring out how to manufacture the engine block. If the manufacturing engineer wants to suggest a change in the design (and the second CAD system includes a STEP output module), he or she can have the CAD system write a STEP AP 203 Part 21 file and send it back to the designer. It is also possible to use STEP to communicate design information at the feature level (AP 224 [1.25]) and manufacturing information at the operation level (AP 238 [1.26]).

It is noted that this chapter is only intended to provide a snapshot of the STEP standards. Discerning readers are referred to the Appendix at the end of this chapter for further information.

1.2 History of STEP

STEP evolved from earlier efforts in building data standards for CAD, particularly the Initial Graphics Exchange Specification (IGES), the first version of which was released in 1980. Many of the organizations and people who worked on IGES and were aware of its weaknesses collaborated in building the next generation product data standard, PDES, the Product Data Exchange Specification, which first appeared in 1984. IGES and PDES were U.S. national efforts, but similar work was being done in France, Germany, and the UK. In 1984, ISO TC184/SC4 (Technical Committee 184, Subcommittee 4 – generally known as SC4) resolved to try to build a standard for computerized product model data. Much of the expertise that had gone into building IGES and starting PDES was transferred to the new international effort that became known as STEP. The first major release of proposed STEP documents occurred in 1988, at which time a large set of models had been assembled into a single "Integrated Product Information Model" (IPIM). Half a dozen documents from the IPIM were adopted as initial drafts of ISO standards at an SC4 meeting in Tokyo in late 1988.

The methodology and architecture of STEP evolved continuously during its early years (and indeed, it is still evolving). By 1989, STEP had focused on the concept of application protocol (AP) as a subset of STEP that would be intended for a specific industrial use and could be implemented and subjected to conformance testing. The architecture for APs was developed in the following few years. To meet industrial needs for information exchange, additional APs have been started from time to time ever since the AP concept came into existence.

The first version of STEP to become an ISO standard was adopted in 1994 and companies such as GE, Boeing, and General Motors began announcing commitments to using STEP in 1995.

It was apparent from the beginning of the use of application protocols that developing and using them was difficult. In addition, it frequently occurred that different application protocols would use the same type of information (particularly geometry and topology). This led to the development of methods of modularizing APs. With modularization, a new AP can be formed by combining existing modules and (usually) adding one or more new application-specific modules. As of June 2008, the SC4 web site (http://www.tc184-sc4.org/) lists 23 application protocols that have become international standards. Four of these were first approved between 2003 and 2007. Eight other APs are still in progress. SC4 meetings continue to be held at which new APs and new editions of existing STEP models of all types are developed.

The STEP modelling methods of EXPRESS [1.9] and Part 21 are broadly usable. Other standards efforts have adopted them. For example, the Parts Library standards, ISO 13584 [1.29], and the standards for Data Model for Computerized Numerical Control, ISO 14649 [1.30], both use EXPRESS and Part 21 files. This enables the integration of other standards with STEP.

Feature-based machining and inspection is covered by STEP Part 238, as is described in detail in Section 1.10. The work has many antecedents in North America, Europe, and Asia extending back over two decades. However, the development in Europe of ISO 14649 in the 1990s was the key to the efforts that culminated in AP 238. Like ISO 10303, ISO 14649 is made up of separate Parts. Several of these Parts were available for comment by 1998. By 2004, second editions of the Parts of ISO 14649 [1.31-1.34] were becoming international standards. Several Parts of ISO 14649 were adopted as conceptual models by the ISO team developing STEP Part 238 in the early 2000s. Development of both ISO 14649 and STEP Part 238 continues today. Both of them are commonly known as "STEP-NC".

1.3 Objectives of STEP

The purpose of STEP, as stated in the document giving the fundamental principles of STEP [1.8], is "to specify a form for the representation and unambiguous exchange of computer-interpretable product information throughout the life of a product. The form is independent of any particular computer system. Later in the purpose section, it is stated that STEP "permits different implementation methods to be used for storing, accessing, transferring, and archiving product data."

We may distinguish "exchange" (which is accomplished by sending a file from one place to another) from "sharing" (which occurs when two or more parties work on the same file or data base at the same time). From the beginning of STEP, many STEP developers have held that supporting sharing is essential. Although sharing is not among the purposes stated in the fundamental principles document, the introduction to that document says that STEP will be "suitable not only for neutral file exchange, but also as a basis for implementing and sharing product data bases."

Additional objectives of the original developers of STEP (as reported in [1.6]) included the creation of a single international standard covering all aspects of computer-aided design and manufacturing (CAD/CAM) data exchange and the implementation and acceptance of this standard by industry in lieu of other methods.

1.4 Overview of STEP Parts

Individual STEP documents are called "Parts". The Parts of STEP may be grouped by type as follows. The Parts are numbered so that all Parts of the same type fall in the same number range. The range is given below after the type. Not all numbers are used. There are several hundred application modules (the last type listed below). The total number of Parts of the other types listed below is about 120 (including those in progress and those which have been withdrawn):

- Overview and fundamental principles (1)
  This is a single document giving an overview of STEP and an exposition of its fundamental principles.
- Description methods (11–19)
  These cover the information modelling language EXPRESS and its graphical form, EXPRESS-G.
- Implementation methods (21–29)
  These cover methods of representing data that has been modelled in EXPRESS.
- Conformance testing methodology and framework (31–39)
  These give the general concepts of conformance testing as well as actual test methods and requirements on testing labs and clients.
- Integrated generic resources (41–59)
  These are EXPRESS information models of widely useful specific subject domains, such as geometry, topology, and tolerances.
- Integrated application resources (101–199)
  These are EXPRESS information models of more narrowly focused specific subject domains.
- Application protocols (201–299)
  These are the Parts intended for implementation in industry. As described in more detail below, each application protocol includes several documents.
- Abstract test suites (301–399)
  These are test suites for testing implementations of application protocols. In principle, for each application protocol numbered 2XX, there is an abstract test suite numbered 3XX.
- Application interpreted constructs (501–599)
  These are narrowly focused modules (written in EXPRESS) that can be combined to make up the implementable part of an application protocol.
- Application modules (1001–1999)
  These are similar to miniature application protocols. An application protocol can be built by including a (usually large) number of application modules. The second edition of Part 203 of STEP has been built this way. Using application modules is a more recent architectural approach than using application interpreted constructs and may replace application interpreted constructs.

1.5 Information Modelling Using EXPRESS and EXPRESS-G

EXPRESS is similar to the data description half of object-oriented programming languages such as C++. That is, EXPRESS supports describing the data structure of an object, but objects do not have any executable functions or methods. In EXPRESS, the definition of a type of object is called an *entity*, rather than a *class* (the term used in C++). A property of an entity is called an *attribute* of the entity. For example, an attribute of a circle is its diameter. Like other object-oriented languages, EXPRESS supports parent/child relationships among entities. The parent is called a *supertype* of the child, which is called a *subtype* of the parent. A subtype entity inherits all the attributes of its supertypes.

In order to support defining attributes, EXPRESS has built-in:

- Simple data types (e.g., STRING and INTEGER)
- Aggregates (e.g., ARRAY, LIST, and SET)

A logically complete set of entity definitions is called a *schema*. In addition to entity definitions, a schema can contain data type definitions and various kinds of constraints on instances of entities. EXPRESS includes a rich set of methods for describing constraints. In addition to making it possible to state rules about the attributes of a single entity, EXPRESS supports stating rules that apply to entire populations of instances of one or more data types.

In order to support stating complex rules, EXPRESS supports writing functions and has built-in:

- Arithmetic operators and expressions (e.g., A+2)
- Logical operators and expressions (e.g., A .AND. B)
- Numerical functions (e.g., cos(x))
- Operators on aggregates (e.g., sizeof)
- Methods of describing a set of objects (e.g., all circles with radius less than 1)
- Entity equality test operators

EXPRESS functions may also be used for computing values of derived attributes.

EXPRESS has additional functionality not described here, but almost all of what appears in a typical EXPRESS file has been mentioned. Figure 1.1 shows an example schema taken from the EXPRESS manual, Part 11 of STEP [1.9]. The schema says that a person must be a male or a female. Each person has a first and last name, date of birth, type of hair, and zero or more children (who are also persons). The age of a person is calculated using the years function. The parents of a person are found as an inverse. A male may be married to a female, in which case the female has an inverse relationship to the male.

```
SCHEMA example;

  TYPE date = ARRAY [1:3] OF INTEGER;
  END_TYPE;

  TYPE hairType = ENUMERATION OF (blonde, brown,
                                  black, red, white);
  END_TYPE;

  ENTITY person
    ABSTRACT SUPERTYPE OF (ONEOF(female, male));
    firstName : STRING;
    lastName : STRING;
    nickname : OPTIONAL STRING;
    birthDate : date;
    children : SET [0:?] OF person;
    hair : hairType;
  DERIVE
    age : INTEGER := years(birthDate);
  INVERSE
    parents : SET [0:2] OF person FOR children;
  END_ENTITY;

  ENTITY female SUBTYPE OF (person);
  INVERSE
    husband : SET [0:1] OF male FOR wife;
  END_ENTITY;

  ENTITY male SUBTYPE OF (person);
    wife : OPTIONAL female;
  END_ENTITY;

  FUNCTION years(past : date) : INTEGER;
    (* This function  calculates the number of years
      between the past date and the current  date *)
  END_FUNCTION;

END_SCHEMA;
```

Figure 1.1. Example EXPRESS schema

Another description method (also given in Part 11) is a graphical form of EXPRESS called EXPRESS-G. An EXPRESS-G diagram shows:

- Entities in solid boxes
- Simple data types in solid boxes with a double line on the right end
- Defined data types in boxes with dashed borders
- Enumeration data types in boxes with dashed borders and a double line on the right end
- Subtypes as a thick solid line connecting a supertype entity to a subtype entity with a circle at the subtype end

- Required attributes as a thin solid line connecting an entity to an attribute of the entity, with a circle at the attribute end and the name of the attribute (and any aggregate description) in text next to the line
- Optional attributes as a thin dashed line connecting an entity to an attribute of the entity, with a circle at the attribute end and the name of the attribute (and any aggregate description) in text next to the line
- Etc.

The EXPRESS-G diagram for the EXPRESS schema shown in Figure 1.1 is shown in Figure 1.2.

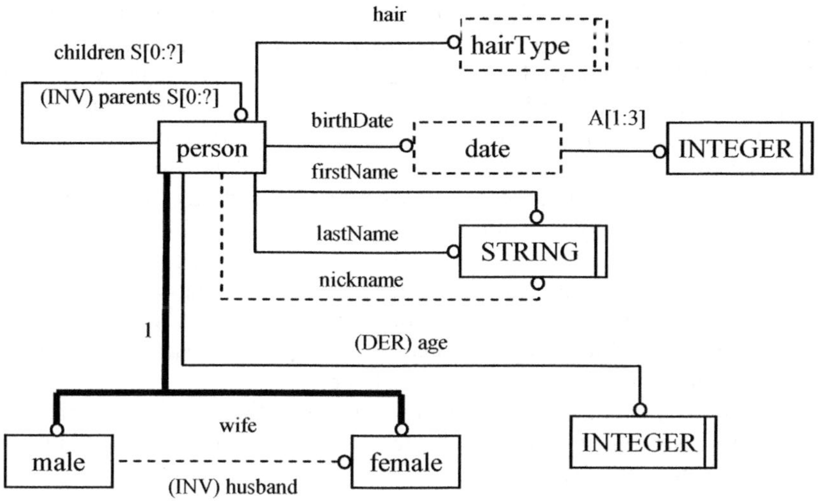

Figure 1.2. Example EXPRESS-G diagram

Several items from (textual) EXPRESS have no representation in (graphical) EXPRESS-G. These include rules, functions, items in an enumeration, and complex subtypes. There is nothing other than graphics in EXPRESS-G that is not in EXPRESS.

1.6 Data Representation

Once we have a method (EXPRESS) of making logical models of entities (such as *male* in Figures 1.1 and 1.2), we need a method of recording and storing instances of entities (such as Tom, Dick, and Harry). This may be done using STEP Part 21 files or XML files, or by using a database with the STEP Data Access Interface (SDAI).

1.6.1 Part 21 Files

Part 21 of STEP [1.10] is the primary method used for making instances of entities in files (computer files or paper files). The title of Part 21 is *Clear text encoding of the exchange structure*, but the files are usually called either "Part 21 files" or “physical files”. Each Part 21 file has a header section and a data section. The data section consists of instances of entities defined in one or more schemas named in the header. The header also includes identifying information about the file.

The Part 21 file format uses a minimalist style in which the same information is not written twice so that there is no room for contradictions in the data. The style assumes that, normally, the data will only be processed by software, and people will look at the data only to create test examples or find bugs, and that making the data more easily readable by these people is less important than eliminating redundancies. This said, it is not grammatically wrong to have identical data lines. In fact when hand-coding a Part 21 file, it is better to allow redundant data.

The Part 21 format is therefore simple and elegant. Each entity instance in a Part 21 file begins with a unique Entity ID and terminates with a semicolon “;”. Entity ID is a hash symbol “#” followed by an integer and has to be unique within the data exchange file. Entity ID is followed by an equal symbol (“=”) and the name of the entity that defines the instance. The names are always capitalized, though EXPRESS is case insensitive. The name of the instance is then followed by the values of the attributes listed between parentheses and separated by commas. Where the data type of an attribute is an entity, the attribute is given using the ID of an entity instance (e.g., #293). Where the data type of an attribute is a simple data type such as integer or string, the value of the attribute is written out literally. Aggregate values such as lists are enclosed in parentheses with commas between entries. Enumeration values are preceded and followed by periods (for example, *blonde* in an EXPRESS enumeration becomes *.BLONDE.* in a Part 21 file).

If an entity is a complex subtype of two or more other entities, a more complex method of representing it called the “external mapping” is used in a Part 21 file. That method is not described here.

Figure 1.3 shows a Part 21 file based on the EXPRESS schema shown in Figure 1.1 and representing a family with four people: Tom, Dick, Harry, and Teresa. Entity #1 is for Tom Thumb, who has the nickname Thumbs, was born on April 4, 1963, has Dick (#3) and Harry (#4) as children, has brown hair, and has Teresa (#2) as his wife. Meanings of #2, #3, and #4 are similar. Note that the derived (age) and inverse (parents and husband) attributes have no representation in the file. It is expected that a system that reads the file will find values for these attributes based on the data that is provided in the file. Comments in the header indicate what each header item means. The special token ("$") is used to represent an object whose value is not omitted. The special token ("*") is similar to ("$") except that the value can be derived from other values according to rules given in the EXPRESS schema. Two consecutive apostrophes ("") are used to encode a null string (string of length zero).

A Family Description

```
ISO-10303-21;
HEADER;
FILE_DESCRIPTION(('A FAMILY'), /*description */
                 '3;1'); /* implementation level */
FILE_NAME('TOMS FAMILY', /* name */
      '2008-06-25',     /* time stamp */
      ('T Kramer'),     /* author(s) */
      ('NIST'),         /* organization(s) */
      'none',           /* preprocessor version */
      'tk',             /* originating system */
      'noman');         /* authorization */
FILE_SCHEMA(('EXAMPLE'));
ENDSEC;
DATA;
#1 = MALE('Tom', 'Thumb', 'Thumbs', (1963, 4, 9),
      (#3, #4), .BROWN., #2);
#2 = FEMALE('Teresa', 'Thumb', 'Terri', (1963, 7, 4),
      (#3, #4), .BLONDE.);
#3 = MALE('Dick', 'Thumb', 'Shorty', (1993, 9, 3),
      (), .BROWN., $);
#4 = MALE('Harry', 'Thumb', $, (1992, 3, 26),
      (), .BROWN., $);
ENDSEC;
END-ISO-10303-21;
```

Figure 1.3. Example Part 21 file

1.6.2 XML Files

Since Extensible Markup Language (XML) [1.2] is a widely used file format and there are many software packages that will manipulate and display it, an XML file format has also been devised for representing STEP data.

XML defines a character-based document format in which rules are given for defining semantic tags that break a document into identified parts. The breakdown is nested so that it forms a tree (like the chapters, sections, and subsections of this book, but without the numbering). The following is a simple XML document defining a milling cutter:

```
<?xml version="1.0"?>
   <MILLING_TOOL>
      MILL 18MM
   </MILLING_TOOL>
```

The first line is a declaration that this is an XML file. It is usually made up of an attribute named “version” and its value “1.0”. Lines two to four define a “MILLING_TOOL” element with “<MILLING_TOOL>” as the start tag and

"</MILLING_TOOL>" the end tag. "MILL 18MM" is the value of the element. XML is flexible because there is no restriction to those tag names. Hence, it is possible to assign human-understandable tag names in an XML document, while computers just interpret an XML document according to a pre-defined formula. Two editions of Part 28 have appeared for specifying how XML may be used to represent STEP data. Edition 1 is only a technical specification. It is retained in STEP to allow existing applications to continue to use it. Edition 2 is a full international standard. It is intended that all new applications should use only the second edition.

Edition 1 of Part 28 specified an XML markup declaration set based on the syntax of the EXPRESS language. The markup declaration sets are intended as formal specifications for the appearance of markup in a conforming XML document. These declarations may appear as part of Document Type Definitions (DTDs) for such a document.

Recognizing the limitations of using DTDs, ISO developed edition 2 of Part 28 [1.15] employing W3C XML Schema. The main theme of the new implementation method is its two-level method. At the lower level, CAD authoring systems can continue to read and write STEP data sets. The only difference at this level is that these data sets can now have an XML format to make them more compatible with the higher level. At the upper level the data sets are modularized by inserting information from the mapping tables into the XML data to explain the meaning of each entity sequence. The new method can open up the definition of an Application Protocol into a series of interconnected XML schemas.

This method is implemented using two languages, a configuration language for describing how to map EXPRESS information into an XML defined form, and the existing STEP mapping table language converted into an XML form.

This new method still allows a wide range of approaches to using XML. To ensure that a unique XML format will be used, some APs (AP 238, for example) include normative annexes unambiguously defining an XML format.

1.6.3 STEP Data Access Interface (SDAI)

The STEP Data Access Interface (SDAI) enables using STEP data in a database system. SDAI does not provide everything needed for implementing real-time sharing of STEP data, but a sharing system can be implemented using SDAI.

Currently, four international standards have been established for SDAI:

- Standard data access interface, Part 22 [1.11]
- C++ language binding to the standard data access interface, Part 23 [1.12]
- C language binding of standard data access interface, Part 24 [1.13]
- Java Programming language binding to the standard data access interface with Internet/Intranet extensions, Part 27 [1.14] (technical specification)

Part 22 provides an application programming interface (API) to data described by an EXPRESS information model. The API allows the separation of systems that store and retrieve STEP data from applications that use STEP data. The API specified in Part 22 is independent of any programming language (since it is written in EXPRESS). Parts 23, 24, and 27 each define a specific way of mapping the generic API to a particular computer programming language.

SDAI allows two different mapping approaches: early binding and late binding. The difference between them is whether an EXPRESS data dictionary is needed by the software applications. There is no data dictionary in an early binding; an EXPRESS model is compiled into an API. In a late binding, the EXPRESS schema definition is needed by late binding applications at run-time. Part 23 (for C++) supports both early and late binding. Part 24 (for C) supports only late binding. Part 27 (for Java) supports both early and late binding. In addition, Part 27 provides for using STEP data in an intranet or on the internet.

The early binding approach generates specific data access functions according to the EXPRESS schemas and the programming language definitions. The entities defined in EXPRESS schemas are converted to C++ or Java classes. The inheritance properties in the EXPRESS schemas are also preserved in those classes. The advantage of an early binding is that the compiler of the programming language can perform additional type checking, and data access can be accomplished with fewer operations. However, because of the complexities of EXPRESS schemas (for example, the number of definitions in a STEP-NC AIM model is around 200 and each definition in the early binding approach needs to have a corresponding class in the programming language), the initial preparation, compiling and linking of an early binding approach was time-consuming in the era in which the SDAI standards were being developed (in 2008, with much faster computers, this is no longer the case).

The late binding approach, on the other hand, does not map EXPRESS entities into classes. It uses EXPRESS entity dictionaries for accessing data. Data values are found by querying those EXPRESS entity dictionaries. Only a few simple functions need to be defined in the late binding approach to get or set values. A late binding approach is suitable for a programming language that does not have strong type checking or an environment that may have multiple EXPRESS schemas (when an EXPRESS schema changes, a late binding application can use a new dictionary without changing the application itself). Late binding is simpler than early binding because there is no need to generate the corresponding classes.

1.7 STEP Integrated Resources

A collection of EXPRESS models, here called the *STEP integrated resources*, provides a fixed set of entities whose instances are allowed to occur in the files and databases that are intended for application. Although the collection is fixed in the short run, over the years it has been modified somewhat. The STEP integrated resources include three types of models:

- STEP integrated generic resources
  These are EXPRESS models for basic capabilities of product data representation. They are Parts 41–59 of ISO 10303, 13 of which are international standards. They include, for example, geometry and topology (Part 42 [1.17]) and product structure configuration (Part 44 [1.18]).
- STEP integrated application resources
- These are EXPRESS models for more specific but widely applicable types of product data, such as draughting (Part 101 [1.19]), kinematics (Part 105

[1.20]), and finite element analysis (Part 107 [1.21]). There are 8 international standards in this group. The range 101–199 is reserved for the Part numbers.
- Generic resources from other ISO standards
These are two EXPRESS models for basic capabilities of product data representation that were developed for other ISO standards and adopted by STEP. They are: arithmetic and algebraic expressions (Part 20 of ISO 13584 [1.29]), and time (Part 42 of ISO 15531 [1.35]).

1.8 Application Protocol (AP)

An application protocol (AP) is focused on a particular application domain. APs are the Parts of STEP intended to be implemented for industrial use. When the AP concept was first introduced in STEP, an AP had three parts:
- Application activity model (AAM)
  - A model of the activities and data flows of the application.
- Application reference model (ARM)
  - A model of the data needed for a particular application.
- Application interpreted model (AIM)
  - An encoding of the ARM in terms of the STEP integrated resources. This is the model that is intended for implementation in systems that use STEP.

APs still have these three parts, but the following documentation strategies have been added in order to modularize APs:
- Unit of functionality (UOF)
  - a set of entities and related constructs from an ARM that perform a single function or a set of closely related functions
- Application interpreted construct (AIC)
  - a UOF reinterpreted using the STEP integrated resources
- Application module (AM)
  - a miniature AP that can be reused

The first cut at modularizing APs was to use UOFs and AICs. Experience showed that these modules were too large to allow reusing modules. The second round of modularizing the STEP architecture introduced AMs. These are like miniature APs and are designed to be reused. It is intended that APs may be constructed from sets of AMs. The second editions of some APs, notably AP 203, use the AM approach. APs constructed in the future will also use it.

In addition, in order to allow for partial implementations of large APs, STEP provides that conformance classes of an AP may be defined that include only a portion of the functionality of the AP.

The following subsections discuss in more detail the items described above.

1.8.1 AAM (Application Activity Model)

The application activity model of a to-be-developed application protocol is a model of the activities and data flows of the application. AAMs are built using IDEF0 [1.39], which is a graphical method of modelling activities and data flows. Activities are represented as boxes, while data, actors, and constraints are represented by arrows. In the IDEF0 approach, an aggregated model is built first to show the big picture with three to six activities. Then one or two rounds of refinement are performed, with each activity at an upper level being expanded into an entire page at the next level down.

AAM models usually have a large number of arrows representing flows of various types of data, but only one or a few of these types are selected for further attention at the next stage (building an ARM). Once the AAM stage is completed and an ARM has been built, the AAM plays no further role. An example of the first page of an AAM is shown in Figure 1.4.

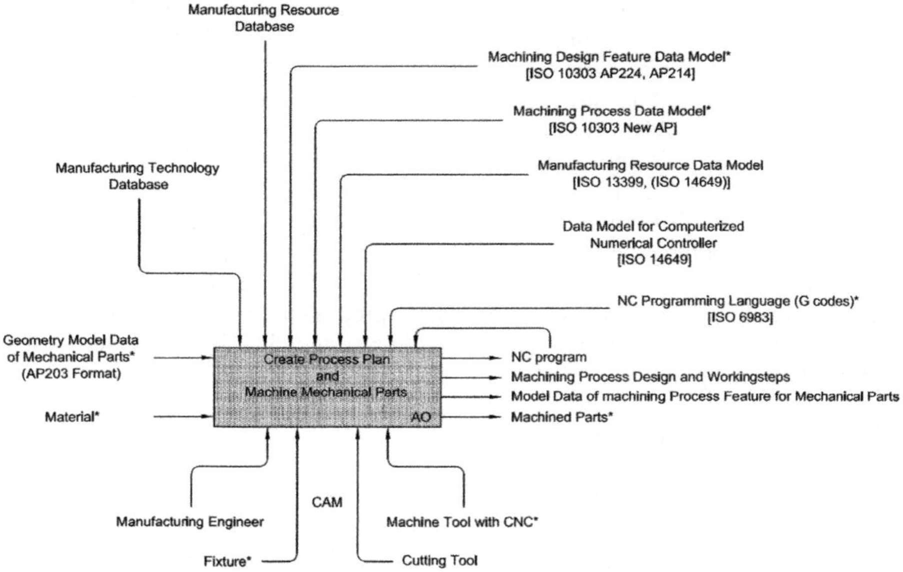

Figure 1.4. An AAM for digital manufacturing based on international standards [1.30]

1.8.2 ARM (Application Reference Model)

The application reference model (ARM) of an application protocol is a model of the data needed for a particular application. The model is given using the terminology of the application so that the model can be understood by practitioners of the application (who are involved in the development of the model). The process of building an ARM usually includes workshops at which domain experts decide what entities should be defined and what their attributes should be.

ARMs may be written in EXPRESS, EXPRESS-G, or IDEF1X. The modelling language is less important than the content.

1.8.3 AIM (Application Interpreted Model)

The application interpreted model of an application protocol is an EXPRESS model of (exactly) the information in an ARM but encoded in terms of the STEP integrated resources. The encoding is done using mapping tables, the format of which is formally defined and is uniform across STEP. Because extensive knowledge of the integrated resources, the format of mapping tables, and strategies for mapping are required to do the mapping, the encoding can only be done by a STEP expert.

For those few APs lucky enough to have an ARM whose structure and semantics are similar to a model included in the integrated resources, the mapping is direct and understandable to a domain expert. The cardinal example of this is STEP AP 203. For most APs, however, the encoding is mostly done using entities from Part 41 [1.16] such as product_definition_relationship or action_relationship, the useful semantic content of which is only that one thing is an attribute of another. The main technique used for getting these extremely generic entities to represent something specific is to write EXPRESS rules in the AIM that require certain strings which are attributes of the generic entities to have specific values. The AIM that results from the encoding is thus usually very verbose and unintelligible to a domain expert. In addition, a Part 21 file written using an AIM is typically five to ten times as long as a Part 21 file containing the same information but written using an EXPRESS ARM.

Because AIMs are so opaquely encoded, to use the data in an AIM, a STEP expert is usually needed to write an application programming interface that can then be used by an applications programmer to build a system that does something useful with the data.

1.8.4 UOF (Unit of Functionality)

As mentioned earlier, a UOF is a subset of the ARM of an AP containing entities and related constructs that support some specific functionality. A number of APs were produced containing explicit UOFs. AP 240, for example, lists 11 UOFs, the first 4 of which are adminstrative, design_exception, library_reference, and manufacturing_machine_tool_resources [1.27]. AP 219 also lists UOFs, for example, administrative_data, dimensional_measurement_analysis, feature_profile, functional_limitations, and part_properties [1.24].

1.8.5 AIC (Application Interpreted Construct)

An AIC is a UOF reinterpreted in terms of the STEP integrated resources. The idea is that an AIC developed for use in the AIM of one AP can be reused in other APs. Over 20 AICs have become international standards.

There is no requirement that an AIC correspond exactly to a UOF (the way an AIM must correspond exactly to an ARM). The functional_limitations UOF of AP 219, for example, corresponds closely but not exactly with Part 519, the AIC for geometric tolerances [1.28]. Because there is no source document to reference, an AIC (unlike an AIM) must include a description of the semantics of each entity it defines.

As mentioned earlier, the UOF/AIC method of modularizing an AP was found to be inadequate to support reuse and is being replaced by the AM method described next.

1.8.6 AM (Application Module)

An application module is like a miniature AP that has only a small bit of functionality [1.3]. Like an AP, an AM has an ARM written in terms of the domain being modelled. In an AM, the thing that would be called an AIM in an AP is called a module integrated model (MIM). Like an AIM, an MIM is a strict reinterpretation of the ARM using the STEP integrated resources. AMs can reference other AMs in order to build complex functionality.

The most conspicuous use of AMs is in edition 2 of AP 203 [1.22], which is comprised of 75 AMs. Several hundred AMs are finished or in progress. AMs have STEP Part numbers in the range 1001–1999 (except that the top level AM for AP 203 is Part 403). AMs are being standardized as technical specifications in order to minimize the time and effort needed to become ISO standards.

1.8.7 Conformance Classes

Because some APs are very large and cover multiple sub-domains, the STEP architecture allows that specific subsets of an AP called conformance classes may be designated as appropriate for implementation. AP 214, for example, has 20 conformance classes [1.23], and AP 238 has four [1.26].

The specification of a conformance class is a listing (or other kind of description) of those entities and related constructs in the AIM of the AP that are included in the conformance class and must be implemented by an application conforming to the class.

1.9 Conformance Testing

STEP provides for an elaborate system of conformance testing in the 30s Parts and the hypothetical 300s Parts, but no formal conformance testing system is currently in place for any application protocol. Early in the development of AP 203, formal testing and certification was done. This has been superseded by a less formal AP 203 vendors forum.

In order for STEP-based systems to interoperate, it is essential for STEP products to conform to the Parts of STEP that they claim to implement. It is not clear, however, what method of obtaining conformance will be most effective. The STEP conformance testing system seems to have been over-engineered compared to what is practical in the relatively small community of STEP implementers.

1.10 STEP-NC

STEP-NC is the application of STEP methods to numerically controlled machines. STEP-NC has been and continues to be developed by global efforts. Within ISO, two different subcommittees (SC1 and SC4) of 184 (TC 184) have been active. SC1 focuses on the control of machines, while SC4 focuses on industrial data. Since numerical control programs for machining a product are product data, there is a natural overlap between SC1 and SC4.

The ISO 14649 set of standards, which are subtitled "Data model for computerized numerical controllers", were developed by SC1. The models are written in EXPRESS and are ARM type models, in that they use domain terminology such as machining. ISO 14649 has the Parts listed below that have become international standards. Other Parts are partially developed (in particular, Part 111: Tools for milling). A round of improving the existing standards is under way.

- Part 1: Overview and fundamental principles [1.30]
- Part 10: General process data [1.31]
- Part 11: Process data for milling [1.32]
- Part 12: Process data for turning [1.33]
- Part 121: Tools for turning [1.34]

These Parts are arranged hierarchically, in that Part 11 uses Part 10 and Part 111, while Part 12 uses Part 10 and Part 121. Part 10 provides a set of basic capabilities for process planning for machined parts. Parts 11 and 12 specialise these capabilities for milling and turning, respectively.

SC1 seems to have the intent that STEP data representation methods be used with ISO 14649, since the Parts of ISO 14649 that include examples use STEP Part 21 files for them. SC1 has not, however, adopted using STEP AIMs. File exchange for industrial use can be accomplished quite well using Part 21 files based on ARM type models. What cannot be done using ARM type models is to store programs in a database that is implemented using only the STEP integrated resources at the lowest level.

The SC4 manufacturing working group has adopted the ARM models built in SC1 as the ARMs for AP 238 [1.26]. Unlike ISO 14649, which has separate Parts as described above, AP 238 incorporates the equivalent of all the Parts of ISO 14649 (except Part 1) in a single, very large model.

Unfortunately, ISO 14649 created its own machining features descriptions rather than using those available in AP 224 of ISO 10303. The two sets of machining features are similar but have many significant differences. In building AP 238, it was decided to modify the ISO 14649 features in order to make them more like those in AP 224. Thus, the real ARM for AP 238 is not exactly the Parts of ISO 14649. There are a few other points at which AP 238 deviates from ISO 14649. AP 214 also has machining features, but they have been made identical to those in AP 224. Inclusion of inspections in AP 238 is an added advantage in that it makes it possible to integrate machining and inspections.

Both ISO 14649 and ISO 10303-238 are thought of as modelling information for process planning at the micro-level; hence the intention of replacing G-code, which is traditionally and still extensively used to program NC machine tools. It is therefore worth mentioning ISO 10303-240 *Process plans for machined products* [1.27], which can model process planning information at the macro-level.

1.11 STEP into the Future

Industrial use of STEP has shown evidence of significant cost savings, higher quality, and reduced time-to-market. It may become a major building block in e-economy, the effort to unite manufacturing businesses among corporate partners, distant suppliers, and across diverse computer environments. Such an important function will only increase in importance in this rapidly changing economy that is increasingly globalised, collaborative, distributed, interoperable and integrated. Because of this, we should see an increased use of the Java programming language in building a standard data access interface with Internet/Intranet (e.g., JSDAI tools). Similarly, representing STEP data using XML is also viewed as an effective means of supporting collaborative projects.

As an essential part of STEP, EXPRESS will continue to be used as a viable means of representing engineering data. Such a trend will be seen in domains outside as well as within STEP related data models. The increased use of EXPRESS will give software developers the incentive to come up with more user-friendly, robust, reliable, and affordable software tools than those we have today (e.g., those listed in the tables in the Appendix at the end of this book).

How STEP will look in 5 or 10 years time is much dependent on a battle to win a balanced trade-off between extensibility of its specification and guaranteed interoperability of applications using the specification. In the recent past, and perhaps still so, a major driver for architectural change in STEP is interoperability between APs. On the one hand, it would be naive to think STEP developers would have the foresight to anticipate all data elements of importance for any significant time span. For example, the increasing use of parametric, feature-based design was not supported in the early parts of ISO 10303. There has also been a desire to represent the designer's intent and design history together with the raw design data. Experimental use of STEP for practices such as product lifecycle management (PLM) and e-commerce, is already taking place. Therefore, there is definitely a need for STEP to be able to expand its current data structures and hierarchy. On the other hand, interoperability between APs will be threatened if expanded data structures are added outside the standard. Past experiences with issuing newer versions of a STEP Part to include extra data (e.g., AP 203) have always led to a significant amount of redevelopment and revamp of the existing tools and systems, which does little to encourage quick adoption of the new standard.

All in all, it is apparent that STEP is much more than an international standard for exchanging product data. It is about enterprise integration, global competitiveness, data archiving, design reuse, and solving challenging manufacturing and business problems.

Appendix Sources of Information about STEP

Further information can be obtained from the following five types of sources: 1 the ISO itself; 2 the Web-site of the ISO Technical Committee, Sub-committee 4; 3 other national and organisational Web sites; 4 books; and 5 other publications.

1 The International Organisation for Standardisation

The ISO holds the STEP standards that can be purchased online via http://www.iso.org/iso/home.htm. Indeed, the first Part of STEP, ISO 10303-1 [1.8] is a must-read document for anyone who has just entered the STEP world.

2 Web-site of the ISO Technical Committee, Sub-committee 4

This ISO Sub-committee is charged with development of the STEP standards. Its Web-site (http://www.tc184-sc4.org) has a great deal of information about STEP, including a complete list of STEP Parts and the famous "STEP on a Page", which lists all the Parts of STEP by type and gives their status.

3 Other national and organisational Web-sites

The Web-sites of national standards bodies and vendors of STEP products have a wealth of information; most of it is available for free download such as FAQs, white papers, examples, and tutorials about STEP. Some also have Web-based STEP applications and online STEP translation services. Some of these frequently visited sites are:

- National Institute of Standards and Technology - http://www.nist.gov
- PDES Inc. http://pdesinc.aticorp.org/
- U.S. Product Data Association (US PRO) - http://www.uspro.org/
- International Industry STEP Centers (ISC) - http://isc.aticorp.org/
- Canadian STEP Centre, http://www.ic.gc.ca/epic/site/ad-ad.nsf/en/ad03581e.html
- Korean STEP Centre, http://kstep.or.kr/
- Italian STEP Centre (CeSTEP), http://xoomer.alice.it/luciano.lauro/
- ProSTEP http://www.prostep.com/
- STEPml - http://www.stepml.org/
- Product Life Cycle Support (PLCS), Inc. – http://www.plcsinc.org
- STEP Tools, Inc. - http://www.steptools.com

4 Books on STEP

There are some books published on STEP standards though most are somewhat dated:

- For the history of STEP and a detailed look at STEP concepts and methods through 1998, the book *STEP the Grand Experience* [1.36] is unsurpassed. This is a book edited by a group of researchers at the US National Institute of Standards and Technology (NIST). Its electronic copy is available at http://www.mel.nist.gov/publications/view_pub.cgi? pub_id=821224.
- The book *STEP for Data Management, Exchange, and Sharing* [1.6] is very readable and gets into fine details of STEP theory.

- The book *Product Data Exchange* [1.5] by Bloor and Owen describes product data exchanges from the early days of computer-aided design and engineering to the development of STEP. The book also gives a detailed account of the development from IGES through to the final publication of STEP.

5 Other publications

The *STEP Application Handbook Version 3* [1.1], dated 2006 (available at http://pdesinc.aticorp.org) is a collection of information about the current state and usability of STEP intended to be helpful to engineering users.

The paper *Fundamentals of STEP Implementation* [1.38] (available at http://www.steptools.com/library) is aimed at savvy software engineers who are using STEP or thinking about doing so (but it does not cover application modules).

The paper *A Modular Architecture for STEP*, available at http://www.nist.gov/msidlibrary/doc/isoma-9977.pdf [1.3] provides a very readable explanation of the AM idea.

The paper *STEP Modular Architecture* [1.4], presents the move to modularize the STEP application protocol architecture.

The paper *On STEP-NC and the Complexities of Product Data Integration* [1.7] discusses both STEP and STEP-NC from the data integration point of view.

The paper *ISO 10303 – STEP: A key standard for the global market* [1.40] gives a business (in this case Boeing) context for STEP.

The report *IDA-STEP (Integrating Distributed Applications on the Basis of STEP Data Models)* [1.37] extensively reviews the implementation of STEP AP214 and AP212. A special focus is on realizing concepts of the digital factory. Some of the software tools from LKSoftWare GmbH are also discussed.

The paper *STEP-compliant NC research: the search for intelligent CAD/CAPP/CAM/CNC integration* [1.43] gives a good overview of STEP-NC research between 1998 and 2005 and discusses current (in 2005) STEP-NC issues.

The chapter Information Sharing in Digital Manufacturing Based on STEP and XML in the book "Collaborative Design and Planning for Digital Manufacturing" [1.41] describes a method for STEP and XML to be combined in presenting product information.

The chapter *STEP-NC to Complete Product Development Chain* in the book Database Modelling for Industrial Data Management: Emerging Technologies and Applications, [1.42] focuses on STEP-NC and how it may close the gap between design and manufacturing for a complete, integrated product development environment.

References

[1.1] Anonymous. 2006. *STEP Application Handbook ISO 10303 Version 3*, SCRA, North Charleston, South Carolina, USA.

[1.2] Anonymous. 2007. *XML (Extensible Markup Language)*, World Wide Web Consortium, USA.

[1.3] Barnard Feeney, A. & Price, D. 1999. A Modular Architecture for STEP, *Proceedings of World Automation Congress 2000.*

[1.4] Barnard Feeney, A. 2002. The STEP Modular Architecture, ASME Transaction, Journal of Computing and Information Science in Engineering. 2(2): 132-153.

[1.5] Bloor, S. & Owen, J. 1995. *Product Data Exchange*. Publisher: CRC, November 30, 1995.

[1.6] Fowler, J. 1995. *STEP for Data Management, Exchange, and Sharing*, Technology Appraisals, Ltd., Twickenham, UK.

[1.7] Hardwick, M. 2004. On STEP-NC and the Complexities of Product Data Integration. *ASME Transaction, Journal of Computing and Information Science in Engineering*. 4: 61-67.

[1.8] ISO 10303-1. 1994 Industrial automation systems and integration -- Product data representation and exchange -- Part 1: Overview and fundamental principles. Geneva, Switzerland: International Organisation for Standardisation (ISO).

[1.9] ISO 10303-11. 2004., Industrial automation systems and integration – Product data representation and exchange – Part 11: Description methods: The EXPRESS language reference manual, Geneva, Switzerland: International Organisation for Standardisation (ISO).

[1.10] ISO 10303-21. 2002. Industrial automation systems and integration – Product data representation and exchange – Part 21: Implementation methods: Clear text encoding of the exchange structure, Geneva, Switzerland: International Organisation for Standardisation (ISO).

[1.11] ISO 10303-22. 1998. Industrial automation systems and integration – Product data representation and exchange – Part 22: Implementation methods: Standard data access interface, Geneva, Switzerland: International Organisation for Standardisation (ISO).

[1.12] ISO 10303-23. 1998. Industrial automation systems and integration – Product data representation and exchange – Part 23: Implementation methods: C++ language binding to the standard data access interface, Geneva, Switzerland: International Organisation for Standardisation (ISO).

[1.13] ISO 10303-24. 1998. Industrial automation systems and integration – Product data representation and exchange – Part 24: Implementation methods: C language binding of standard data access interface, Geneva, Switzerland: International Organisation for Standardisation (ISO).

[1.14] ISO 10303-27. 1998. Industrial automation systems and integration – Product data representation and exchange – Part 27: Implementation methods: Java™ programming language binding to the standard data access interface with internet/intranet extensions, Geneva, Switzerland: International Organisation for Standardisation (ISO).

[1.15] ISO 10303-28. 2007. Industrial automation systems and integration – Product data representation and exchange – Part 28: XML representations of EXPRESS schemas and data, using XML schemas, Geneva, Switzerland: International Organisation for Standardisation (ISO).

[1.16] ISO 10303-41. 2000. Industrial automation systems and integration – Product data representation and exchange – Part 41: Integrated generic resource: Fundamentals of product description and support, Geneva, Switzerland: International Organisation for Standardisation (ISO).

[1.17] ISO 10303-42. 2003. Industrial automation systems and integration – Product data representation and exchange – Part 42: Integrated generic resource: Geometric and topological representation, Geneva, Switzerland: International Organisation for Standardisation (ISO).

[1.18] ISO 10303-44. 2003. Industrial automation systems and integration – Product data representation and exchange – Part 44: Integrated generic resource: Product

structure configuration, Geneva, Switzerland: International Organisation for Standardisation (ISO).
[1.19] ISO 10303-101. 1999. Industrial automation systems and integration – Product data representation and exchange – Part 101: Integrated application resources: Draughting, Geneva, Switzerland: International Organisation for Standardisation (ISO).
[1.20] ISO 10303-105. 1996. Industrial automation systems and integration – Product data representation and exchange – Part 105: Integrated application resource: Kinematics, Geneva, Switzerland: International Organisation for Standardisation (ISO).
[1.21] ISO 10303-107. 1996. Industrial automation systems and integration – Product data representation and exchange – Part 107: Integrated application resource: Finite element analysis definition relationships, Geneva, Switzerland: International Organisation for Standardisation (ISO).
[1.22] ISO 10303-203. 2005. Industrial automation systems and integration – Product data representation and exchange – Part 203: Application protocol: Configuration controlled 3D design of mechanical parts and assemblies (modular version), Geneva, Switzerland: International Organisation for Standardisation (ISO).
[1.23] ISO 10303-214. 2003. Industrial automation systems and integration – Product data representation and exchange – Part 214: Application protocol: Core data for automotive mechanical design processes, Geneva, Switzerland: International Organisation for Standardisation (ISO).
[1.24] ISO 10303-219. 2007. Industrial automation systems and integration – Product data representation and exchange – Part 219: Application protocol: Dimensional inspection information exchange, Geneva, Switzerland: International Organisation for Standardisation (ISO).
[1.25] ISO 10303-224. 2006. Industrial automation systems and integration – Product data representation and exchange – Part 224: Application protocol: Mechanical product definition for process planning using machining features, Geneva, Switzerland: International Organisation for Standardisation (ISO).
[1.26] ISO 10303-238. 2007. Industrial automation systems and integration – Product data representation and exchange – Part 238: Application protocol: Application interpreted model for computerized numerical controllers, Geneva, Switzerland: International Organisation for Standardisation (ISO).
[1.27] ISO 10303-240. 2005. Industrial automation systems and integration – Product data representation and exchange – Part 240: Application protocol: Process plans for machined products, Geneva, Switzerland: International Organisation for Standardisation (ISO).
[1.28] ISO 10303-519. 2000. Industrial automation systems and integration – Product data representation and exchange – Part 519: Application interpreted construct: Geometric tolerances, Geneva, Switzerland: International Organisation for Standardisation (ISO).
[1.29] ISO 13584-20. 1998. Industrial automation systems and integration – Parts library - Part 20: Logical resource: Logical model of expressions, Geneva, Switzerland: International Organisation for Standardisation (ISO).
[1.30] ISO 14649-1. 2003. Industrial automation systems and integration – Physical device control – Data model for computerized numerical controllers - Part 1: Overview and fundamental principles, Geneva, Switzerland: International Organisation for Standardisation (ISO).
[1.31] ISO 14649-10. 2004. Industrial automation systems and integration – Physical device control – Data model for computerized numerical controllers - Part 10:

General process data, Geneva, Switzerland: International Organisation for Standardisation (ISO).

[1.32] ISO 14649-11. 2004. Industrial automation systems and integration – Physical device control – Data model for computerized numerical controllers - Part 11: Process data for milling, Geneva, Switzerland: International Organisation for Standardisation (ISO).

[1.33] ISO 14649-12. 2005. Industrial automation systems and integration – Physical device control – Data model for computerized numerical controllers - Part 12: Process data for turning, Geneva, Switzerland: International Organisation for Standardisation (ISO).

[1.34] ISO 14649-121. 2005. Industrial automation systems and integration – Physical device control – Data model for computerized numerical controllers - Part 121: Tools for turning, Geneva, Switzerland: International Organisation for Standardisation (ISO).

[1.35] ISO 15531-42. 2005. Industrial automation systems and integration – Industrial manufacturing management data - Part 42: Time model, Geneva, Switzerland: International Organisation for Standardisation (ISO).

[1.36] Kemmerer, S. (editor). 1999. STEP The Grand Experience, NIST Special Publication 939, National Institute of Standards and Technology, Gaithersburg, Maryland, USA.

[1.37] LKSoftWare. 2004. IDA-STEP (Integrating Distributed Applications on the Basis of STEP Data Models), EDAG Engineering & Design AG, LKSoftWare GmbH.

[1.38] Loffredo, D. 1999. Fundamentals of STEP Implementation, STEP Tools, Inc., Troy, New York, USA. [online] available: http://www.steptools.com/library/fundimpl.pdf. [31 July, 2008].

[1.39] NIST. 1993. Integration Definition for Function Modelling (IDEF0), Draft Federal Information Processing Standards Publication 183. December 21. National Institute of Standards and Technology, MD, USA.

[1.40] Mason, H. 2002. ISO 10303 – STEP: A key standard for the global market. ISO BULLETIN, April 2002. pp 9-13.

[1.41] Qiu, X. & Xu, X. 2009. "Information Sharing in Digital Manufacturing Based on STEP and XML", in Collaborative Design and Planning for Digital Manufacturing, edited by L. Wang & A.Y.C. Nee, in press. Springer Veralag.

[1.42] Xu, X. 2006. STEP-NC To Complete Product Development Chain, In Database Modeling for Industrial Data Management: Emerging Technologies and Applications, edited by Z. Ma, pp. 148-184, Idea Group Publishing.

[1.43] Xu, X., Wang, H., Mao, J., Newman, S.T., Kramer, T.R., Proctor F.M. & Michaloski, J.L. 2005. STEP-compliant NC research: the search for intelligent CAD/CAPP/CAM/CNC integration, International Journal of Production Research, 43(17): 3703-374

2

Feature-based Process Planning Based on STEP

Thomas Kramer[1] and Frederick Proctor[2]

Intelligent Systems Division, MS8230, National Institute of Standards and Technology
100 Bureau Drive, Gaithersburg, MD 20899, USA
Email: [1] thomas.kramer@nist.gov, [2] frederick.proctor@nist.gov

Abstract
This chapter begins by describing characteristics a process planning language should have. Then it discusses the extent to which four process planning languages have these characteristics. The first two of these are STEP standard process planning languages (all of AP 240 and part of AP 238). The third is part of ISO 14649. The fourth, called FBICS-ALPS, is a version of the ALPS language adapted for FBICS, a system that does automatic feature-based process planning. Next the chapter summarizes the machining features available in AP 238, ISO 14649, and AP 224. The way in which FBICS does feature-based process planning is presented and a system is described that translates FBICS plans into ISO 14649 plans (which are conceptually identical to AP 238 plans). Finally, the chapter is summarized, and improvements in FBICS needed to make it industrially useful are presented.

2.1 Introduction

ISO 10303 and ISO 14649 include several parts that deal with process plans and features and, hence, are used or could be used in STEP-NC. In addition, a Feature-Based Inspection and Control System (FBICS) has been built at the U.S. National Institute of Standards and Technology (NIST) that uses STEP methods and standards. This chapter discusses process plans, features, and process planning (the act of making a process plan that uses features). For each of these three topics, issues are discussed, the relevant sections of ISO 10303, ISO 14649, and FBICS are presented, and these sections are compared.

This chapter uses many STEP concepts introduced in Chapter 1. Readers who are not familiar with STEP should read Chapter 1 before tackling this chapter. Throughout this chapter, it should be borne in mind that (with a few exceptions) AP 238 and ISO 14649 have identical semantics, since ISO 14649 provides the ARM for AP 238.

2.2 Process Plans

The first subsection of this section defines process plan and discusses the desirable characteristics of a process planning language. The other four subsections are devoted to presenting three STEP-based process planning languages and discussing the extent to which each of them has the desirable characteristics.

2.2.1 Definition and Desiderata

A process plan, in general, is a recipe for transforming some input materials or partially finished products into finished or (more) partially finished products. Process plans may be for continuous processes (as in oil refining) or discrete processes. Discrete process plans consist of individual operations. Since we are interested in manufacturing using NC machines, which is done with individual operations, we deal here only with discrete process plans, primarily plans for machining by milling or turning. FBICS can also write and execute plans for inspection using a coordinate measuring machine (CMM) or a machining centre equipped with a touch trigger probe, but few details of that aspect of FBICS are given here.

Since process plans must be stored, a file format for representing them is required. The STEP way to get a file format, as explained in Chapter 1, is to write an EXPRESS schema giving the semantic content of an information model and then to use either the Part 21 rules or the Part 28 rules to obtain a file format from the schema. We will assume in the rest of this chapter that EXPRESS is used to define information models for process plans, and Part 21 is used for writing process plan files. The combination of EXPRESS and Part 21 constitutes a *process planning language*. If Part 28 or some other format were used for writing process plans according to the same EXPRESS model, that would constitute a different language. It would have the same semantics but different grammar and syntax.

Several characteristics of a process planning language for feature-based manufacturing are desirable. Namely, the language should:

- Be machine-readable
- Be machine processable into a serviceable application programming interface (API)
- Have a generic core suitable for all types of machining (and other discrete processes)
- Be vertically extensible to be suitable for several levels of a hierarchical control system
- Be horizontally extensible so that it is suitable for various types of machines
- Be capable of refinement in stages

Machine-readable

Since we have assumed the language is modelled using EXPRESS and Part 21, the language is sure to be machine-readable. Several of the STEP systems described in Chapter 1 are able to read EXPRESS and Part 21 files. Many of those that read Part

21 files are able to save a usable representation of what they read and to provide access to the data.

Machine Processable into a Serviceable API

By *serviceable* we mean that the API produced from a model of the language by an automatic EXPRESS processing system should be readily usable by a programmer who is also a domain expert.

Several of the STEP systems listed in Chapter 1 can read an EXPRESS file and produce an API from it in a language such as C++ or Java. Although these tools can produce an API from any EXPRESS schema, the API may fail to be serviceable for either of two reasons.

First, the data in which an applications programmer is interested may have been atomized so that it is spread across several instances of entities from the STEP integrated resources, and, individually, the instances have very little meaning. This happens when the EXPRESS model being processed is an AIM for which the ARM entities and types do not correspond closely to any entities and types found in the STEP integrated resources. The domains for which this is a problem include almost everything except CAD, for which topology and geometry in AIMs (or AMs) correspond very closely to topology and geometry in Part 42. When an API fails to be serviceable for this reason, it is necessary to buy or write a set of software that implements a serviceable API on top of the correct but unserviceable API produced by an automatic EXPRESS processor.

Second, the ARM model itself may be generic by the virtue of requiring text strings in many places. When this is done, an API produced from the EXPRESS model will often return a string to the application programmer who will then have to produce a routine manually to deal with the string. Such an API is not serviceable.

Generic Core

Many functionalities of a discrete process planning language are essential or desirable regardless of the domain for which the language is designed. These include functions that:

- Identify what the plan acts on
- Identify resources used in the plan
- Represent individual plan steps
- Specify the order in which step may be executed, possibly including loops
- Allow alternative sets of steps to be executed
- Refer to external documents (most especially process plans for the next hierarchical level down)
- Provide methods of making decisions in order to choose between alternatives (usually this means having Boolean and numerical expressions)
- Provide variables for use in expressions or steps

For feature-based planning, of course, it is essential that a method of describing features be added to the core.

The generic core should be extended (using EXPRESS!) as needed for defining process planning languages needed for specific purposes. Using the same core for all types of plans needed in a system will speed system development, minimize the amount of relearning system programmers must do, enable reuse of basic execution routines, and make the overall system more maintainable.

Vertically Extensible

Hierarchical control is a proven method of operating a shop that produces discrete parts. While control hierarchies may be seven or more levels deep, here we consider only three: cell, workstation, and task, as shown in Figure 2.1. The figure shows superiors above their subordinates. Arrows indicate commands. The figure has two subordinates for each superior, but each superior may have any number of subordinates.

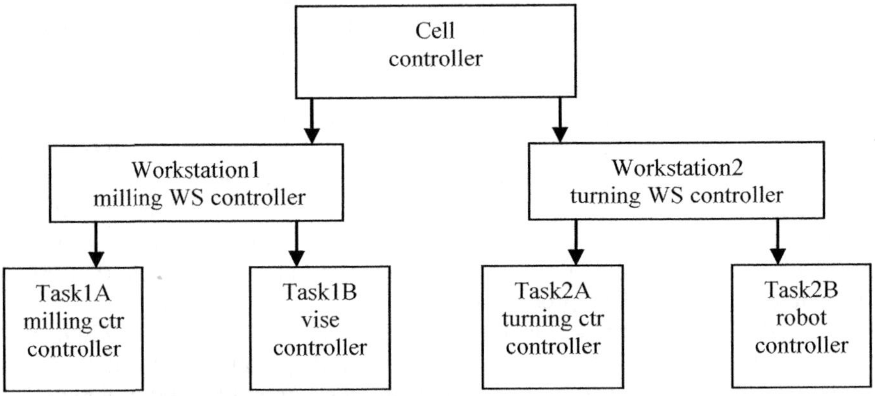

Figure 2.1. Prototypical controller hierarchy

The cell level controls a whole shop or a large portion of a shop. A cell has a collection of machines that, among them, can produce a finished part. At the cell level, a process plan is made to produce a finished part from stock or a near net shape such as a casting. A cell process plan says which features to make or inspect on which machines in which fixturings, and in what order the workpiece in process should be sent to the workstations of the cell.

The workstation level is the next level down from the cell level. A workstation is part of a cell and contains a single principal machine, possibly attended by ancillary machines such as powered fixtures, roller tables, or material handling robots. A workstation process plan says which features of the part design or the workpiece are to be made in a single fixturing. A single step in a feature-based workstation plan for a milling centre or turning centre identifies the feature to process and the operation to perform on the feature. The step usually indicates a specific tool or type of tool to use and may include feed and speed rates and strategies for using the tool.

The task level is the next level down from the workstation level. A step in a task level process plan for machining typically says what tool path, feed, and speed to use. In current practice outside of STEP, an NC programming language such as RS274 [2.4] is used at the task level.

In the hierarchy, each superior controller can give either planning commands or execution commands to its subordinates.

The core process planning language should be capable of being extended vertically to multiple levels of the control hierarchy.

Horizontally Extensible

At the workstation control level, where a single principal machine is used, different languages are needed for different types of workstations (those using a milling centre, a turning centre, or a CMM, for example). The core process planning language should be capable of being extended by defining specific types of plan steps for each different kind of machine. Since the workstations are all at the same hierarchical level, this is horizontal extension.

Refinable in Stages

Process plans are usually made so that some or most resources that can enter and leave a workstation (such as cutting tools) are specified by type, without distinguishing between instances of things of the same type. When it is time to execute a process plan, specific instances of these resources must be selected and linked to the plan. It is also possible that two workstations are almost the same but have slight differences in their fixed resources or that resources in a single workstation become unavailable (from breakage or maintenance, for example). It is useful, therefore, to make a process plan that includes alternate methods of doing the same thing, where the way the alternatives are done depends on the resources available.

In the operation of a shop, jobs are usually scheduled. In order to control the shop, timing information (such as earliest start and latest completion) may be inserted in the process plan, and actual times may be reported while the plan is being executed.

A process planning language should be such that process plans may be readily refined from (1) the stage at which generic resources and resource-dependent alternatives are specified to (2) the stage at which specific resources are identified but jobs are not scheduled to (3) the stage at which specific resources are identified and a schedule has been made.

Process plans may also be generated containing alternate operations for which the resources are all expected to be available. In order to optimize the plan, it may then be useful to simulate a number of different ways of executing the plan, select the best alternatives, and discard the other alternatives. Doing this is another way in which a plan may be refined, and process planning languages should support it.

The shelf life of a process plan gets shorter and shorter as the plan becomes more and more refined. Where a plan has been optimized based on conditions that may change, some record of those conditions should be saved with the plan so that, at execution time, it can be checked that the conditions still hold.

2.2.2 ISO 14649 and AP 238 Process Planning Languages

ISO 14649 and AP 238 (of ISO 10303) use process planning languages that have the same semantics, but AP 238 adds a level of encoding in order to use the STEP integrated resources. ISO 14649 models the core language in Part 10 [2.9], a language for milling machine process plans in Part 11 [2.10], and a language for turning centre process plans in Part 12 [2.11]. Language models for other types of machines are being developed in ISO 14649. AP 238 [1.7] puts everything in one giant model.

Machine processable into a Serviceable API

The EXPRESS schemas for Parts 10 and 11 of ISO 14649 are readily processable into a serviceable API. This was done by the authors for building an ISO 14649 interpreter [2.18] with functionality sufficient to interpret Example 1 given in Annex E of Part 11 of ISO 14649. The API was generated using STEP Tools ST-Developer. Although Part 12 of ISO 14649 was not processed, we believe it could be processed readily and would give a serviceable API.

The EXPRESS schema for AP 238 is automatically processable (because all EXPRESS schemas are automatically processable), but the API that results is not serviceable. In order to get a serviceable API from the automatically generated code resulting from processing AP 238, it is necessary to add a large, complex layer of manually written code on top of the automatically generated code. This is because the EXPRESS schema for AP 238 is an AIM type schema in which (as in almost all AIMs) data from several AIM entities must be recombined in order to obtain data that would be provided directly by an API built from the ARM schema.

The authors built an AP 238 interpreter with the same functionality as the ISO 14649 interpreter. The output of both interpreters was calls to functions implementing canonical machining commands [2.19]. An implementation of the canonical machining commands in which each command just prints itself was compiled into both interpreters. An AP 238 Part 21 file representing the same information as the ISO 14649 Part 21 file for the example was written by hand. When the ISO 14649 file was processed by the ISO 14649 interpreter and the AP 238 file was processed by the AP 238 interpreter, the files of canonical commands produced by the two interpreters were identical. Identical output files were also produced when other pairs of equivalent test files were used.

Generic Core

The core process planning languages defined in Part 10 of ISO 14649 and in AP 238 are almost adequate. They do everything a core language should do except the following:

- The plan structure is innately hierarchical in that an executable may be a Workplan, and a Workplan includes a list of executables, but no provision was made for the system executing a plan to be part of a hierarchy of controllers. In particular, there is no way to reference a plan for another controller.

- The only kind of expression provided is a Boolean expression. Arithmetic expressions are not defined. This omits some important functionalities. For example, although numeric variables are provided, and a "while" loop is defined with a Boolean test for getting out of the loop, there is no way to increase or decrease the value of a variable, so that the results of a test can change. The most common way of running a loop is to increment a counter variable each time around the loop and stop when the counter reaches a particular value. This cannot be done in ISO 14649.

Vertically Extensible

The process planning languages defined in Part 10 of ISO 14649 and in AP 238 are not currently vertically extensible, but they would be if the problems noted immediately above were fixed.

Horizontally Extensible

The process planning languages defined in Part 10 of ISO 14649 and in AP 238 are horizontally extensible and have been extended for milling and turning. The languages were constructed so as to be horizontally extensible.

Refinable in Stages

There is nothing in ISO 14649 that discusses the notion of refinement in stages, but ISO 14649 will support refinement. Chung and Suh [2.2] reported implementing refinement by first generating ISO 14649 process plans containing alternatives whose feasibility depends on machine capabilities and then refining the plan by selecting from among the alternatives. The unrefined plan is suitable for a variety of turning machines. Its structure is called a neutral process sequence graph. The refined plan is suitable for a specific turning machine and is called a hardware-dependent process sequence graph.

The authors of Part 111 of ISO 14649, "Tools for milling machines" [2.12], clearly intend that process plans be refinable by describing tools in a plan only in terms of nominal tool parameters and then selecting specific tools at execution time. This might be implemented either by refining the plan immediately before it is executed or by selecting specific tools as the plan is executed. Part 10 gives each tool a tool_id that is a label and says the tool_id must be unique, but it does not identify the collection among which uniqueness should apply. If the tool_id is intended to identify a specific tool in a tool inventory, then specifying a tool_id can be the method of going between planning stages. If the tool_id is intended to identify a type of tool in a tool catalog, then it is not useful for going between planning stages.

Workstation Process Plan Contents

The main content of an ISO 14649 or AP 238 workstation level process plan, excluding administrative and flow of control items, consists of instances of the following types of entity:

- Workingstep: each Workingstep points to a feature and an operation; the operation makes the feature or partially makes the feature (in the case of roughing operations).
- Feature: each feature should be referenced by at least one Workingstep.
- Operation: each operation should be used in at least one Workingstep and may be used in several Workingsteps. An operation usually has four major components: (1) machine_functions, (2) a tool, (3) a technology, and (4) either a list of tool-paths or one to three strategies (tool-path and strategies are both optional). Most operations also have one or more parameters specific to the type of operation (such as overcut_length for milling).
- Machine_functions: coolant, chip removal, axis clamps, etc. – things the machine can do other than turning the spindle and moving the tool relative to the part.
- Tool: each tool should be used in at least one operation.
- Technology: each technology should be used by at least one operation. A technology specifies feed rate, spindle speed, and other motion control items.
- Strategy: each strategy should be used by at least one operation. A strategy specifies the nature of a portion of a tool-path. For example, to start cutting a pocket, the strategy might be (1) pass through air (if a starter hole in the pocket has been made), or (2) plunge straight down into the material, or (3) spiral down into the material.
- Tool-path: each tool-path should be used by at least one operation. A tool-path gives the path the tool should follow. It may also give (1) the feed rate of the tool at each point on the path, (2) machine functions on the path, and (3) technology on the path. The path of the tool may be specified by pointing at explicit geometry or by giving a rule for generating explicit geometry.

The general idea behind having a list of tool-paths for a single operation is that it gives full control over everything the machine can do at each point on the path of the tool while performing the operation. If a tool-path is not specified, the technology and the machine_functions are constant throughout the operation. With a tool-path, they can be varied. Whenever a list of tool-paths is given for an operation and it is given high precedence, there is no use for a strategy, and the operation's machine_functions and technology may be overridden.

Conceptually, tool-paths are at the task hierarchical level and do not belong in a workstation level plan. However, if tool-paths were not provided at the workstation level, in order to allow the user complete control over machining operations (which is definitely needed now and is likely to be needed for the foreseeable future), it would be necessary to have task level process plans. Such plans would be on the same level as M and G codes traditionally used for machine control programs. It is not clear whether having plans at both the workstation level and the task level would be worth the trouble.

2.2.3 AP 240 Process Planning Language

Machine Processable into a Serviceable API

The EXPRESS schema for AP 240 is automatically processable (because all EXPRESS schemas are automatically processable), but the API that results is not serviceable. As with AP 238, this is because AP 240 uses an AIM schema. Also, as with AP 238, it is possible to build an API that provides access to ARM type data by writing a large, complex library of functions that extract ARM-like data from AIM data. Unlike AP 238, however, even doing that is not enough to produce an API usable in a specific field. This is because many of the attributes of AP 240 entities are strings, and it is intended that the strings have meaning in a specific field. To use AP 240 in a specific field, an additional document must be written that explains which strings may be used and what they mean. Then, to make a usable API, it is necessary for a programmer to read and understand the document and add a third layer of code that knows what to do when faced with particular values for the strings.

Generic Core

The core process planning language defined in AP 240 has some of the things a core language should have. It can:

- Identify what the plan acts on
- Identify resources used in the plan
- Represent individual plan steps
- Specify the order in which steps may be executed. Only sequential execution is provided
- Allow alternative sets of steps to be executed. This is supported in only the following two ways: (1) a process plan may identify a list of alternate process plans, and (2) alternate_activity is defined (though it is not clear how to use it)
- Refer to external documents

The core language defined in ap 240 does not provide:

- Expressions of any sort. Thus, no execution time decisions can be made automatically by executing steps specified in an ap 240 plan
- Variables of any type

Vertically Extensible

AP 240 explicitly identifies the work_cell, workstation, and machine hierarchical levels. The introduction says a process plan is to be "used by programmers to generate machine tool control programs", which implies that only the workstation level is supported. AP 240 defines enough types of data, however, to support both cell level and workstation level process plans. In most current discussions of AP 240, it is said to be intended to be useful for "macro" process plans, and macro appears to mean the cell level.

Horizontally Extensible

AP 240 is horizontally extensible by using agreed-upon sets of strings in the right places. Extending AP 240 horizontally by EXPRESS subtyping could be done only by ignoring the values of many existing attributes.

Refinable in Stages

It is possible to refine an AP 240 process plan in stages by eliminating alternatives. It may also be possible to refine an AP240 plan by specifying id numbers for resources such as cutting tools. The Id numbers, however, are not optional, so to refine by going from specifying a type of tool to specifying a particular instance of a tool, a bogus id number would need to be used in the stage one plan indicating that any tool of the type described will do. Then a real tool id would be used in the refined plan.

AP 240 does not appear to have been constructed with refinement in mind. There is no data element for indicating the stage of a plan.

Workstation Process Plan Contents

It is difficult to determine by studying AP 240 what the contents of a workstation level process plan are intended to be. The standard does not provide any examples of plans at any level, and the authors of this chapter have not been able to find any examples. It appears that a workstation process plan could be built by having data of the sort shown in Figure 2.2, which is a pseudo-Part 21 file snippet. Each entity instance in the figure contains zero, one, or two of the values that would be required in a real Part 21 file, and the names of attributes (which would not appear in a real Part 21 file) are shown in *italics*. The value of each named attribute is separated from the name by an equals sign (which also would not appear in a real Part 21 file). Moreover, the entities in the figure are ARM entities, not AIM entities.

Note that the names of the processes and their parameters are simply strings. No specific processes or process parameters are defined in AP 240. Thus, to make use of an AP 240 workstation level process plan of the sort shown in Figure 2.2, it would be necessary to write a document for the particular type of workstation similar to Part 11 or Part 12 of ISO 14649, explaining what the various terms and values mean. This same problem would arise with an AP 240 process plan for any other hierarchical level.

```
#1 = PROCESS_PLAN_VERSION
        (activities_to_produce_part = (#2, #3));
#2 = MANUFACTURING_PROCESS
        (assigned_feature = #4, assigned_operation = #6);
#3 = MANUFACTURING_PROCESS
        (assigned_feature = #5, assigned_operation = #7);
#4 = MANUFACTURING_PROCESS_FEATURE ( );
#5 = MANUFACTURING_PROCESS_FEATURE ( );
#6 = PROCESS_ACTIVITY
        (process_parameters = (#8), uses_to_perform = #10);
#7 = PROCESS_ACTIVITY
        (process_parameters = (#9), uses_to_perform = #11);
#8 = PROCESS_PROPERTY
        (process_name = 'drill', process_characteristics = (#12, #13));
#9 = PROCESS_PROPERTY
        (process_name = 'countersink', process_characteristics = (#14, #15));
#10 = TOOL_ASSEMBLY ( );
#11 = TOOL_ASSEMBLY ( );
#12 = NUMERIC_PARAMETER
        (parameter_name = 'speed', parameter_value = 2000);
#13 = NUMERIC_PARAMETER
        (parameter_name = 'feed', parameter_value = 507);
#14 = NUMERIC_PARAMETER
        (parameter_name = 'speed', parameter_value = 1254);
#15 = NUMERIC_PARAMETER
```

Figure 2.2. AP 240 workstation level process plan file snippet

2.2.4 FBICS-ALPS Process Planning Languages

The process planning languages used in FBICS have a modified version of ALPS (A Language for Process Specification [2.1]) as their core. The language is called FBICS-ALPS and is modelled in the FBICS_ALPS schema. ALPS itself does not define expressions, but FBICS_ALPS imports a separate EXPRESS schema for expressions written for FBICS. The FBICS_ALPS core has nothing specific to a particular hierarchical level (cell, workstation, or task).

Separate sections of another EXPRESS schema written for FBICS, the FBICS_COMBO schema, define additions to the ALPS core for three specific types of stage 1 plans: for a cell, for a machining workstation, and for an inspection workstation. Since machining and inspection are both in the same schema, hybrid plans containing both machining and inspection tasks may be written. Since interpreters for the RS274 and DMIS languages (for machining and inspection, respectively) were available to the FBICS project, no attempt was made to replace those languages with ALPS-based EXPRESS schemas. Those languages are used at the FBICS task level.

Machine Processable into a Serviceable API

The FBICS process planning languages have all been processed into usable APIs using commercially available STEP tools. The APIs have all been extensively used and tested.

Generic Core

FBICS-ALPS is the generic core of FBICS stage 1 plans for the cell and for inspection and machining (or both) workstations. FBICS-ALPS is entirely generic.

FBICS-ALPS has all of the desirable characteristics listed in Section 3.2.1 plus other functionalities that may be useful in some situations. These include allocating and deallocating resources during execution, and synchronizing the execution of plan steps. These other functionalities are not used in FBICS, however.

The resources used in FBICS are limited to cutting tools and touch trigger probes. Cutters and probes available for planning are given in a Part 21 file tool catalog while those available for execution are given in a Part 21 file tool inventory. An EXPRESS cutting tool catalog model developed at NIST was enhanced for use in FBICS by adding entities for probes and tool instances. The enhanced model is used for the FBICS tool inventory and tool catalog Part 21 files.

Vertically Extensible

FBICS-ALPS was constructed so as to be vertically extensible and is fully vertically extensible. Vertical extensions are made in other schemas by defining sets of subtypes of primitive_task_node.

The version of ALPS with which the FBICS project started was a souped-up version designed for production management. It had one impediment to being extended using EXPRESS. That is, although primitive_task_node had only one attribute (named subtask), the attribute got in the way of gracefully building subtypes of primitive_task_node. The value of subtask was a work_element, an entity that had only a name. If primitive_task_node had been used as a supertype, the subtask attribute would have become useless.

In the FBICS_ALPS schema, the subtask attribute of primitive_task_node has been removed. It is intended and implemented that sets of subtypes of primitive_task_node should be defined in EXPRESS schemas that extend the core language. Each subtype of primitive_task_node (defined in other schemas) adds whatever attributes it needs.

Horizontally Extensible

FBICS-ALPS was constructed so as to be horizontally extensible and is fully horizontally extensible. Horizontal extensions, like vertical extensions, are made in other schemas by defining sets of subtypes of primitive_task_node. Indeed, FBICS grew horizontally from its original implementation, FBICS, which covered only machining, to include inspection.

Refinable in Stages

The production management version of ALPS with which the FBICS project started was carefully designed to support three stages, which it called process plan, production-managed plan, and production plan. This was simplified in FBICS-ALPS to only two stages. A stage one plan is called a process plan, has alternative execution paths, and has various different types of steps. A stage two plan is called an operation plan, has only one linear execution path, and has only one kind of step, which is called one_operation. A one_operation has two attributes: an operation type identifier and the name of a file to be used at the next hierarchical level down that will carry out the operation. The simplicity of stage two plans made it very easy to implement building and executing them, but it introduced yet another information model that an application programmer would need to understand. It would probably have been a better idea to use the same format for stage two plans as for stage one plans, even though that would have required more code for generating and traversing stage two plans.

Workstation Process Plan Contents

The main content of an FBICS stage one workstation level process plan, excluding administrative and flow of control items, consists of instances of operations. Each operation specifies the index in the associated features file of the feature to operate on, the tool catalogue name of a tool to use, the feed rate, the spindle speed, the coolant use, and values of parameters associated with the specific kind of operation (such as pass depth for a peck drilling operation).

An FBICS stage one workstation level process plan points to a number of other files whose data is implicitly included. These other files include:

- The feature-based design of the workpiece before the plan is executed
- The feature-based design of the workpiece as it should be after the plan is executed
- The feature-based design of the fixture used to hold the workpiece while the plan is executed
- The features referenced in the plan
- A setup file giving the locations in machine coordinates of the four previous items

Each FBICS workstation level process plan is generated using a tool catalog that is selected when FBICS is initialized. Since a stage one plan may be executed or refined long after it is generated, the name of the tool catalog used by a plan really should be added to the plan.

As mentioned in the previous subsection, stage two FBICS plans consist of ordered lists of one_operations, each of which identifies a type of operation and names the file a subordinate should use to carry out the operation. In the case of the workstation level, the subordinate's file is either a DMIS file for inspection [2.3] or an RS274 file for machining [2.15].

2.2.5 Summary Table

Table 2.1 gives a summary of the extent to which each of the four languages just described has the desirable characteristics of a process planning language.

Table 2.1. Characteristics of four process planning languages modelled in EXPRESS

Language → Characteristics ↓	ISO 14649	AP 238	AP 240	FBICS-ALPS
Machine-readable	Yes	Yes	Yes	Yes
Machine-processable into serviceable API	Yes	No	Doubly no	Yes
Generic core	Almost	Almost	Almost	Yes
Vertically extensible	Almost	Almost	Yes	Yes
Horizontally extensible	Yes	Yes	Somewhat	Yes
Refinable in stages	Somewhat	Somewhat	Somewhat	Yes

2.3 Features

Across all domains of discourse, the word *feature* has many different meanings. In this chapter a feature is a type of shape (such as a pocket or hole) associated with a process plan or with the design of a piece part. In FBICS, the definition of a feature is narrowed so that a feature must be a closed volume in space.

A discussion of the issues to be considered in building a suite of features for material removal may be found in [2.13]. One of the issues is how a feature relates to a machining operation acting on the feature. In many systems the relation is not stated explicitly. The relation used in FBICS is this: when the operation is finished, all of the material in the feature must be removed, and the operation may not remove any material outside of the feature.

Many feature models intended to be useful for machining have been proposed. In this section we deal with only those features defined in AP 224 of STEP and Parts 10 and 12 of ISO 14649. FBICS uses AP 224. Part 11 of ISO 14649 (for milling) uses only the features defined in Part 10 of ISO 14649. Part 10 defines general features plus specialized feature types for milling but not for turning. Part 12 of ISO 14649 (for turning) defines and uses turning features. These are subtypes of general features defined in Part 10. AP 238 uses all the features defined in Parts 10 and 12 of ISO 14649, but some of them are modified so as to be more like AP 224 features.

The reasons that stereotyped features are useful are:

- Humans think in terms of stereotyped features
- Automatic recognition of stereotyped features from a boundary representation may be feasible
- Sterotyped operations may usually be defined for cutting stereotyped features
- Automatic process planning for parts with stereotyped features may be feasible
- Automatic tool-path generation may be feasible for stereotyped features

Figure 2.3 shows a part with a variety of stereotyped features. Part 224 features and ISO 14649 features are similar in shape but not identical. The EXPRESS supertype-subtype hierarchies are also similar but not identical. Both define features in a native coordinate system and locate each feature by locating the native coordinate system of the feature in a setup or part coordinate system. Both have all the features listed in Figure 2.4, but not by the names shown below or in the hierarchical arrangement shown below. In particular, AP 224 does not separate features into milling features and turning features.

Other similarities between AP 224 features and ISO 14649 features include:

- AP 224 and ISO 14649 both allow tolerances to be applied to parameters representing distances and angles
- AP 224 features and ISO 14649 features both have a lot of ambiguities that should be fixed

AP 224 features and ISO 14649 features have some fundamental differences:

- AP224 allows protrusions and bosses that add material to a base shape. ISO 14649 has no protrusions and allows bosses only as material inside a step, pocket, or planarFace that should not be milled away when the rest of the feature is milled away
- AP224 allows fillet as a transition feature. Fillets are not a type of transition feature in ISO 14649 (probably because fillets, like protrusions, add material to the part). ISO 14649 allows fillet shapes on the interior of features by providing for bottom corner and side corner radii. Curiously, neither provides for a flat fillet (which is easily machined at the bottom of a feature with the right type of endmill)
- ISO 14649 has tool-path feature, which is a dummy feature with no geometry that may be associated with an operation for which the tool-path is specified

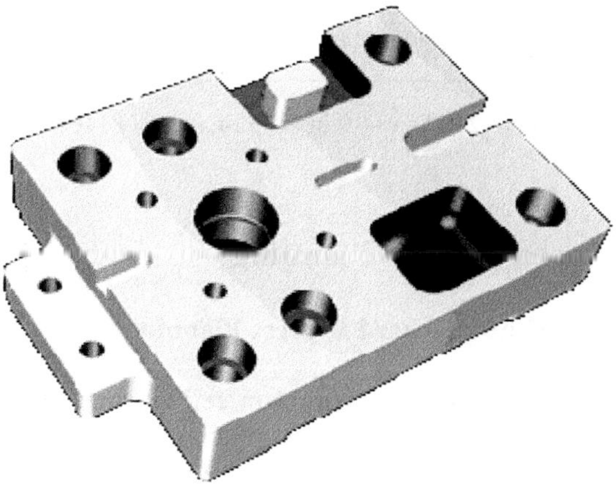

Figure 2.3. Some stereotypical features: holes, pockets, slots, planar faces, steps, bosses, and freeform surfaces

```
featureForMachining
   advancedBrepFeature (for complex shapes)
   transitionFeature (for chamfers and edge rounds)
   stereotypedFeature (for milling and turning)
      millingFeature (pockets, planarFaces, and steps may have bosses)
         hole (with various bottom conditions)
         pocket (with flat bottom, no bottom, or complex bottom)
            closedPocket (entirely surrounded by material on sides)
               rectangularClosedPocket (profile is rectangle)
               generalClosedPocket (profile is closed but arbitrary)
            openPocket (one side open)
               rectangularOpenPocket (profile is partial rectangle)
               generalOpenPocket (profile is open but arbitrary)
         slot (with various cross sections and end conditions)
         planarFace
         thread
         step
         sphericalCap (could also be a turning feature)
         roundedEnd
         generalOutsideProfileFeature (removal at the boundary of a part)
         replicateFeature (patterns of features, with omissions and offsets)
            circularPatternOfFeatures (on a plane)
            rectangularPatternOfFeatures (may be skew)
            randomPatternOfFeatures
      turningFeature
         outerRound
            outerDiameter
            outerDiameterToShoulder
         revolvedFeature
            revolvedFlat (like a chamfer on the end of a circular boss)
            revolvedRound (like an edge round on the end of a circular boss)
            groove
            generalRevolution
         knurl
   compoundFeature (two or more features combined)
```

Figure 2.4. Features in AP 224 and ISO 14649

2.4 Feature-based Process Planning

2.4.1 Overview of Feature-based Process Planning

Process planning is the activity of creating or refining a process plan (or a set of process plans). Here we focus on process planning for machining a part using machining features. Process planning may be done automatically, manually, or partially automatically and partially manually. In current industrial practice, most feature-based process planning is primarily manual at the cell and workstation levels but primarily automated using a computer-aided machining (CAM) system at the task level (where tool-paths are created).

2.4.2 Features and Process Planning

The sequence of activities for making a part of a given design starts with recognizing machining features from the design. Part designs are usually given as boundary representations, but may be given as design features (which may not be usable as machining features), or via constructive solid geometry. Feature recognition may be entirely automatic, done manually with a CAM system, or partly automatic and partly manual. Describing a part entirely in terms of features is most feasible when the features are either rotational (i.e., of the sort that can be made on a turning centre) or have one or a few characteristic directions (i.e., of the sort that can be made on a 3-axis machining centre in one or a few fixturings).

In some descriptions of feature-based process planning, it is assumed that once features have been identified, they cannot be changed during process planning. This is not a practical approach. It is common to start process planning and realize that a better plan can be built by changing the features. In an industrially usable system, it must be possible to change features during the construction of a process plan:

- If one of the features is the outside boundary of the part, it may be obstructed by clamps. In this case, the closed profile of the outside boundary may be divided into two or more open profiles, each of which is part of the boundary. The clamps can be moved in between cutting the two open profiles.
- If a large pocket with a complex profile is to be machined, and there is a narrow gap in the pocket through which only a small diameter cutter can pass, then it will be best to separate the pocket into three parts: two large pockets (one on either side of the gap) and one small pocket that passes through the gap. The large pockets can be cut efficiently with a large diameter tool while the gap pocket can be cut with a small diameter tool.
- If the profile of a pocket has concave corners with small radii, it may be useful to define a similar pocket with larger radii in the concave corners and add small pockets that fit into the corners with small radii. Then the large pocket may be machined with a large tool, and the small pockets may be machined with small tools.
- If two pockets intersect and their bottoms are on the same plane, it may be desirable to combine them into a single pocket with a more complex profile.

2.4.3 Feature-based Process Planning in FBICS

FBICS implements automatic two-stage hierarchical feature-based process planning starting from a design described in a STEP Part 21 file based on the AP 224 ARM EXPRESS model. FBICS plans for 3-axis machining. Only the round_hole, counterbore_hole, and rectangular_pocket AP 224 features are implemented in FBICS. The equivalent of an AP 224 step, however, may be made in FBICS by using a rectangular_pocket to remove the same volume as the step. FBICS makes plans for the cell level, the workstation level, and the task level, with only one controller at each level. Each of the three controllers is in a separate computer process. An interprocess communications system is used to connect the controllers with each other and with other FBICS processes. At the cell and workstation levels, stage one plans are built in FBICS-ALPS with operation types appropriate to the level and the capabilities of the equipment.

The architecture of FBICS as implemented is shown in Figure 2.5.

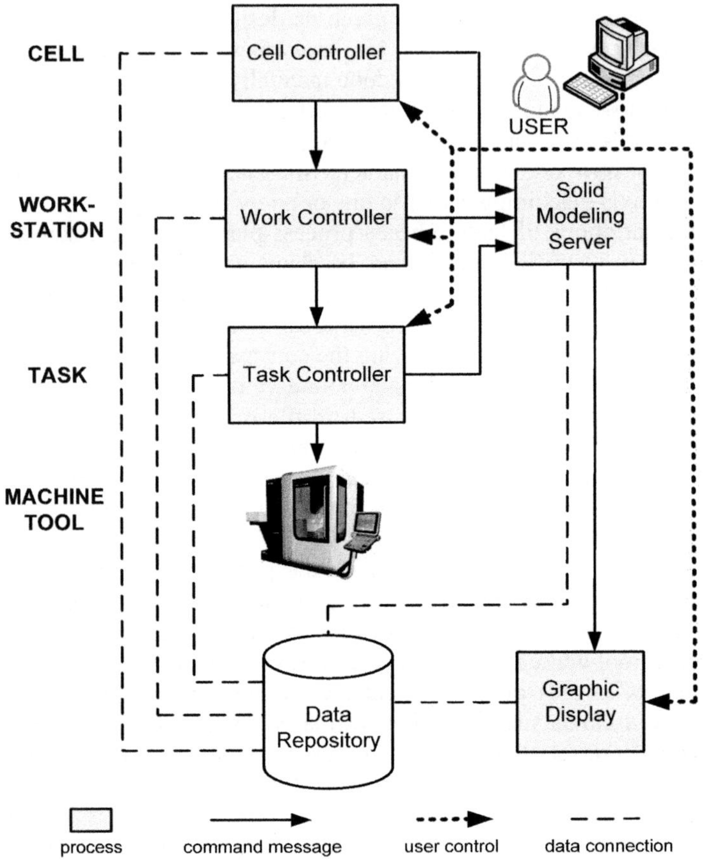

Figure 2.5. FBICS architecture

FBICS can plan for machining alone, machining with intermittent on-machine inspection, or inspection alone. How it does each of these is described in detail in [2.17]. In this chapter, only FBICS planning for machining is described.

FBICS uses a solid modeller in its own process. Each of the three controllers can use the modeller. A set of FBICS interface functions for the modeller was written to make using it independent of the underlying commercial solid modeller.

FBICS has a user interface to each of the three controllers. Each user interface can transmit a command from the user to make a plan or to execute a plan. The most complex activities occur when the user commands the cell to make a plan and then commands the cell to execute the plan, so that is described here. FBICS can plan automatically off-line for one, two or all three hierarchical levels. In the example that follows, we will assume the cell has been commanded to plan two levels deep.

User Gives Planning Command

When the user gives a planning command to the cell, the user specifies:

- The name of a Part 21 file containing the AP 224 design of the part. This file describes a block and the features that must be removed from the block to make the final intended shape.
- Whether a file giving the shape of the incoming workpiece exists. If the file exists, FBICS will read it. If the file does not exist, FBICS will write it; in this case the workpiece design will be the block shape given in the AP 224 design of the part.
- The name the workpiece file to read or write (according to whether it exists or not). This is also an AP 224 Part 21 file.
- The base name to use for the Part 21 plan files FBICS will write.
- The base name to use for the Part 21 AP 224 feature files FBICS will write.
- The base name to use for the Part 21 setup files FBICS will write.
- The number of hierarchical levels for which FBICS should plan (two in this example; one and three are the alternatives).

When FBICS executes, it generates a large number of files (often in the hundreds). Many of these files refer to other files, so it is necessary for FBICS to have an elaborate system of generating file names. To simplify this discussion of how FBICS works, in the remainder of this section, file names will be ignored.

Cell Makes Cell Level Stage One Plan

When the cell gets the user command, the cell reads in the design of the part and calls on the modeller to make a solid model of the part. The solid model is made by creating a solid from each of the features in the design, creating a solid from the block in the design, and Boolean subtracting all the features from the block.

The cell tells the modeller to make a solid model of the workpiece, which is done the same way as making a model of the design. In order to reason about how to make the part and to produce a graphical display of what it is thinking about, the cell tells the modeller to make a copy of the workpiece to use for the part-in-progress. The part-in-progress model changes as planning proceeds.

With the help of the modeller, the cell generates a lot of information about each feature in the design.

The cell checks whether each of the design features intersects the workpiece. Those design features that do not intersect the workpiece do not need to be machined. The other design features are put on a list of features to make.

The cell checks that the designed part is contained within the workpiece. If not, that means material needs to be added to the workpiece in order to make the part. Adding material is not possible by machining, so planning stops if the part is not contained in the workpiece.

The cell then determines what setups of the part-in-progress are needed and which features should be machined in each setup. To do this, the set of features to be made is divided into subsets called direction_sets. The features in a direction_set all have their Z axes pointing in the same direction (so that they are all machinable from the same direction). The features in a direction_set cannot necessarily all be made in the same fixturing, since other features may be blocking access to them. Hence, each direction_set is separated into two subsets, makeables and unmakeables. The makeables are initially those features that are not blocked in the Z direction by any other features (i.e., the makeables are those features in the direction_set that are immediately accessible for machining on the workpiece). The unmakeables are the rest of the features in the direction_set. After the initial assignment of makeables and unmakeables, any feature in the unmakeables that is blocked only by features in the makeables is transferred to the makeables (since it will be makeable in the same fixturing after the features that block access to it are made). The features in the unmakeables are all repeatedly checked until no more features can be transferred.

Setup_sets (sets of features that can be made in a single setup) and a partial ordering for making the setup_sets are then constructed concurrently by the following iterative procedure:

1. If one or more direction_sets has no unmakeables, those direction_sets are used as a group of setup_sets representing setups that can be made in any order. Otherwise, if one or more direction_sets has non-empty makeables, those makeables are used as a group of setup_sets representing setups that can be made in any order. The features in the setup_sets that have been identified are considered to have been made, and the part-in-progress is updated by boolean subtraction of those features.
2. The detailed information about all the features that have not yet been made is recomputed. Any direction_set with no remaining features is deleted from the direction_sets. The makeables and unmakeables of each direction_set are recomputed.

The two steps above are repeated until all features have been considered to have been made. This produces a total ordering of groups of setup_sets in which the setup_sets of each group may be made in any order. This ordering is captured in a cell-level stage one FBICS-ALPS process plan. The steps in the plan are all run_setup steps, one for each setup_set. Each step gives the name of a setup file.

For each setup, a features file is generated and saved, a fixture is selected to hold the part during the setup, and the names of the files describing the features and the fixture are inserted in the setup file. In FBICS, the selection of fixtures is naive – really just a placeholder for more intelligent fixture selection routines yet to be developed.

If it is possible to execute setups in more than one order, for some of the setups, it will not be known what the shapes of the incoming workpiece and outgoing workpiece are because it is not known what machining will have been done previously. Thus, the setup files in stage one plans refer to incoming and outgoing workpiece designs that do not exist. When a stage one plan is executed or refined, however, when any setup is about to be run, it will be known what the shapes are, and files describing the shapes will be available.

It is possible that the design has features that are not accessible. In this case planning fails. If planning seems to have succeeded, the modeller is used to check that the part-in-progress is now the same shape as the design.

Cell Makes Stage Two Plan

The cell has now made a stage one plan. It could execute this plan directly, but it has been commanded to plan two levels deep. The cell could traverse its stage one plan in every legal order and tell the workstation to make a plan for each possible traversal of the cell's stage one plan, but that may cause a combinatorial explosion of plans. One of the test parts used in FBICS required seven setups, and they could be done in any order. There would be 7! (= 5040) different ways to do that. To optimize the set of plans being generated, it would probably be useful to explore some of the different ways in which the stage one plan could be executed, but that has not yet been implemented. What has been implemented is to pick one legal execution order arbitrarily and use it.

The stage two cell plan will be linearly ordered and will consist of run_setup operations, each of which names a workstation level process plan that the workstation should execute. Since the cell has been commanded to plan two levels deep, the workstation level plans named in the cell stage two plan will be stage one plans. If the cell had been commanded to plan three levels deep, the workstation plans would be stage two plans.

Now that the cell has fully determined the order in which setups will be run, the shape of the workpiece as it leaves each setup and enters the next is found, and Part 21 AP 224 files describing these shape are saved. The setup files are modified by inserting the names of these intermediate shape files.

Cell Tells Workstation to Plan

Now the cell traverses the stage two plan it just made. For each of the setups it has decided to use, the cell gives a command to the workstation to plan for that setup. Each command contains only the name of the setup file to read, the name of the plan to write, and the number of levels deep to plan. In the scenario we are describing, each command tells the workstation to plan one level deep since the cell was commanded to plan two levels deep.

Workstation Makes Stage One Plans

Upon receiving a planning command from the cell, the workstation reads files for the setup, the features to make, the fixture, the incoming workpiece, and the outgoing workpiece. It then calls on the modeller to make solid models of the features, fixture, and workpieces in the positions given in the setup file. A part-in-progress is built by copying the incoming workpiece. A check is made that the outgoing workpiece is contained in the incoming workpiece.

The workstation knows that all the features in the setup can be made, but it does not know in what order they can be made. Therefore, the workstation goes through the same kind of accessibility analysis the cell performed in dealing with the makeables and unmakeables of a setup-set, but the workstation forms a partial ordering of the features as it does so.

For each feature to be machined, the workstation selects a cutting operation to make the feature and a type of tool to cut with. Values are selected for tool use parameters (feed, speed, etc.) by using tool usage rules. These rules are loaded when FBICS is initialised from a file prepared by the user. Thus, the user has local control over how tool use parameters are found. The tool usage rules use variables and may contain logical and arithmetic expressions. The variables may be any of ten variables whose names are hard-coded in FBICS. They include, for example: "material" (meaning the material the workpiece is made from) and "diameter" (meaning the diameter of the tool). The values of the hard-coded variables are set automatically in various ways. For example, the value of material is set when the workpiece file is read. Allowed values of non-numerical variables such as material must be given in the tool usage rules file.

The workstation writes a stage one process plan containing the operations that have been planned with the partial ordering of operations that has been determined.

That completes the planning commanded by the user.

Cell Executes Stage Two Plan

Now that planning is finished, the user gives the cell a command to execute a stage two plan. The command from the user gives only the name of the cell stage two plan to execute. To execute the cell plan (the gist of which is to run each of a list of workstation stage one plans) for each workstation plan, the cell sends the workstation a command to run the named plan. After sending each command, the cell waits until the workstation says it is done before sending the next command.

Workstation Executes Stage One Plans

Each time the workstation gets a command from the cell to execute a stage one plan, the workstation reads the plan and the files named in the plan (setup, features, etc.) and commands the task controller to open the setup also. Then the workstation repeatedly decides what plan step to execute next, decides what operations are necessary to carry out the plan step, and, for each operation, writes a file describing the operation in detail, commands the task controller to write a segment of NC code to carry out the operation described in the file, and (after the task controller reports that it has written the code) commands the task controller to execute the segment of code it has just written.

Some of the workstation stage one plan steps may require more than one operation. For example, steps for changing the tool and turning coolant on and off are not included in a workstation stage one plan because they may or may not be needed, according to the order in which the cutting steps are executed. But changing the tool and turning coolant on and off must be done from time to time, so for each plan step it executes, the workstation inserts these operations if they are needed.

When traversal of the workstation stage one plan is finished and all required operations have been carried out, the workstation closes its view of the setup and commands the task controller to close its view of the setup.

Task Controller Writes Code Segment

Each time the workstation commands the task controller to write a segment of NC code to carry out an operation, the task controller reads the file that the workstation wrote describing the operation and calls a code generating function that knows how to generate code for the particular kind of operation. The code is placed in a file.

Task Controller Executes Code Segment

Interpreters for both the RS274NGC language [2.15] and the DMIS language [2.14] are built into the FBICS task controller. Each time the workstation commands the task controller to execute a segment of NC code (contained in a named file), the task controller calls on the appropriate interpreter to interpret the code. Interpreting the code produces calls to canonical machining functions or canonical inspection functions, which are then executed to do the required cutting or inspection.

2.5 FBICS to ISO 14649

FBICS workstation level stage one machining plans are conceptually very similar to ISO 14649 plans for milling. They are similar enough that it has been possible to build a translator for Part 21 files. The translator reads a Part 21 file based on the FBICS_ALPS and FBICS_COMBO schemas and writes a Part 21 file based on the MILLING_SCHEMA and MACHINING_SCHEMA of ISO 14649. The input files may be generated automatically by FBICS or they may be generated any other way, as long as only those entities and types are used that could be in a schema generated by FBICS.

FBICS stage one plan files are shorter than ISO 14649 plan files because FBICS plan files:

- Refer to external setup files
- Include the data from an iso 14649 operation and technology in a fbics operation
- Refer to tools by tool type id

Hence, to make an iso plan file from a fbics plan file, in addition to handling two plan files, the translator must:

- Read the fbics setup file and use it to build a setup entity in the 14649 plan file
- Create a 14649 operation and a technology from each fbics operation
- Have on hand a tool catalog data base and a tool inventory data base, which when combined with the tool type id from the FBICS plan, will enable the construction of a 14649 tool from each tool type id mentioned in the FBICS plan

Other reformatting is done, but it is more straightforward than the details just mentioned.

2.6 Conclusion

This chapter has identified the desirable features of a process planning language: be machine-readable, be machine-processable into a serviceable API, have a generic core, be vertically extensible, be horizontally extensible, and be capable of refinement in stages.

Four languages for writing feature-based process plans have been developed using the STEP methodology of an EXPRESS schema plus Part 21. Two of these (AP 238 and AP 240) are Parts of STEP (ISO 10303). The third (ISO 14649) is an ISO standard. The fourth (FBICS-ALPS) is not a standard.

The AP 238 and AP 240 process planning languages are difficult to use because their EXPRESS schemas are AIM type schemas, making it necessary to build complex software that decodes semantics. AP 240 suffers from the additional liability of making extensive use of strings, so that interoperability will be possible in a particular application domain only if documents are written specifying what strings may be used in the domain and what they mean.

The ISO 14649 process planning languages are usable now for workstation level plans for machining and turning, but ISO 14649 has not been built to support hierarchical control.

The architecture of FBICS and the approach to feature-based process planning implemented in FBICS provide the kernel of an industrial strength automated machining and inspection system, but many improvements are needed to make FBICS workable in an industrial setting, as follows.

- Currently, FBICS handles only a few feature types. The FBICS methods need to be extended to cover a fuller set of features. This should be straightforward, but a large amount of code of various types is needed for each type of feature.
- FBICS does not include making an analysis of whether machining features that have been recognized should be subdivided or combined. This kind of analysis is an essential part of cost-effective process planning and should be added.
- FBICS does not include an analysis of how a part is to be located and held while it is being machined, and it does not include process plan steps for moving clamps in a single setup. It also does not have a method of planning for clamps and locating surfaces automatically. Those items are needed.
- FBICS does not optimize the plans that it makes. Where there are not many ways in which plan steps can be executed, it would be feasible to find the cost of every possible way to execute a set of plans and pick the cheapest, but combinatorial explosion is quite likely (there are N! orders in which to drill an array of N holes, for example). To optimize effectively, the scenarios in which explosions are likely would have to be identified and heuristics developed for each scenario.
- FBICS has a fancy method (not described here) of using tolerances on parameters of pockets to determine when to do in-process inspection. FBICS uses the results of the inspection to modify the tool path used for final finish milling. FBICS does not make any use of geometric tolerances, however. The use of tolerances in planning for machining must be expanded. For example,

if a hole that has been drilled has a tight location tolerance, it might be planned to centre drill before drilling. Or, if the hole has a tight cylindricity tolerance, it might be planned to ream the hole after drilling.

- FBICS currently does not make particularly efficient plans. For example, FBICS does not plan to do rough milling. The planning methods should be changed to make more efficient plans.
- Currently, FBICS planning rules are hard-coded. The planning rules should be changed to be in loadable files, so that each shop could use the planning rules it prefers.
- Humans are much more flexible and innovative in problem solving than any automated process planning system is now or will be for the foreseeable future. An effective system must allow for human intervention at each major stage of process planning. To provide this opportunity, FBICS includes many interfaces where files are used for data. For each of these data interfaces, a friendly, graphical user interface is needed so that a human can intervene in the planning process easily and intuitively.

References

[2.1] Catron, B., Ray, S.R. 1991. *ALPS – A language for process specification*, International Journal of Computer-Integrated Manufacturing, Vol 4, No 2, 105-113.

[2.2] Chung, D-H., Suh, S-H. 2008. *ISO 14649-based nonlinear process planning implementation for complex machining*, Computer-Aided Design 40, 521-536.

[2.3] Dimensional Measurement Standards Consortium. 2007. *Dimensional Measurement Interface Standard 5.1*, ANSI/DMIS 105.1-2007, Part 1.

[2.4] Electronic Industries Association. 1979. *Interchangeable Variable Block Data Format for Positioning, Contouring, and Contouring/Positioning Numerically Controlled Machines*, EIA Standard EIA-274-D.

[2.5] ISO 10303-11. 2004. *Industrial automation systems and integration – Product data representation and exchange – Part 11: Description methods: The EXPRESS language reference manual*, Geneva, Switzerland: International Organisation for Standardisation (ISO).

[2.6] ISO 10303-224. 2006. *Industrial automation systems and integration – Product data representation and exchange – Part 224: Application protocol: Mechanical product definition for process planning using machining features*, Geneva, Switzerland: International Organisation for Standardisation (ISO).

[2.7] ISO 10303-238. 2007. *Industrial automation systems and integration – Product data representation and exchange – Part 238: Application protocol: Application interpreted model for computerized numerical controllers*, Geneva, Switzerland: International Organisation for Standardisation (ISO).

[2.8] ISO 10303-240. 2005. *Industrial automation systems and integration – Product data representation and exchange – Part 240: Application protocol: Process plans for machined products*, Geneva, Switzerland: International Organisation for Standardisation (ISO).

[2.9] ISO 14649-10. 2004. *Industrial automation systems and integration – Physical device control – Data model for computerized numerical controllers - Part 10: General process data*, Geneva, Switzerland: International Organisation for Standardisation (ISO).

[2.10] ISO 14649-11. 2004. *Industrial automation systems and integration – Physical device control – Data model for computerized numerical controllers - Part 11:*

Process data for milling, Geneva, Switzerland: International Organisation for Standardisation (ISO).

[2.11] ISO 14649-12. 2005. *Industrial automation systems and integration – Physical device control – Data model for computerized numerical controllers - Part 12: Process data for turning*, Geneva, Switzerland: International Organisation for Standardisation (ISO).

[2.12] ISO 14649-111 (FDIS). 2004. *Industrial automation systems and integration – Physical device control – Data model for computerized numerical controllers - Part 111: Tools for milling machines*, Geneva, Switzerland: International Organisation for Standardisation (ISO).

[2.13] Kramer, T.R. 1994. *Issues Concerning Material Removal Shape Element Volumes (MRSEVs)*, International Journal of Computer Integrated Manufacturing, Vol. 7, No. 3, 139-151.

[2.14] Kramer, T.R., Proctor, F.M., Rippey, W.G., Scott, H. 1998. *The NIST DMIS Interpreter – Version 2*, NISTIR 6252, National Institute of Standards and Technology.

[2.15] Kramer, T.R., Proctor, F.M., Messina, E. 2000. *The NISTRS274NGC Interpreter – Version 3*, NISTIR 6556, National Institute of Standards and Technology.

[2.16] Kramer, T.R., et al. 2001. *A Feature-based inspection and machining system*, Computer-Aided Design 33, 653-669.

[2.17] Kramer, T.R., et al. 2004. *Feature-based Inspection and ControlSystem*, NISTIR 7098, National Institute of Standards and Technology.

[2.18] Kramer, T.R., Proctor, F., Xu, X., Michaloski, J.L. 2006. *Run-time Interpretation of STEP-NC: Implementation and Performance*, International Journal of Computer Integrated Manufacturing, Vol. 19, No. 6, 495-507.

[2.19] Proctor, F., Kramer, T.R., Michaloski, J.L. 1997. Canonical Machining Commands, NISTIR 5970, National Institute of Standards and Technology.

[2.20] Wallace, S., et al.. 1993. *Control Entity Interface Specification*, NISTIR 5272, National Institute of Standards and Technology

3

A Heuristic STEP-NC Based Process Planning Tool for Sequencing NC Machining Operations

Ulrich Berger[1], Ralf Kretzschmann[2] and Klaus-Peter Arnold[3]

Chair of Automation Technology, Brandenburg University of Technology
Siemens-Halske-Ring 14, Cottbus, Germany
Email: [1] ulrich.berger@ tu-cottbus.de
[2] ralf.kretzschmann@ tu-cottbus.de
[3] automatisierung@ tu-cottbus.de

Abstract
Nowadays, the process planning for sequencing NC (numerical control) machining operations is still done manually in principle. In the last decade several approaches for automatic process planning based on Artificial Intelligence (AI) and heuristic algorithms have been developed. Additionally, new technologies such as CAD/CAM (Computer Aided Design/Computer Aided Manufacturing) systems, feature-oriented specifications, and interfaces (e.g., STEP-NC) were introduced. Nevertheless, the process planner still has to modify and acknowledge the generated Workplans manually. In order to overcome this problem, an approach for enabling the automatic preparation of STEP-NC based Workplans with methods known from graph theory is introduced in this chapter. Therefore a STEP-NC Workplan is mapped into a directed graph in a mathematically defined way. Based on that, it is possible to use algorithms to find the shortest path and a Hamiltonian Path (HP) inside this directed graph as optimal sequenced solution under given requirements. Thus, the Workplan will be structured and reordered. Finally, the corresponding NC machining code will be generated and distributed to the machinery. Hence in this chapter, the requirements, the investigation, and the selection of suitable knowledge structuring and processing concepts, the mathematical fundamentals, and the work flow of sequencing a system are investigated. The focus of this chapter is the investigation of heuristic algorithms in order to sequence the STEP-NC machining operations. Finally, a preliminary demonstrator is introduced.

3.1 Introduction

Today, the current situation for manufacturing enterprise is characterized by increasing requirements from the customer side on quality and individualization of products. In contrast to that, there is also pressure on product prices at the same time. Thus, the enterprises are forced to shorten the product lifecycles of their products in order to react faster to the customer demands. As a consequence, the design and manufacturing process planning phase and the manufacturing process in general become more and more critical stages in the product lifecycles. The machining of components is an essential process in the scope of manufacturing. In this process step the components are machined to the given requirements from the production defining and designing phase. Therefore, the cost, the manufacturing time, and the product quality are strongly depending on the used manufacturing technology. In this scope, a lot of innovations were introduced in the last few decades. New manufacturing technologies such as the multi-axis machining and hybrid combination of milling and laser ablations for high precision finishing are already introduced and established [3.43]. Furthermore, enhanced CAD/CAM system allows very complex calculation for optimized cutter location paths to assure a high surface quality of components. As a consequence, more machining strategies have to be considered in the process planning phase. Thus, the NC process chain is getting more complex. More information has to be handled and exchanged between the different phases of the process change. However, the effective NC programming is still a bottleneck within this process chain. The reason is the mangled information exchange caused by prevalent organizational and technical obstacles such as the use of the ISO 6983 as NC programming interface [3.5]. It was shown by Brunnermeier in 2002 that the US automotive spend $ 1 billion to fix data error caused by interoperable interfaces [3.38]. A solution to overcome these problems is the introduction of STEP-NC (ISO 14649) and the feature technology. They enabled the process planning and the exchange of NC Workplans to the machinery [3.23, 3.26]. As a result, the process planning knowledge can be included in the corresponding machining features. Nevertheless, the sequencing (selecting and ordering) and acknowledgement of suitable machining features and machining operations to given design feature as a mapping function is still done manually with the help of the experience of the process planner. In contrast to that, STEP-NC enables the automatic process planning. Thus, the objectives of automatic process planning are the time savings comparing with a manual planning and the use of the whole available machinery. A lot of different automatic process planning methods and approaches are introduced in the last decades. The introduction of automatic rule-based process planning algorithms is successful in preparing Workplans in small machining application fields. However, these so-called expert systems are limited, because of small fields of application and difficult handling. The knowledge acquisition is also a bottleneck.

To overcome these limitations, a new innovative approach for using heuristic algorithms known from the graph theory for sequencing machining operations is presented in this chapter. With the help of knowledge-based machining operations as the elemental part of an NC Workplan, the selection and reordering of these elements can be executed in a mathematically defined way. With the help of mathematical defined mappings, it is possible to transfer a Workplan of an NC program in a directed graph [3.8] for further processing. As a consequence, the well-known algorithm (Floyd-Warshall and Travelling-Salesman) from the graph theory and combinatorics can be used to process and optimize the graphs under given requirements as objective functions e.g., time reduction, in a traceable way [3.1]. As a consequence, requirements from the industry (e.g., more transparency and reducing the processing time) can be achieved.

The chapter is organized as follows. The state-of-the-art is presented in the following section. At first, arising problems and challenges of the NC process chain will be introduced. Then, STEP-NC as solution to overcome this issue is introduced and the automatic process planning of a STEP-NC based process chain is outlined. As a result, this section concludes with the statement of problem formulation of the sequencing problem. The third section deals with objective and requirements of a new approach based on the actual state of the art. This new approach for sequencing of machining operations based on algorithm known from the graph theory is introduced in the fourth section in a detailed way. The evaluation of the demonstrator of the approach will be done by using a carefully selected benchmark part in the fifth section. Finally, the conclusion and outlook complete this chapter.

3.2 State of the Art and Related Work

The actual state of the art of the NC process chain is presented in this section by using carefully selected literature. First, the three phases of the NC process chain will be introduced. Afterwards, actual challenges and problems in the process chain are discussed. Then, STEP-NC is introduced as an enabler to overcome the discussed problems. Based on that, the process planning methods to generate NC Workplans and NC programs are investigated in detail. Finally, the problem statement of the sequencing problem is defined.

3.2.1 NC Process Chain

The NC process chain consists of three fundamental steps. First of all, the part (workpiece) including manufacturing requirements has to be designed with powerful CAD software systems [3.7].

Afterwards, the following planning phase consists of further detailed sub-process steps. Thus, the main sub-step is the preparation of the Workplan in the scope of NC machining. The Workplan will be used in order to compile the instructions for the NC machine controller. Therefore, the Workplan includes all information about the machining task. That includes in detail the raw part and the workpiece geometry, the shop-floor equipment (tools, machine, clamping), and the sequence of the machining operations. The NC programming tool is used to create the Workplans. The NC programming method depends on the complexity of the workpiece and the Workplan. Commonly used methods are shop-floor-oriented programming procedure (SOP) and powerful CAPP (Computer Aided Process Planning) and/or CAM software applications. Afterwards, the Workplan is compiled into NC programs based on ISO 6983 based NC instructions or ISO 14649 data structure [3.5, 3.7, 3.26].

Finally, the part is machined by NC machines executing these generated instructions in the shop-floor as the last step of the process chain [3.7]. The NC programs have to be distributed to the dedicated machine by a DNC (Distributed Numerical Control) system at the right time according to the production, planning, and scheduling results. These shop-floor scheduling tasks are supported by MES (Manufacturing Execution System) and/or PPS (Process Planning System) [3.7,

3.11]. The determining factors for shop-floor scheduling are the machinery, the tools, the operators, and constraints regarding to the capacity and time [3.7].

3.2.2 Challenge and Problems in the Process Chain

The introduced traditional NC machining process is characterized by permanent enhancements in the last decades. There are five factors which influence these enhancements and the complexity of the process chain as seen in Figure 3.1 [3.1]. The traditional ISO 6983 based process chain is facing the following facts.

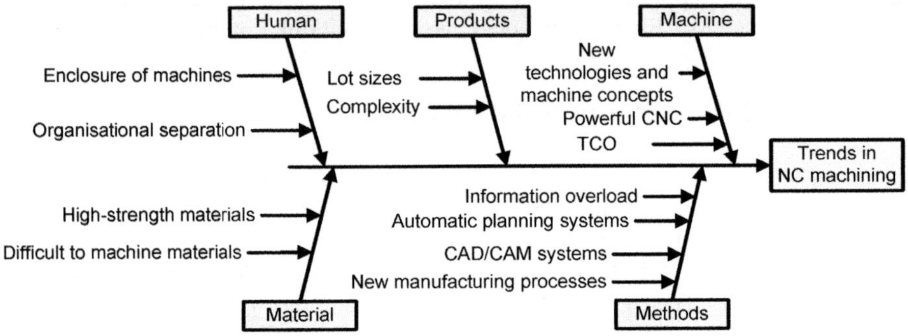

Figure 3.1. Complexity along the NC process chain [3.1]

The first factor deals with the operator on the machine and the programmer/planner in the process planning phase. The organizational separation of the shop-floor and the process planning departments lead to information deficiencies due to the use of interfaces in order to exchange information between the process planer and the operator. Furthermore, the manual process planning is very time consuming, because a lot of different information have to be handled by the operator and programmer. The planner is faced with an information overload caused by the availability of new machines, technologies, materials, and methods. The second and the third factors are focusing the manufacturing of complex products with variable lot sizes and the machining of new resources such as expensive materials which are difficult to machine (e.g., nickel for turbine blades). As a result, complex programs have to be generated. Therefore, more potential programming errors have to be avoided in the CAM system. But these systems do not support all information concerning the available machinery (such as dynamic and kinematics machine models). The result is that the programs are not 100% verified [3.21]. Thus, a modification of the non-human-readable programs is necessary for setting up the machine. The fourth factor is the use of new machining cost models (such as TCO – total cost of ownership) [3.34]. New technologies such as multi-axis-machining and hybrid technologies (e.g., machining with dry ice blasting and laser ablation) are also applicable now. Therefore, the NC process chain comprehends more knowledge intensive processes. A new knowledge exchange is enabled within these phases with the help of new methods (fifth factor). These are the use of integrated design and process planning by joining CAD and CAM systems and introducing the feature technology paradigm [3.20]. The introduction of automatic rule-based process planning algorithms is only successful in preparing Workplans in small machining application fields. However, these so-called expert systems are very difficult to handle as mentioned before. Nevertheless, the information and knowledge exchange from and to the shop-floor is still handicapped by using the ISO 6983 exclusively for generating NC programs for a fixed setup. Thus, the short-term shop-floor scheduling is prevented. As a consequence, it is hard to modify these NC programs, if the allocation will change because of unpredictable states in the machinery. An information feedback of these modified NC programs fails because of missing or insufficient information feedback possibilities back to the design and process planning. As a result, the knowledge is kept on the one hand in the design and process planning phase and on the other hand in the shop-floor [3.22]. Due to the missing link to the CAM information, recurrent problems occur (broken die and collision), while setting up a machine in the shop-floor with a new program [3.21].

As discussed, the traditional NC process chain is facing new complexities. Therefore Xu sums up the required capabilities for a new generation of CNC machining in order to meet the increasing costumer demands [3.23]. Comparing the ISO 6983 interface base process chain to the requirements, the commonly used NC process chain fails for the following reasons:

- Seamless information flow: ISO 6983 fails because of information lacks between the CAM model and the NC programs
- Feature-based machining: ISO 6983 fails because of computing all machining feature information into cutter locations
- Autonomous CNC: ISO 6983 does not support this feature
- Fault tolerant: ISO 6983 does not support this feature
- Networked and distributed: ISO 6983 does not support this feature because of non interoperability between different controller
- Modular, scalable: ISO 6983 does not support this feature, because of defining fixed tool movement
- Portable, adaptive: ISO 6983 does not support this feature because of non interoperability between different controllers

Due to the presented facts, the ISO 6983 based NC programming is thwarted. This aspect also influences the process planning activities. New solutions for the process planning and NC programming are needed. Therefore the STEP-NC based process chain is introduced in order to increase dramatically the interoperability in the process chain. As a consequence, a new method for enabling the STEP-NC process planning is presented in the current chapter.

3.2.3 STEP-NC – STEP Compliant Numerical Control

The ISO 6983 based NC process chain suffers from the "historic" programming language. The disadvantages of the "G/M" code have been discussed extensively in the literature [3.19, 3.23]. Thus, a new NC programming language called STEP-NC was introduced [3.23]. STEP-NC is a high-level and feature-based NC interface. The target is to provide the CNC controller with more structural information about the machining process as the ISO 6983 does. Consequently, a bi-directional information flow is enabled as shown Figure 3.2.

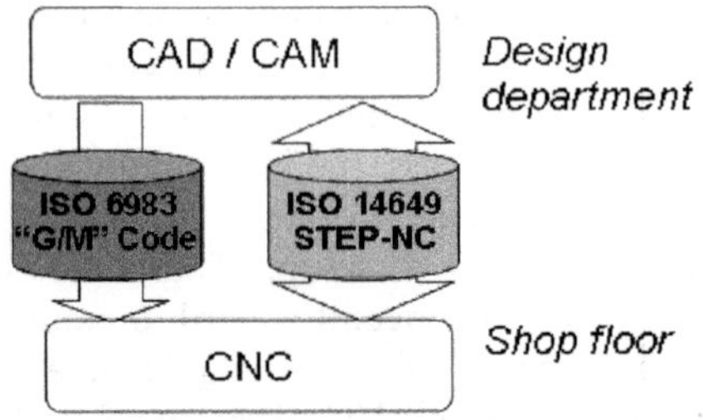

Figure 3.2. Comparison of the old ISO 6983 and new ISO 14689 [3.23]

STEP-NC is based on the STEP standard and was developed in several projects around the world [3.23, 3.24, 3.37]. STEP-NC provides an object-oriented data model for CNC controller and was introduced as an ISO 14649 and STEP Part 238 standard [3.26, 3.36]. First demonstrators, prototypes of STEP-NC controllers and software solutions were developed in the last few years. They support the different phases of the STEP-NC process chain and they provided excellent results [3.23]. Therefore, the STEP-NC based process chain is introduced in this chapter in order to increase dramatically the interoperability in the NC process chain. As a consequence, a new method for enabling the STEP-NC process planning will be presented in the next section. The process planning in STEP-NC based process chain is based on STEP. Therefore several STEP AP (application protocols) are used. A survey of STEP and the used AP can be found in Chapter 1 of this book.

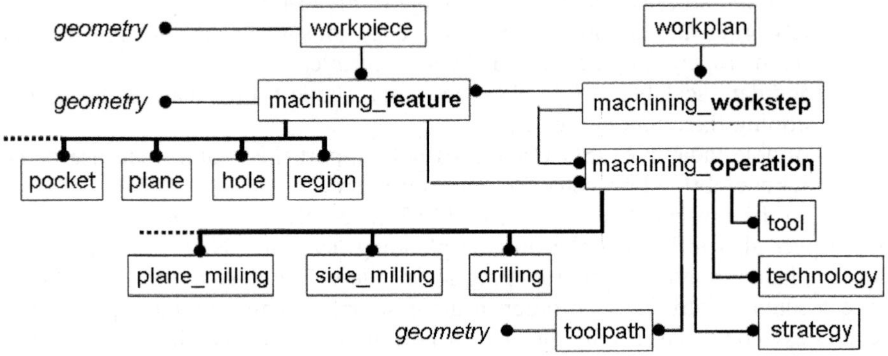

Figure 3.3. ISO 14649 data model for 2.5D and 3D feature milling

As seen in Figure 3.3, a workpiece and a Workplan are associated with the help of machining features and machining workstep. That means a workpiece consists of machining features (based on mapped and/or transferred design features). The machining features are represented by one machining workstep, which consists of several machining operations. The machining operations represent the activities of the machine in order to machine the operations. An operation includes the tool, the technology (e.g., feed rate, spindle speed, the strategy, and the tool path). Thus, a machining operation has alternative machining operations that are doing the same machining feature but with alternative parameters.

There are two questions arising in the scope of handling machining operations:

- Which machining operations sequence will be selected to represent a machining feature?
- How will the selected machining operation be ordered to meet the manufacturing objectives in an optimized way?

To answer these questions, different approaches are presented in the following sections.

3.2.4 Process Planning in the STEP-NC Process Chain

The process planning is supported by CAM, CAPP, and MES systems. These systems help the planner to select corresponding machining operations in order to machine the given workpiece. Furthermore, these systems support the planner to schedule machining jobs at the shop-floor. Nevertheless, these activities are very difficult and time consuming. Therefore a need for introducing automatic process planning, in order to decrease the time consuming programming and planning activities in the STEP-NC process chain, is identified. The process planning in the scope of NC manufacturing can be divided into two different levels with two different planning horizons as seen in Figure 3.4.

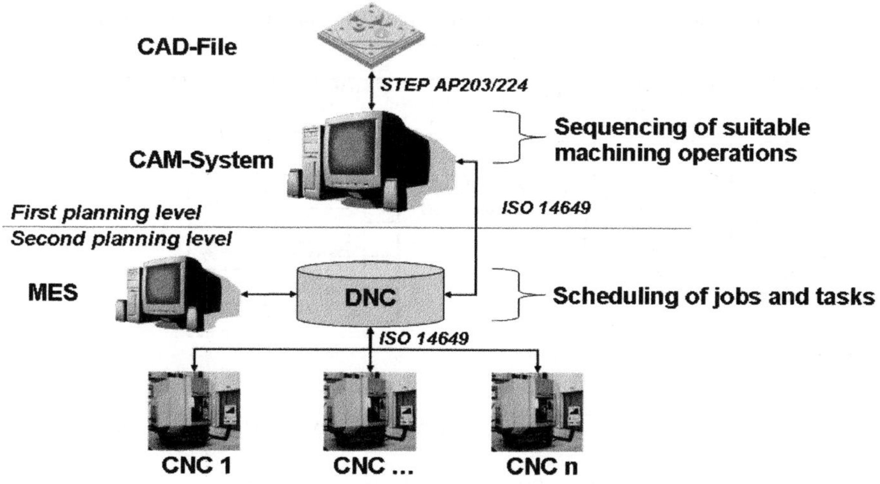

Figure 3.4. Identifying different planning levels in the STEP-NC process chain

The first planning level describes the selection and ordering of machining operations including the already mentioned parameters (such as tool, machine, technology, path, etc.). The planning horizon is the Workpiece with all geometrical and technical elements. Additional available information about the machining operation strategies and the technologies belong to the planning horizon.

The second planning level schedules the generated NC programs based on the Workplans into current manufacturing situation at the shop-floor. This task is performed by MES and/or PPS [3.7, 3.11]. The planning horizon in this case is the current job scheduling and capacity planning.

The following section deals with the first planning level. Therefore different planning methods are introduced that can be considered as potential planning approaches and methodologies for STEP-NC process planning. For the sake of completeness, a survey of methods for the scheduling of machining jobs in the second planning level is outlined.

Investigation of Process Planning Approaches

It was proved in the late 1980s that the automatic process planning can be done with the help of the computer [3.9]. Nevertheless, this problem and especially the scheduling job problem is an NP-hard (nondeterministic polynomial-time hard) problem [3.27]. Thus, different approaches were introduced dealing with this problem in scope of the generation of Workplans for the NC machining and of scheduling the jobs into the shop-floor. In Figure 3.5, a selection of suitable methods for sequencing of machining operations is presented. In the following, one rule-based planning method and one heuristical approach are outlined as related work that can be used for STEP-NC machining operations. Finally, a short survey of AI based approaches is presented.

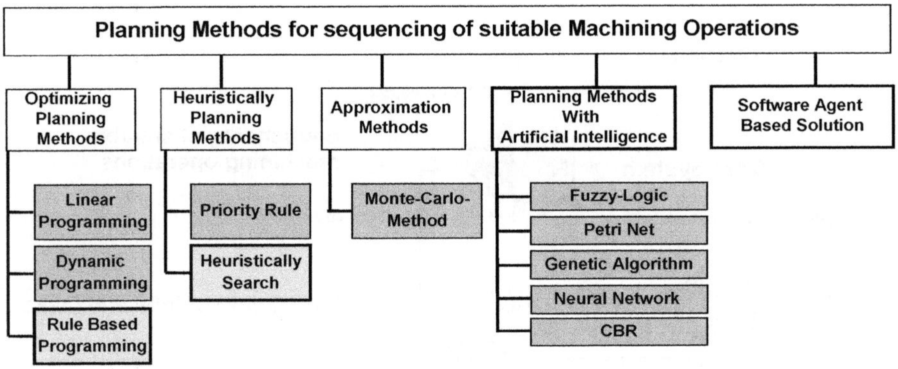

Figure 3.5. Planning methods for the sequencing of suitable machining operations [3.44]

Sormaz developed a system for rule- and feature-based process selection. This system uses 32 rules that map the design feature of a workpiece into the suitable machining feature supported by several machining technology databases. The knowledge representation is XML based and all selected process parts are validated with the help of virtual simulation before generating the NC program code [3.28]. The limit here is the complex handling of conflicting rules.

Hellberg introduced a heuristic algorithm, which used so-called "workpiece intermediate states" as kind of planning history for the step-by-step planning of the Workplan. The target is to overcome the limitations of the direct rule-based planning approaches. Therefore, different rules are also introduced which determine the application of the machining operations. With the help of the planning history, it is possible to select and compare different planning alternatives in the ordering phase. Thus, a backtracking in planning process is possible in the case a planning branch is not successful [3.10].

Finally, in the field of the AI, there are existing different algorithms and methods for the sequencing of machining operations [3.12]. These are Fuzzy-Logic, Petri-Net, Genetic Algorithm (GA) [3.13], and a Neural Network (NN) for selecting technology parameters [3.6,3.14] and agent-based process planning [3.32]. Finally, Case-Based-Reasoning (CBR) is used in order to map new design feature to already given design feature by comparison. Afterwards the assigned machining features are selected and manually sequenced [3.17].

For the sake of completeness, a survey of methods for the scheduling of machining jobs in the second planning level are investigated by Blazewicz [3.2]. Common used methodologies are multi-agent systems for reacting for dynamic changes and disturbances on the shop-floor [3.27], Neural Network for scheduling different jobs on parallel machines [3.29], optimization function for scheduling jobs on one machine considering preventive maintenance using dynamic programming algorithms [3.30], and finally using a combination of simulated annealing (SA) and Genetic Algorithm for job scheduling [3.31].

In conclusion, a lot of different process planning methods for sequencing of (STEP-NC based) machining operations and scheduling of jobs were introduced in the shop-floor. The knowledge representation and processing is done with methods know from the AI such as Neural Networks and Genetic Algorithms. Nevertheless, software agents-based systems and optimization approach based on heuristic algorithm and dynamic programming are often used too. The selection of the suitable knowledge representation and processing depend on several criteria.

Mapping to Practical Software Solutions

After the introduction of process planning approaches for STEP-NC based machining operations, the investigation of market surveys outlines the practical potential of these approaches. Actual market surveys and investigation by CIMdata and CAD-CAM-Report showed that the offered CAM-Systems (in 2006) do not support the automatic process planning in a satisfactory way [3.3, 3.4]. It was shown that feature-based programming is state-of-the-art. Furthermore, the STEP interface for sharing CAD model is supported. Nevertheless, the CAM systems often suggest different alternatives for the selection and sequencing of the machining operations given by features. Thus, the planner still has to select one alternative out of them as the machining operation selection. Afterwards, several CAM systems support the planner in ordering the selected machining operation to given fixed requirements. Thus, the process planning at the first planning level is still done manually. As a consequence, the process planner has to acknowledge the selection of the machining feature and the machining operations corresponding to the design feature of the workpiece manually. This task consumes a lot of time. The objective is the automatic sequencing of STEP-NC based machining feature and machining operations regarding a given Workplan and the given machining requirements. The benefits will be the time saving in the generation of Workplans.

3.2.5 Mathematical Formulizing of the Statement of Problem

The statement of problem for sequencing machining operations can be formulized as follows.

Let *E* be a function that calculates the effort of a machining feature *MF* regarding to a given set of assessment weights selection *AWS* (e.g., time, cost, etc.):

$$E(MF, AWS) \rightarrow \Re \tag{3.1}$$

The overall objectives of the process planning in the first planning level are the time saving by computing an optimal Workplan and the effort reduction of machining the complete Workplan. Thus, the following objective function (3.2) with given assessment weights selection *AWS* and *l* machining features have to be computed:

$$Min \sum_{i=0}^{l} E(MF_i, AWS) \tag{3.2}$$

Let *EM* be a function that calculates the machining effort of a machining operations *MO* to a given sets of assessment weights selection *AWS* (e.g., time, cost, etc.) and let *EC* be a function that calculates the effort to change a setup of *MO1* to the next *MO2* in the Workplan:

$$EM(MO, AWS) \rightarrow \Re \tag{3.3}$$

$$EC(MO_{1,} MO_2, AWS) \rightarrow \Re \tag{3.4}$$

Let *ST* be a function that each machining feature *MF* maps to a sequence of *p* machining operations *MO* as an ordered set under the limitations *L*:

$$ST(MF, L) \rightarrow (MO_1, \ldots, MO_p) \tag{3.5}$$

For a Workplan with *q* machining features which were mapped with (3.4) into *m* machining operations and given assessment weights selection *AWS* (e.g., time, cost, etc.), the total effort for machining the Workplan can be calculated with (3.6). The objective will be the reduction of the total effort:

$$Min \sum_{k=1}^{m} (EC(MO_{k-1}, MO_k, AWS) + EM(MO_k, AWS)) \tag{3.5}$$

The total effort can be reduced by the selection and ordering of machining operations as described in (3.6). For the sequencing of m non-parallel machining operations the total number of solutions is *((m)!)*. Therefore two aspects are important. At first, suitable machining operations that meet the given requirements has to be selected (3.5) and second, the machining operations have to be ordered in a way that the lowest changing effort in total can be achieved (3.6).

For the sake of completeness, the following equation has to be defined. Let SE be a function that maps each design feature DF to an ordered set of r alternative MF.

$$SE(DF) \rightarrow (MF_1, \ldots, MF_r) \tag{3.7}$$

3.3 Objectives and Requirements for Sequencing of Machining Operations

The role of the objective function (3.6) is to optimize the effort of machining Workplans. Thus, the need of a new solution for that function going beyond the actual state of the art is explained. This leads to identifying and defining the objectives and requirements for a need solution.

3.3.1 Reasoning for a New Solution

The presented state-of-the-art methods and approach are not developed for a STEP-NC based solution. All presented methods used their own knowledge representation which is independent from the STEP-NC standard. Therefore, additional information exchange between a STEP-NC data model and the internally used knowledge representation of each presented methods is necessary. As a consequence, information lacks are possible. A STEP-NC based solution is possible, because of the availability of STEP-NC based controller [3.23]. Furthermore, the presented solutions are not implemented in the CAM systems today. The introduced automatic rule-based process planning algorithms are exclusively successful in preparing Workplans in small machining applications fields. However, these so-called expert systems are very difficult to handle as mentioned before. As a result, the process planner selects the machining operation due to his subjective experience. Neural Network based approaches are common today [3.6, 3.14]. Nevertheless, the result computation is not traceable. Furthermore, it was shown by Hellberg that Genetic Algorithms are not usable to find such good sequencing result like heuristic algorithms are [3.11]. The CBR based algorithm by Gerken is also not usable, because it is based on subjective validation. As a consequence, the planner is using an often used but not optimal sequencing proposal. To overcome this gap between research and application, an innovative and easy-to-use approach has to be introduced by considering the following requirements.

3.3.2 Objectives and Requirements of a New Solution

The organizational objective of a new solution is to select and reorder the Workplan in a traceable and customized way. The fundamental basis is the use of process knowledge provided by the personnel involved in the NC process chain and process monitoring. As a result, the use of feature technology and STEP-NC along the whole NC process chain is an objective. These features have to be used to describe the physical design of the workpiece (CAD model) and the assigned best-practice machining operations to machine the workpiece [3.9]. A feedback from the shop-floor has to be established to guarantee and validate an optimal assignment of machining operation to design features. An essential aspect is to enable the feedback of operators back to the process planning about unsuccessful machining operations. Summing up the objective and requirement are the following [3.44]:

- Shorten the programming time by introducing automatic process planning tools and algorithms
- Enhancement of the utilization and consideration of the whole available machining equipment (machine, clamping/fixtures, tools) and the functions of the CAM system
- Preparation of Workplans to given user defined design requirements (surface quality, roughness) and manufacturing criteria (cost, time, quality)
- Introducing an objective effort calculation procedure in order to assess machining operations and machining features
- Generating Workplans according to (3.5)
- Enabling knowledge acquisition and representation to develop a knowledge database (KB) for gathering process planner experience for re-use in automatic process planning
- Using a suitable knowledge representation for the knowledge database
- Developing a traceable knowledge processing for generating Workplans
- Implementing the approach on a knowledge-based NC programming system [3.1]

The following assumption has to be defined in order to develop the approach. In order to machine a design feature, the utilization of several various machining features and machining operations are possible. These machining features are specified by CNC machines, clamping/fixtures, tools, technologies, and strategies settings based on STEP-NC data model machining operations. The machining operations are assessable with several assessment criteria such as machining time, surface quality, roughness, collision potential, etc.

3.4 Approach for STEP-NC Machining Operations Sequencing

3.4.1 Methodology and Keynote

The development methodology of the approach for sequencing of STEP-NC machining operations revolve around the designing of an expert and/or knowledge-based system [3.39]. The main idea of an expert system is to empathise the decision making of a human being based on knowledge representation and processing.

Therefore, different knowledge representation and processing methods are highlighted in order to find suitable methods for designing a sequencing system.

The keynotes of the considered approach are the computation of assessed and alternative machining operations based on STEP-NC. Hence, these machining operations have to be stored in a central database in order to gather the planner knowledge in a structural manner and to re-use it. The machining operations are rated through metrics that are dependent upon their machining characteristics. The assessed machining operation will be utilized later in order to select and order suitable machining operations for a given Workplan in an optimized way. STEP-NC and a directed graph that shows the alternative of a Workplan are used as knowledge representation. Furthermore, algorithms from the graph theory are used to process the information that is stored in the directed graph to generate the Workplan. Heuristics are used to overcome the NP-hard problem of scheduling the machining operations in an optimal way in shortening the run-time [3.27]. This approach has to be integrated in the current NC process chain. As a consequence of the approach, benefits are feasible such as time-saving in the process planning and cost saving by using all available machining technology.

3.4.2 Functional Principle and Application Scope

In general, the focus of this approach is the machining process planning for feature-based workpiece based on ISO 10303 Part 224 "Mechanical product definition for process plans using machining features" [3.25]. The workpiece will be machined with geometrical defined cutting edge regarding to the DIN 8589 [3.33].

The functional principle is explained with the help of Figure 3.6 (top). Nowadays, it is state-of-the-art that the process planner selects the corresponding machining features and the machining operations for each design feature. This selection is done by means of the experience of the planner. The deficit is the loss of planner expertise in case the planner leaves the department. To overcome this problem, the proposed approach maps a list of design features into a list of machining operations in a mathematically defined way. The mapping is done into three sub-steps in detail and is outlined in Figure 3.6 (bottom).

The first step imports the CAD model of the workpiece which consists of design features. This data is provided by a STEP based CAD interface. The CAD features will be represented by a list of design features with constraint connections within them. Afterwards, the design features are grouped (I) by defined sorting criteria such as the same kind of feature. An example is given in Figure 3.6 (graph a).

The next step is the application of the structuring function (II). Therefore, the design features have to be mapped into suitable machining features and machining operations. The knowledge database is used to store such mappings. The database includes all already made mappings from previous Workplans. Thus, one machining feature can be represented by several alternative (various) sequences of machining operations (see Figure 3.6, graph b). Now, the alternatives have to be reduced in each horizontal level. This task is done by the structuring algorithm that selects the combination of machining operations which has the lowest effort in total as shown in Figure 3.6 (graph b, thicker links). Afterwards, the order of the selected machining operations has to be changed, due to the fact that operations on the same machining, using the same clamping and/or tool, can be machining one after the other in order to save effort for changing setups. Therefore additionally constraint has to be added to the graph (see Figure 3.6, graph c).

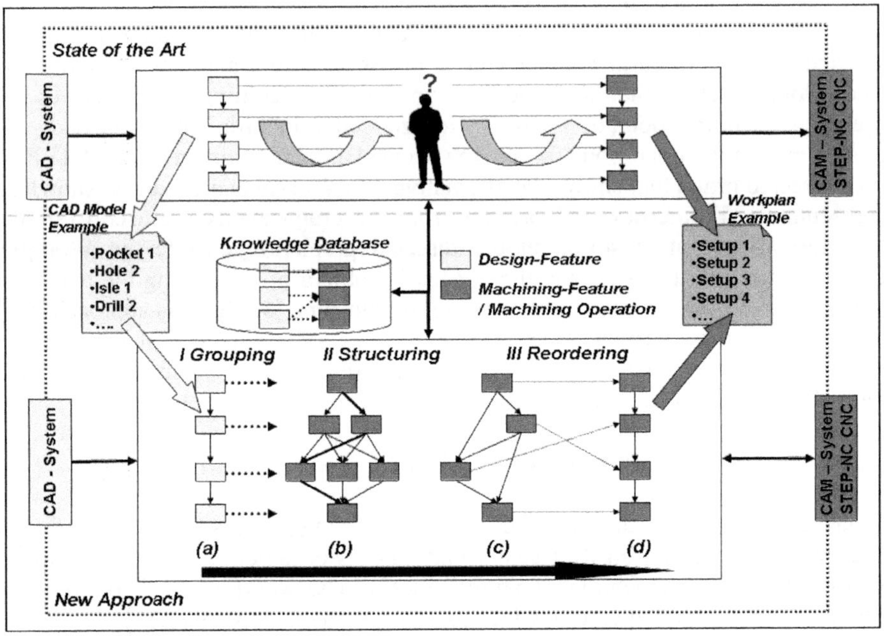

Figure 3.6. Functional principle of approach in comparison with the state-of-the-art [3.44]

The reordering (III) algorithm as the third step changes the sequence of the operation due to the mentioned fact of effort saving. The result is a sequence of suitable machining operations corresponding to the given design features and to the process planner defined criteria (see Figure 3.6, graph d). Finally, the machining features list is provided to the CAM-System that generates the corresponding NC program based on ISO 14649. In order to meet interoperability requirements concerning given ISO 6983 based controller, the transfer to a CAM system in order to calculate "G/M" code is possible optionally.

Based on the functional principle, there are several issues to be defined in order to realize this approach. First, the transfer of the Workplan into a suitable knowledge representation has to be determined. Furthermore, the database structure and assessment of machining operations has to be defined. The next step is introduction of effort and effort calculation rules based on the assessment to enable the selection of alternative machining features and machining operations. The last step is the investigation of suitable algorithms in order to process the desired structuring and reordering functions. Finally, an overall architecture of such an approach has to be determined and a work flow has to be specified in order to transfer it into a real application as a demonstrator.

3.4.3 Architecture and Modules

The architecture of the approach is modular and consists of four main components. As already mentioned, the Workplan has to be sequenced to the given criteria. This task will be done by the first module called "Process Planning Tool". As seen in Figure 3.7, this module has access to the CAD system in order to receive the workpiece description. Furthermore, an access to the STEP-NC based CAM-System (e.g., provided by RWTH Aachen [3.23]) is enabled to process the sequencing of machining operations. Afterwards, the Workplan is sent to the machinery at shop-floor level. The second module "Process Visualization Tool" is introduced in order to meet the requirement (e.g., traceable Workplan generation). This tool illustrates the current discussed Workplan in a graphical way. Thus, the process planner can easily interact with the system.

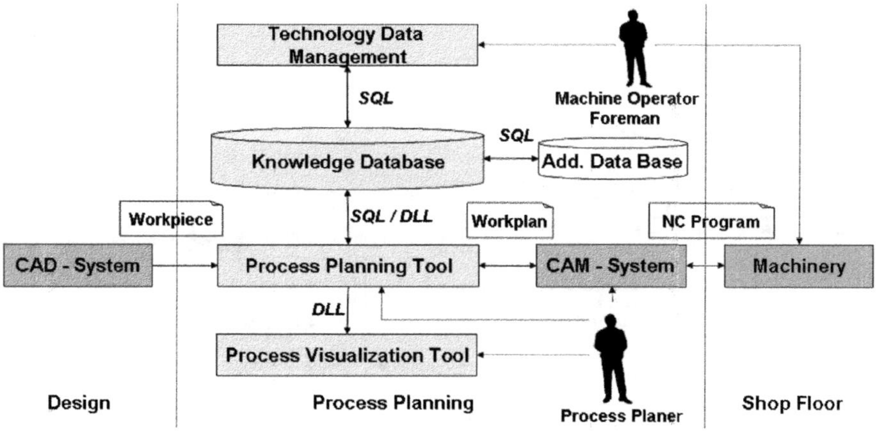

Figure 3.7. Architecture and modules of the approach

Furthermore, the Workplans are generated with assessed machining operations as objective. Therefore, two additional modules have to be introduced. First, these operations have to be stored in the third introduced module called "Knowledge Database". This database has additional access to additional databases. These databases include information and parameter settings about tools and machines that can be used in the machinery. Second, the machining operations have to be assessed with the help of the introduced fourth module "Technology Data Management". The machine operators can now access the individual machining operations.

3.4.4 Model for Workplan Representation and Processing

Investigation of Workplan Representation

The selection of a model for the representation of Workplans is discussed in this section. The result is the selection of a suitable representation.

Feature-based concepts are well known established structures in order to define a Workplan [3.16, 3.17, 3.20, 3.22]. Therefore, Cai and Gerken defined the machining process plan elements in a hierarchical way as shown in Figure 3.8 (right). In comparison to that, the ISO 14649 defined fewer process plan elements as shown in Figure 3.8 (left). Both definitions include an element called "Machining Operation". Nevertheless, the concept level of both elements is equal. Thus, the STEP-NC based machining operation is used in order to represent the basic element of a machining Workplan. Furthermore, a machining feature is represented by further machining operations to realize (3.5). Additionally, the VDI2218 introduced feature mapping for assigning features from different applications scope [3.20]. This methodology is addressed to connect the design and the corresponding machining features to realize (3.7). An additional entity "Assessment" is linked to the element "Machining Operations" in order to store the assessment of machining features and machining operations. In conclusion, the STEP-NC data model is used as the Workplan representation.

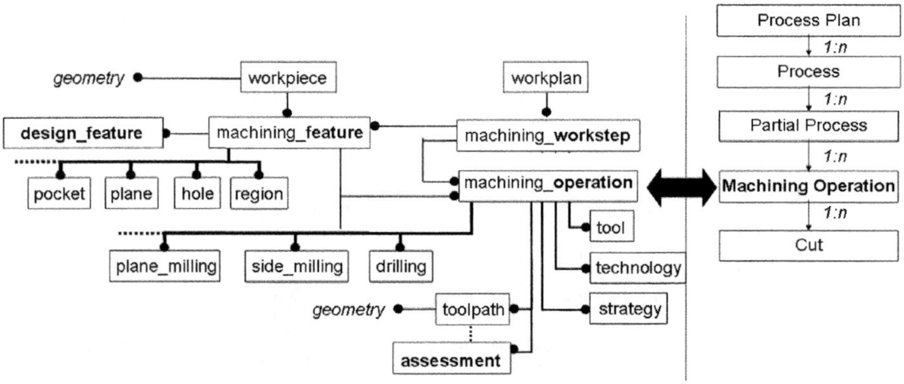

Figure 3.8. Linking between design, machining feature and machining operation

Investigation of Workplan Processing

A suitable concept that addresses the operations of grouping, structuring, and reordering has to be found in order to process the information of a Workplan. Therefore requirements have to be determined and suitable available concepts have to be benchmarked. The knowledge processing structure should meet the following requirements. Alternative Workplans are used to machine the same machining tasks. Therefore, the requirements for the concept are handling time dependency, outlining different alternatives in a Workplan, and handling and assessing machining operations. Furthermore, the application of traceable and well known algorithms for grouping, structuring and reordering alternative machining operation is required. Thus, the following knowledge representation and knowledge processing concepts are investigated: Neural Networks, Genetic Algorithm, formal rules-based languages, and methods based on the graph theory [3.18].

Genetic Algorithms are based on a very interesting concept to generate sequences of machining operations by combination and mutation of machining operation orders. A suitable method is the character coding. Conversely, it was shown by Hellberg that GA based processing is not efficient in the scope of sequencing machining operations [3.10].

Neural Networks are a set of "neurons", which are organized in an input layer, a processing layer and an output layer. NN have to be trained with special training algorithms. The disadvantage of this "black box" concept is that the Neural Networks are processing information in a non-traceable way [3.18]. Thus, Neural Networks do not conform to the requirements.

Formal and rule-based languages are used to formalize a given well structured domain under discussion [3.18]. Nevertheless, algorithms for processing structures such as Workplans are missing. Thus, this concept is not suited, too.

The graph theory fulfils the given requirements as the last concept. First, it is possible to model time dependency with time conditioned connector between the nodes. Second, the different paths inside the graph are utilized to represent the alternative Workplans. By introducing efforts/costs for passing paths, the assessment of sub-Workplans and Workplans are enabled. Finally, the use of algorithms introduced in the graph theory such as the Floyd-Warshall (FW) algorithm and the TSP (Travelling Salesman Problem) solving algorithm enable the processing of the Workplan [3.15]. The methods from the combinatoric are used to transfer the task of generating an optimized Workplan as operations research problem.

Defining the Workplan Processing

A Workplan with all alternative machining operations can be transferred into a directed graph. A directed graph DG is defined as an arrangement of a set of nodes V and a set of links E. These links are connecting two nodes out of the set of nodes [3.8]. By mapping the Workplan into the directed graph, the set of machining operations are represented by the set of nodes. The links between the nodes are representing time-related dependences between two machining operations. A link $e = (a,b)$ means that the machining operation a has to be machined before the machining operation b. Two different time dependences have to be modelled. On the one hand, the machining dependence is caused by the use of different machining technologies. For example, the roughing of a pocket must be done before finishing can be applied. On the other hand, the geometrical dependency for machining operations is modelled. That means that drilling inside a pocket has to be machined first. An example of an alternative Workplan is illustrated in Figure 3.10. In this graph, different alternative machining features for design features are shown. These design features are ordered in a vertical way. Consequently, the alternative machining features and the corresponding machining operations are ordered in a horizontal way. Each line represents alternative machining operations, which have to be applied to generate the Workplan.

Three aspects have to be introduced in the directed graph in order to assign and process the alternative Workplans. First, the machining operations have to be assessed with criteria that have to be determined. Based on that, effort calculation rules (in order to realise (3.3.x)) have to be introduced for the machining of an operation and the changing of setups between two machining operations, second. Finally, the algorithms have to be defined in such a way that they realize the function of grouping (I), structuring (II), and reordering (III) with the help of the assessed machining operations.

First, two criteria set are introduced for assessing machining operations, that are represented by C (3.8). The first set A includes criteria for assessing the machining characteristics of the machining operations. This set A includes criteria such as tool wear, run time, machining costs, and surface quality and roughness. Some of these criteria are feature-size independent (e.g., roughness, surface quality) and some depend on the geometrical size of the feature. As an example, the run time has to be calculated by an additional linked CAM-System that provides this information. The second set D includes criteria for assessing the setup changing between two setups of sequenced machining operations. This set D includes criteria such as geometrical position of the machining features, the used tool, the used clamping and the used machine.

Second, additional weights W (27.2) have to be introduced in order to define calculation rules of the effort in the Workplan. The main idea is to customise the effort calculation corresponding to the given manufacturing scenarios. A manufacturing scenario includes objectives and weighting factor for the given overall manufacturing metrics (time, cost, quality). In detail, the weights MW define the weight of these manufacturing criteria. The additional weights SW define the sub weights, due to the fact that the assessed and normalised criteria A and D are combined in the effort calculation. Each criterion from A or D influences only one overall manufacturing metric. Thus, an effort can be calculated by adding the product of the corresponding assessed criterion with the corresponding sub-weight SW and corresponding manufacturing weight MW to a total sum. Therefore, the already introduced assessment weights selection AWS is represented by the set C and W now (3.10). Thus, the effort for changing between two machining operations MO setups as link is defined as shown (3.11). The effort for machining a given machining operation MO is described by (3.12):

$$C = (D, A) \tag{3.8}$$

$$W = (MW, SW) \tag{3.9}$$

$$AWS = (C, W) \tag{3.10}$$

$$EC(MO_1, MO_2, AWS) = \sum_{k=1}^{o} (MW_1 * SW_{1,k} * D_{1,k}(MO_1, MO_2)) \tag{3.11}$$

$$EM(MO_1, AWS) = \sum_{i=1}^{n} \sum_{k=1}^{m} (MW_i * SW_{i,k} * A_{i,k}(MO_1)) \tag{3.12}$$

The multi-factor effort calculation enables multi-parameter decisions such as selection of machining operations. As a result, the given Workplan consisting of machining operations can be assessed with the help of this effort equation. Thus, the assessment helps to select the best suitable Workplan under several alternative Workplans, which fulfilled (3.5). A further advantage is that the multi-parameters decision of sequencing machining operations is mapped with the help of weights into a one parameter (effort) decision. Thus, the effort calculation depends on the criteria which the process planner can determine. Now, it is possible to change the value of an effort corresponding to a given scenario which is described by a concrete specification of these criteria.

Finally, it is possible to use algorithms known from the graph theory in order to compute significant paths through the directed graph with the help of the effort calculation rules. The overall objective is to find a solution of (3.6). Thus, two major algorithms can be identified that are suitable for the application in the field of NC machining [3.1]. On the one hand, the Floyd-Warshall algorithm calculates the shortest distance (as a total sum of link efforts) between two nodes (machining operations) with the help of the transitive closure [3.8]. With the help of this algorithm, different alternative machining operations for the same machining feature can be rated. Thus, the function structuring (II) of the approach is realized. The time complexity is $O(n^3)$, where n is the amount of nodes. On the other hand, the TSP solving algorithms can be executed in order to reorder (III) all machining operations in a Workplan in order to get the lowest effort in total to machine all machining operations in a row (as Hamiltonian Path). The TSP is NP-hard [3.8]. Consequently, an optimal Hamiltonian Path can be calculated by comparison of all machining operations combination. Therefore, the time complexity is $O(n!)$, where n is the amount of nodes. This high time complexity of the algorithm is not acceptable. Several algorithms were introduced (Christofides algorithm, nearest/farthest insert algorithm, greedy algorithm, all nearest neighbours algorithm, double minimum-spanning-tree, space filling curve heuristic) in order to find a heuristic solution for a Hamiltonian Path in acceptable runtime. The use of these heuristic algorithms shortens the runtime of the combinatoric algorithm [3.8, 3.40, 3.41] drastically. As a consequence, the function of reordering is realized with such a heuristic algorithm. At least, the first function grouping (I) is done by a common sorting algorithm. Table 3.1 sums up the explored and used algorithms. It is shown, that a common sorting algorithm for the grouping is used due to the low complexity. Furthermore, the FW and TSP are used to realize the functions (II) and (III). Nevertheless the complexity of the TSP is too high. Therefore a heuristic algorithm will be introduced in the next section.

Table 3.1. Exploring different algorithm to realize the approaches functions (I)–(III)

Algorithm	Result	Time complexity
Function (I) Grouping		
Sorting algorithm (Quicksort)	1 order	$O(n^2)$, *n nodes*
Permutation	n order	$O(n!)$, *n nodes*
Function (II) Structuring		
FW algorithm	1 order	$O(n^3)$, *n nodes*
Variation	a^b order	$O(a^b)$, *a alternatives per each nodes, b nodes,* $n = a * b$, $b = const.$
Function (III) Reordering		
TSP algorithm	1 sequence	$O(n!)$, *n nodes; without heuristics*

3.4.5 Workflow for the Knowledge-based NC Programming System

The workflow of the proposed approach is specified based on the mathematical background and also the methodology of the approach. A specimen Workplan as shown in Figure 3.10 (left) is used in order to exemplify the approach.

As already mentioned, the process planning tool is responsible for the sequencing of machining operations. Thus, the tool has access to the knowledge database, the visualization tool, and the CAD/CAM systems. In Figure 3.9 the workflow of the approach is presented in a detailed way. The workflow is divided into three operations (I)–(III). The objective of the approach is the generation of a Workplan for machining a given workpiece. Therefore, a feature-based CAD model with design features has to be investigated. At first, the design feature has to be imported and grouped. The ordering is done by a sorting algorithm that groups the design feature list by similar feature types. The first step of grouping is done as shown in Figure 3.6 (graph a).

The second step is the mapping of the design feature list into suitable alternative machining features and operations. Afterwards, the alternative operations have to skip. Thus, an example for a Workplan is used to describe this second and the third step. An example of CAD model description and Workplan is given in Figure 3.10 (left), The CAD model consists of three design features (DF) which can be mapped into six machining features (MF). Each of the machining features is represented by machining operations. These mappings are stored in the knowledge database. Going into the detail, the machining features MF1a and MF1b have a shared machining operation MO1 (e.g., roughing) and two alternative steps MO2a (e.g., finishing 3-axis) and MO2b (e.g., finishing 5-axis). The machining feature MF2 is represented by only one machining operation. The last design feature can be mapped into three alternatives machining features (MF3a, MF3b, and MF3c). These MF are sharing one machine operation MO4 (e.g., roughing) and three alternative machining operations (MO5a, MO5b, and MO5c). Two sets have to be defined in order to map the introduced list of design features into the corresponding alternative machining features and to apply the effort calculation rules (3.11) and (3.12). At first, the limitations L for the technology and strategy parameters of each machining features and operation have to be determined. The process planner can define these limitations given to the available resources at the machinery. The second set is the weights W in order to define the used manufacturing scenarios for calculating the effort of Workplans. Now the suitable machining operations corresponding to the given machining features are selected by (3.5) for each mapped machining features. The mappings are stored in knowledge database.

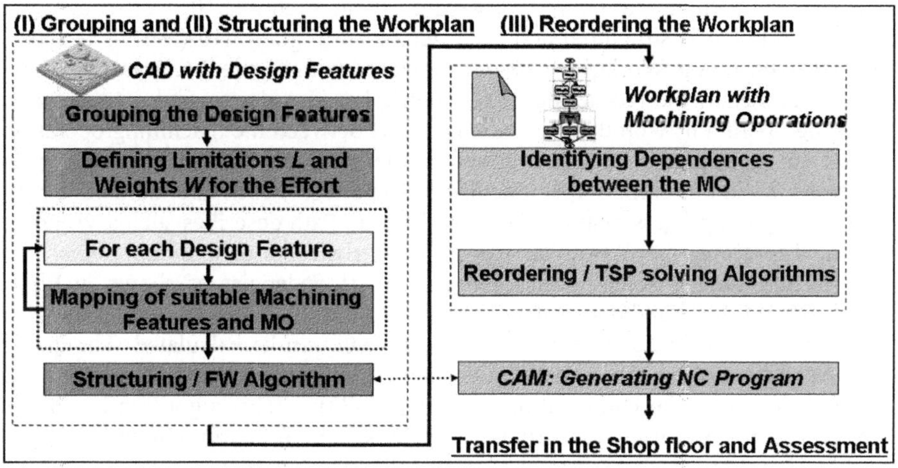

Figure 3.9. Workflow of the approach

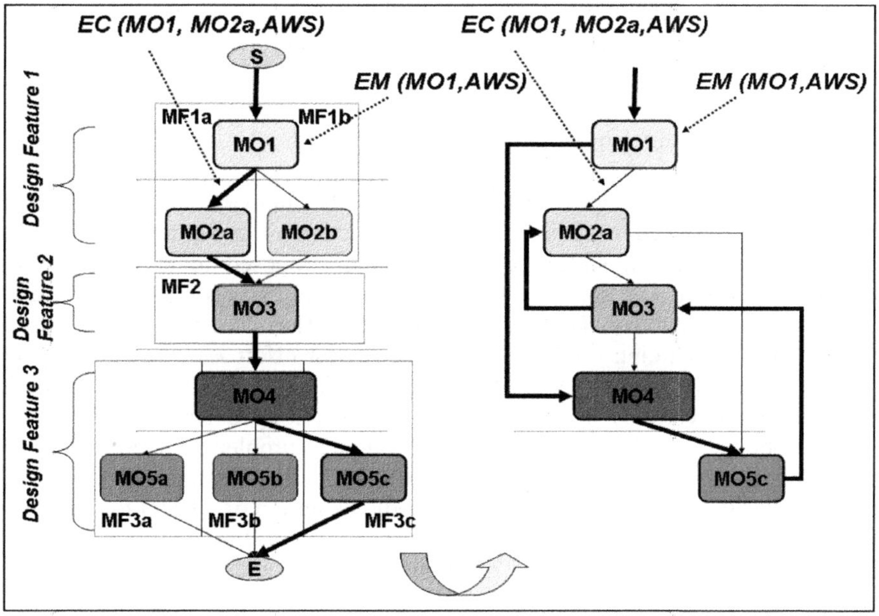

Figure 3.10. Specimen Workplan as directed graph (*left*) with Hamiltonian Path (*right*)

The next step is the calculation of the effort for each node (machining operation) and link (dependencies) with the help of a CAM system for the feature-size dependent criteria and the introduced calculation rules (3.11) and (3.12). Afterwards, the process planer selects the best-fit alternative machining operations with the FW algorithm that calculates the "shortest path" inside the graph. Therefore, the additional start node "S" and end node "E" have to be added. As a result, the marked machining operations in a horizontal line represent the operations with the lowest effort to the given limitation L and weights W in Figure 3.10. Consequently, a suitable Workplan is structured and bold marked in Figure 3.10 (left). The second step of structuring the Workplan is done (compared with Figure 3.6, graph b).

The last step is the reordering of the Workplan in order to find a sequence of machining operations with the least sum of effort between the machining operations in total to meet (3.6). This Workplan is the optimal Workplan to the given limitations L and specified manufacturing scenarios, which are defined by the weights W by the process planner. The Hamiltonian Path describes such a Workplan as path. Therefore, the Workplan has to be enriched with additional links that represent additional machining and geometrical dependencies as shown in Figure 3.10 (right). All links and nodes are assessed with the calculation rule EC (3.11) and EM (3.12). Afterwards, a suitable Hamiltonian Path will be calculated. The quality of the path depends on the executed algorithm. As already discussed, the best path can be achieved by the combination of all possibilities. This algorithm has a high time complexity. The uses of heuristic algorithms shorten the runtime drastically. A possible Hamiltonian Path through the example is shown in Figure 3.10 (right, thicker marked). The last step is done as shown in Figure 3.6 (graph d). Finally, the optimized NC Workplan is found. Optionally, a CAM system generates the NC program corresponding to the given Workplan by translating the machining operations. The scheduled NC programs will be dispatched to the corresponding machine. Thus, the operator has now the possibility to assess the machining operations for re-using them.

3.4.6 Approach for Sequencing Algorithm

Three different functions have to be assigned with automatic executed algorithm based on the presented workflow. The first function of grouping (I) the design features is done with a sorting algorithm. It is proposed to use the common Quicksort algorithm. The second function of structuring (II) is realized by the FW algorithm presented above. The FW minimized the effort of machining features by selecting these machining operations with the lowest effort EM. Thus, the second part of the objective at equation (3.6) is minimized. The third function of reordering the machining operations has to be done by computing a Hamiltonian Path as described above. Calculating a HP is a NP-complete problem. Therefore the TSP solving algorithm is used to compute the HP [3.8, 3.42]. As already discussed, a heuristic algorithm reduces the runtime for finding solution of a Hamiltonian Path in an acceptable runtime [3.8, 3.40, 3.41]. Heuristics algorithms use assumptions to shorten the amount of investigated solution. As a consequence, the complexity is decreased [3.41]. It is shown by Turau that it is not possible to validate the quality of the HP of a heuristic result in a mathematical way [3.40]. But it was shown by

Reinelt by empirical experiments that heuristic algorithms have fixed quality borders to validate the difference to the optimal solutions.

For the presented approach, the "All nearest neighbour" (ANN) algorithm will be used, due to the fact that the ANN is a simple heuristic algorithm. The main idea is to sequence machining operations (nodes) in such a way that the effort between two operations is as low as possible. This means that the setup changing effort is as low as possible for each operation pair. However, the first part of (3.6) is minimized in such a way that the effort function *EC* is minimized for each operation pair. Deadlocks are possible due to the fact that the nearest node (machining operation) can influence other nodes. Therefore, different dependencies are added between the nodes by geometrical and machining dependencies. Thus, a three step heuristic algorithm will be used as shown in Figure 3.11.

```
// Init. of the Variables
allNodes[]
path[] = EMPTY, paths[][] = EMPTY
restNodes[] = EMPTY, possibleNodes[] = EMPTY
lastNode = EMPTY, startNode = EMPTY
// Selecting each Node in the Graph as a starting nodes of a HP
FOR EACH Nodes startNode FROM IndependentNodes (allNodes[],allNodes[])
   path[] = startNode,
   restNodes[] = allNodes[] – startNode,
   lastNode = startNode
   // Computation of the next Node which will be visited after the last note in current sequence
   WHILE restNodes[] NOT EMPTY
      // (1) Determining all possible Notes with a connection of the last note in current sequence as a Set
      possibleNodes[] = ConnectingNodes (lastNode, restNodes)
      // (2) Determining all Notes that are also independent from the Set
      possibleNodes[] = IndependentNodes (possibleNodes[], restNodes[])
      // (3) Determining the Note with the shortest Distance now from the Set
      lastNode = ShortestDistance (possibleNodes[], lastNode)
      path[] = path[] + lastNode
      restNodes[] = restNodes[] – lastNode
   LOOP
   // Storing the HP corresponding to the starting Node
   paths[startNode][]=path[]
LOOP
// (4) Selecting the shortest path
Path[] = shortestPath (paths[][])
```

Figure 3.11. Heuristic algorithm ANN [3.44]

In general, the algorithm selects one shortest path by calculating different HP with regard to all possible starting nodes in the directed graph. The central idea of this algorithm is to reduce the desired possible nodes which can be selected as next node in a loop. Therefore three functions are used. The first function (1) *ConnectingNodes* selects all possible nodes which are accessible from the last selected planning node. Afterwards, this selection is reduced to all nodes which are independent in the scope of machining and geometric by the function (2) *IndependentNodes*. The results are all nodes, which can be accessed without having a deadlock. The last step is now the selection of node with the shortest connection (via the effort) to the last planning node. This task is done by the function (3) *ShortestDistance*. As a result, the nearest node was selected for each inner loop. The inner loop for each starting node is finished when all nodes in the graph are included. Afterwards, the shortest HP from all HP is selected with the function (4) *shortestPath*. This HP represents the sequence of the machining operations.

3.5 Technical Realization and Evaluation

3.5.1 Technical Realization

The approach is implemented as an ongoing process in a demonstrator. The demonstrator consists of the four individual software modules, which are introduced as modules of the approach. The module "Process Planning Tool" is implemented using Visual Basic. The current state of the tool allows the importing of CAD models in order to map the design features into machining features and machining operations. Afterwards, the reduction of alternatives is computed by the FW function. Actually, the ANN algorithm as presented above is finalized. Thus, all three functions of grouping, structuring and reordering are implemented. The further three modules are already implemented, too. The "Process Visualisation Tool" is used to represent the alternative Workplans as a net in MS Visio and MS Excel. Thus, the process planner can easily retrace the process planning with the help of the charted Workplan and Gantt chart. The "Knowledge Database" is also implemented in MS Access and includes 21 tables. These tables are storing all relevant information for the mapping of design features into assessed machining features and operations. Furthermore, the created Workplans are stored. An interface to the current STEP-NC data model of machining operation is planned and in the current designing phase. The last software module "Technology Data Management" allows the machine operator to assess the generated Workplan with the machining operations with a form-based interface to the knowledge database in a comfortable way.

3.5.2 Evaluation of the Approach

The evaluation will be done by the process planning for the machining of the NCG (NC Gesellschaft) Benchmark Part 2004 (see Figure 3.12, left). The benchmark park consists of important and representative features that are used in industry. Furthermore, the future validation will be done with the help of a real industrial part. This part is provided by "Nisaform, s.r.o." (see Figure 3.12 right) and includes representative features of die and mould-maker industry.

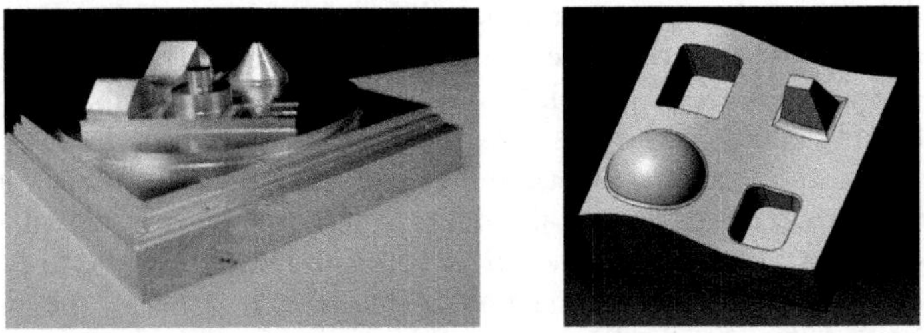

Figure 3.12. Machined NCG Benchmark Part [3.35] and Nisaform benchmark part

The used NCG benchmark part consists of several design features. The corresponding machining features are already assessed and stored in the database of the demonstrator. The demonstrator has applied with an example Workplan consisting of six design features. In Figure 3.13, this example is shown. At first the grouping (a) of the design feature was done. Afterwards the six design features were mapped to the machining features (b) with the help of the stored mappings. That means each design feature was mapped into all possible alternative machining features. As seen in Figure 3.13, the alternatives machining features are illustrate by a directed graph. Afterwards, theses machining features were mapped (c) into several machining operations corresponding to the content of the knowledge database and to given limitations (e.g., using five axes and all tools). Afterwards, the weights *W* for the effort equation were defined. With the help of the given criteria sets *A* (tool wear, machining costs, collision potential, surface quality, roughness) and *D* (machine, clamping, tool, geometric position) the alternative machining operations were benchmarked. Then the structuring was done. The results are the highlighted machining operations in the morphological box and in the directed graph. These machining operations are divided into roughing and finishing operation.

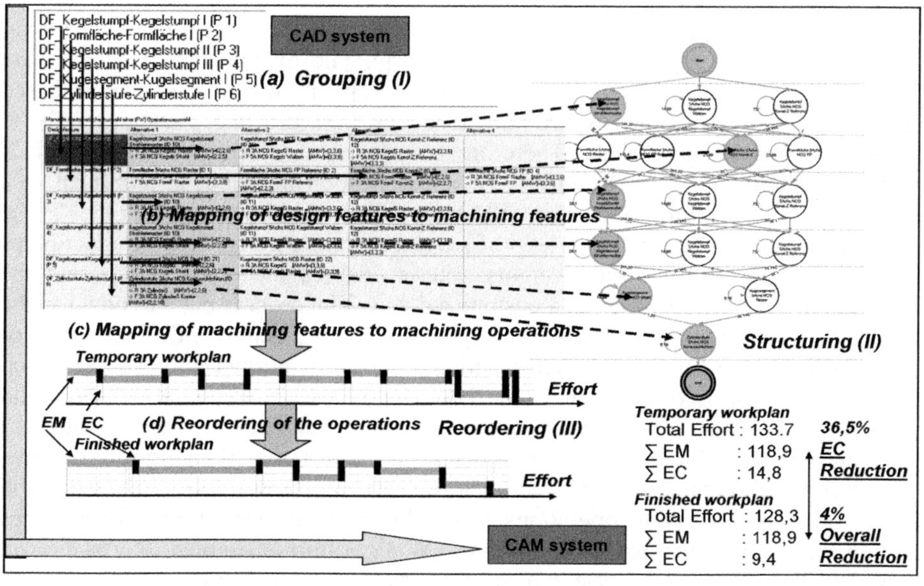

Figure 3.13. Example evaluation for effort saving

Afterwards, a temporary Workplan represented by a Gantt chart as seen in Figure 3.13 was created. Here, the grey boxes represent the effort of machining the operations and the black boxes represent the effort of changing the setup between two machining operations. Finally, the temporary Workplan was reordered (d). This function is realised by the help of the ANN. As shown in Figure 3.13, the finished Workplan now comprehends a new order of machining operations. All the independent operations are grouped together. An effort saving (for changing the setups – *EC*) of over 36% and in total of 4% was achieved to the given limitations *L*, and weight *W* of the manufacturing scenario in this simple example. The first trials showed the applicability of the directed graph to model different alternative Workplans. The applicability of the FW and the ANN is also validated. Further test follows in order to confirm the first evaluation results.

3.6 Conclusion and Outlook

Nowadays, there are significant deficiencies in the information flow along the NC process chain. The reasons are the use of insufficient interfaces within the process chain. As a consequence, knowledge is kept in these uncoupled process chain phases. Nevertheless, the deficits could be solved by common integrated CAD/CAM software solutions and particularly by the introduction of the STEP-NC interface. Thus, modifications in the shop-floor are supported in an adequate way. Nevertheless, there are still deficits in automatic preparation of Workplans for STEP-NC-based NC machining. Therefore, the need for a new time saving solution for automatic re-using of expert knowledge was pointed out. The objective and requirement were formulated.

The approach presented deals with the automatic sequencing of machining operations under a given manufacturing scenario. The fundamental concept is the use of feature-based design and assessment. By introducing a feature-based knowledge representation, expert knowledge as best practice strategies will be stored in a knowledge database. Therefore, a mathematical description of Workplans based on STEP-NC was introduced. This specification describes a Workplan as a sequence of machining features with alternative machining operations. These operations are assessed with the help of several certain criteria. The introduction of manufacturing scenarios with weights criteria enabled the planner to assess the different machining operation in a customized way. The process planner now has the possibility to select different alternatives parameters for machining different areas of a workpiece regarding to a given machining scenario. Furthermore, the selection process is automated by using the three functions grouping, structuring and reordering. For that reason, different mathematical descriptions were investigated and rated. Finally, the graph theory was selected with the applied algorithm. Now, the machining operations based Workplan will be transferred into a directed graph with nodes as machining operations and links as quantified efforts dependences within them. Thus, the application of the FW algorithm is enabled. It was shown, that the FW calculates the best choice of alternative machining operations for each machining feature in order to reduce machining effort. Additionally, a heuristics algorithm for the TSP computes the optimal sequence of the selected machining steps in three step algorithm in order to reduce the effort for changing setups.

The advantages of the approach are the traceable and easy-to-handle generation of Workplan with sequenced machining operations. The benefits are the time saving and process planning for diverse machinery equipments. Thus, it is possible by modifying the limitations *L* and manufacturing scenarios to try out different alternative Workplans. As a result, the need for new machinery equipment can be felt and finally rated in relation to the effort and benefit of the new equipment. The next advantage is the high interoperability due to that fact that STEP-NC is supported as interface. Furthermore, the planning knowledge of the planner in selection technology and strategy for each machining feature and operations is stored in a central knowledge database in a structural way. The re-use of them are enabled.

Nevertheless, the approach has several areas for improvements that have to be discussed. Unknown design features are not processable because of missing mappings. Thus, these features have to be created. Furthermore, the approach is usable for commonly used feature based Workplans. Free-form features are very difficult to handle. Additionally, the approach needs a CAM system as a backbone for calculating the run time of each individual machining operation. This function can require a high effort when considering complex and huge Workplans. The ANN algorithm used a heuristically assumption to reorder the machining operations. The equation (3.6) will be solved by minimising the "local" effort between two machining operations pair. In general, the algorithm assumed that the global minimum of all effort is a sum of all minimised local efforts. Finally, the approach has not considered the current job scheduling at the shop-floor up till now, but integration into MES is possible.

The demonstrator showed in the first tests with the NCG benchmark part that the automatic process planning based on the FW and heuristic TSP algorithm is possible. A time reduction in the process planning and machining effort of over 36% for *EC* and in total 4% in the example Workplan underlines the supposition. Further tests, and also with further real industry benchmark parts (Nisaform), are to be followed to confirm the first evaluation results.

A next step is finalising the implementation of the demonstrator. Several databases (e.g., machinery) will be added. Further algorithms for mapping of design features into machining features have to be identified. Finally, the approach will be included in job scheduling methods related to an actual machinery situation such as MES. All in all, the approach supports the planner to automatically select suitable machining operations corresponding to a given manufacturing scenario. All the objectives and requirement which were addressed are fulfilled.

References

[3.1] Berger, U., Kretzschmann, R., Noack, J. 2008. An approach for a knowledge-based NC programming system, In *Proceedings of the 5th International Conference on Informatics in Control, Automation and Robotics*, Portugal, May 12-16, 2008.

[3.2] Blazewicz, J., Domschke, W., Pesch, E. 1995. The job shop scheduling problem: Conventional and new solution techniques, *European Journal of Reselinkh 93.*

[3.3] CIMdata, Inc. 2008. *NC Software and Services Market Assessment Report Version 16.* . [Online]. Available: http://www.cimdata.com/. [27 July 008].

[3.4] CAD-CAM-Report2006. CAM-Systeme werden intelligenter *CAD/CAM-Report Nr. 7+8*, 25. Jahrgang.

[3.5] ISO 6983. 1983. *Numerical control of machines -- Program format and definition of address words*, ISO (International Organization for Standardization).

[3.6] Duran, O., Rodriguez, N., Consalter, L., A. 2008. End Milling: A Neural Approach for Defining Cutting Conditions", In: *Proceedings of the 4th International Workshop on Artificial Neural Networks and Intelligent Information Processing,* Portugal, May 12-16.2008.

[3.7] Eversheim, W. 1996. *Organisation in der Produktionstechnik – 1. Band,* VDI-Verlag, Düsseldorf.

[3.8] Jungnickel, D. 1990. *Graphen, Netzwerke und Algorithmen,* Wissenschaftsverlag, Mannheim.

[3.9] Hamelmann, S. 1995. *Systementwicklung zur Automatisie-rung der Arbeitsplanung,* VDI-Verlag, Aachen.

[3.10] Hellberg, K. 1993. *Methoden zur automatischen Erzeugung von Arbeitsgangfolgen,* VDI-Verlag, Düsseldorf.

[3.11] Kletti, J. 2007. *Konzeption und Einführung von MES-Systemen,* Springer, Berlin.

[3.12] Leung, H., C. 1996. Annotated Bibliography on Computer Aided Process Planning *International Journal of Advanced Manufacturing Technology,* Vol. 12, No. 5: pp. 309-3297, 1996.

[3.13] Sandou, G., Font, S., Tebanni, S. 2008. Feeding a genetic algorithm with an ant colony for constrained optimization, In *Proceedings of the 5th International Conference on Informatics in Control, Automation and Robotics,* Portugal, May 12-16, 2008.

[3.14] Khalifa, R., B., Yahia, N., B., Zghal, A. 2006. New neural networks approach for optimal machines tools selection In: *12th IFAC Symposium on Information Control Problems in Manufacturing,* France, Sant-Entienne, May 17-19.

[3.15] Berger, U., Kretzschmann, R., Aner, M. 2007. Development of a holistic guidance system for the NC process chain for benchmarking machining operations, In: *Proceedings of the 12th IEEE Conference on Emerging Technologies and Factory Automation,* Greece, Patras, September 25-28, 2007.

[3.16] Cai, J. 2007. *Development of a Reference Feature-based Machining Process Planning Data Model for Web-enabled Exchange in Extended Enterprise,* Dissertationssschrift, Shaker Verlag, Aachen.

[3.17] Gerken, H., 2000. *Management von Erfahrungen mit einem Assistenzsystem für die Arbeitsplanung,* Dissertationssschrift TU-Berlin.

[3.18] Görz, G. 2003. *Handbuch der Künstlichen Intelligenz,* Oldenbourg, München.

[3.19] Pritschow, G., Heusinger, S. 2005. *STEP-NC-basierter Korrekturkreis für die Schlichtbearbeitung von Freiformflächen,* Jost-Jetter Verlag, Heimsheim.

[3.20] VDI. 2003. *Information Technology in Product Development - Feature Technology,* VDI-Gesellschaft Entwicklung Konstruktion Vertrieb.

[3.21] Warnecke, G., Valous A. 1993. *Informationsrückkopplung zwischen NC-Fertigung und Arbeitsplanung,* Schnelldruck Ernst Grässer, Karlsruhe.

[3.22] Weck, M., Wolf, J., Kiritsis, J, D. 2001. STEP-NC – The STEP compliant NC Programming Interface, *IMS Forum,* Ascona Schweiz.

[3.23] Xu, X., Wang, L., Rong, Y. 2006. STEP-NC and Function Blocks for Interoperable Manufacturing *IEEE Transactions on Automation Science and Engineering,* Vol. 3, No. 3, July 2006.

[3.24] WZL Aachen. 2007. *Welcome to the official STEP-NC page.* [Online]. Available: http://www.step-nc.org. [20 July 008].

[3.25] ISO 10303-224. 2006. *Industrial automation systems and integration* – Product data representation and exchange – Part 224: Application protocol: Mechanical product definition for process planning using machining features, Geneva, Switzerland: International Organisation for Standardisation (ISO).

[3.26] ISO 14649-1. 2003. *Industrial automation systems and integration* - Physical device control - Data model for computerized numerical controllers - Part 1:

Overview and fundamental principles, ISO (International Organization for Standardization).

[3.27] Weiming, S., Hao, Q. 2007. Agent-based Dynamic Scheduling for Distributed Manufacturing, In: *Process Planning and Scheduling for Distributed Manufacturing*, Springer, London.

[3.28] Sormaz, D., N., Arumugam, J., Ganduri, C. 2007. Integration of Rule-based Process Selection with Virtual Machining for Distributed Manufacturing Planning, In: *Process Planning and Scheduling for Distributed Manufacturing*, Springer, London.

[3.29] Yu, A., Gu, X., Jiao, B. 2008. A Coupled Transiently Chaotic Neural Network Approach for Scheduling Identical Parallel Machines with Sequence Dependent Setup Times, In *Proceedings of the 17th IFAC World Congress,* Seoul, Korea, July 6-11.

[3.30] Allaoui, H., Artiba, A., Goncalves, G., Elmaghraby, S. E. 2008. Scheduling n jobs and preventive maintenance in a single machine subject to breakdowns to minimize the expected total earliness and tardiness costs, In *Proceedings of the 17th IFAC World Congress,* Seoul, Korea, July 6-11.

[3.31] Zhang, C., Li, P., Rao, Y., Li, S. 2005. A New Hybrid GA/SA Algorithm for the Job Shop Scheduling Problem". *Evolutionary Computation in Combinatorial Optimization*, pp. 246-259.

[3.32] Wang, L., Sehn, W. 2003. DPP: An agent-based approach for distributed process planning" *Journal of Intelligent Manufacturing*, Vol. 14, pp. 429-439.

[3.33] DIN 8589. 2003. *Manufacturing processes chip removal - Part 0: General; Classification, subdivision, terms and definitions*, DIN Deutsches Institut für Normung e. V. - Normenausschuss Technische Grundlagen (NATG), Bezugsquelle Beuth Verlag GmbH.

[3.34] Reidelbach, H. 2005. Anforderungen und Empfehlung eines großen Kunden an den deutschen Maschinenbau, In: *1. Deutscher Maschinenbau Gipfel*, Berlin, October 11-12, 2005.

[3.35] NC-Gesellschaft. 2000. *Anwendung neuer Technologien: NCG 2004-Prüfrichtlinien und Prüfwerkstücke für hochdynamische Bearbeitungen (HSC), Teil1: Fräsmaschinen und Bearbeitungszentren*, 2000.

[3.36] ISO 10303-238. 2007. *Industrial automation systems and integration* – Product data representation and exchange – Part 238: Application protocol: Application interpreted model for computerized numerical controllers, Geneva, Switzerland: International Organisation for Standardisation (ISO).

[3.37] ESPRIT Project EP 29708. 2001. *STEP-Compliant Data Interface for Numerical Controls (STEP-NC) – Final Report*. [Online]. Available: http://www.step-nc.org/html/..%5Cdata%5Ceu1_final_report.pdf. [25 October 007].

[3.38] Brunnermeier, S., Martin, S. 2002. Interoperability costs in the US automotive supply chain, *Supply Chain Management: An International Journal*, Volume 7, Number 2: pp. 71-82(12).

[3.39] Lenz, A. 1991. *Knowledge Engineering für betriebliche Expertensysteme*, 1. Auflage, Deutscher Universitätsverlag Wiesbaden.

[3.40] Turau, V. 2004. *Algorithmische Graphentheorie*, 2. Auflage, Oldenbourg Wissenschaftsverlag München.

[3.41] Schmitting, W. 2006. *Das Travelling-Salesman-Problem - Anwendungen und heuristische Nutzung von Voronoi-/Delaunay-Strukturen zur Lösung euklidischer, zweidimensionaler Travelling-Salesman-Probleme*, Dissertationssschrift, Münster.

[3.42] Reinelt, G. 1994. *The Travelling Salesman Computational Solutions for TSP Applications*, Springer, Berlin Heidelberg.

[3.43] Eßmann, J. 2008. *Hymould,* [Online]. Available: http://www.hymould.eu. [08/08/2008].

[3.44] Berger, U., Kretzschmann, R., Arnold, K-P., Minhas. S. (2008): Approach for the Development o a heuristic Process Planning Tool for Sequencing NC Machining Operations. In: APPLIED COMPUTER SCIENCE (ACS), Vol. 4, No 2, 2008, pp. 17-42

4

STEPNC++ – An Effective Tool for Feature-based CAM/CNC

John Michaloski[1], Thomas Kramer[2], Frederick Proctor[3], Xun Xu[4], Sid Venkatesh[5] and David Odendahl[6]

[1–3] Intelligent Systems Division, MS8230, National Institute of Standards and Technology
100 Bureau Drive, Gaithersburg, MD 20899, USA
Email: [1] john.michaloski@nist.gov, [2] thomas.kramer@nist.gov, [3] frederick.proctor@nist.gov,

[4] Department of Mechanical Engineering, University of Auckland,
20 Symonds Street, Auckland, New Zealand
Email: x.xu@auckland.ac.nz

[5,6] Boeing Aircraft, Seattle, WA
Email: [5] sid.venkatesh@boeing.com, [6] david.j.odendahl@boeing.com

Abstract
This chapter discusses the realization of direct translation of feature-based CAM files into feature-based CNC part program files. The information infrastructure that allows this to happen is STEP-NC, as described by ISO 14649 Parts 10 and 11. Among the many benefits cited for STEP-NC, the elimination of the costly and inefficient process of post-processing using one standard CNC definition is most commonly cited. However, this chapter argues that a much greater benefit can be found using STEP-NC to resolve the CAM/CNC impedance mismatch where CAM comprehensive process information is reduced into motion primitives. The CAM/CNC impedance mismatch and resulting lack of process information make intelligent machining difficult. Given this perspective, a demonstration system was developed to show that feature-based CNC is possible with STEP-NC, which preserves more feature-based CAM process knowledge to make intelligent machining possible. The demonstration system incorporates: (1) part representation using STEP-NC Part 21 files, (2) reading and analyzing feature-based elements of the STEP-NC, (3) translation into a CNC feature-based representation, and (4) generation of actual CNC programs relying on conversational programming "Canned Cycles". Cutting and simulation tests have confirmed the advantages of the approach. Overall, the demonstration system shows that a standard information infrastructure such as STEP-NC is essential for advancing manufacturing to enable improvements in efficiency, product quality, life-cycle cost, and time-to-market.

4.1 Introduction

In design and manufacturing, organizations depend on information technology to handle the tasks necessary to make products. Over the last two decades, Computer-

Aided Design (CAD) and Computer-Aided Manufacturing (CAM) information technology has greatly benefitted manufacturing industries everywhere. CAD/CAM benefits include cost-savings, more rapid and improved designs, higher productivity, and better return on investment. Today, it is commonplace to see complex manufacturing done with global CAD/CAM collaboration and virtualization that includes numerous partners located locally and internationally, from the same and different companies, scattered throughout the world. This globalization vision for manufacturing has resulted in streamlined supplier chains, optimized resources, and distributed collaborative design.

Nevertheless, despite the positive contributions of CAD/CAM to the design and manufacturing processes, barriers still exist for achieving seamless end-to-end manufacturing. Sharing product information throughout the job stream within the entire manufacturing life-cycle is still a major problem. A key contributor to the loss of information within production is the lack of a product exchange standard. The "Interoperability Cost Analysis of the U.S. Automotive Supply Chain" study performed by the Research Triangle and commissioned by National Institute of Standards and Technology (NIST) estimates that imperfect interoperability imposes at least $1 billion per year on the members of the U.S. automotive supply chain [4.1]. Most significant are the costs of repairing or re-entering of non-standard upstream design product data files that are not usable for downstream production applications. Additionally, re-entering and/or reformatting information results in duplication of the data, which can introduce substantial semantic problems maintaining consistent values between two different copies of the same information. Duplicate representation requires more manpower and computational costs and important information can often be lost during translation. A similar problem in data base technology is known as the impedance mismatch. In electronics, impedance mismatch is intended to describe the mismatch between output and input signals, but has been borrowed by the software community to describe a mismatch between incompatible software technologies.

STEP-NC, as defined by ISO, the International Organisation for Standardisation, 14649 [4.2, 4.3] or ISO 10303-AP238 [4.4], attempts to ameliorate the productivity losses due to the impedance mismatch between the CAM machining specification and the actual Computer Numerical Control (CNC) machining. STEP-NC is the colloquial term that refers to both ISO 14649 and ISO 10303-AP238, where ISO 14649 is the Application Reference Model (ARM) schema, and ISO 10303 AP238, is the reinterpretation of ISO 14649 into the fully-integrated Application Integrated Model (AIM) schema. In this chapter, STEP-NC will refer to the ISO 14649 family of machining standards.

STEP-NC offers accurate and complete product definition data from product design all the way to the machine tool. Current discrete part manufacturing does not produce parts based on all the product information that has been described upstream by the design process. Instead, the part fabrication process starts with the CAD part design, incorporates CAM feature-based machining operations, then translates the machining operations into simple machine tool motion primitives, or, "M" and "G" codes. Consequently, much of the important design and fabrication knowledge is lost and the machine tool has little knowledge to work with regarding the part features, operations, or resources used necessary to optimize and safeguard the machining. This loss of process knowledge is clearly detrimental, as today's machine tool CNCs have the capability to synthesize many more aspects of the machining process in order to realize intelligent control.

This chapter describes STEPNC++, an open-source software package for translating feature-based STEP-NC CAM files into feature-based CNC programs in a portable, standard manner. Key to improving this translation is minimizing whenever possible the use of CNC primitive motion tool-paths. Two critical information infrastructure technologies made featured-based CAM/CNC possible – STEP-NC and C++ meta-programming techniques. For the metal working industries, STEP-NC offers a variety of feature-rich machining specifications that can be used for sharing part design and process data up/downstream within the manufacturing process. Recent advances in C++ compilers have made meta-programming templates, which can be thought of as compile-time execution, a powerful technology in generating succinct but powerful code.

4.2 Feature-based CAM to Feature-based CNC

Conceptually, computer integrated manufacturing (CIM) is a production model by which all elements of the factory (i.e., people, equipment, materials, and computers) are organized and integrated around common data repositories. Figure 4.1 shows the CIM sequence of production stages, commonly referred to as "Art to Part", within a modern manufacturing process for producing parts. In spite of CIM advances, from the perspective of Computer Aided Engineering (CAE), CAD systems, CAM systems, and CNC machines, the process still requires better data integration at all levels and along multiple dimensions (design, analysis, process planning, fabrication) of the manufacturing enterprise.

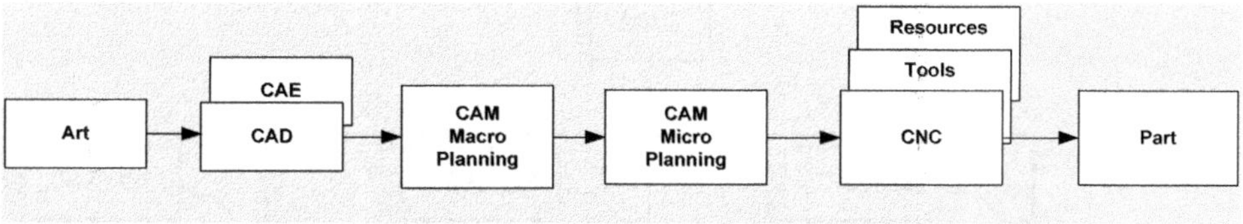

Figure 4.1. Computer integrated manufacturing Art-to-Part production

The goal of "Art-to-Part"" is to manufacture high quality products in the most cost-efficient manner. Initially, the "Art" stage involves an idea or customer need for a discrete part. The CAD stage uses software packages and solid modellers to design the part geometry and tolerances in a concurrent engineering environment that incorporates part intent, and relationships such as mating, geometric fit, or assembly. CAE analysis studies the related parts for structural weakness with regard to loads, strains, and stresses. The CAM macro stage accepts a CAD geometrical part definition and produces a process plan applying software-derived or domain expert's manufacturing knowledge. The CAM micro stage produces a detailed manufacturing NC process plan including NC code that is verified in simulation prior to manufacturing.

On the shop-floor, the machining stage uses a machine tool to cut a metal casting or raw stock to create the "Part." A CNC computer directs the machining operations. The machining stage employs a CNC machine tool with sufficient accuracy and precision to meet part tolerances and associates the necessary raw stock or casted material, fixtures, and tooling as identified in the detailed manufacturing process plan. The CNC runs a "Part Program" that is a sequence of CNC operations to machine the part.

In today's Art-to-Part process, most CNC part programs are generated by CAM modelling software systems. However, due to inadequate integration, CAM files cannot run natively on any given CNC machine tool, and so require a translation process. This CAM to CNC translation is known as post-processing, which has traditionally been the principal means of translating CAM files into CNC programs. Post-processing takes the CAM feature-based operations and generates a machining part program, not as a list of features and operations, but as ISO 6983/RS274D [4.5, 4.6], or "M & G" code motion primitives – lines or arcs. Henceforth, this chapter will use the ISO 6983 standard when referring to "M & G" motion.

Unfortunately, CNC vendors have augmented and extended the ISO 6983 programming languages as they have seen fit to meet functional needs and user requirements. The proliferation of modifications to the standard has led to hundreds of vastly different ISO 6983 dialects. Consequently, CAM software vendors maintain hundreds of CNC-specific post-processors to provide some form of interoperability. Figure 4.2 outlines the generation of a legacy machine part program, where CAM software, using a different Post-Processor per machine, converts the CAD part models into the required motion-based ISO 6983 NC part program.

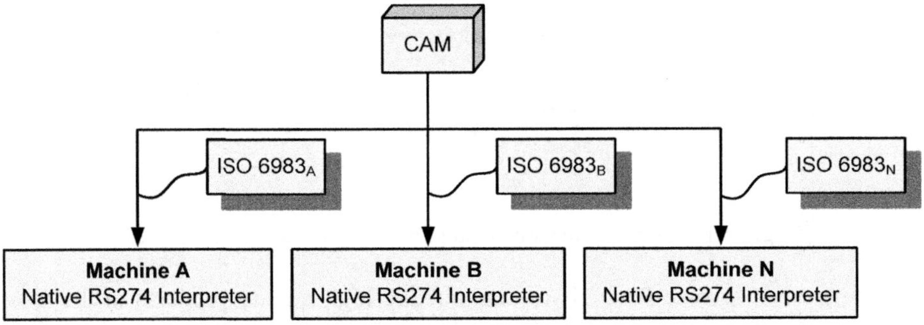

Figure 4.2. Post-processing

Post-processing leads to productivity losses due to the impedance mismatch of the process knowledge between the CAM system and the CNC machine. Presently, most CNC machines receive ISO 6983 data defining each axis movement as motion primitives in order to manufacture a part. This geometric data is referred to as machine control data (MCD). MCD provides a very low level of instruction: tool, axes positions, feed, and speed. The problem with MCD programs is that they are not portable or adaptable. Portability is a problem in five axis milling, since unique axes-position data must be generated for each machine control combination (part, tool, and machine configuration) on which the part is to be run [4.7]. Adaptability is a problem because feature-based information is not provided to allow adaptive real-time servo changes in machining dynamics (feed and speeds) based on machining conditions. Overall, the CAM-CNC post-processing has discarded all of the feature-based part and process information that could help yield higher quality parts faster.

A shift to feature-based CNC provides a new opportunity for manufacturers to streamline the entire "Art-to-Part" process by directly mapping a feature-based CAM file into a feature-based CNC program. The term "feature-based" refers to a description of the shape of the object to be made as a volume whose shape is an instance of some members of a predefined set of types of shape. Examples of machining features include faces, pockets, slots, bosses, and holes, some of which are shown in Figure 4.3.

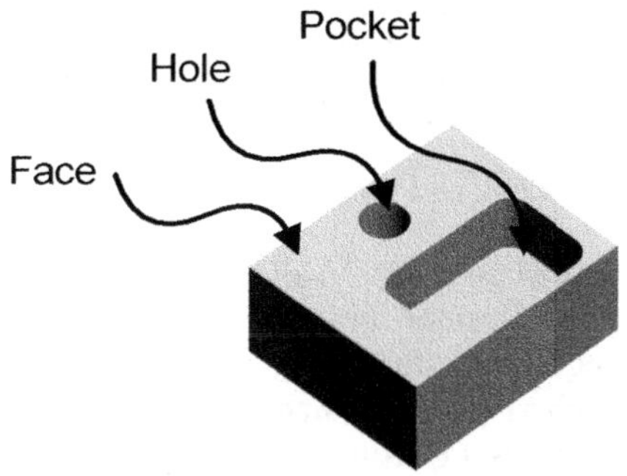

Figure 4.3. Example machining features

A machining process plan is a sequence of feature-based operations. CAM users generate plans by associating machining features to the part geometry. For each part feature, a sequence of machining operations must be defined. For example, given a hole as a machining feature, the CAM system can associate a series of drilling operations – center-drill, drill, and then ream.

Figure 4.4 shows the new feature-based CAM-CNC infrastructure. The enabler of feature-based CNC is "conversational programming". Conversational programming allows parts to be programmed using high-level features that can be run directly on the CNC. The conversational programming functionality is part of the ever expanding body of "canned cycles" that CNCs support. Traditionally, CNC provided higher-level canned cycles to handle more difficult machining requiring tight synchronization of tooling and motion. For example, common ISO 6983 milling canned cycles include drilling, boring, tapping, and peck drilling sequences. Modern CNCs typically provide support for roughing and finishing canned cycles to machine rectangular and circular pockets, slots, threading, and engraving, with a list of ever-growing feature-based operations. Further, in feature-based CNC, tolerance information can be coupled with the feature-based canned cycles to allow better production cycle times, accuracy, consistency, predictability, and process reliability. These are just some of the manufacturing factors that can be improved, leading to better parts at lower cost.

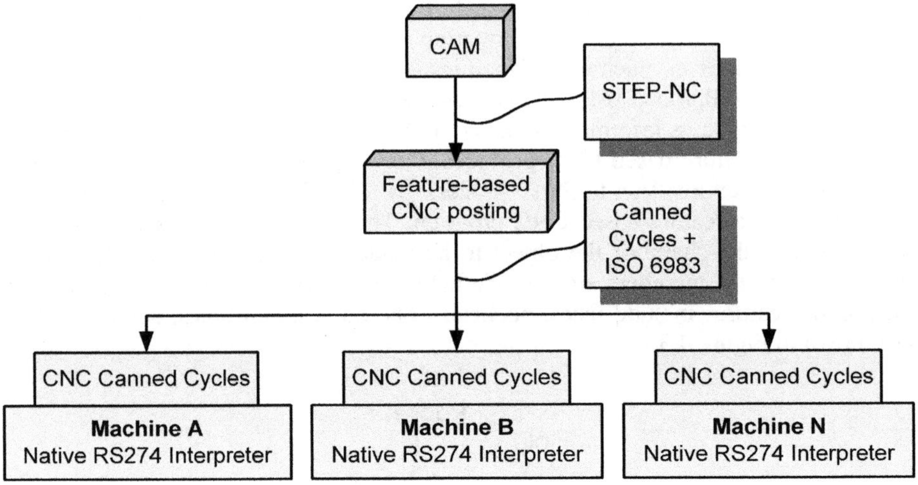

Figure 4.4. Feature-based CAM-CNC

Even though direct translation of CAM features into CNC machining features seems intuitive, this is not common practice. This feature-based CAM posting to motion-based CNC integration strategy is clearly deficient, but over time has become a familiar, if not cherished, manufacturing practice. In contrast, STEP-NC is a standard developed by the ISO to communicate feature-based CAD/CAM product data to the CNC. STEP-NC is intended to replace ISO 6983 with a richer product data description, including geometry, features, and tolerances. With STEP-NC, accurate and complete feature-based product data flows all the way from product design, down to the machine tool. Therefore, STEP-NC offers the opportunity to maintain a part's actual feature representation and not use lines and arcs to approximate features.

In addition to CAM-CNC feature-based data exchange, STEP-NC offers a number of potential manufacturing advantages. Foremost, STEP-NC contains a comprehensive model that describes complete process and machining data that can be made available to the CNC machine tool. Thus, STEP-NC allows for better vertical integration between CAD/CAM and CNC with a full complement of production knowledge such as tooling, stock, materials, and setup. Assuming STEP-NC compliance to conformance classes spelled out in the standard, STEP-NC allows easier horizontal integration through addition (or substitution) of an equivalent class of STEP-NC compliant CNC. STEP-NC itself is a unified life cycle engineering description of the part to allow "design anywhere, build anywhere, or maintain anywhere" in the virtual enterprise. Ergo, distributed machining with STEP-NC is possible since all the relevant information is available in one file as one unified life-cycle model – part program, graphics, raw stock, program origin, workpiece setup, tools – so that design and manufacturing data for a part can be distributed to machine shops around the world via the Internet. Finally, adaptive machining is possible now that a complete STEP-NC product model is made available to the CNC, using real-time CNC adaptive servoing and trajectory control based on process state and tolerances.

There is a wealth of opportunities open to manufacturers should they use STEP-NC for feature-based CAM/CNC. A perfect product data exchange between CAM and CNC interchange would be ideal, but is probably not realistic. So, there is a need to balance the strengths of CAM against the strengths of CNC, and leverage the best technology whenever possible. CAM systems currently provide a better holistic view of the machining process and offer a wider range of machining strategies. CNC systems understand machine dynamics to offer more precise control of the actual motion, but may not offer the best total tool-path optimizations. Overall, selection of appropriate combination of CAM/CNC cutting strategies will have the most positive effect on the part precision, intermediate surface roughness, cycle time, and surface finish of the final part. This chapter will look at a couple of specific feature-based CNC opportunities that offer great benefits.

Our first objective was to look at feature and part tolerance information in a feed-forward tolerance scheme so that roughing cycles can have improved cycle times, and finishing cycles can have tighter tolerance and better surface finish. The next CAM/CNC functionality we will investigate is the verification and the correctness of the CAM machining parameterization based on CNC feature-based performance. Due to the complexity of manufacturing processes, specifying process parameters, such as machining speeds, feed rates, and tool selection, is often ad hoc and empirical, resulting in sub-optimal solutions. Often, the "final" tooling, feeds and speeds parameters have been determined by costly trial-and-error prototype machining. This chapter will show how a STEP-NC system simplifies logging higher-level, lossless, feature-based CNC process knowledge for use as feedback in CAM machining parameterization.

An even more compelling use of tolerance and data logging occurs when used with feature-based CNC simulation. Currently, CAM can "prove-out" or verify a CNC program through CAM simulations to detect such errors as gouging and tool collisions. However the simulated motions only approximate the actual CNC motions. Instead, if the CAM is verified against the actual CNC motion controller, this could guarantee that CAM features match the CNC machined features. Feature-based verification could then perform experiments with tolerances or feeds/speeds to optimize the process in the CNC background or with a CNC simulator, while at the same time saving valuable machine time and extending tooling and machine life.

4.3 Feed-forward Tolerancing

In manufacturing, part quality is measured by the conformity of manufactured parts to a specified tolerance. To start, CAD models the "nominal" or "exact" part geometries (shape, size, and form of the part) and adds dimensional tolerances to describe the allowable variation for the sizes, locations, orientation, and associativity relationships between tolerances using standards such as ASME Y14.5 [4.8] or ISO 1101 [4.9] that describe geometric dimensions and tolerances (GD&T). CAD design tolerances are then mapped into CAM feature-based machining tolerances. Traditionally, the inspection process measures part features in a specified sequence to determine feature tolerances and overall part tolerance. The inspection compares the machined part against the tolerance dimensioning to determine part quality – but only after the part has been machined. Since the determination of a bad part has been done after the part has been machined, it is very difficult to incorporate the inspection feedback into the CAM software to improve faulty part quality.

Taguichi observed that quality should be designed into the part and not inspected into it [4.10]. In today's CNC, the opportunity exists to perform feed-forward feature-based tolerancing to describe more precisely the range of machining tolerances while the part is being machined. In a traditionally feature-based CAM to motion-based CNC, tolerance information may be recorded in the CAM file, but this is rarely done even though the CNC has the capability to modify motion generation during contouring based on axes' tolerances. Without tolerance information, CAM users must limit entire tool-paths containing contouring motion to the maximum speed suitable for any cornering because of high loads during cornering. The lack of tolerance information within the RS 274 standard has also contributed to its lack of CNC adoption.

In conversational CNC machining, tolerances for the axes motion can be established for the basic classes of machining – roughing, finishing, and semi-finishing or rough-finishing. For roughing, the emphasis is on speed to reduce cycle times while balancing this with tool life. For finishing, the emphasis is on accuracy, minimizing tool deflection, and attaining the proper surface finish quality. Semi-finishing provides a cut to achieve the proper surface quality (e.g., remove scallops) particularly for machining tough materials before the finishing pass.

Of issue is the difference between maintaining machining tolerance and surface finish in MCD vs feature-based machining. MCD part programs take 3D Cartesian space geometrical data (with the possibility of additional orientation dimensions for four and five axes motions), almost exclusively linear or circular paths, and dynamic data (feeds and speeds) to produce motion trajectories. The dynamic model of a machine defined by axes limits for speed, force and jerk are a major factor in motion trajectories. Trajectory motion of the cutter along the linear or circular path is achieved by interpolating between the programmed points along the path to generate intermediate Cartesian positions with spacing between points based on the programmed feed-rate. These intermediate trajectory points are then transformed into axes-space to control the machine.

However, to approximate non-circular contour paths, a "cloud of linear points" along the path is programmed, and the path is interpolated linearly by "blending" a series of short trajectory segments. Trajectory motion based on small linear paths can often result in instantaneous changes in direction at the transitions between the short segments, causing step changes in axis velocity and discontinuities of acceleration or jerk. These instantaneous changes can result in reduced surface finish quality, part accuracy errors, undesirable machine vibrations, and excessively jerky trajectory motion producing wear/tear on the axes.

Currently, in a feature-based CAM to motion-based CNC world, most CAM systems' use of tolerance information is special-case, not routine. This is due to the proliferation of vastly different ISO 6983 dialects leading CAM vendors to invest in support of hundreds of post-processors. With such a myriad of non-standard dialects and arbitrary CNC functionality, CAM providers address the largest potential set of CNC implementations with the minimal set of CNC functionality, so that they choose to embed functionality in the CAM software and then use linear approximations to achieve contouring within the CNC.

In contrast, CNC feature-based machining can use the machine dynamics to its advantage, since the CNC is calculating trajectories based on instantaneous machine dynamics. Combining geometrical paths with dynamic motion control can greatly improve machining accuracies. Figure 4.5 shows that when machining a contour, such as the 90° right-angle corner, the tolerance-enabled CNC maintains the cutter path within the specified tolerance limit of the ideal cutter path, even with sparser intermediate trajectory points.

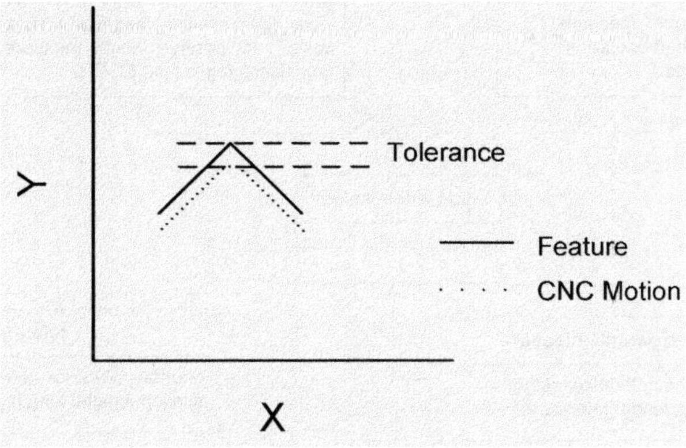

Figure 4.5. Feature-based NC tolerance

STEP-NC contains part and feature tolerance data, which can be used to adjust the machining process to accommodate the desired part tolerance. Figure 4.6 shows the Unified Modelling Language (UML) [4.11] class relationship for the tolerance primitives in the STEP-NC standard. A `toleranced_length_measure` is the basic entity to describe a length measure with tolerance. The `toleranced_length_measure` contains a theoretical length as a real value but also has an `implicit_tolerance` encapsulating the type of tolerance. STEP-NC has the "`plus_minus_value`" type of tolerance to describe the upper and lower limits valid for a scalar dimension. STEP-NC also defines a "`limits_and_fits`" type of tolerance that describes an ISO 286 [4.12] tolerance system to specify the difference between a measured actual size and the corresponding basic size as a quality or the accuracy grade of a tolerance as it applies to a shaft or to a hole. STEP-NC also declares a generic "`shape_tolerance`" entity, which is defined as a length.

The STEP-NC tolerance capabilities will be explored for one of the most common operations in machining metal parts – pocket milling. Pocket milling is a machining operation to remove all the material inside some arbitrary closed boundary on a flat surface of a workpiece to a fixed depth [4.13–4.18]. The geometry of the pocket is defined by its contour on the outer face of the workpiece and its depth.

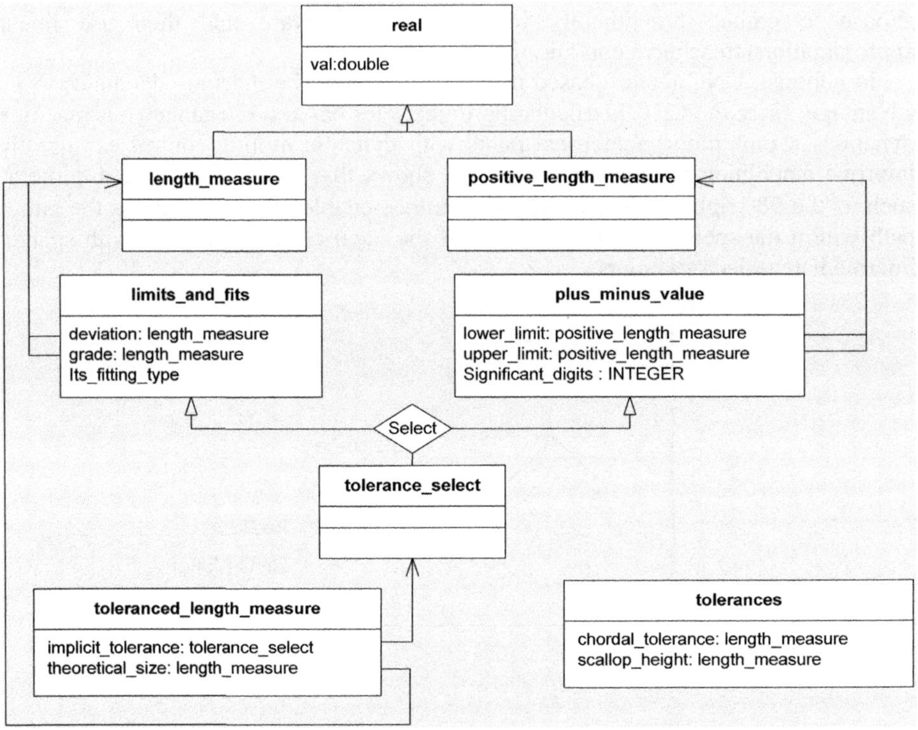

Figure 4.6. UML representation of tolerance information

It is common machining practice to form a pocket by milling it using one or more end mills. If the inside boundary of a pocket is large in comparison to the cross section of the end mills suitable for cutting it, it is common practice to make a "roughing cut" to remove the bulk of the material by making a slightly smaller pocket. This leaves a thin layer of material that is removed later by a finish end mill. A successful finishing operation depends on matching or exceeding the tolerance information given for the pocket.

As currently defined, STEP-NC provides pocket tolerances using the STEP-NC tolerance primitives previously discussed. Figure 4.7 shows the STEP-NC pocket nomenclature. STEP-NC feature tolerances exist for most aspects of the machining features and operations, but frequently a tolerance is tagged as optionally reflecting a pragmatic view of GD&T in the machining world. All STEP-NC 2½D manufacturing features have a depth, defining the feature bottom. For pockets, the bottom is defined as an elementary plane without tolerance (basically an absolute Cartesian depth placement, and vectors defining a direction and reference direction). Instead, we use the STEP-NC "`global_tolerance`", valid where no other tolerances are specified, which defines a tolerance for the workpiece as the STEP-NC `shape_tolerance` primitive.

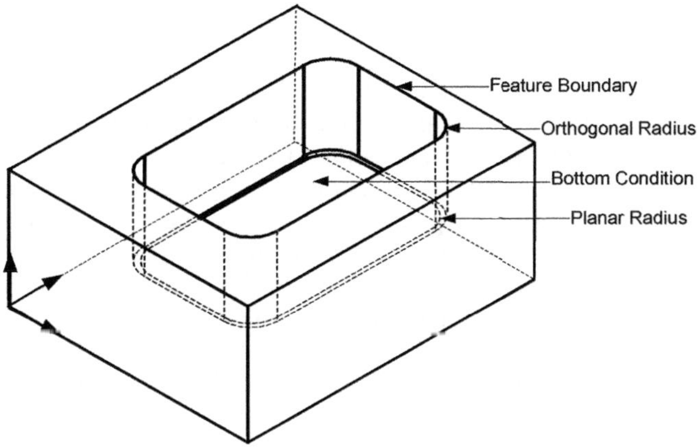

Figure 4.7. STEP-NC pocket nomenclature

All other STEP-NC pocket feature attributes include a tolerance of some kind. The corners of the profile of a pocket have an orthogonal_radius attribute, which is an optional `tolerance_length_measure`. Between the floor of the pocket and its walls there is a `planar_radius` attribute and an optional `tolerance_length_measure`. The planar and orthogonal values with tolerance are optional, but should not be optional if feedforward tolerancing is required. For rectangular pockets, the `feature_boundary` is described by the pocket length and width, both supporting toleranced dimensions. For circular pockets, the diameter can have a tolerance.

Clearly, the basic tolerance strategy in manufacturing is to select a CNC machine with sufficient precision and accuracy to meet or exceed the required tolerances. However, tolerances can positively or negatively impact the cost of producing the part. Manufacturers lose time and money when they over-tolerance or under-tolerance the machining operation. Bad machining strategies lead to longer cycle times and difficulty in maintaining process parameters such as surface finish. Even if CAM produces small increments of linear data to approximate the contour, machining strategies must use special cornering tool-path feed-rates that still may not meet the desired tolerances. Thus, incorporating feed-forward tolerance knowledge into the feature-based CNC process plan is imperative to improving cost and part quality.

4.4 Smarter Machining Process Parameterization

It is the job of the manufacturing engineer to choose CAM machining operations and specify the machining parameters. The current state of CAM finds process planning generally relying on human expertise and knowledge. Typically, manufacturing engineers and shop-floor machinists use their years of experience to define the machining operations and specify the "correct" parameters. In most cases, the planning decisions are based on standard reference documents [4.19], company-specific "crib sheets", or rule-of-thumb. CAM systems offer some simple lookup assistance that is starting to tap into the realm of expert systems, but even this can be complicated by many issues. First, the same process can have different results even when running on identical brand and model of machine tools. Additionally, this CAM process planning includes evaluating complex tradeoffs, such as proper tools, short cycle times vs long tool life, and other parameterization.

A survey by the Kennametal Corporation illustrates the difficulty in properly choosing machining operation parameters, finding that U.S. industry chooses the correct tool less than 50% of the time, uses cutting tools at their rated cutting speed only 58% of the time, and uses cutting tools to their full life capability only 38% of the time [4.20]. These sub-optimal practices are estimated to cost U.S. industry $10 billion per year. A fully integrated CAM-CNC process planning system with feedback from an actual CNC would clearly help in removing the guesswork from CAM parameterization.

One of the problems contributing to this poor parameterization machining dilemma is the difficulty not only in understanding the machining process itself, but also in assessing how well parts have actually been made by the machine tool. Inspection can tell us what features are out of tolerance, but cannot tell us what process aspects contributed to any problems. Was it the wrong tool, tool wear, or incorrect feed and speeds? In assessing the machined part, the only way to understand truly the process is to do in-situ process monitoring. Unfortunately, after the CAM model has been posted, all the feature-based knowledge has been stripped out of the program, making it difficult to understand the "bigger picture" inside the program. Since the CNC is programmed at the motion primitive level, correlating any process feedback to the associated CAM feature is very difficult. Linking downstream motion primitives to upstream CAM features can be done after the fact with much time and effort, but does not seem to be cost-effective to industry.

Figure 4.8 illustrates the use of collecting machining data in the "Art-to-Part" scenario. Feature-based data collection can be used for reusing or refining macro-planning level CAM feature operations and at the micro-planning level CAM for scientifically quantifying machining operations. Given this scenario, the CNC machining can simply associate with each feature a timestamp, alarms, mismatched actual vs programmed feed rates, excessive lag or following errors, or perform on-machine feature inspection to quantify machining performance. Stochastic analysis of the machining data would instill a more scientific and knowledge-based approach to feature-based machining. This CNC machining analysis can be used to make more informed decision in the CAM machining parameterization. With the millions of machining data points generated for a milling process, manufacturers need science-based help managing the process to understand the problems that confront them. CNC feature-based machining offers a better way in which machining can be scientifically evaluated.

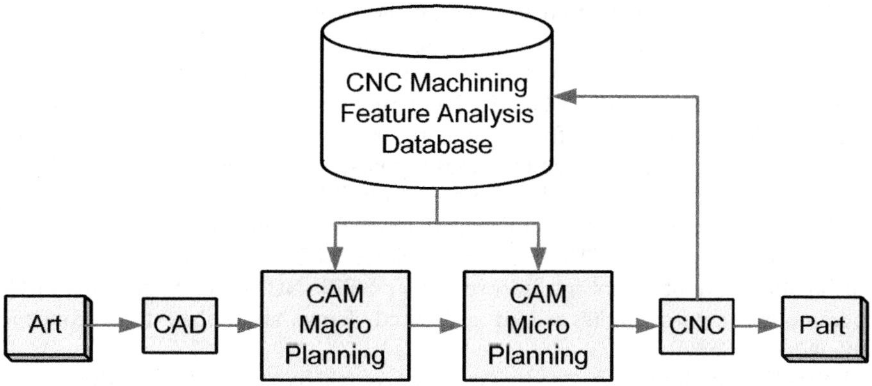

Figure 4.8. Feature-based CNC machining feedback

4.5 STEPNC++ Implementation

Mitigating all the potential benefits of using STEP-NC is the fact that no CAM or CNC vendors provide native STEP-NC support. The general lack of STEP-NC acceptance is due to several issues. First, STEP-NC is different and the machine tool industry is rather conservative, preferring to focus on improving "cycle time" rather than taking risks on unproven new features. Second, it is not directly apparent how standards contribute to a company's bottom line for the next quarter or the next fiscal year. Third, companies are profit-driven, and therefore reluctant to invest in solutions that benefit everyone. Finally, although STEP is powerful, it is complex, with a relatively narrow manufacturing-centric appeal.

STEP-NC is complex for a number of reasons. STEP-NC is part of the ISO 10303 family of standards known as STEP. The STEP standards community developed their own data modelling language, EXPRESS (ISO 10303-11) [4.21], to describe specifications. An EXPRESS model definition is contained in one or more constructs called EXPRESS "schemas". STEP Part 21 defines an exchange file format for transmitting instances of data that has been modelled in EXPRESS schemas. ISO 14649 Parts 10 and 11, and STEP Part 21 are specified in EXPRESS.

The STEP standards community uses EXPRESS to model all the ISO 10303 STEP Parts. However, the narrow scope of the application domain (manufacturing) has placed a limit on the amount of software development to a small number of dedicated, EXPRESS related developers. The limited number of EXPRESS software tools is due in part to the complexity of developing EXPRESS software as well as the rather (comparatively) small market for manufacturing software. Contrast this to the ubiquitous deployment and pervasive adoption of Extensible Markup Language (XML) [4.22] used for Internet communication and business applications, where numerous large software companies provide both commercial and free language tools. The lack of STEP EXPRESS development tools and the cost of those that exist make entry into the STEP and STEP-NC environment difficult.

STEPNC++ implementation focuses on the evolution and promotion of STEP-NC development technologies within the manufacturing community by providing a set of library frameworks, and standards validation implementations. The goals of the STEPNC++ feature-based CAM-CNC software system include low-cost entry, novel solutions, ease of software integration, speed, and standard component libraries requirements. To achieve these goals, we adopted C++ [4.23] meta-template programming tool and techniques to build an ISO 14649 Parts 10/11 and ISO 10303 Part 21 parser. By comparison, much experimental STEP work at NIST has used a commercial STEP software package [4.24, 4.25], but the code is not distributable without a runtime license. Other STEP-NC native feature-based CNC programming environments are of a limited scope and relate to a conceptual framework [4.26].

EXPRESS is defined in "a derivative of Wirth Syntax Notion", which is very similar to Extended Backus Naur Form (EBNF), an ISO standard representation [4.27], a formal meta-language in which to describe languages. To achieve our objectives, we used Boost [4.28], a collection of platform-neutral C++ libraries of reusable, easily-customizable code that features the Spirit parser library, which was used extensively for parsing EXPRESS schemas. Figure 4.9 shows the feature-based CAM/CNC data flow of our STEPNC++ system parsing CAM files to produce a CNC feature-based part program:

1 First, create the CAD/CAM models and then outputs a STEP-NC ISO 14649 Part 21. This could be hand-generated using the STEPNC++ class hierarchy.
2 Read the Part 21 file using the Boost's Spirit parser framework to produce a traversable parse tree, which is a tree whose nodes are labelled by the matched EBNF productions.
3 Create a STEPNC++ instance hierarchy by traversing the Spirit parse tree twice – once to extract all the STEP-NC Entities defined, and a second time to resolve all the Entity attributes. Given a STEPNC++ instance hierarchy, extract the features and operations from the ISO 14649 plan into a CNC feature-list representation. Verify the STEP-NC plan using rule-based knowledge.
4 Generate a vendor-specific ISO 6983 feature-based CNC part program based on a generic feature-based posting paradigm.
5 Run (or simulate) the feature-based CNC part program. Note, at this point, feature-based CAM/CNC does not remove the entire aspect of CNC-specific posting from the Art-to-Part process.

Spirit is an object-oriented, back-tracking, recursive descent parser implemented using C++ template meta-programming (TMP) techniques [4.29, 4.30]. C++ TMP is a programming abstraction in which templates are used by a compiler to generate temporary source code, which is merged by the compiler with the rest of the source code and then compiled [4.31]. The use of meta-programming templates can be thought of as compile-time execution. An important part of code generation with meta-programming is expression templates [4.32]. The technique of expression templates allows entire expressions to be passed to functions as parameters and inlined into the function body. Expression templates have been used to achieve faster execution performance for the Blitz++ scientific computations library [4.33], where the goal of Blitz++ is to attain Fortran execution speed using C++.

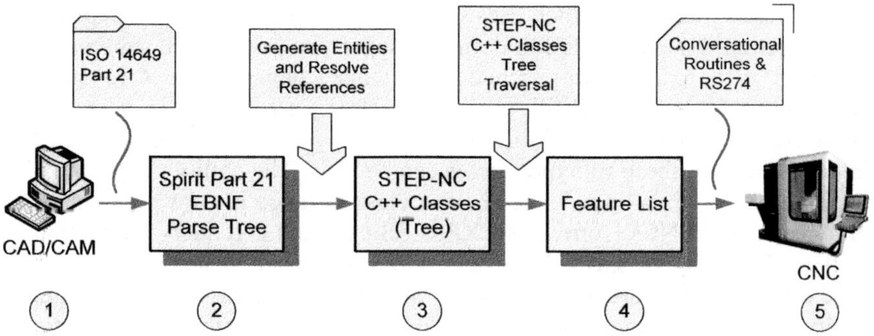

Figure 4.9. Overview of the STEPNC++ feature-based CAM/CNC data flow

Spirit can simplify the development of experimental and commercial STEP-NC applications and should be expected to minimize the start-up costs for smaller STEP-NC programs. First, the Spirit code is embedded directly into the STEPNC++ system. There is no "compiler-compiler" phase, and no translation of EBNF into BNF is required. Second, translating EBNF grammars into Spirit code that produces a parse tree is straightforward. Figure 4.10 shows the transformation from an entity input described in the EXPRESS grammar into a parse tree. A Spirit parse tree is a tree representation of the input where a branch or leaf node corresponds to a matched EBNF rule. Parse trees allow multiple passes to be done over the data without having to re-parse the input since a semantic action need not be tied directly to a matched rule. By comparison, Yacc/Lex [4.34] generally intertwine code handlers while parsing the grammar.

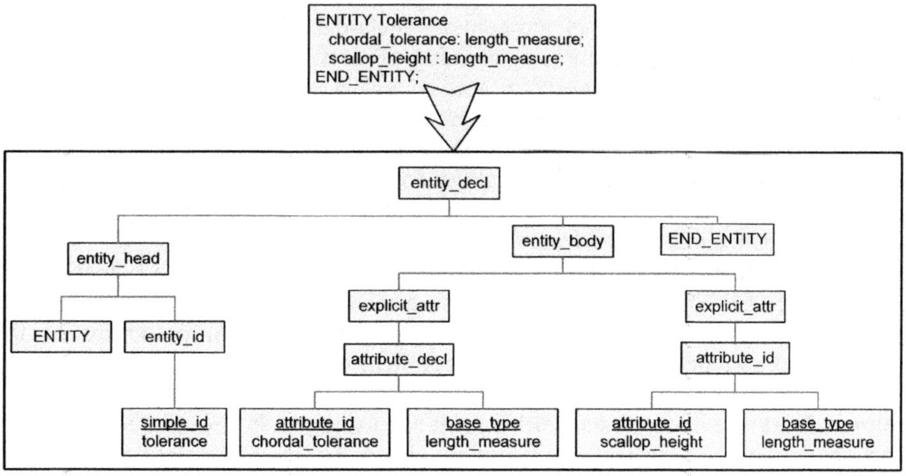

Figure 4.10. Entity parse tree

The STEPNC++ classes allow one to create, read, and write ISO Part 21 files based on ISO 14649 Parts 10 and 11 schemas. However, the critical STEPNC++ functionality is the ability to traverse the STEPNC++ instance hierarchy and extract the desired feature-based CNC knowledge. EXPRESS relies heavily on the inheritance class hierarchy, which uses supertype and subtype relationships to refine parameterization. Often, EXPRESS defines a supertype from which subtypes derive, many times in a long, multi-pronged inheritance chain. Figure 4.11 shows the Milling_Cutter UML EXPRESS class hierarchy. Each cutter is declared as a separate class, so that when an abstract "`milling_cutter`" is used as an Entity attribute, any one of these subtype classes can be used. The ability programmatically to inspect and use class knowledge about a class instance is known as "metaclass" programming. STEPNC++ provides EXPRESS tree path and query navigation tools as well as Entity metaclass knowledge, such as the class name, and super and subtype inheritance hierarchy. STEPNC++ also provides access to knowledge about the entity variables and methods as well as operations for creating new objects of the class.

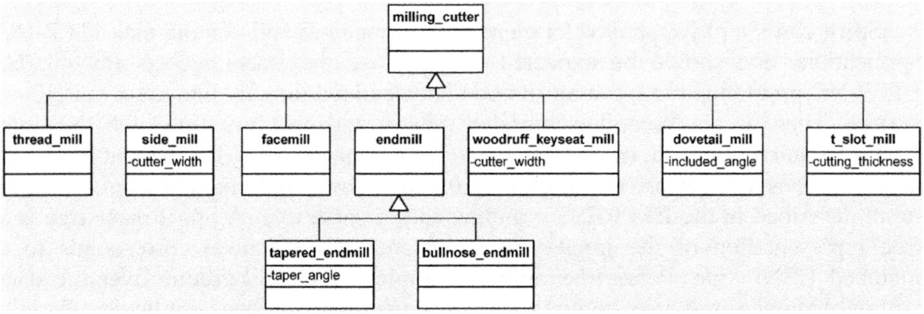

Figure 4.11. Milling cutter UML class hierarchy

The EXPRESS model provides a tree representation of data that often contains many levels of nested elements and contains meta-data throughout. Navigation is a key element in simplifying the programming process. STEPNC++ Query is a tree traversal mechanism developed to access STEP-NC parts' data. STEPNC++ provides a textual tree query mechanism similar to XPath, where the result of a STEPNC++ Query is the selection of nodes or meta-information about the nodes (e.g., array size). Query path expressions consist of a series of attribute names, separated by the slash character ("/"). The returned value of the path expression is the node sequence that results from the last step in the path.

For example, the STEPNC++ Query provides a convenient and efficient way to access the STEP-NC workpieces using the attribute names of EXPRESS entities. Below, two methods are shown for accessing the material properties of the workpiece. First, metadata about a tree-node item size is retrieved from the STEP-NC data. A positive item size establishes that at least one workpiece has been defined. Next, the material standard is retrieved using purely textual tree navigation starting at the root (project) and moving down the branches until the material standard node (EXPRESS type string) is found. By comparison, a hybrid text query and C++ pointer navigation is shown accessing the material id. Pointer navigation is especially hard in STEP-NC because the data can be sparse, where tree-nodes can be optional and have no data. Null pointers are used to signify entities with missing attribute data in STEPNC++. This can lead to C++ exceptions if not properly tested for NULL, making textual query even more appealing.

```
CString szStandardMaterialId;
CString szMaterialId;
int m= StepQueryItemSize("/project/its_workpieces");
if(m>0)
{
  szStandardMaterialId =((EStringPtr) StepQuery("/project/its_workpieces[0]/"
                                  "itc_material/standard_identifier")->GetVal();

  EworkpiecePtr workpiece=(Eworkpiece*) StepQuery("/project/its_workpieces[0]" ;
  szMaterialId = workpiece->its_material->material_identifier-> GetVal();
}
```

In summary, the STEP-NC Query mechanism allows one to extract features and operations from the ISO 14649 plan for translation into a vendor-specific feature-based CNC representation. Class metadata is available to differentiate object instance inheritance chains. Tree traversal techniques similar to XPATH simplify accessing the multi-layered object-oriented STEP-NC plan representation. However, pure pointer traversal through the tree is also available.

4.6 Validation and Analysis

Using STEPNC++, a series of tests were conducted on a Deckel Maho Guildermeister (DMG) DMU 70 eVolution running a Siemens 840D controller to validate feature-based CAM to feature-based CNC. In combination with the DMG, a Siemens 840D Simulator was used for testing and proof of concept. The DMG is a high-speed machine with feed ranging to 20,000 mm/min and speeds up to 30,000 rpm. The Siemens 840D is an open architecture CNC that offers functionality for high-speed and 5-axis machining. The 840D has extensive conversational programming comparable to other CNC and machine tools in industry.

To understand feature-based CAM/CNC machining better and the implications of the feed-forward tolerance knowledge, STEPNC++ was used to test the milling of the "NIST Logo" part, which is constructed entirely using simple rectangular, closed pocket features. A software tool using STEPNC++ was developed to read the NIST Logo STEP Part 21 file, and then translate the pocket knowledge into 840D compliant pocketing knowledge to produce a CNC part program. Using UML notation, Figure 4.12 shows major aspects of the STEP-NC feature and side/bottom machining operation class hierarchy related to pocketing that were used.

Several Siemens 840D canned cycles are available for milling pockets (Figure 4.13). "Pocket1" is the milling routine we will discuss, as it is the simplest. Pocket1 requires a milling cutter with an "end tooth cutting across centre" [4.17, 4.35]. Pocket1 uses contouring tool-paths, which are good for attaining a consistent surface finish. Contour milling uses successive offsets of the pocket contour as the cutting paths of the tool.

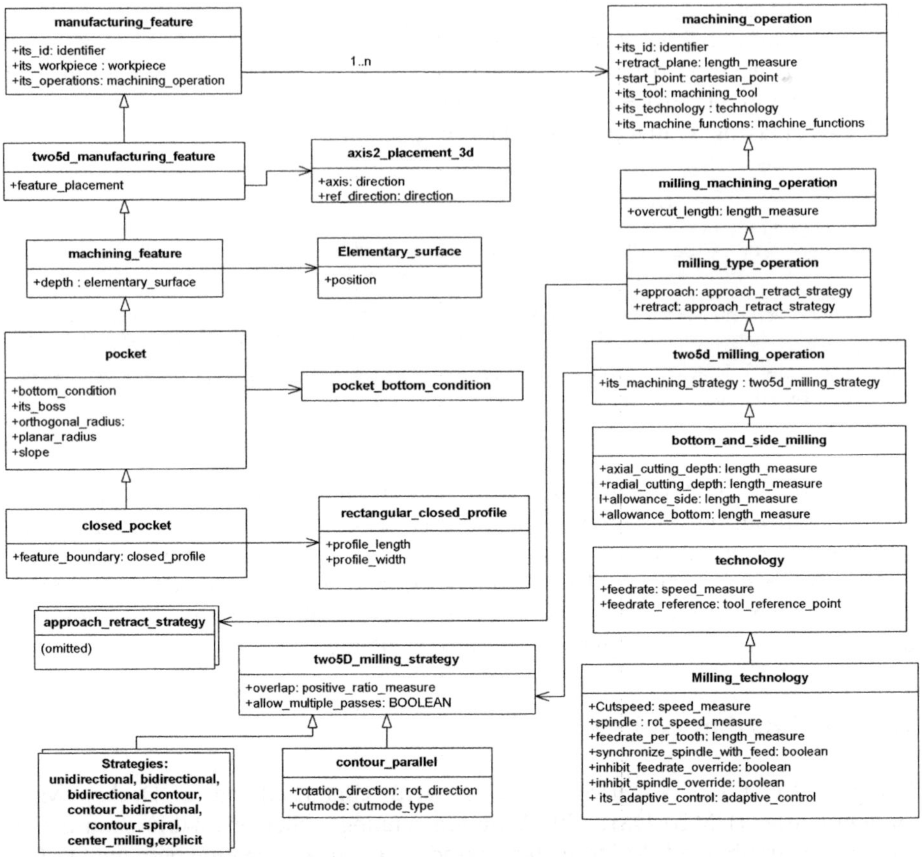

Figure 4.12, UML representation of rectangular pocket related STEP-NC classes

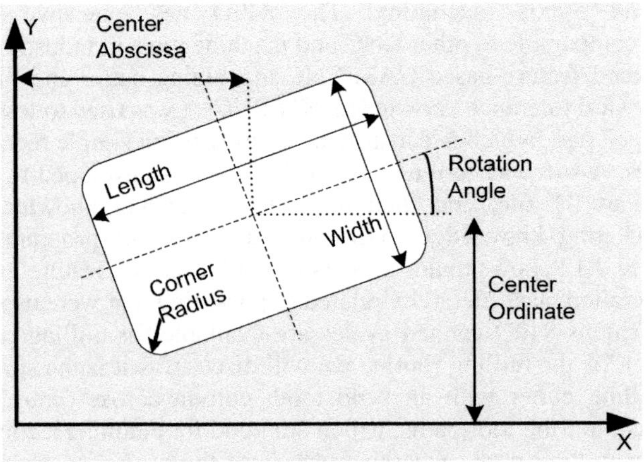

Figure 4.13. Overview Pocket1 parameters

Table 4.1 shows the mapping from STEP-NC knowledge into the 840D "Pocket1" canned cycle. In the table, "its_feature" refers to the closed_pocket class. "MillingOp" refers to a milling operation done by traversing from the feature to the "bottom_side_milling" operation.

Table 4.1. STEP-NC to canned cycle pocket mapping

840D Pocket1 Parameters	STEP-NC Pocketing Parameters
Retract plane	"MillingOp"/retract_plane
Reference Plane – i.e.,Top of pocket	its_feature/feature_placement/location/coordinates[z]
Safety clearance plane	Hard coded as 2 mm
Absolute depth (relative depth available - not used)	its_feature/depth/position/location/coordinates [z]
Length	its_feature/feature_boundary/profile_length/theoretical_size
Width	its_feature/feature_boundary/profile_width/theoretical_size
Center abscissa	its_feature/feature_placement/location/coordinates[x]
Center ordinate	its_feature/feature_placement/location/coordinates[y]
Corner radius	its_feature/orthogonal_radius/theoretical_size
Rotation angle	Angle between longitudinal axis and abscissa. Computed using: its_feature/feature_placement/ref_direction/direction_ratios
Cycle type: Roughing, Finishing, Combination	Use classname of /its_feature/its_operations[i] to derive rough or finish
Feedrate for depth infeed	Not addressed in approach_retract_strategy
Milling direction: Climb vs conventional	"MillingOp"/its_machining_strategy/cutmode Assuming contour_parallel machining strategy. STEP-NC default is conventional.
Not used.	Stepover
Finish feed	"MillingOp"/its_technology/feedrate In: technology, (requires units conversion)
Finish speed	"MillingOp"/its_technology/spindle In: milling_technology
Feedrate	Feedrate for surface machining "MillingOp"/its_technology/feedrate In: technology, (requires units conversion)
Spindle speed	"MillingOp"/its_technology/spindle In: milling_technology
Spindle direction	"MillingOp"/its_technology/spindle sign
Tool type	Classname of "MillingOp"/its_tool/its_tool_body
Tool radius	"MillingOp"/its_tool_body/dimension/diameter
Finish Allowance Side (for roughing)	"MillingOp"/allowance_side
Finish Allowance Bottom (for roughing)	"MillingOp"/allowance_bottom
Max one infeed depth	"MillingOp"/axial_cutting_depth

Below is a sample of the STEPNC++ generated feature-based NC block sequence to mill a pocket. The Workingstep name is "mainpocket". Line 1 is a comment (denoted by leading ‘;’) indicating this is a “closed_pocket” type feature. Line 2 is a comment describing the mainpocket operation “bottom_and_side_finigh_milling”.

```
;WS(1) mainpocket Feature=closed_pocket
;WS(1) mainpocket Operation(0)=bottom_and_side_finish_milling

; Tool=endmill (120), length=100 mm, radius=3mm
N010 $TC_DP1[1,1]=120 $TC_DP3[1,1]=100 $TC_DP6[1,1]=3

; Canned cycle to set tolerances - finishing
N020 CYCLE832(0.01,102001)
N030 T1
N040 M6
N050 G0  F2000 X38.100 Y25.400 Z48.100
N060 F4M03S1000

POCKET1(48.100,  38.100,  2.000,  -7.620,  ,  66.040,  40.640,  7.620,
38.100, 25.400, 0.000, 4.480, 4.480, 12.700, 20.000 ,1)
```

In the feature-based CNC operation, a tool change operation is performed, which corresponds to a T1 (select tool 1) and an M6 (tool change). For simulation purposes, the tool is defined within the program so that the CNC can determine the relationship between the tool and contouring tolerances. $TC_DP[i, j] defines tooling parameters, where i corresponds to T, the tool number and j corresponds to D, the cutting edge number. $TC_DP1 defines the tool type – an end-mill, $TC_DP3 defines the tool length – 100 mm, and $TC_DP6 defines the tool radius – 3 mm. The block N020 uses the 840D CYCLE832 canned cycle and defines the tolerance information for a finishing operation, with the emphasis on accuracy. CYCLE832 sets the tolerance band for contours and provides additional machining support for smoothing, block compression, spline interpolation, feed-forward control and jerk limits. The block N060 defines feed F to be 4 m/min, M3 calls for a counter-clockwise spindle rotation, and S1000 instructs a spindle speed of 1000rpm. Pocket1 calls the pocketing canned cycle with the parameters as outlined in Table 4.1.

Figure 4.14 shows the output for milling 11 pockets to form the NIST Logo: “Main pocket”, “stem of T”, “top of ST”, “left of S”, “middle of S”, “right of S”, “bottom of S”, “I”, “right of N”, “middle of N”, and “left of N”.

World-class manufacturing depends on the ability to understand and maintain outstanding machine tool performance. Trouble arises when machining produces out-of-tolerance parts. Circumstances that can contribute to part defects include tool wear, temperature variations, fixture stability, spindle loads, tool runout, improper cutting depth, uneven workpiece surface, and variations in material hardness.

Empirical knowledge and machining wisdom are helpful but accurate measurements are advantageous in analyzing root causes. Part of the difficulty in determining root causes is the large amount of data that must be continually collected and analyzed. CNC process traceability and data logging provides a systematic mechanism to enable informed manufacturing. The idea to perform CNC data monitoring for machining enhancements is not a new concept, and is found within industry and academia [4.36–4.38]. CNCs typically offer data logging canned cycles, but these are limited so as not to interfere with synchronized machining.

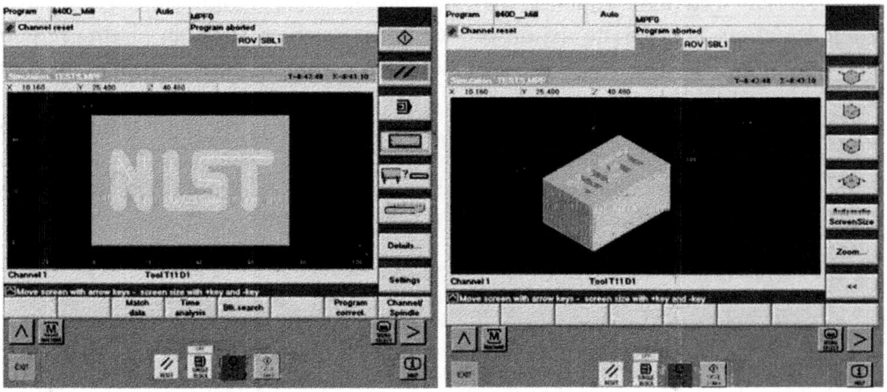

Figure 4.14. Simulated feature-based CAM/CNC cutting NIST logo part

The STEPNC++ feature-based traceability demonstration system monitored CNC process data using an Ole for Process Control (OPC) [4.39] data server. The CNC VM Data Logging mode was adapted from familiar logging approaches [4.40, 4.41] and the architecture is shown in Figure 4.15 and discussed below:

- *CNC Data Server* provides CNC data collection and external communication mechanism. In a standard NC logging process, an open architecture CNC with some turnkey or in-house data collection mechanism must be available. When one or more values change, the CNC Data server will notify the logging client of the changes and exchange the new data values.
- *Data Logging Client* is a separate application running in the background of the CNC computer or on a remote computer that may be running one or more data collection operations.
- *Data Collection* manages the communication issues related to logging data and handles CNC Data Server alarm notification.
- *Data List* defines the list of data to collect from the CNC.
- *Data Handler* manipulates the data after collection according to a given configuration. It is assumed each logged data list can be time stamped or incrementally marked.
- *Filter* provides more concise data logging with the ability to configure the save options of the data collection. The default filter is to log all the data items whenever one or more of the data items change.
- *Destination* provides for saving the log to a Persistent Data Store, in some common data format. Typical formats include Text, Comma Separated values (CSV), XML, or data base. Supporting a limited number of import/export file formats with broad industry support has been deemed preferable to having a separate language to describe formatting and analysis of the data. The None

format allows data logging without physical file saving to allow monitoring for potential machine faults.

- *Formatter* provides support for transforming the data to a corresponding Destination format.
- *Trigger* provides support for conditional event monitoring. Upon event detection, appropriate action can be taken, such as issue an alarm. Triggers can play a key role in part integrity and in implementing machine safeguarding rules.
- *Data Analysis* generates reports, provides traceability, and trouble shoots machining performance, or ties directly into the CNC and provides safeguarding machining operations – such as machine halting or feed rate override reduction should analysis reveal a problem.

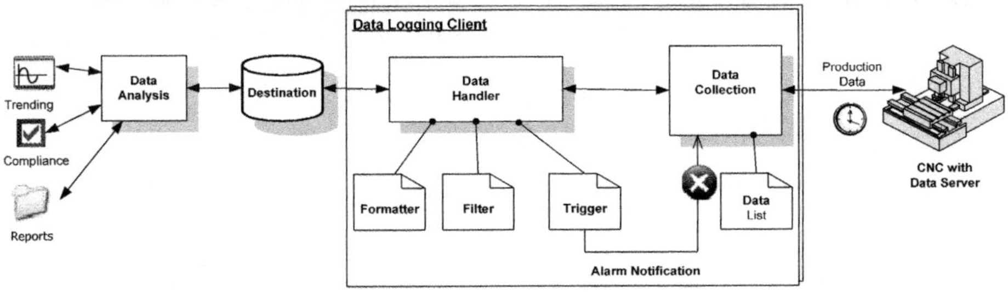

Figure 4.15. System architecture for feature-based CNC traceability

Initially, data was logged using the CSV data logging format, as it is the most common import and export format for spreadsheets and databases. Data collected for the "NIST Logo" machining included Timestamp, Feature, Program Name, Line Number, X/Y/Z Position and Following Error, Feed-rate, Spindle Speed, Spindle Load, and Tool Wear Offsets. Data points were logged at 10Hz (100 ms updates). Given data collection rates at 10Hz, data logs can easily generate 100 Mb of data especially for parts taking hours of machining.

The STEPNC++ data logging has filters to reduce the potentially enormous amount of collected data. However, detecting abnormal data is the key to improving machining, such as excessive lags or spindle load. For example, filters were applied to track excessive following errors, and for a complex part with concave walls, excessive lag was found as expected. Another helpful data concentration mechanism would be to use statistical analysis to summarize the variability of the actual vs programmed process values, such as for feed-rate, spindle speed position, and tool wear.

Real-time data logging software running in parallel with the CNC operations can help analyze complex problems. Long term it would be nice to have a machining knowledge base or expert system to assist those machinists lacking in years of experience when they need to troubleshoot a problem. The next step in the traceability is to incorporate an EXPRESS model to log traceability data such that, we will use STEP-NC as the feature description language to associate the logged data to the machining feature information.

4.7 Discussion

In CAM/CNC program exchange, the use of ISO 6983 causes an interface impedance mismatch, whereby the descriptive process knowledge is reduced into a motion-control CNC world. STEP-NC offers a more complete machining process model in which CAD/CAM systems and CNC machines can exchange process knowledge. By including valuable process information that can be utilized by the CNC, feature-based CAM/CNC part program exchange shows many potential benefits from the additional STEP-NC process knowledge – faster roughing cycle times, better intermediate surface roughness and final surface finish, as well as improved part geometrical accuracy and precision.

The future of feature-based CAM-CNC part program exchange depends on using STEP-NC in a reasonably easy and cost-effective manner. The STEPNC++ system demonstrates feature-based CAM/CNC that leverages the conversational canned cycles of modern CNCs. The STEPNC++ system is based on the Boost C++ meta-template programming library and contains software for reading, translating, posting, and manipulating ISO 14649 STEP-NC files. The core of STEPNC++ consists of Spirit EBNF parser implementations of various STEP related grammars. The grammars can be used via a C++ library interface and provide an object-oriented interface for feature-based machining and milling capabilities. Commercial STEP solutions could be used in place of much of STEPNC++ and should perform just as well if not better to achieve feature-based CAM/CNC.

In this chapter, STEPNC++ explores the use of tolerance-based CNC machining to improve finishing accuracies and roughing cycle times. Even though CNCs support tolerance-based machining, the concept is only in its infancy. The programming CAD/CAM tolerance information is typically confined to quality and inspection use. Recently, ISO 10303-203.E2, the "Configuration Controlled Design of 3D Mechanical Parts and Assemblies – Edition 2" standard [1.42], has added tolerance knowledge to the standard which should further advance the manufacturing industry toward tolerance-based machining.

The new "enlightened" feature-based CNC machining also exposes new machining challenges, for example, what should a CNC (or translator) infer if optional data is missing? Contrary to popular opinion, conformance and interoperability of STEP-NC will now play an even larger role as the universe of CNC machining competence has broadened. Adding powerful functionality can be expected to add complexity. Case in point is the open-ended questions of "correctness" and "completeness" in a STEP-NC program.

Correctness can be an issue given an unconstrained STEP-NC program. In order to accommodate the potentially large combinations of features to machining operations, the CNC will have to establish some conformance profile of "intelligence". Detection of over-specified, under-specified, or missing data through STEP-NC "Part Program" compiling can be considered an "early binding" specification, in that all the relevant CAM/CNC feature exchange knowledge is fully quantified and correct, and the CNC does not make any high-level planning decision. STEP-NC also offers "late-binding" specification that allows the CNC to make process planning decisions, such as allowing the CNC to select tools as well as generate feature-based paths. In this case, questions arise as to what decisions the

CNC should be allowed to make and under what circumstances should the CNC be allowed autonomous decision making.

Looking at a feature-by-feature CAM-CNC implementation of STEP-NC shows that it could be extremely sensitive to "completeness" issues. Theoretically, a minimal, but acceptable, STEP-NC program could contain just a stock and material description, a pocket feature with tolerances but no operations or tooling that could be sent to the CNC. An "intelligent" CNC could handle this, as it supports a corresponding roughing/finishing canned cycle to machine a pocket and could select tooling to satisfy the feature programming and tolerance requirements. As a general rule, not all CNCs may have the capability to support this intelligent feature-based approach. This implies the need for a CNC profile against which STEP-NC part program are evaluated in order to guarantee machining suitability. The issues of completeness and suitability are not necessarily as ominous as they sound as the CNC is already making such decisions, such as default tolerances, that have been fixed within the CNC by the machine tool vendor, that are being done generally unbeknownst to the users. However, programming CNC decision knowledge opens up the broad spectrum of learning and data storage computer science technology. The STEPNC++ demonstration system is working on incorporating KD Tree to assist the CNC in decision making [1.43]. KD Trees are spatial learning data structures for indexing points in k-dimensional space where each axis is treated equally.

Although enlightened feature-based STEP-NC CNC machining appears both promising and foreboding, the reality of feature-based CNC machining at this time is limited to "early binding", such that any STEP-NC program must be fully instantiated and correct to run on a commercial CNC. But this in itself is a commanding step forward, especially since the CNC has a native understanding of the machine dynamics and control/trajectory algorithms of the machine and can theoretically better manage tolerance-based feeds and speeds. Consequently, to increase the popularization of STEP-NC and intelligent tolerance-enabled machining in general, elevating the mantle of feature-based CAM-CNC should be a top priority.

Disclaimer

No approval or endorsement of any commercial product by the National Institute of Standards and Technology is intended or implied. Certain commercial equipment, instruments, or materials are identified in this report in order to facilitate understanding. Such identification does not imply recommendation or endorsement by the National Institute of Standards and Technology, nor does it imply that the materials or equipment identified are necessarily the best available for the purpose.

References

[4.1] S. Brunnermeier and S. A. Martin, Interoperability Cost Analysis of the U.S. Automotive Supply Chain: Final Report DIANE Publishing, 2000.

[4.2] International Organization for Standardization, "ISO 14649: Industrial automation systems and integration - Physical device control - Data model for computerized numerical controllers - Part 11: Process data for milling," Geneva, Switzerland,2000.

[4.3] International Organization for Standardization, "ISO 14649: Industrial automation systems and integration - Physical device control - Data model for computerized numerical controllers - Part 10: General process data," Geneva, Switzerland,2004.

[4.4] International Organization for Standardization, "ISO/DIS 10303-238: Industrial automation systems and integration û Product data representation and exchange û Part 238: Application Protocols: Application interpreted model for computerized numerical controllers," Geneva, Switzerland,2004.

[4.5] Electronic Industries Association, "ANSI/EIA-274-D-1980, Interchangeable Variable Block Data Format for Positioning, Contouring, and Contouring/Positioning Numerically Controlled Machines," Washington, D.C.: 1979.

[4.6] International Organization for Standardization, "ISO 6983: Numerical Control of machines - Program format and definition of address words - Part 1: Data format for positioning, line and contouring control systems," Geneva, Switzerland,1982.

[4.7] S. Venkatesh, D. Odendahl, X. Xu, J. Michaloski, F. Proctor, and T. Kramer, "Validating Portability of STEP-NC Tool Center Programming," in IDETC/CIE 2005 25th Computers and Information in Engineering Conference (CIE) Long Beach, CA: ASME, 2005.

[4.8] ASME, "ASME Y14.5M-1994 Standard on Dimensioning and tolerancing," 1994.

[4.9] International Organization for Standardization, "ISO 1101-2004 Indication of special specification operators for straightness, roundness, flatness and cylindricity,"2004.

[4.10] S. H. Park, Robust design and analysis for quality engineering Springer, 1996.

[4.11] Object Management Group (OMG), "Unified Modeling Language," http://www.uml.org/: 2008.

[4.12] International Organization for Standardization, "ISO 286-1:1988, ISO system of limits and fits - Part 1: Bases of tolerances, deviations, and fits," 1998.

[4.13] Z. Bouaziz and A. Zghal, "Optimization and selection of cutters for 3D pocket machining," International Journal of Computer Integrated Manufacturing, vol. 21, no. 1, pp. 73-88, 2008.

[4.14] H. S. Choy and K. W. Chan, "Machining tactics for interior corners of pockets," International Journal of Advanced Manufacturing Technology, vol. 20, no. 10, pp. 741-748, 2002.

[4.15] A. Hatna, R. J. Grieve, and P. Broomhead, "Automatic CNC milling of pockets: geometric and technological issues," Computer Integrated Manufacturing Systems, vol. 11, no. 4, pp. 309-330, 1998.

[4.16] M. Held, G. Lukacs, and L. Andor, "Pocket Machining Based on Contour-Parallel Tool Paths Generated by Means of Proximity Maps," Computer-Aided Design, vol. 26, no. 3, pp. 189-203, 1994.

[4.17] T. R. Kramer, "Pocket Milling with Tool Engagement Detection," J. of Manufacturing Systems, vol. 11, pp. 114-123, 1992.

[4.18] Y. S. Tarng, Y. Y. Shyur, and B. Y. Lee, "Computer-Aided Generation of the Cutting Conditions in Pocket Machining," Journal of Materials Processing Technology, vol. 51, no. 1-4, pp. 223-234, 1995.

[4.19] "Machinery's Handbook 25th Edition," Industrial Press Inc,2008.

[4.20] "NIST Predictive Process Engineering Program," http://www.mel.nist.gov/msid/ppe.htm: 2008.

[4.21] International Organization for Standardization, "ISO 10303-11: 1994, Industrial automation systems and integration - Product data representation and exchange - Part 11: Description methods: The EXPRESS language reference manual," Geneva, Switzerland,1994.

[4.22] The World Wide Web Consortium, Extensible Markup Language (XML) 1.0 (Fourth Edition) 2006.

[4.23] B. Stroustrup, The C++ Programming language (Third Edition) Addison-Wesley, 1997.

[4.24] T. R. Kramer, H. m. Huang, E. Messina, F. M. Proctor, and H. Scott, "A feature-based inspection and machining system," Computer-Aided Design, vol. 33, no. 9, pp. 653-669, 2001.

[4.25] F. M. Proctor and T. R. Kramer, "A Feature-based Machining System using STEP," vol. 3518, pp. 156-163, Sept.1998.

[4.26] S. H. Suh and S. U. Cheon, "A Framework for an Intelligent CNC and Data Model," The International Journal of Advanced Manufacturing Technology, vol. 19, no. 10, pp. 727-735, June2002.

[4.27] International Organization for Standardization, "ISO/IEC 14977: Information technology - Syntactic metalanguage - Extended BNF," Geneva, Switzerland,Aug.2001.

[4.28] "The Boost C++ libraries," www.boost.org: 2002.

[4.29] "Spirit," http://spirit.sourceforge.net/links.html: 2008.

[4.30] J. d. Guzman and D. Nuffer, "The Spirit Parser Library: Inline Parsing in C+," C/C++ Users Journal, Sept.2003.

[4.31] D. Abrahams and A. Gurtovoy, C++ Template Meta-programming : Concepts, Tools, and Techniques from Boost and Beyond (C++ in Depth Series) Addison-Wesley Professional, 2004.

[4.32] T. L. Veldhuizen, "Expression Templates," C++ Report, pp. 26-31, 1995.

[4.33] T. L. Veldhuizen, "Arrays in Blitz++," in ISCOPE 1998: Proceedings of the Second International Symposium on Computing in Object-Oriented Parallel Environments London, UK: Springer-Verlag, 1998, pp. 223-230.

[4.34] S. C. Johnson, "Yacc: Yet Another Compiler Compiler," in UNIX Programmer's Manual, 2 ed New York, NY, USA: Holt, Rinehart, and Winston, 1979, pp. 353-387.

[4.35] K. Tang, "Geometric Optimization Algorithms in Manufacturing," Computer-Aided Design & Applications, vol. 2, no. 6, pp. 747-758, 2005.

[4.36] R. Aronson, "In-Process Gaging is Cost Effective," Manufacturing Engineering, 2007.

[4.37] H. Kunsoo and P. Changho, "Unmanned turning force control with selecting cutting conditions,", 3 ed Proceedings of the American Control Conference: 2003, pp. 2602-2607.

[4.38] J. Mou, M. A. Donmez, and S. Cetinkunt, "Integrated error correction system for machine performance improvement," Proceedings of the 1994 American Control Conference: 1994.

[4.39] OPC Foundation, http://www.opcfoundation.org.

[4.40] Sun Microsystems, "Java Logging Overview," http://java.sun.com/j2se/1.4.2/docs/guide/util/logging/overview.html: 2001.

[4.41] SAP, "Tutorial - Logging and Tracing Mechanism in SAP Version 1.2,".

[4.42] International Organization for Standardization, "ISO 10303-203 Industrial automation systems and integration - Product data representation and exchange - Part 203: Application protocol: Configuration controlled design,"1994.

[4.43] A. Moore, "An introductory tutorial on kd-trees," Robotics Institute, Carnegie Mellon University, Technical Report No. 209 , 1991

5

A STEP-Compliant Approach to Turning Operations

Yusri Yusof[1] and Keith Case[2]

[1]Department of Mechanical and Manufacturing Engineering, University of Tun Hussein Onn Malaysia (UTHM), Parit Raja, 86400 Johor, Malaysia
Email: yusri@uthm.edu.my

[2] Wolfson School of Mechanical and Manufacturing Engineering, Loughborough University, Leicestershire LE11 3TU, UK
Email: k.case@lboro.ac.uk

Abstract
There is no doubt that today manufacturing is more competitive and challenging than ever before in trying to respond to "production on demand". Due to the complexity of programming there is a need to model their process capability to improve the interoperable manufacturing capability of machines such as turning centres. This chapter focuses on the use of the new standard; ISO 14649 (STEP-NC), to address the process planning and machining of discrete turned components. It explores how ISO 14649 can be used to combine turning and milling operations to support interoperable CNC manufacturing of rotational asymmetric components at a single turning centre. The major contribution of this chapter is the creation of a computational environment for a STEP-NC compliant system for turning operations, namely SCSTO. SCSTO is the experimental part of the research, supported by specification of information models and constructed using a structured methodology and object-oriented methods. SCSTO was developed to generate a Part 21 file based on machining features to support the interactive generation of process plans utilising feature extraction. An important aspect was the need to overcome the complexities of component geometry including milling features so as to have the ability to manufacture turn/mill components.

5.1 Introduction

In the last few decades the economic priorities of manufacturing have shifted from being based on low cost of a standard product without compromise on consistency and quality, to the use of modern industrial manufacturing facilities with a *"production on demand"* concept. That concept has been adopted in order better to meet the challenges and take advantage of the opportunities of economic globalisation. This concept found expression in DABA, or "Design Anywhere, Build Anywhere" [5.1], where the changing business environment leads to a need to collaborate beyond geographic boundaries supported by the rapid advancement of information technology associated with manufacturing technology. In a traditional approach design and manufacturing are considered to be separate but in a modern global and modern environment that separation becomes a major weakness leading to slow and costly production cycles. This happens because design has its own team and so does manufacturing. The design focus is to design the product and pass information to engineers to decide how to make the product or how to realise the design. If anything happens and there is a need to redesign the product, the engineer will pass information back to the design team. Today, with the use of computer technologies and communication technologies in the manufacturing industries, the methods mentioned above are largely being replaced by CAD and CAM to implement concurrent engineering. Widespread CADCAM systems will reduce human interaction and the result should be increased production, reduced costs and better quality of product. The fundamentals of planning a machining process in a numerically controlled environment lie with the control and quality of operation planning and that planning time represents 50–80% of the actual machining time for single parts or small batches [5.2]. It becomes more critical for complex situations and new manufacturing technologies tend to extend the time further. Process planning has been defined by [5.3] as a function within the manufacturing environment which deals with the selection of manufacturing processes and parameters to be used to create the final product [5.3]. Investigations by Younis showed that an efficient CAPP system could result in reduction of the manufacturing costs by up to 30% and would also reduce the manufacturing cycle and the total engineering time by up to 50% [5.4]. Hence, the focus has been on process planning as the task of the determination of manufacturing processes, which for instance can determine whether or not a product should be manufactured through turning operations.

5.1.1 Standard Product Data Exchange

During the design process and production, both teams have to work closely until the final product is realised. One of the issues in that process is data transfer and data exchange. If the product is very complex such as those found in aerospace and automotive industries, the final product comes with varieties of sub-components that have been designed by different departments using a variety of tools or software. For example the part design information might be transferred using any one of a number of industry data exchange standards such as IGES, DXF and other formats. To eliminate the problem resulting from a variety of standards, new standards have been developed. ISO 10303 (informally known as STEP) and more recently ISO 14649 (informally known as STEP-NC) have been introduced as part of an international effort aimed at achieving fully interoperability and bi-directional information exchange [5.5]. To date, two different ISO subcommittees are working towards such a STEP-NC standard with two different focii; ISO TC 184/SC1 is working on ISO 14649, termed the ARM whereas ISO TC 184/SC4 is developing STEP AP-238, termed the AIM. Both models represent the data model information to program intelligent CNC controllers, but the AIM is fully STEP compliant, whereas the ARM contains the information required to program a CNC machine. The ARM is to be used in an environment in which CAM systems have exact information from the shop-floor, whereas AIM is more suitable for a complete design and manufacturing integration [5.6].

5.1.2 STEP-NC Environment for Manufacturing

ISO 14649 is referred to as STEP-NC due to its interaction with ISO 10303 (STEP) and was initiated to provide a data model for a new breed of intelligent CNC controller that is well-structured with *Workplans* and *Workingsteps.* ISO 14649 aims to model the complete information requirement that must exist in a controller to control a machine tool by defining "what-to-make" and plans "how-to-make". STEP-NC has been developed as a result of several research projects carried out by companies and academic institutions. In terms of international research and development into these standards, projects such as OPTIMAL [5.7] largely overcame the legacy standards of ISO 6983. OPTIMAL is one of the earliest STEP-compliant systems and is based on feature information and machining strategies. The research does not stop there, because the researchers now focus on identifying and defining interoperable manufacturing and STEP-NC compliance in the context of concurrent engineering. In particular, information reviews of STEP-NC, manufacturing processes and manufacturing resources are also major foci in this research area. The new NC programming data model purports to support a well structured hierarchical interface, and object-oriented and two way communication from the CAD environment down to the shop-floor [5.8]. STEP-NC is an improved interface between the CAD world and the manufacturing arena. It is recognized as such since it provides process information at the time and place of the manufacturing activity. The proposed STEP-NC data format supports accurate and timely adaptive control of the production equipment and provides feedback for information back to the planning activity. The current standard of programming NC namely G & M codes or ISO 6983 has had no significant change since the format of NC machines was developed at MIT in 1952 [5.2, 5.8-5.12]. Figure 5.1 shows the evolution of NC machines from using hardwired configurations to the current fully-integrated systems that can be found almost everywhere, from small job shops in rural communities to multi-national companies in large urban areas. During the pre-CNC epoch the program language had been modified by vendors and controller developers who added their own commands. Since the 1970s significant developments have been made towards more automatic and reliable computer numerically controlled machines with new machining processes. Today's highly sophisticated CNC machines utilise a variety of cutting technologies such as multi-turret and multi-spindle in complex axial configurations and this machine capability increases the level of flexibility and capability compared to the previous decade [5.13]. A large number of Computer Aided Systems (CAx) have been developed and implemented in recent years to support all stages of product life by computer systems and many can simulate virtual CNC machining with the complete machine toolpath [5.14]. Since the first NC machine was introduced, various process plan packages have been developed and each system tried to interpret the part data format more reliably. Most of these systems are specialised to support certain applications, and are based on an information model that handles the application specific view of the product. These current trends are aimed at open systems but they are predominantly used in retrofitting applications for conventional NC machines.

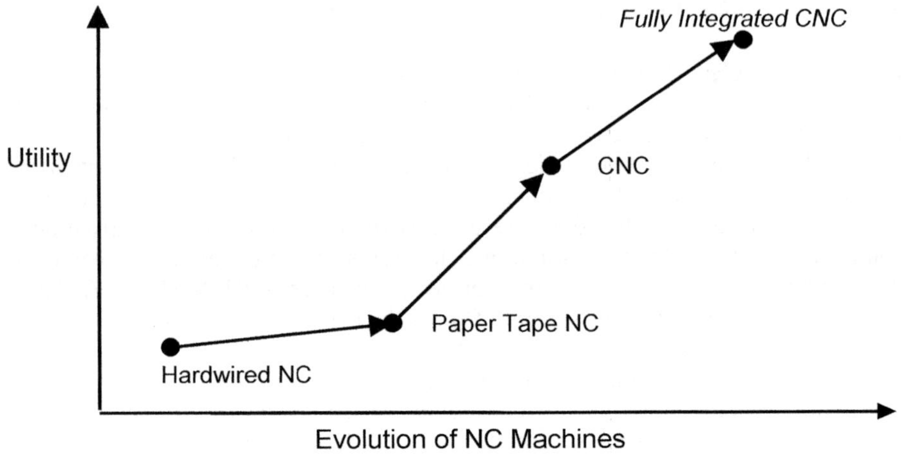

Figure 5.1. Evolution of NC machines [5.15]

5.2 Related Work

STEP-NC has been developed as a result of several research projects carried out by companies and university institutes which is focused on machining and inspection. Recently a number of projects involving the areas of STEP-NC based interoperability and research and development for various CNC manufacturing processes have been started. Overall research activities in specific areas of STEP-NC based on manufacturing technology and processes focus more on milling rather than turning activities, due to ISO 14649, Part 11 for milling operation being established before Part 12 (Turning). Research activities in milling operations have been more common than turning in the last 5 years. Work from Shimamura is recognised as one of the earliest pieces of research to address an alternative for enhancing the capability of the existing NC machines economically using a PC-based retrofitting scheme for the manufacture of free form surfaces [5.16]. In 2002, research and development in terms of manufacturing technology and processes began with a proposal for the conceptual framework for designing and implementing an intelligent CNC system by Suh and Sheon [5.17], followed by Hardwick providing the first outlook on STEP-NC compliant manufacturing [5.18]. Lee and Bang have successfully developed and built a five-axis milling machine that is run by STEP-NC in XML [5.19] and another prototype system has been proposed by Newman et al. for a STEP-compliant CAD/CAM system based on one of these frameworks using the new ISO 14649 standard for milling components [5.20]. Finally, test and validation methods have been proposed for testing data for numerical control [5.21]. It is noticeable that, in 2006, researchers were extremely focused on this particular area, and details can be found in a special issue edition of the International Journal of Computer Integrated Manufacturing (IJCIM) for STEP-Compliant Process Planning and Manufacturing [5.22]. Kumar introduced a STEP-compliant framework that makes use of self-learning algorithms that enable the manufacturing system to learn from previous data and results in error elimination and consistent quality products. It has been tested and certified for pocket and hole features for milling [5.23]. The latest achievement in 2007 is the successful development of a system called ST-FeatCAPP for prismatic parts based on ISO 14649 by [5.24]. The system maps a STEP AP224 XML data file, without using a complex feature recognition process, and produces the corresponding machining operations to generate the process plan and corresponding STEP-NC in XML format. Liu et al. also proposed an NC programming system for prismatic parts to be machined using STEP-NC machine tools, and the system consisted of three functional modules, namely (1) a feature-based modeller, (2) a process planner and (3) a part program generator. The system can read the STEP-NC file and calculate the toolpath automatically compared to current systems that only produce low level control information [5.25]. One of the aims for the next generation of CNC machines is to be interoperable and adaptable so that they can respond quickly to changes in market demand and the manufacturing needs of customized products [5.22]. As part of this, 2006 was a time when researchers were particularly focused on proposing a framework for turning. Most of the researchers proposed prototype systems to support data interoperability between the various CAx systems based on ISO standard 14649 that provided the first data exchange format used in the operation of NC machines [5.26–5.32]. Among these systems, G2STEP is the latest system to cover the machine functioning from pre-processor to STEP-NC part program generation including part program verification [5.33]. This development of a future manufacturing platform to enable different processes and capability such as milling applications, multi-axis and complex components as the basis of the integration of CAD/CAPP/CAM and CNC will be a major research task for years to come. The changing business environment over the past decades including globalisation resulted in the standards ISO 10303 and ISO 14649 (STEP and STEP-NC) being introduced to solve the interoperability issues. For the time being many obstacles come from software/hardware vendors as the current approaches give them many opportunities to maintain their market, but the new standards can provide the platform for the future of global interoperable manufacturing [5.34]. The Shop-floor Programming System (SFPS) introduced by Suh is the first system fully compliant with ISO 14649 [5.35] and, to date, only this system has been patented (US patent references; 6400998, 65112961, 6556879, 6650960 and 6671571). SFPS and other systems related to STEP compliance that have been developed by academia all over the world are shown in Table 5.1.

TurnSTEP clearly defines the number of setups as either one setup or two setups dependent on the independent machine format [30]. TurnSTEP has some weaknesses such as that threads cannot be automatically generated but need to be defined and the process plan graph edited by the user manually. The output of this system can be in text and XML file formats [30]. As reported, TurnSTEP is at a prototype stage and the implementation of another part, which is intelligent and autonomous is still under development. In terms of implementation of bi-directional information flow, none of the systems show how it would work and do not make it clear how the functionality is supported in prototype systems. So far the test components used contain only simple turning operations with z and x axes and do not cover multi-axis machining. The author strongly agrees with the suggestion by Heusinger and Rosso-Jr, for the STEP-NC compliant information structure to support the milling capability of the NC turning centre to meet industrial needs mapped by ISO 14649 Part 11 and 12 (milling and turning) [5.31, 5.37]. The author has noticed that all the proposed systems use a feature recognition approach and feature based techniques to allow the user to edit the part program. Xu has stressed that the commercial software, namely ST-Plan, can create STEP AP 224 machining features from CAD files (AP 203 or AP 214) [5.12]. All the proposed systems comply with ISO 14649 and this is the first stage to develop the universal manufacturing platform for CNC machining as proposed by [5.29, 5.34].

Table 5.1. Review of STEP-compliant systems

No	Systems	Input	Output	Domain
1	SFPS (Milling) [5.35]	STEP AP203 & AP214	Part program physical file (text)	Prismatic
2	STEPTurn [5.31, 5.36]	STEP AP203	Part program physical file (text)	Rotational
3	TurnSTEP [5.29, 5.30]	STEP AP	ISO 14649 physical file and extensible mark-up language (XML)	Rotational
4	G-Code Free for lathe [5.6, 5.26]	STEP AP 203	Native CNC language program	Rotational
5	G2STEP (2-axis turning machining) [5.33]	G-codes	STEP-NC part program	Rotational

5.3 Design of a STEP Compliant System for Turning Operations (SCSTO)

This section proposes a system framework for a STEP Compliant System for Turning Operations (SCSTO) which considers both informational and functional perspectives of the system. From a functional perspective the proposed system has been designed to be a semi-automatic process planning system, meaning that it does not automatically generate manufacturing information directly from the CAD model. It is aimed at the creation of feature-based process plans for manufacturing processes such as turning operations. The proposed system is for turning operations and is based on a STEP compliant environment. It consists of several elements that define turning features and generate STEP code compliant with ISO 10303 Part 21 [5.38]. The system is based on feature-based design and begins with the selection of the workpiece followed by the choice of turning manufacturing features and finally the choice of the tools. The output of the system is a physical file complying with Part 21 [5.38]. The aim of this work is to address the process planning and machining of rotational components and to propose a STEP Compliant NC structure for generation of ISO 14649 code which can be used for turning component manufacture. Interoperability within this context is a significant objective. Interoperability is defined as the ability to integrate STEP-NC compliant information in the product life cycle including CAD, CAPP, CAM and CNC, combined with feasible information structures to represent various configurations of turning machining centres. The overall framework is illustrated in Figure 5.2 and is based on the Java programming language. The prototype has been developed using JBuilder 2005 [5.39] to provide a suite of integrated development tools related to STEP standards. This concept has been used to generate Java classes from the EXPRESS schema and to handle the STEP Part 21 physical file format. The data model for manufacturing of turned components was based on the ISO 14649 standard. Part 10 is the backbone of the standard covering the common data structure. In the standard, the part is defined as a *workpiece* while the task is defined as a *Workplan* consisting of a series of *machining_workingsteps* to carry out the *machining_operation* on a *manufacturing_feature*. In turning operations it becomes a Workplan with a series of *turning_workingsteps* to carry out the *turning_operation* on a *turning_feature*. The *turning_operation* itself is supported by *turning_technology*, *turning_machine_function* and *turning_strategy*.

The UML was developed representing the representation and model diagrams, the constraints and the extension mechanisms. UML is the most widely known and used standardized notation for object-oriented analysis and design. The most useful standard UML diagrams are use case diagram, class diagram, sequence diagram, state chart diagram, activity diagram, component diagram and deployment diagram. For the purposes of this chapter, only class diagrams and their notation have been used. In order to develop the system, a set of computing tools will be used. The Java programming language is used for the actual development of software components based on the object oriented methodology and UML is utilised as the modelling language. The manufacturing models refer to the process that deals with production such as operation and strategies. The UML represents the various objects for the SCSTO manufacturing environment and the relationships between these objects as shown in Figure 5.3 [5.40, 5.41]. Each data type in these models is based on ISO 14649 part 10 [5.42] and part 12 [5.43] and UML diagram for the turning feature is shown in Figure 5.4.

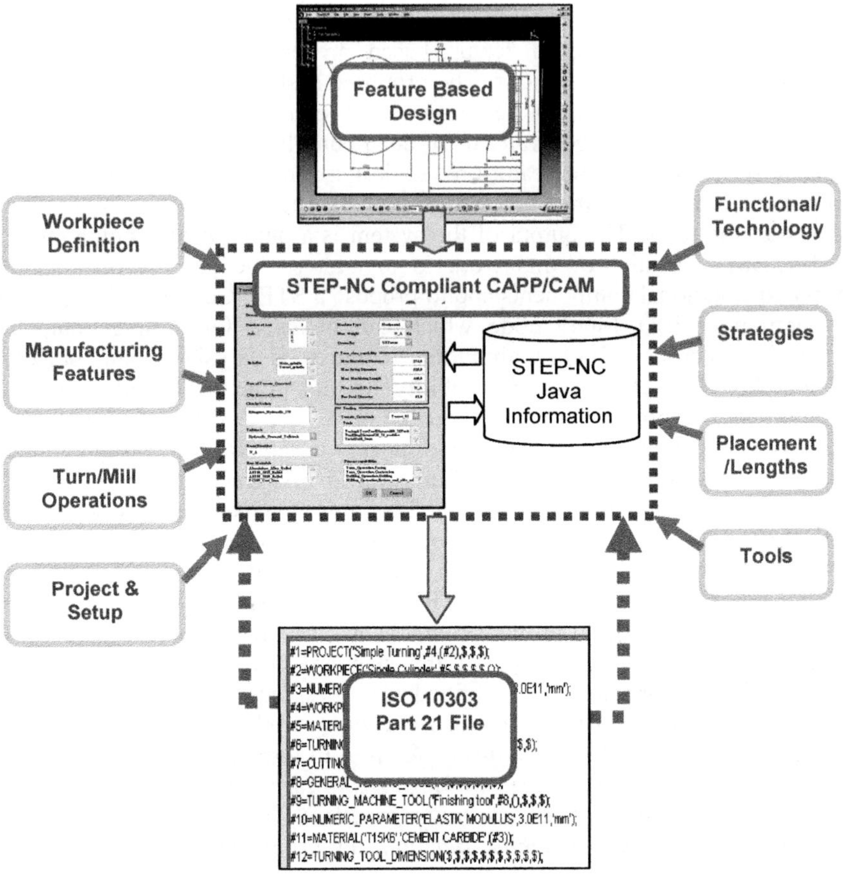

Figure 5.2. Proposed SCSTO system

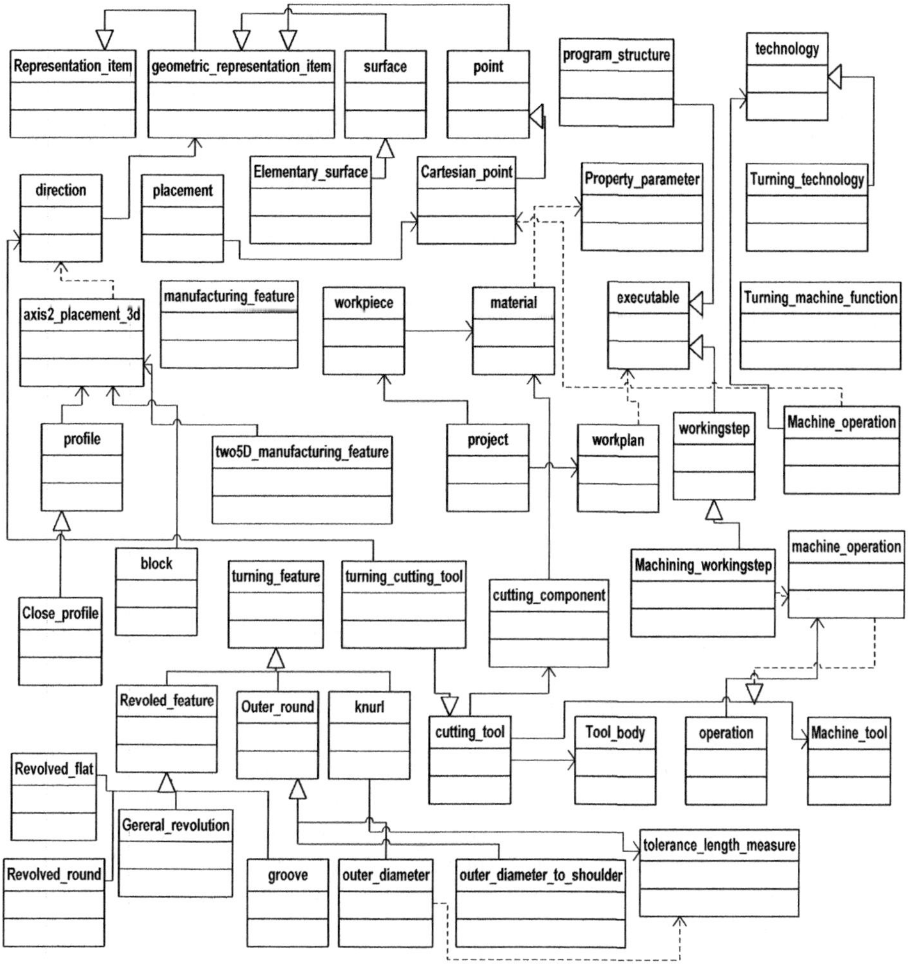

Figure 5.3. UML Diagram for SCSTO [5.40, 5.41]

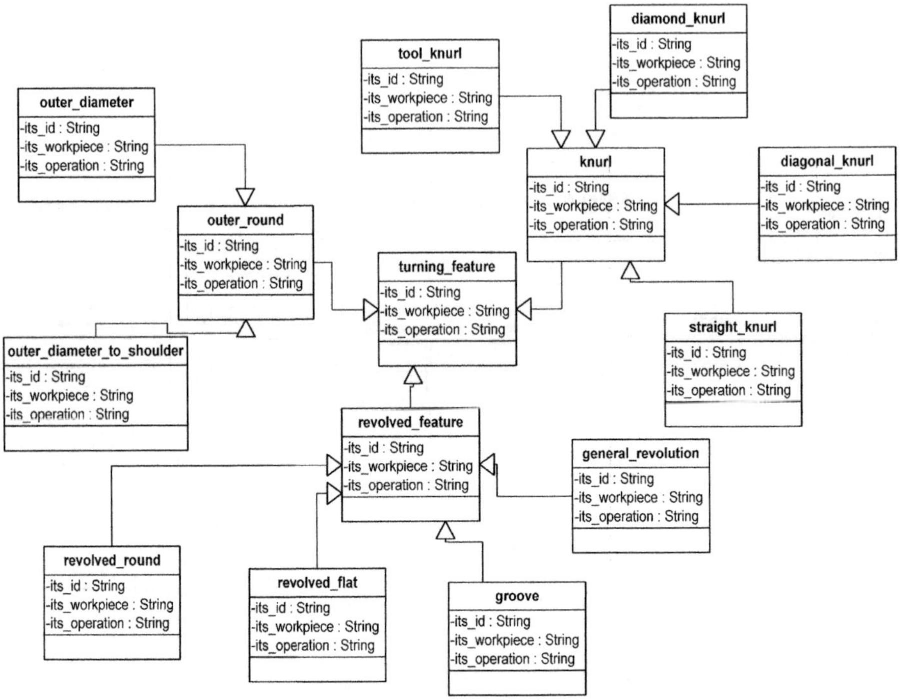

Figure 5.4. UML diagram for the turning feature

It starts with gathering the information related to SCSTO, focuses on the product and manufacturing models to provide additional input into the construction of the model and forms the basis of the conceptual information. A sample of code containing the implementation and structure of one the most general class turning feature library is given in Figure 5.5 and the code consists of public class, constructors and methods.

```
package iso14649;
// Imports
import java.util.ArrayList;
public class turning_feature extends two5D_manufacturing_feature {
 // Constructors
 public turning_feature() { }
 public turning_feature(String _its_id, workpiece _its_workpiece, machining_operation[]
_its_operations, axis2_placement_3d _feature_placement) { }
 // Methods
 public ArrayList get_prop_names() { return null;}
 public void set_attributes_from_part21_line(String part21line, ArrayList A) { }
 public ArrayList get_attributes() { return null;}
 public ArrayList get_output() { return null;}
}
```

Figure 5.5. Sample code for general class turning feature library

Finally, after considering the product data model and the manufacturing data model referring to manufacturing resources, processes and strategies, the author has developed a model for SCSTO using STEP-NC schemas. This model becomes a platform for developing the SCSTO prototype based on functional and information referring to resources, processes and strategies. The overall system as shown in Figure 5.6 consists of three main subsystems; manufacturing features creator, manufacturing operations and program generator. In SCSTO, the user has a choice either to create a new project with new features or to open a project from a CAD file. Geometry described in the AP 224 format defines the features but not their location and orientation. Feature geometry is defined in ISO 14649 Part 12 [5.43] and more formally described in terms of UML diagrams. The alternative is to create the feature by user definition which is limited to features within the feature library. If the user needs to define a new feature, the first step in designing a part is to specify the base part shape and the dimensions associated with the shape to define its size. In this thesis the base part shape is limited to cylindrical, so only this shape can be recognised for further processing. The base part shape is considered as the initial shape of the material before machining the features. Cylinder length and diameter are the only parameters needed to define the base part that is positioned with the z-axis parallel to the longitudinal axis of the shape. The x and y axes are orthogonal to the z axis. The axis origin is positioned at the centre of the circular profile forming the bottom of the cylindrical base part. The second subsystem consists of five major components:

1 Integration and preparation of part design data
2 Selection of machining strategy
3 Selection of machining technology
4 Selection of machine functions
5 Selection of cutting tools

The third subsystem is the Generation of Process Plans to generate a physical STEP-NC process plan file. This file consists of information such as machining operations, cutting tools, machining parameters, etc. If the user is not satisfied with this part program, it can be edited to modify either turning features or turning operations. The turning machining concept is based on 2D profiling except for drilling perpendicular to the z axis. Figure 5.7 shows the raw material and the working area which is defined as the Boolean difference between the raw material and the finished part. The product model is contained within a set of interconnected objects. The class definition of these objects is based on the entity definitions that exist within the ISO 14649 standard. From the standard, classes are written in the Java language and an example of the structure of these classes is shown in Figure 5.8. The SCSTO system adheres to the Windows standard for user interface design.

All functions can be accessed from the pull-down menus, and common functions are accessible via toolbar icons. Figure 5.9 shows the SCSTO user interface window. It consists of several pull-down menus and toolbars. Figure 5.10 shows an example activity for project dialog. One of the major functions of the proposed system is machining operations, to manufacture the part. Generally, there are two types of machining operations: roughing and finishing. Roughing is used to remove material from the original raw material by multiple surface passes down to the finishing allowance. Finishing then removes the finishing allowance to yield the final form of the feature. The operation is one of the following: facing, grooving, contouring, threading, or knurling and for milling, drilling, boring, centre drilling and reaming. Due to special machining the proposed system covers all types of turning machining and only drilling under milling operations as defined in ISO 14649 Part 11 [5.44]. The drilling operations have been combined together with turning operations.

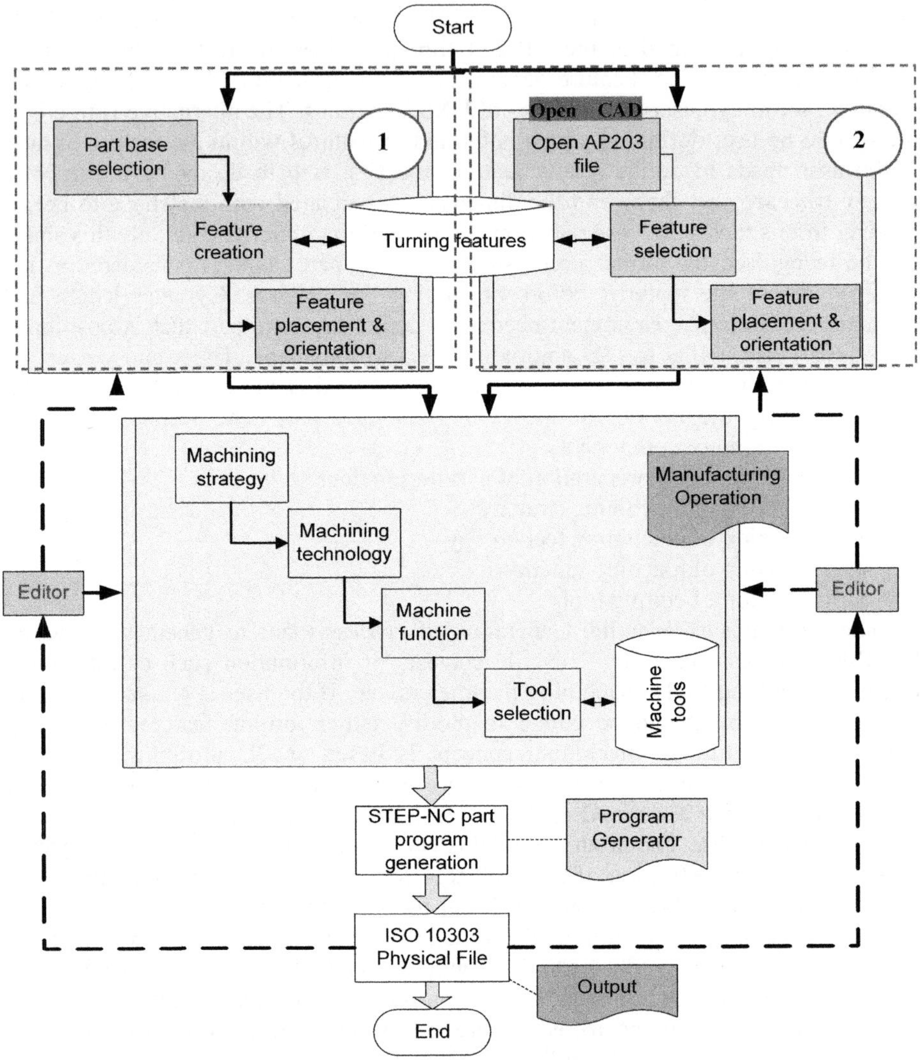

Figure 5.6. Architecture of SCSTO

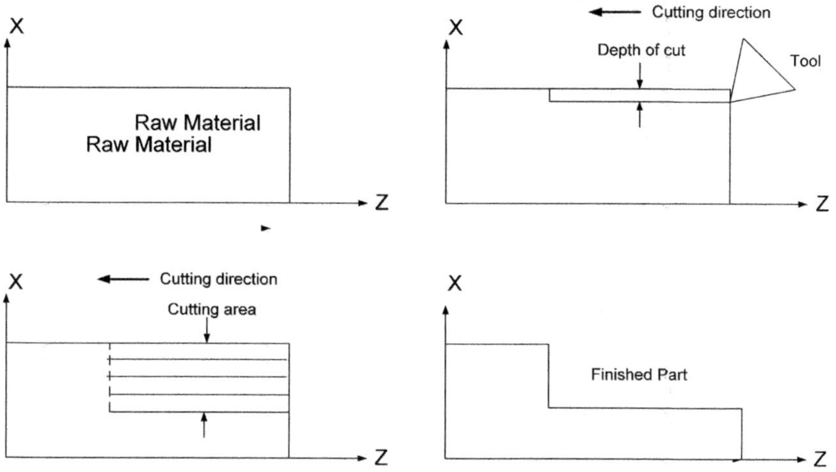

Figure 5.7. 2D turning concept

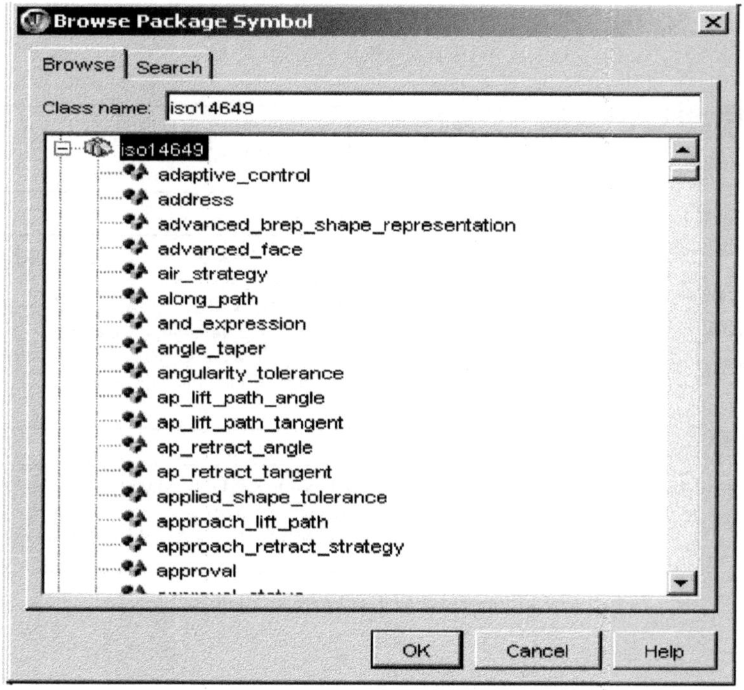

Figure 5.8. Java classes based on ISO 14649 standard

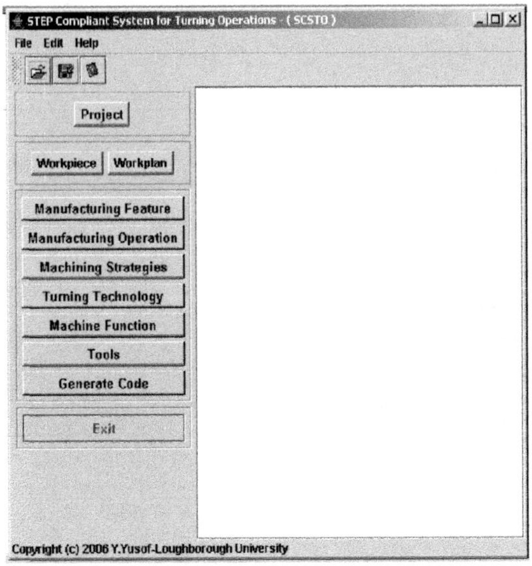

Figure 5.9. The SCSTO user interface window

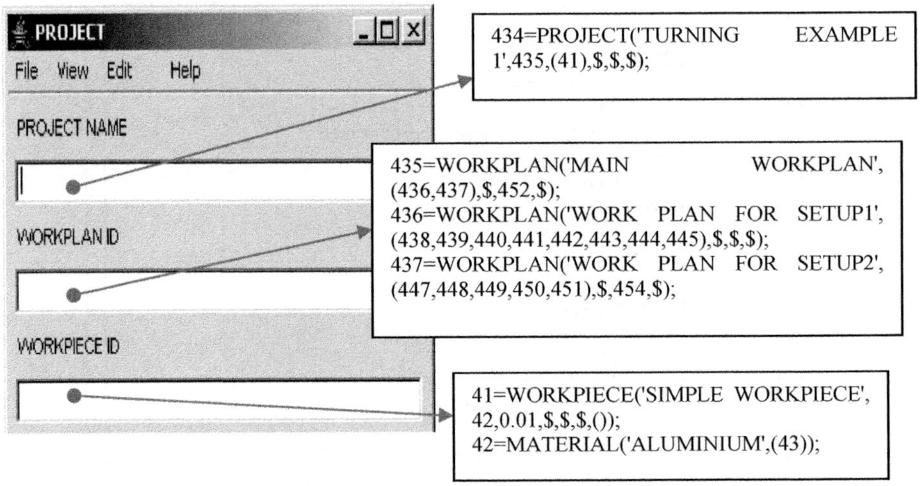

Figure 5.10. SCSTO project dialog

5.4 Case Study Component

Machining operations for the case study component are turning, grooving, threading, off centre drilling and milling on the side face. Its general attributes are double-sided and asymmetrical. The main contribution of this case study is to show the ability of SCSTO to read from a CAD file and recognise all the features with definition of turning technology, turning strategies, defining tools, Workplan and Workingsteps. This description focuses on the imported file and the recognition of the features. Figure 5.11 presents the 3D solid and isometric views of the component to provide a better understanding of the 3D geometry.

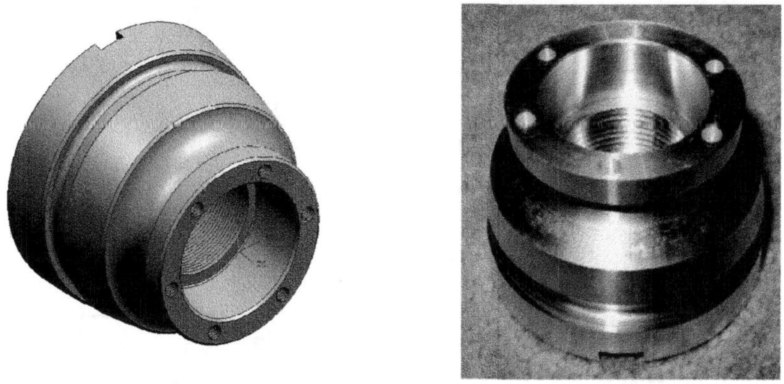

Figure 5.11. 3D drawing for component - final product (top-view) for component

The main purpose of this case study is to investigate the prototype system's capability of creating features directly from a CAD file. The case study component has been designed in Unigraphics software, exported as a STEP AP 203 file and then imported into SCSTO. Case study component has a minimum of one setup if the machine has a counter spindle and two-sided machining. The complete machining process involves 23 processes and also depends on the machine configuration as shown in Figure 5.12. The figure clearly shows this type of machining should have a linear axis and a rotary axis (c axis). This demonstration for case study component starts with the user beginning the modelling process by selecting the base part as the cylinder type. This shape is considered as the starting raw material for modelling the component. Then the user creates a step feature and attaches it to the base part. The remaining features are created and attached to the base part. The component was created in Unigraphics (UG) version NX3 to export file based on UG environment.

Finally, an ISO 14649 part program is generated by clicking the Generate Code button. The program is based on workpiece and *machining_workingsteps* in a physical file text format. This text file can be saved to a selected directory folder. As mentioned the part program can be edited by the user based on manufacturing features, strategies, tools, etc. When the user has finalised the part program, it can be sent to the machine controller. The manufactured product after the finishing process is shown in Figure 5.11.

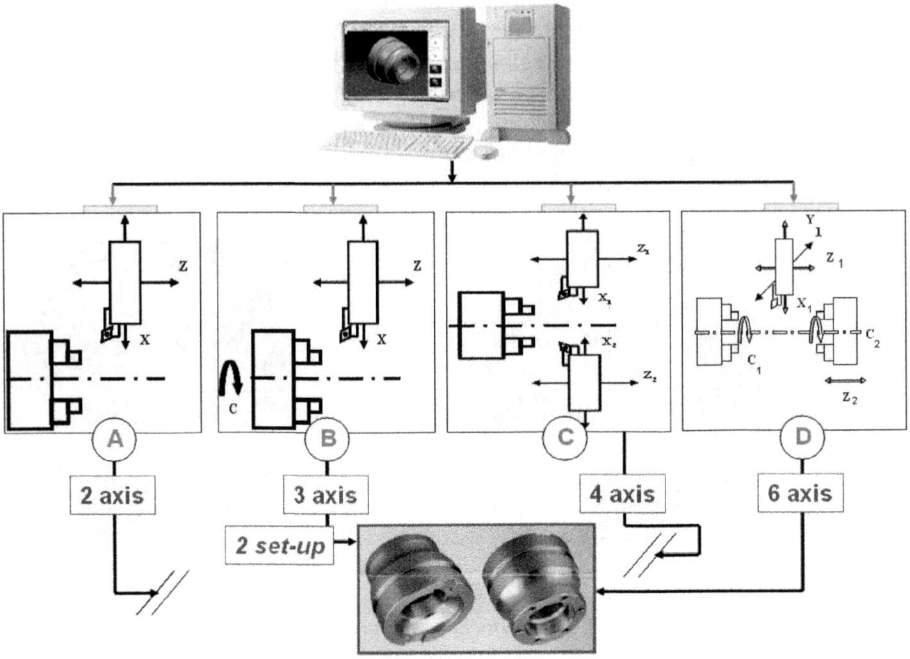

Figure 5.12. Process planning for case study component

5.5 Conclusion

This research work was motivated by industrial requirements of concurrent engineering, standard product data models and an integrated manufacturing environment, which are further reflected in process planning. The conclusions of the work can be elaborated as the information modelling in this area addresses the problem of capturing and representing manufacturing information related to resources and processes. The interaction between the different types of models could provide a description of the products, how they should be manufactured, and what manufacturing resources should be used. This would provide an information platform upon which several different computer-based tools to support the innovation process can be built. This will allow the provision of reliable manufacturing information to assist in the performance of product development life cycle activities and related decisions. SCSTO was developed to generate a Part 21 file based on machining features to support the interactive generation of process plans utilising feature extraction. The system was constructed using a structured methodology for its planning and object-oriented methods for its implementation. A case study component was tested to show that the new approach (STEP-NC) can generate code which is comparable to the currently used G-code with some benefits such as the elimination of the post-processor. Efforts are under way to fulfil the STEP-NC challenge by combining Parts 11 and 12, for turn/mill operations. STEP-

NC forms a possible basis to satisfy the latest requirements and demands with respect to a bi-directional CAx process chain for machining. STEP-NC forms a possible basis to satisfy the latest requirements and demands with respect to a bi-directional CAx process chain for machining. In addition its development as a future manufacturing platform to enable different process models to be integrated for the adaptable integration of CAD/CAPP/CAM and CNC will be a major avenue of research for years to come. It is expected that the recommended future extensions will enhance the usefulness of this chapter, and will meet the requirements for global interoperable manufacturing for real-life parts. There is no doubt that so far none of the proposed systems are fully capable of machining turn/mill components. Work to date has focused on the separate parts of ISO 14649 using Part 11 for milling operations including drilling and Part 12 for turning. No significant work has been done on combining the two parts for turn/mill components. However, the author and some researchers [5.31] believe that this industrial requirement could be achieved through research and development involving collaboration by researchers, users, manufacturers, academia and the ISO committee. If developers look from the business perspective, and academia focuses on theoretical aspects, the objective of combination turning and milling machining compliance with the new standard (STEP-NC) can be realised.

References

[5.1] D. Normile, K. Tyrka, D. Bulkeley, and D. Bak, "Design anywhere, build anywhere-Product Data Management and the Internet are redefining the relationships between manufacturers, customers, and suppliers," in *Global Design News*, vol. November, 2000.

[5.2] F. Ahlquist, "A Methodology for Operation Planning," in Department of Mechanical Engineering, Licentiate thesis, 2002, 89

[5.3] L. Alting and H. Zhang, "Computer aided process planning: A state of the art survey.," *International Journal of Production Research*, vol. 27, pp. 553-585., 1989.

[5.4] M. A. Younis and A. M. A. Wahab, "A CAPP Expert System for rotational components," *Computers and Industrial Engineering*, vol. 33, pp. 509-512, 1997.

[5.5] ISO, "International Standard 14649-1: Part 1 : Industrial automation system and integration - Physical device control - Data model for computerised numerical controllers - Part 1: Overview and fundamental principles," 2003.

[5.6] X. W. Xu, H. Wang, J. Mao, S. T. Newman, T. R. Kramer, F. M. Proctor, and J. L. Michaloski, "STEP-Compliant NC Research: The search for Intelligent CAD/CAPP/CAM/CNC Integration," *International Journal of Production Research*, vol. 43, pp. 3703-3743, 2005.

[5.7] ESPRIT, "Project 8643, Optimised preparation of manufacturing information with multi-level CAM-CNC coupling (OPTIMAL): final report for publication.," 1997.

[5.8] P. Muller, "STEP-NC – New data interface for NC programming," in *STEP-NC Newsletter, Issue 1, April 2000*. Erlangen, Germany, 2000, pp. 1-2.

[5.9] E. Fortin, J. F. Chatelain, and L. Rivest, "Improving Information Flow from CAM to CNC : an Innovative Architecture Based on BNCL Virtual Machine," presented at Computer & Industrial Engineering International Conference (29th ICC&IE), Montreal, Canada, 2001

[5.10] R. D. Allen, J. A. Harding, and S. T. Newman, "The application of STEP-NC using agent-based process planning," *International Journal of Production Research*, vol. 43, pp. 655-670, 2005.

[5.11] S. T. Newman, "Integrated CAD/CAPP/CAM/CNC Manufacture for the 21st century," presented at International Conference in Flexible Automation and Intelligent Manufacture (FAIM2004), Toronto, Canada, 2004

[5.12] X. Xu and S. T. Newman, "Making CNC Machine Tools More Open, Interoperable and Intelligent," *Computers in Industry*, vol. 57, pp. 141-152, 2006.

[5.13] A. Nassehi, S. T. Newman, and R. D. Allen, "STEP-NC compliant process planning as an enabler for adaptive global manufacturing," *Robotic and Computer Integrated Manufacturing*, vol. 22, pp. 456-467, 2006.

[5.14] S. Newman, "Integrated manufacture for the 21st century Development of the STEP-NC standard and its implications for manufacturing processes worldwide," *Metalworking Production*, vol. 148, pp. 13-16, 2004.

[5.15] J. Michaloski, "Case Study in the Challenges of Integrating CNC Production and Enterprise Systems," presented at ISA EXPO, Chicago, Illinois, US, 2005

[5.16] A. Shimamura, M. Moriyama, and K. Kasuga, "Data detection method for curve and surface data in STEP data exchange system," *Seimitsu Kogaku Kaishi/Journal of the Japan Society for Precision Engineering*, vol. 62, pp. 701-705, 1996.

[5.17] S. H. Suh and S. U. Cheon, "A framework for an intelligent CNC and data model," *International Journal of Advanced Manufacturing Technology*, vol. 19, pp. 727-735, 2002.

[5.18] M. Hardwick, "Digital manufacturing using STEP-NC," *Technical Paper - Society of Manufacturing Engineers. MS*, pp. 12, 2002.

[5.19] W. Lee and Y. B. Bang, "PC Based STEP-NC Milling Machine Operated by STEP-NC in XML Format," *Journal- Korean Society of Precision Engineering*, vol. 19, pp. 185-193, 2002.

[5.20] S. T. Newman, R. D. Allen, and R. S. U. Rosso-Jr, "CAD/CAM solutions for STEP Compliant CNC Manufacture," presented at Proceedings of the 1st CIRP(UK) Seminar on Digital Enterprise Technology, School of Engineering, University of Durham, 2002

[5.21] A. B. Feeney and S. Frechette, "Testing STEP-NC Implementations," presented at Proceedings of the 5th Biannual, World Automation Congress, 2002.39-44

[5.22] X. W. Xu, P. Klemm, F. M. Proctor, and S. H. Suh, "STEP-Compliant Process Planning and Manufacturing," *International Journal of Computer Integrated Manufacturing*, vol. 19, pp. 491-494, 2006.

[5.23] S. Kumar, A. Nassehi, S. T. Newman, R. D. Allen, and M. K. Tiwari, "Process control in CNC manufacturing for discrete components: A STEP-NC compliant framework," *Robotics and Computer-Integrated Manufacturing*, 2007.

[5.24] S. Amaitik and Kilic, "An intelligent process planning system for prismatic parts using STEP features," *The International Journal of Advanced Manufacturing Technology*, vol. 31, pp. 978-993, 2007.

[5.25] R. Liu, C. R. Zhang, A. Nassehi, and S. T. Newman, "A STEP-NC programming system for prismatic parts," *Materials Science Forum*, vol. 532-533, pp. 1108-1111, 2007.

[5.26] X. Xu and J. Wang, "Development of a G-Code Free, STEP-Compliant CNC Lathe," presented at Proc of the 2004 International Mechanical Engineering Congress and Exposition (IMECE),, Anaheim, California, U.S.A., 2004

[5.27] Y. Wei, K. T. Chong, T. Takahashi, S. Liu, Z. Li, Z. Jiang, and J. Y. Choi, "A framework for CNC turning system based on STEP-NC," presented at ICMIT 2005: Mechatronics, MEMS, and Smart Materials, 2005

[5.28] S.-J. Shin, S.-H. Suh, and I. Stroud, "Reincarnation of G-code based part programs into STEP-NC for turning applications," *Computer-Aided Design*, vol. 39, pp. 1-16, 2007.

[5.29] I. Choi, S.-H. Suh, K. Kim, M. Song, M. Jang, and B.-E. Lee, "Development process and data management of TurnSTEP: a STEP-compliant CNC system for turning," *International Journal of Computer Integrated Manufacturing*, vol. 19, pp. 546-558, 2006.

[5.30] S.-H. Suh, D.-H. Chung, B.-E. Lee, S. Shin, I. Choi, and K.-M. Kim, "STEP-compliant CNC system for turning: Data model, architecture, and implementation," *Computer-Aided Design*, vol. 38, pp. 677-688, 2006.

[5.31] S. Heusinger, R. S. U. Rosso-Jr, P. Klemm, S. T. Newman, and S. Rahimifard, "Integrating the CAx Process Chain for STEP-Compliant NC Manufacturing of Asymmetric Parts," *International Journal of Computer Integrated Manufacturing*, vol. 19, pp. 533-545, 2006.

[5.32] Y. Yusof, R. S. U. Rosso-Jr, S. T. Newman, and K. Case, "The Design of a STEP-NC Compliant CAD/CAPP/CAM System for the Manufacture of Rotational Parts on a CNC Turning Centre," presented at Proceedings of the 23rd International Manufacturing Conference (IMC23), University of Ulster, Northern Ireland, UK, 2006.19-28

[5.33] S.-J. Shin, S.-H. Suh, and I. Stroud, "Reincarnation of G-code based part programs into STEP-NC for turning applications," *Computer Aided Design*, vol. 39, pp. 1-16, 2007.

[5.34] S. T. Newman, A. Nassehi, X. W. Xu, R. S. U. Rosso-Jr, L. Wang, Y. Yusof, L. Ali, R. Liu, L. Zheng, S. Kumar, P. Vichare, and V. Dhokia, "Interoperable CNC for Global Manufacturing (Keynote paper)," presented at Flexible Automation and Intelligent Manufacturing, FAIM2007, Philadelphia, USA, 2007.1-13

[5.35] S. H. Suh, B. E. Lee, D. H. Chung, and S. U. Cheon, "Architecture and implementation of a shop-floor programming system for STEP-compliant CNC," *Computer-Aided Design*, vol. 35, pp. 1069-1083, 2003.

[5.36] X. W. Xu, "Realization of STEP-NC enabled machining," *Robotics and Computer-Integrated Manufacturing*, vol. 22, pp. 144-153, 2006.

[5.37] R. S. U. Rosso-Jr, S. T. Newman, and S. Rahimifard, "The adoption of STEP-NC for the manufacture of asymmetric rotational components," *Proceedings of the Institution of Mechanical Engineers Part B: J. Engineering Manufacture*, vol. 218, pp. 1639-1644, 2004.

[5.38] ISO, "International Organization for Standardisation - ISO 10303-Part 21 Industrial automation systems and integration - Product data representation and exchange - Part 21: Implementation methods: Clear text encoding of the exchange structure," 2002.

[5.39] M. Landy, S. Siddiqui, and J. Swisher, *JBuilder Developer's Guide*, 2003.

[5.40] Y. Yusof, K. Case, S. T. Newman, and X. W. Xu., "A STEP Compliant System for Turning Operations," presented at 17th International Conference on Flexible Automation and Intelligent Manufacturing (2007 FAIM), Philadelphia, USA, 2007.140-147

[5.41] Y. Yusof, "A STEP Compliant Approach to Turning Operations " in Wolfson School of Mechanical and Manufacturing Engineering Doctoral Thesis, 2007, 248

[5.42] ISO, "International Standard 14649-10: Part 10 : Industrial automation system and integration - Physical device control - Data model for computerised numerical controllers - Part 10 : General process data," 2004.

[5.43] ISO, "International Standard 14649-12: Part 12 : Industrial automation system and integration - Physical device control - Data model for computerised numerical controllers - Part 12 : Process data for turning," 2005.

[5.44] ISO, "International Standard 14649-11: Part 11 : Industrial automation system and integration - Physical device control - Data model for computerised numerical controllers - Part 11: Process data for milling," 2004.

6

Circular Sawblade Stone Cutting Technology Based on STEP-NC

Julio Garrido Campos

Automation and Systems Engineering Department, University of Vigo, E.T.S.I. Industriales, 36200 Vigo, Spain
Email: jgarri@uvigo.es

Abstract
Computer Numerical Control (CNC) stone cutting technology with circular sawblades has been used to machine stone construction parts such as flagstones, rebates, mouldings, columns, balusters, etc., from dimensional ashlars. There are major differences between conventional metal cutting and stone cutting processes using circular sawblades, not to mention the fact that each process is governed by specific technology parameters. Circular sawblade machines used for stone cutting can be quite complex, with five or more axes. In some cases, they are automated using CNC systems with the same programming technology as the CNC machines for metal cutting. The need to increase productivity, quality and effectiveness in stone cutting has challenged the stone processing industry to make technological adaptations of successful concepts applied in other industries. While the STEP-NC standard has shown its benefits for conventional metal cutting, the main issue discussed in this chapter is whether STEP-NC can also bring benefits to the stone cutting sector, where new trends in regard to outsourcing, collaboration and even shop-floor integration have not traditionally been major business issues. A secondary issue is about how standards could accommodate sawblade stone cutting technology. In an attempt to address both of these issues, the chapter reviews some particularities of sawblade stone machining processes and proposes an extended STEP-NC compliant model.

6.1 Introduction

Several stone cutting machine technologies exist, such as stone cutting by diamond wire, milling [6.1], abrasive water-jet cutting [6.2], linear blades with reciprocating motion [6.3], circular sawblade cutting machines [6.4], etc. They vary in terms of efficiency and cost, quality of the finished stone part, and the set of shapes or features they are able to machine.

Among these stone cutting processes, sawing with diamond-impregnated segments mounted on a circular steel core is most widely used for granites and stone [6.5]. Even though the term sawing is commonly used [6.6], it is, in fact, a grinding process in which stone constituents are worn away by passing rigid grits over the machined surface. The diamond crystals, acting as cutting edges, are firmly held in a matrix, which progressively erodes the granite, exposing fresh particles, while those protruding sufficiently for cutting undergo mechanical degradation until they are finally dislodged.

Processing stone is a multidimensional and complex task in which, in order to obtain the economically best sawing conditions, the ideal balance has to be obtained between tool life, cutting rate and quality. Factors such as physical material properties, sawblade characteristics, sawing conditions, cooling efficiency, etc. are all interrelated and influence process efficiency and quality. Different conditions during the cutting process is a relevant aspect of this technology, for example, tool cutting power variations, changes in the stone structure, etc. [6.7]. To ensure an optimised process, the control system or/and operator should be able to make changes in the middle of the process. Stone cutting machines may, therefore, be very complex, both in terms of mechanical structure and also in terms of the control system, designed to lead to minimum tool and machine vibrations for high workpiece accuracy and also low noise, optimum productivity and minimum tool wear rate regardless of the conditions of the process [6.6].

From the axis motion control point of view, using circular blades limits the number of axes which can be moved when the tool is in the stone (only movements in the cutting plane are possible). However, to make complex features such as mouldings, additional axis movements may be needed and performed when the blade is out of the stone. Some machines have five or more axes (e.g. six axes in the machine as shown in Figure 6.1), controlled with programmed systems capable of performing up to three interpolated axis movements using, in some cases, the same technology as metal CNC machine counterparts, namely, the G&M codes (ISO 6983).

Figure 6.1. Six-axis circular sawblade stone cutting machine

ISO 6983 (G&M codes) is still widely used for CNC machines, although it has become a bottleneck for some machining technologies because the information passed to the CNC controller is limited to how-to-make information (excluding information about the actual workpiece) and also because the data model non-compliance throughout the process chain of CAD, computer-aided process planning (CAPP), CAM and CNC [6.8].

The new STEP-NC standard [6.9] defines a richer interface or information model for data transfer between CAD/CAM programming systems and CNCs than the ISO 6983 standard does. It makes available higher information content to the CNC, including information describing a set of complex and structured tasks (Workingsteps) for machining the features of a workpiece [6.10]. All this information on the control side can be used to implement high-level intelligent controllers. Another key potential advantage of STEP-NC is the implementation of interoperability, both at the shop-floor level (providing production flexibility) and between the various components of the CAx chain [6.11].

As Sokolov suggested [6.12], STEP-NC represents an obvious improvement for widely used manufacturing methods such as milling and turning, and it also has major benefits for developers of new machines and processes as well as for specialised processes. The first question we raise, however, is whether STEP-NC also represents a potential improvement for technologies where new trends in outsourcing, collaboration and even shop-floor integration are—or have traditionally been—not so obvious, such as, for instance, in sawblade stone cutting. A second question is how the standard can accommodate sawblade stone cutting technology.

From the point of view of STEP-NC, it would be worthwhile exploring as many processes as possible. A wide range of CNC manufacturing processes have already been addressed by research initiatives and projects with a view to extending the STEP-NC standard [6.13]. There have also been research initiatives with regard to stone machining. Stroud et al. have addressed stone milling operations and have defined new architectural stone features [6.14]. The draft of Part 15 of the ISO 14649 standard—although aimed primarily at wood and glass processing—defined some stone sawblade operations. The approach is very general, though. This Part has been withdrawn subsequently by SC1. Therefore, no detailed research into STEP-NC and circular sawblade stone cutting operations has as yet been reported.

Section 6.2 summarises certain particularities and current needs of the stone cutting process, while Sections 6.3 and 6.4 refer to model development. Section 6.3 reviews the particular characteristics of the different sawblade cutting processes and the main process data. This description should not be seen as an exhaustive review of all possible stone cutting operations, but rather as a brief presentation of general and common processes. Section 6.4 translates these technology peculiarities to a data model for circular sawblade stone cutting, reviewing the main feature definitions from different sources (ISO 14649-10, ISO 14649-11, ISO 10303-AP238 and Stroud et al. [6.14]) and defining and detailing new machining operations and technology parameters. Finally, Section 6.5 is a brief conclusion with some remarks about the model implementation and future work.

6.2 Stone Cutting Process Needs

Stone cutting machines are typically isolated machines with embedded CAM systems. Outsourcing, CAD/CAM integration and automated process planning have not traditionally been main business issues in this sector. Machines are typically easy to use, with user-friendly interfaces and flexible control systems capable of dynamically adapting to machining process changes through on-line operator orders. However, the need to increase productivity, quality and effectiveness in stone cutting construction parts poses a challenge for the stone processing industry in terms of technological adaptations of concepts successful in other industries [6.15]:

- Stone cutting shop-floors have different kind of machines (diamond wire, sawblade, water jet, etc.), often from different commercial brands and with different capabilities (rough saw, stone planner, precision saw, etc.). The same part could be made by different machines. For instance, a planar bounded shape could be made either by a water-jet machine or by a circular sawblade machine. Also, different machines performing different types of operations may be used for one stone part. Process planning would improve production efficiency [6.16] and machine interoperability would increase the flexibility of production across the shop-floor with parts capable of being interchanged. STEP-NC has provided the manufacturing community with the basis for defining resource-independent process planning and interoperability between standard platforms [6.11].
- Although outsourcing is not a major issue in the stone business, markets globalisation is an important factor. Stone parts for building façades machined in Vigo, Spain are shipped, for example, to New York or Tokyo, and the same could be said about stone processed in other regions. CAD/CAM integrated solutions to perform shop-floor process planning and machine control directly from a CAD building façade design already exist in the stone cutting business. STEP-NC has shown the way forward for automatic interoperable CNC manufacture of parts from CAD models.
- Automation approaches for stone cutting machines share key functionality concepts with modern intelligent CNC controllers. One is online machining process adaptation to changing conditions, which is an important issue in stone cutting technology. CAM-CNC interfacing with the ISO 6983 standard limits its implementation [6.17], as this implies a full pre-calculation of the toolpaths [6.18]. Instead, a common approach for stone machines is embedded CAM systems that directly and continuously generate tool-path axis control movements from feature definition while taking into account online parameters and online operator orders. Stone cutting machine software control systems are, in many cases, feature-based, where the controller program offers a range of information such as the feature to be machined, tool types to be used, etc., while machine-specific decisions are left to the CNC and its operator. This is the same view for future STEP-NC controllers [6.19].

All this suggests that the stone processing business in general, and circular sawblade machining in particular, would indeed benefit from STEP-NC technology.

6.3 Understanding and Modelling Stone Cutting Processes

The STEP method, which drives standard data specification and modelling processes [6.20], is closely focused on the development of application protocols (APs) (Section 1.8). It starts by defining scope and requirements for industry needs, characterised in an application activity model (AAM) according to product types, product data types and where and how the data is used (explained in Sections 6.3.1 and 6.3.2).

The second phase of the modelling process is a definition of the detailed information requirements that are to be fulfilled. These requirements are documented through the development of an application reference model (ARM) (see Section 1.8.2). As mentioned in Section 1.10, the SC4 manufacturing working group has adopted the ARM models built by SC1 in a single AP-238 model. While a new technology would represent a new part for the ISO 14649 standard, in STEP it would become an AP-238 extension model [6.21]. Section 6.4 will present the ARM definition from an ISO 14649 perspective.

6.3.1 Stone Cutting Processes

Obtaining a final stone construction part from a quarry stone block is based on a fairly standard sequence of activities applied worldwide. These activities are summarised in Figure 6.2 as an AAM.

Stone processing commences in quarries where cubic blocks are extracted. Although block dimensions vary from quarry to quarry, typical dimensions are 180–300 cm long, 150 cm wide and 100–200 cm wide.

The blocks arriving at the stone yard from the quarry are irregular in size and shape. Process planning decisions focus on the most efficient way to cut rough blocks into slab form while minimising wastage (the machines to use, tools, fixtures, cutting parameters, schedules, etc.). Blocks are sawn into slabs using block cutter machines (activity A1 in Figure 6.2). These large machines have one or many tools (frame blades, diamond-wire cutting machines and even large diameter circular saws) for cutting blocks in a single operation into a number of slabs. The next step is secondary sawing to cut the slabs into dimensional ashlars. The machines used in this case are usually computerised bridge saws with large circular blades (activity A2 in Figure 6.2).

The result of these two steps is a set of stocks from which finished stone parts will be machined in more complex, precision cutting machines (activity A3 in Figure 6.2). These multi-axis machines use different technologies: diamond-wire contouring machines, milling machines for small parts and sawblade machines. For stone construction parts, circular sawblades are the most popular cutting mechanism and it is also the main subject of this chapter. To obtain a good surface-finish, a surface treatment may be applied, whether polish, bush-hammer, flamer, etc. (activity A4 in Figure 6.2).

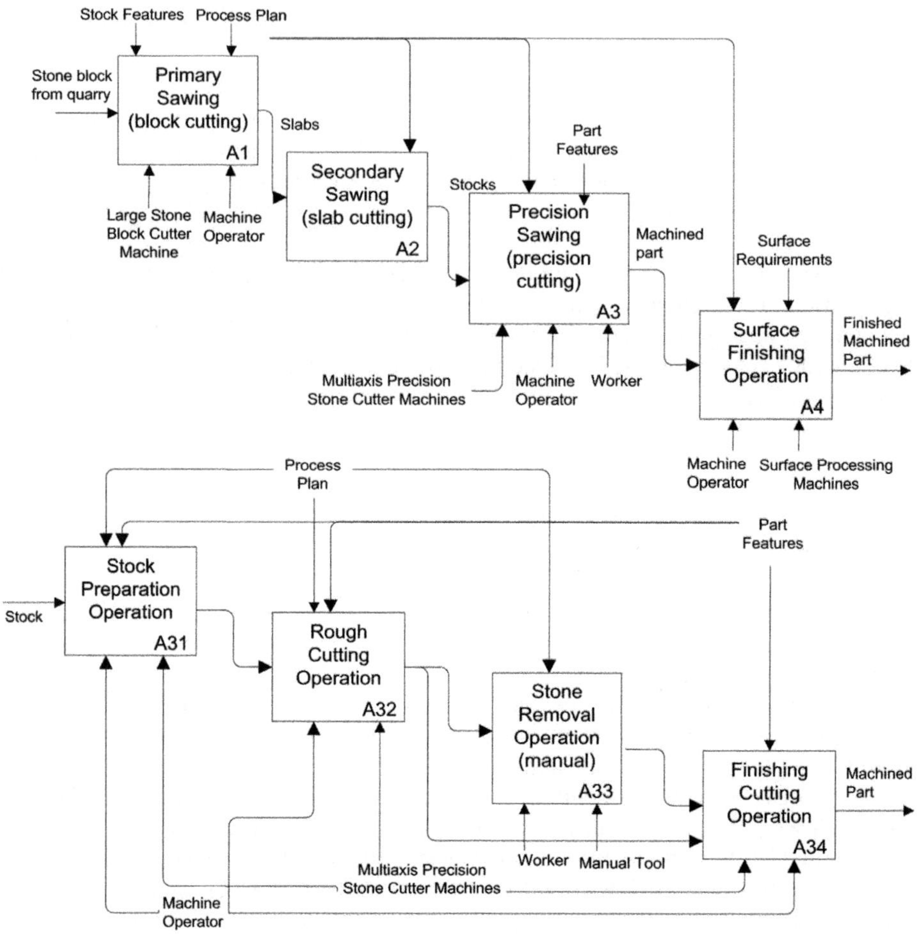

Figure 6.2. Application Activity Model for stone cutting construction parts

6.3.2 Automatic Stone Cutting Machines

Automatic sawblade cutting machines for construction parts are very similar, despite the fact that there are many machine manufacturers in many different regions, and despite the fact each manufacturer machine is equipped with particular solutions, special options and custom functionalities. Machines do vary in terms of mechanical configuration and in terms of control. Simpler machines make single or repetitive cuts, and some are equipped with several blades or several diamond wires to make more than one cut at a time. Other machines are designed and automated to be capable of performing complex tasks. Although each machine manufacturer uses specific hardware and automation solutions, the cutting process is similar (e.g. the activities described in Figure 6.2 as A32, A33 and A34):

- The first phase of performing a set of rough cuts (activity A32 in Figure 6.2)

- The second phase of eliminating the stone between the cuts (activity A33 in Figure 6.2)
- The third phase of finishing the part to obtain a smooth surface (activity A34 in Figure 6.2)
- Depending on the task to be performed and machine capabilities, there may be previous additional phase of preparing the stock through large cuts aimed at obtaining a surrounding roughly patterned shape in order to reduce the amount of material to be removed in the next precision cutting phase (activity A31 in Figure 6.2).

In the first phase, cuts are made progressively with cut depth following the outline or profile of the final part (the desired surface) and leaving a specific distance between cuts (Figure 6.3). The sawblade makes a complete longitudinal cut and goes downward step by step, making several parallel cuts at different depths (passes) (Figure 6.4). Once it reaches the desired depth, the sawblade disc exits the stone and moves to start another parallel cut at a given distance from the previous cut. The number of passes to make a cut and the cutting distance between cuts depend on parameters such as stone hardness, feed rate, sawblade speed, etc. A specific distance is maintained between the depth of the cuts and the desired surface in order to avoid scratching the stone when performing the next phase—traditionally done by hand using tools as mallets, chisels, etc. —of removing the remaining material slots (stone slides) so as to obtain a first rough shape. This process is improved if cut distance is reduced in the first phase. The stone slides would be thinner and precision higher, although processing would take longer.

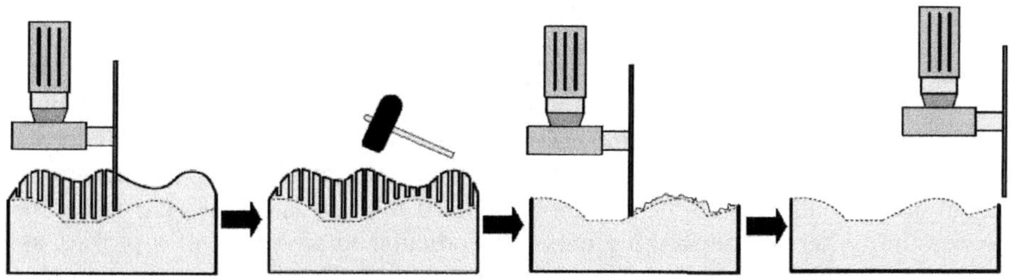

Figure 6.3. Circular sawblade stone cutting process phases (*left*) and sawblade path of a moulding cut process (*right*)

Once the material between cuts (slides) is removed, a finishing process produces the final surface. The rough and terraced profile obtained from the previous phase is smoothed by making overlapping cuts. These cuts are performed in several passes, with each cut gradually approaching the final surface.

There are additional parameters and several cutting strategies. For instance:

- If the machine can swing and rotate the sawblade disc, this provides for other cutting strategies. With a swing axis, different angles and reach points are feasible that would not be possible for parallel cuts.
- If the sawblade swings, cuts may be made perpendicular to the feature's final surface (as explained in the next example), instead of in parallel, vertically or

horizontally. This has the advantages of reducing the number of cuts needed and of producing a better quality final surface with a reduced terrace effect.

- To avoid lateral stone cracking, care has to be taken when doing a first pass of an open boundary and when doing the last pass of a through-end.
- The sawblade speed/feed rate is a critical parameter because the diamond matrix erosion of the tool is very dependent on this rate.
- The direction of sawing—whether upward or downward (down-cutting or up-cutting mode)—is selected depending on the blade rotation direction. These directions are quite different in terms of the forces produced and the way the diamond-wire works.

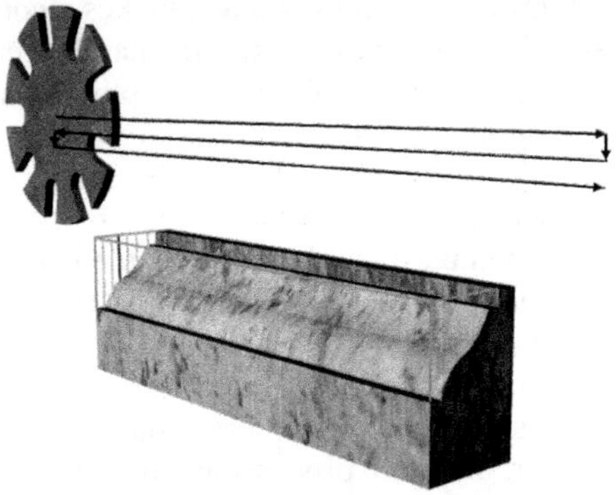

Figure 6.4. Single cut process

Ultimately, the choice of tools and adequate machining parameters depends on know-how and experience and also on in-process parameter change decisions.

Some machines have an additional axis to rotate the stock and hence enable turned parts (revolution parts and indexed parts) to be machined. The cutting plane has to be perpendicular to the revolution axis (if the part is revolving), and there is no possibility here of performing cuts perpendicular to any point of a profile, as in operations for plane mouldings (see the next example). The finishing phase may also be different in revolution operations, as the sawblade may perform a contour movement following the profile in the direction of the revolution axis in order to remove a thin layer of remaining material. The same kind of machine may also make indexed parts which, in appearance, are very similar to revolution parts, except that the process is not a turning process (because the part cannot be rotating when the blade is cutting), although the full perimeter is machined through step-by-step rotating movements of the part.

6.3.3 An Example

Figure 6.5 shows an example of a linear moulding like the one in Figure 6.4. Note that only some cutting positions are illustrated to view the process better. On the left are some of the positions of the sawblade disc when performing the first rough cutting phase. To make a cut-out, the sawblade moves from its retracted position to the contact point, where it makes a first pass. The blade then feeds downward a predefined distance and makes several passes until the disc reaches the *allowance* depth (a depth of material which shall be left on top of the surface defined by the associated manufacturing feature, in the direction of the surface normal). On the right are some sawblade positions in the finishing phase, in which cuts are made much closer to one another and with smaller depth passes. In this case the sawblade moves from the new retracted position to the contact point (now the *allowance* depth), where it starts the passes until the sawblade reaches the desired final surface.

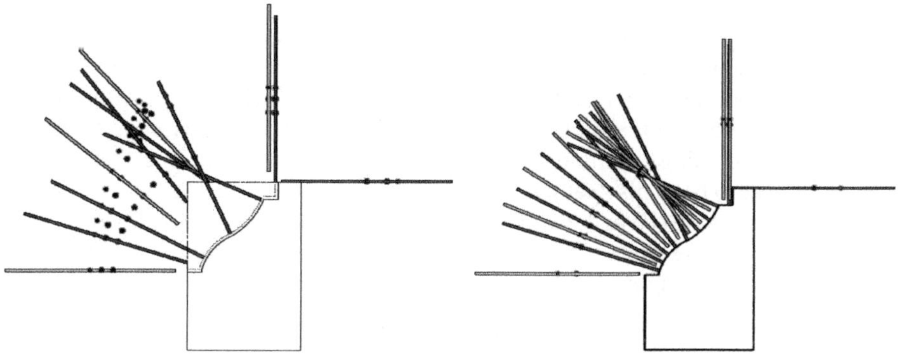

Figure 6.5. Moulding features: rough cutting phase (*left*) and finishing phase (*right*)

6.4 STEP-NC Data Model for Sawblade Stone Cutting

In order to expand the STEP-NC standard, two main strategies may be adopted: to extend manufacturing process standardisation and to extend the range of applications that can use the STEP-NC approach [6.14]. The first strategy has been applied to EDM [6.22] by defining a complete extended model of STEP-NC that became ISO 14649-13, to closed loop manufacturing processes for inspection that became a new ISO 14649-16, and in ISO 14649-15 to define process data for contour cutting of wood and glass, etc. [6.23]. The second strategy consists of using STEP-NC current models (turning and milling) for new processes. One example is the dry high-speed milling of marble and industrial ceramic (the LITHO-PRO project), in which the STEP-NC milling model has been used, but defining a new range of process-specific features [6.14].

For the case of sawblade stone cutting, STEP-NC would need to be expanded to account appropriately for its particularities. In this section we describe a STEP-NC extension model being used to automate a STEP-NC-compliant sawblade stone machining prototype. The data model is built upon the basic STEP-NC process model (ISO 14649-10) and in accordance with Part 11 for milling and Part 12 for turning. The model defines technology-specific data types representing stone cutting processes following the basic STEP-NC approach of separating geometric and technological information.

The extended model uses features already defined in the standard for milling and turning and in other research projects [6.14], to be described in Section 6.4.1. In accordance with other technological parts (for example, Part 11 for milling) and with new developments (for example, Part 16 for inspection [6.24]), the model defines so-called sawbladecutting_workingsteps, including machining operations and features to be machined. The main operations proposal and technology parameters and strategies are described in Section 6.4.2.

6.4.1 Disc Sawblade Cutting Features

Features specify the information necessary to identify shapes of interest in a mechanical product. These shapes represent volumes of material either removed by machining operations or resulting from a series of machining operations (ISO 14649-1). Features are therefore directly related to the set of machining operations for a given machining technology. The sawblade stone cutting basic feature is the cut-out (Figure 6.6). This feature may be described with the slot feature already defined in ISO 14649-10 and also in ISO 10303 AP-224. A slot is defined as a type of multi_axis_feature that is a channel or depression with continuous direction of travel. To the left of Figure 6.6 is a slot feature as defined in ISO 14649-10.

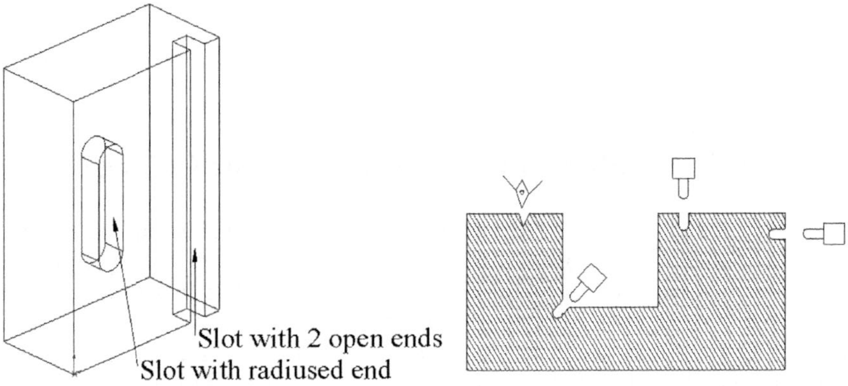

Figure 6.6. ISO 14649-1 slot feature (*left*) and ISO 14649-12 cut_in feature (*right*)

ISO 14649-12 defines a special feature which matches the sawblade single cut for turning machines: the cut_in, defined as a slot or groove where the geometrical shape of a cut_in is similar to that of groove (defined as a narrow channel or depression that is swept through one complete revolution about an axis). As a cut_in shape is identical to the shape of the used tool (ISO 4649-11), a cut_in feature for sawblade turning would have the shape of the sawblade edge.

The turning cut_in feature (ISO 14649-12) has a tool_direction parameter to allow cuts with different angles to be performed (as seen to the right of Figure 6.6). While only vertical cuts are allowed for turning (revolving) sawblade cutting, for surface sawblade cutting operations, cuts other than vertical would be needed.

Therefore, a tool_direction parameter would also be needed for the planar slot to become a planar_cut_in feature.

However, complex stone sawblade cutting machines can produce more complex stone construction parts such as mouldings, columns, etc. Two main kinds of complex features are considered: planar features for mouldings (Figure 6.7) and turning features for columns and balusters (Figure 6.8).

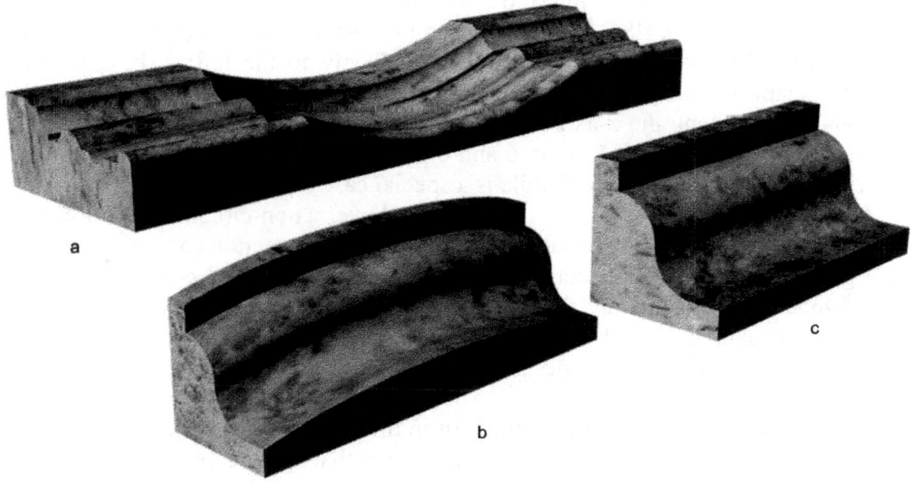

Figure 6.7. Bonded curve path moulding (a), arc path moulding (b) and linear path moulding (c)

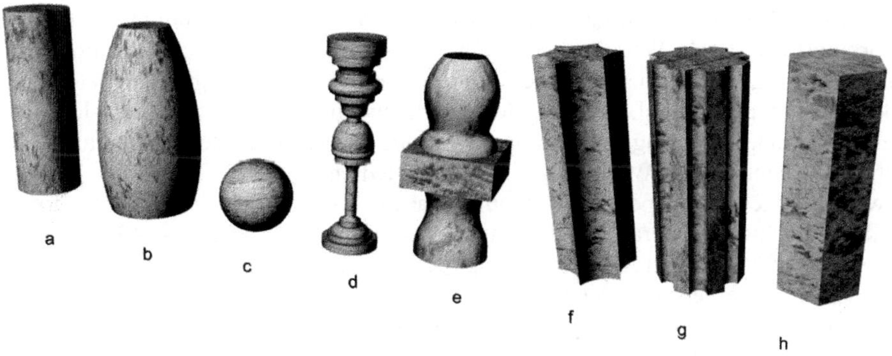

Figure 6.8. Turning features: revolution features (*the five parts on the left*) and indexed features (*the three parts on the right*)

Stroud et al. [6.14] have defined a set of aesthetic features to be produced by stone milling. Among them, rebates (extruding a general profile along a path) match sawblade moulding or plane features. Moulding may refer to different profiles (simple arc, simple square, general profile, etc.) and these may be extruded along a linear path (Figure 6.7c), along an arc path (Figure 6.7b), along a bounded curve path made of lines, arcs, etc. (Figure 6.7a) or even along a general path.

Columns and balusters are very common stone shapes in construction. They are turning features, made in turning machines with an additional axis to rotate the stock. Depending on how this rotation is performed, there are two kinds of features: revolution features and indexed features. Revolution features have a circular_closed_profile (ISO 14649-10) section at each point of the column/baluster axis (z-axis) (Figure 6.8, the five parts immediately to the left: a, b, c, d, e). A cylinder shape has the same circular sections along the z-axis (Figure 6.8a). The sections change along the z-axis according to shapes as illustrated in the other four rounded shapes in Figure 6.8 (b, c, d and e parts); note, however, that the part with its original square shape in the middle is a special case (Figure 6.8e).

Indexed features (Figure 6.8f,g,h) have non-circular sections: a rectangular_closed_profile section, a non_closed_profile section (polygonal section) or a general_closed_profile section. In addition to these closed profiles defined in ISO 14659-10, there is an ellipse_closed_profile as a cross-section for stone columns and balusters. Additional features can also be included in turning features to complete a column/baluster design, such as negative or positive semi-cylindrical pattern features [6.14].

Both revolution and indexed features fit in the turning_features described in ISO 14649-12, defined as classical geometric shapes that can be obtained by turning cylindrical workpieces in operations with two axes (x and z) or three axes (x, z and c, the rotator axis). However, features obtained in 3-axis operations are not as yet included in the standard, and so a new turning feature—indexed_feature—would need to be specified.

6.4.2 Sawblade Cutting Operation Data

A STEP-NC program is organised around a main Workplan containing a series of Workingsteps. Each Workingstep applies a specific machining operation to a manufacturing feature using specific tools, sets of technology parameters and specific strategies [6.25]. Workingsteps represent the essential building blocks for describing manufacturing or handling operations that involve interpolating axes. The actual content of a Workingstep is specified in Entity Operation. Thus, a sawblade_machining_operation would have a similar structure to a milling_machining_operation for milling (defined in ISO 14649-11) or a turning_machining_operation for turning (defined in ISO 14649-12).

Figure 6.9 represents a simplified model for sawblade stone cutting operations, while Figure 6.10 describes machining strategies for some of the operations in Figure 6.9. Two main kinds of operations have been defined: surface operations and turning operations.

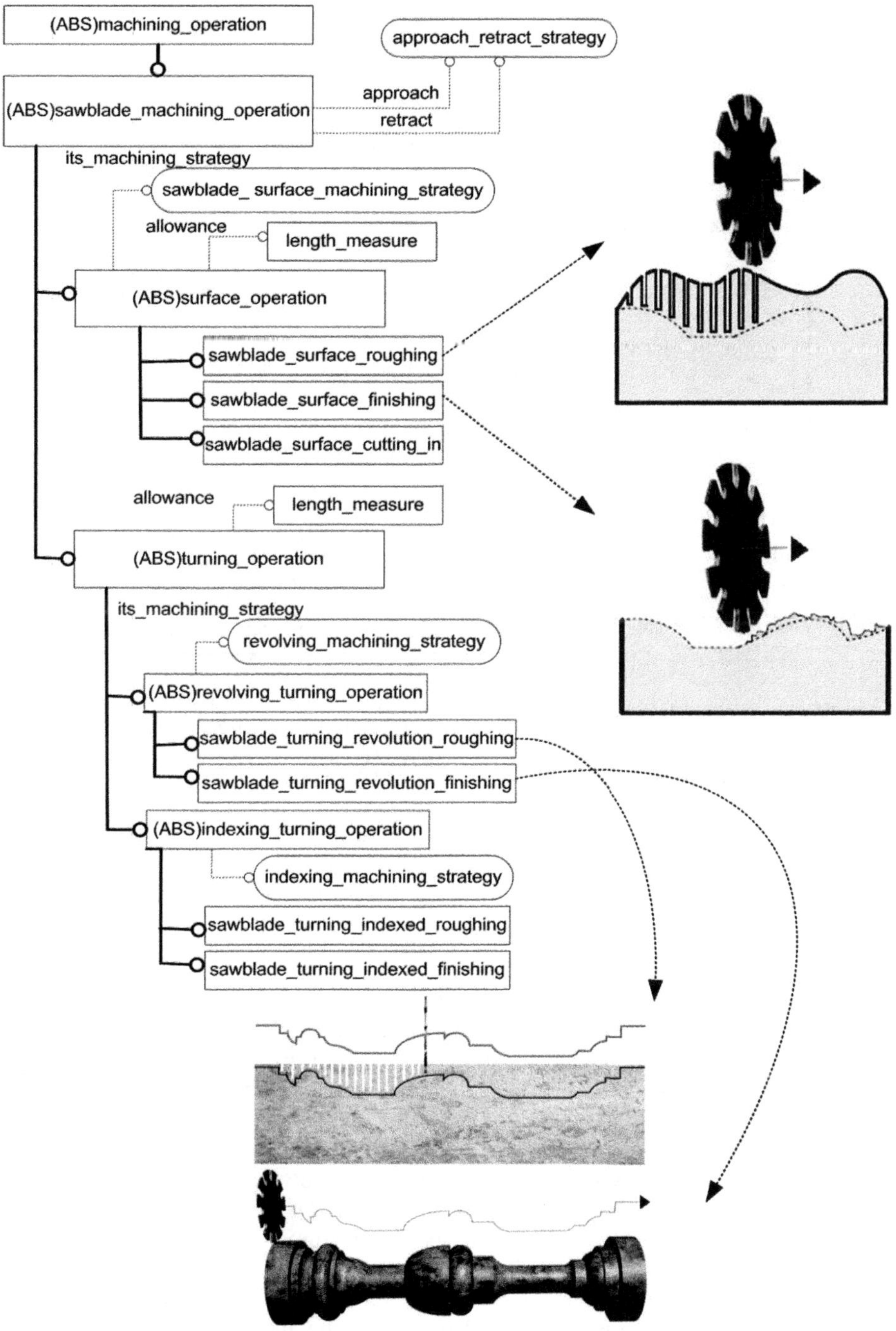

Figure 6.9. Sawblade cutting operations

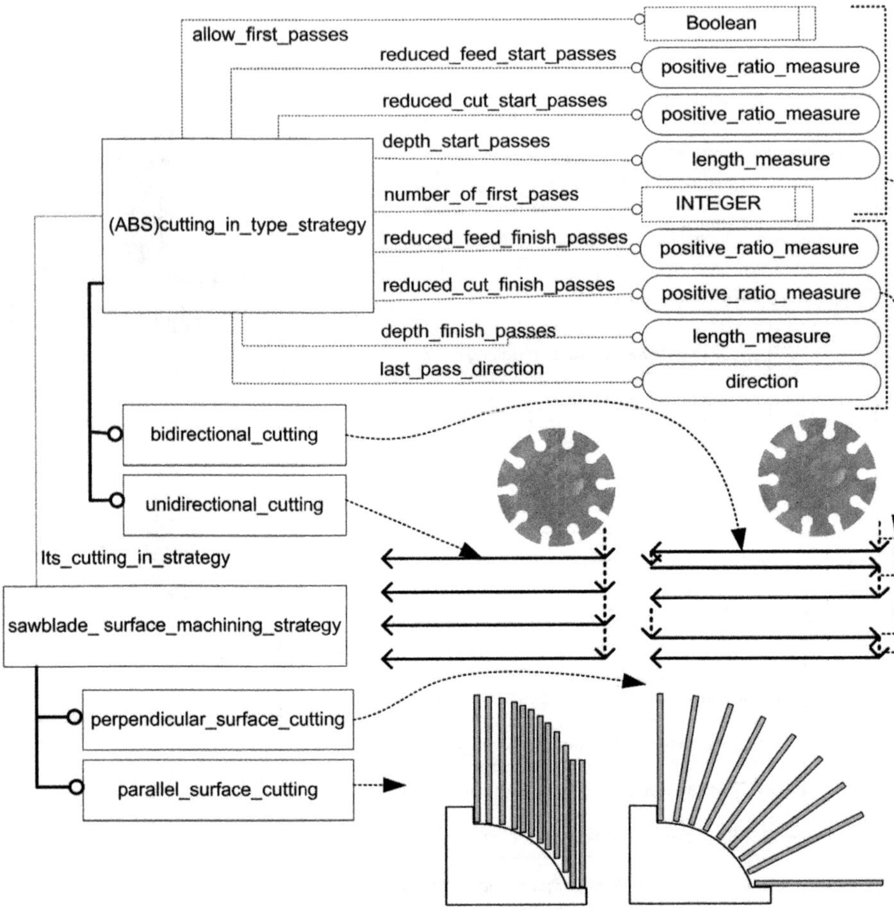

Figure 6.10. Sawblade cutting strategies

Surface operations are used to machine features for a fixed stock (single cut_in feature, moulding features) and also to machine planar faces in an indexed feature. To perform a cut_in, the sawblade_surface_cutting_in operation would be invoked, while to perform a moulding, the sawblade_surface_roughing and sawblade_surface_finishing operations would be invoked in sequence.

When performing a sawblade_turning_operation, the stock has to rotate, either continuously to obtain revolving features through revolving_turning_operations, or step-by-step to obtain indexed features through indexing_turning_operations. In both cases—and as happens with the surface features—roughing and finishing operations are executed.

In all cases, roughing operations perform several cuts to the appropriate depth (the allowance depth of material to left on top of the final surface defined by the manufacturing feature in the direction of the surface normal). For each single cut, several passes are needed to obtain the final allowance depth. Finishing operations remove the remaining material between the allowance surface and the surface defined by the associated manufacturing feature.

In turning operations, rough cuts are performed perpendicular to the rotator axis as the stock rotates. The process is replicated several times along the axis direction, with cut distance as an important parameter. The finishing phase may be implemented in a way similar to the sawblade_surface_operation, but with the vertical cuts much closer to each other. Alternatively, a contour cutting operation can be implemented, in one or in several passes, to obtain progressively the final contour. In this case the sawblade is moved transversally (in the rotator axis direction) along the path and the cuts are made by the lateral part of the sawblade.

Different strategies may be adopted in each operation. Figure 6.10 represents some of the strategies associated with the sawblade_surface_operation, while other strategies (not illustrated) can be defined for turning operations.

There is a common set of parameters for all surface operations that are associated with the single cut approach. Summarised in the cutting_in_type_strategy, they specify, among other things, whether each cutting pass is done in the same direction (when the blade finishes a pass, it moves back without cutting and without going deeper into the stone) or whether it is implemented in zig-zag, with cuts in both directions (a two-directional cutting strategy). Different parameters may also be selected for the first and last passes of a cut—to prevent losing the correct direction in the first pass and to avoid scratching the stone in the last pass, respectively (Figure 6.10). Also, upward or downward cutting may be selected in the last_pass_direction parameter.

To machine a moulding (sawblade_surface_operation), several cuts are performed with a cutting_in_type_strategy for all the cuts. Other strategies are related to the approach to calculating cut orientation for the desired profile. Cuts may be parallel to a plane, but with a depth following a profile, or they may be perpendicular to the tangential at one point of a profile (Figure 6.10).

As with the standard for milling and turning, the model is completed with other information definitions—technology_parameters, approach_retract_strategies, machine_functions, etc. For example, a technology entity would include technology parameters as feedrate value, sawblade speed value, Boolean values to allow or disallow feedrate override, etc.

Information would also be needed on the main tool parameters, for example, sawblade diameter, maximum allowed blade depth in the stone, etc.

6.4.3 Data Model Implementation

A first STEP-NC prototype CAM-CNC system was implemented with the extended model described above and based on a commercial circular sawblade stone cutting machine. The machine has six axis movements and is able to perform turning (Figure 6.1), although, to date, only a STEP-NC interface for surface features has been implemented.

The machining system architecture (Figure 6.11) is composed of a proprietary CAD/CAM-embedded system for selecting feature and machine parameters, although profiles for features may be imported from CAD files in DXF format. The CAM information is translated to XML (eXtensible Markup Language) and communicated to the MMI (Man Machine Interface). Both CAD/CAM and the MMI run in the same embedded PC under Windows CE, but the first could well be run in another computer as the MMI system can also import the XML files. The MMI adds shop-floor information to the XML file (for example, current sawblade diameter). The resulting file is the input for the machine low-level control module, responsible for axis motion control, alarm management and input/output management. This module runs in a TwinCAT PLC Run-Time system with TwinCAT NC axis control and is programmed in IEC 1131.

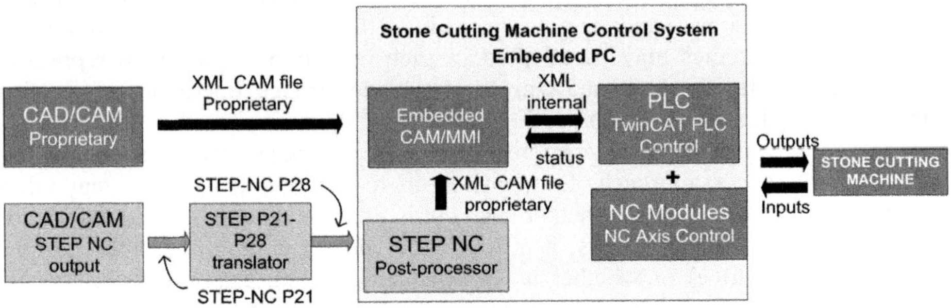

Figure 6.11. Implementation architecture

The STEP-NC control architecture follows the CAM-embedded STEP-NC controller structure defined by Zhang et al. [6.19], where the controller is a combination of a STEP-NC code interpreter, a basic CAM and an NC controller (the interpreter translates the physical file into internal data format and the CAM makes decisions on machine-specific details and generates low-level control commands that are executed by the NC control).

In the prototype, a CAM application allows features and operation parameters to be selected and generates a STEP Part 21 file following the extended model. This file is transformed to a STEP Part 28 format by a STEPPart21-to-STEPPart28 translator and is communicated to the machine. The machine-embedded CAM/MMI system first processes the XML Part 28 file—taking advantage of XML file access libraries—and a post-processor translates the data into the machine information system, matching feature operation parameters coming from the STEP-NC file to the feature operation parameters of the machine.

6.5 Conclusions

Several approaches may be adopted in order to take circular diamond sawblade machining processing of stone parts into account in STEP-NC. A less ambitious approach is to use the current state of the standard, where a CAM system calculates the tool-paths and expresses them in the form of cutter location data. The resulting CAM file is then post-processed adding shop-floor parameters, such as current sawblade diameter, to produce axis movement data. In this approach, standard tool-path data are generated offline, which limits online machining process adaptation to changing conditions—an important issue in stone cutting technology. Moreover, sawblade cutting technology for stone parts have several specific parameters with no clear equivalent technologies as defined in milling, turning, etc. To consider them explicitly in the STEP-NC model, new technology entities will have to be added probably to become a new ISO 14649-6.

Although only surface features have been tested, the prototype implementation demonstrates that the model described above provides the required data for a circular sawblade stone cutting machine to be able to make construction parts. In the implementation, stone cutting STEP-NC data give the machine a comprehensive description of the part (what-to-make) and specifies technological requirements. Detailed tool-path calculations are made by the machine online, thereby maintaining its original process adaptation capabilities. Much more work needs to be done in order to cover the complete process chain from design to process planning and manufacturing. To reach this point, all CNC automated processes in a stone processing plant would need to fit in STEP-NC. In addition to circular sawblades, a large number of profiling and finishing tools and technologies are also needed to consider: milling tools, water jet cutting, diamond wire cutting, frame saw blades cutting, etc. Some of them could be treated using conventional metal cutting approach (milling [6.14], water jet). The way of considering other processes as frame saw blade cutting, diamond wire cutting, surface finishing, etc., is an open issue: whether adding new technology operations or reusing existing ones, for instance, to investigate a possible matching between wire EDM technology and diamond wire stone cutting technology.

From the implementation point of view, stone cutting machine control software tools are, in many cases, feature-based and constructed around software modules for performing specific features [6.26]. It was not difficult to convert a feature-oriented sawblade machine—working with an embedded CAM proprietary system—to an open STEP-NC-compliant machine, as the existing feature control algorithms in the original machine can be used.

Finally, in traditional machining processes, software/hardware vendors see the current lack of standards as an opportunity to maintain their market advantage [6.11]. However, small or medium sized machine vendors in more specific sectors like stone cutting would be interested in machines (standard or otherwise) with new interpretability capabilities and with the ability to develop and interact with new process planning systems. STEP-NC should be presented as a technology to achieve this aim, provided that moving to the new paradigm would not raise difficult technological issues.

Acknowledgments

This work is supported by the MCYT (Ministerio de Ciencia y Tecnología) under Project DPI2006-05772, Spain.

References

[6.1] Polini, W. and Turchetta, S., 2004, "Force and specific energy in stone cutting by diamond mill", In *International Journal of Machine Tools and Manufacture*, **44**, pp. 1189 – 1196.

[6.2] Li, X.H., Liao, Y., Lei, X.Y., Lei, Y.Y., Lu, Y.Y. and Jiao B.Q, 2003, "Numerically Controlled Water Cutter and its Applications in the Machining of Natural Rock Materials", In *Key.Engineering Materials*, **250**, pp. 274-280

[6.3] Wang, C.Y. and Clausen, R., 2002, "Marble cutting with single point cutting tool and diamond segments", In *International Journal of Machine Tools and Manufacture*, **42**, pp. 1045 – 1054.

[6.4] Brook, B., 2002, "Principles of diamond tool technology for sawing rook", In International journal of rock mechanics & mining science, 29, pp. 41 – 58.

[6.5] Konstanty, J. 2002, "Theoretical analysis of stone sawing with diamonds", In Journal on Material Processing Technology, **123**, pp. 146 – 154.

[6.6] Tonshoff, H.K., Hillmann-Apmann, H. and Asche, J, 2002, "Diamond tools in stone and civil engineering industry: cutting principles, wear and applications", In *Diamond and Related Materials*, **11**, pp. 736 – 741.

[6.7] Wang, C.Y. and Clausen, R., 2003, "Computer simulation of stone frame sawing process using diamond blades", In *International Journal of Machine Tools and Manufacture*, **43**(6) , pp. 559 – 572.

[6.8] Xu, X.W. and He, Q., 2004, "Striving for a total integration of CAD, CAPP, CAM and CNC". In *Robotics and Computer-Integrated Manufacturing*, 20, pp. 101-109.

[6.9] Hardwick, M. and Loffredo, D., 2001, "STEP into NC", In *Manufacturing Engineering*, **126**, pp. 38 – 50.

[6.10] Xu, X.W. and Newman, S.T., 2006, "Making CNC machine tools more open, interoperable and intelligent - a review of the technologies". In *Computers in Industry*, **57**(2), pp. 141 – 52

[6.11] Newman, S.T., Nassehi, A., Xu, X.W., Rosso, R.S.U., Wang, L., Yusof, Y., Ali, L., Liu, R., Zheng, L.Y., Kumar, S., Vichare, P. and Dhokia, V., 2008, "Strategic advantages of interoperability for global manufacturing using CNC technology", In *Robotics and Computer-Integrated Manufacturing*, **24**, pp. 699 – 708.

[6.12] Sokolov, A., Richard, J., Nguyen, V., Stroud, I., Maeder, W. and Xirouchakis, P. "Algorithms and an extended STEP-NC-compliant data model for wire electro discharge machining based on 3D representation", 2006, *International Journal of Computer Integrated Manufacturing*, **19**(6), pp. 603 – 613.

[6.13] Xu, X., Wang, H., Mao, J., Newman, S.T., Kramer, T.R., Proctor F.M. and Michaloski, J.L., 2005, "STEP-compliant NC research: the search for intelligent CAD/CAPP/CAM/CNC integration". In *International Journal of Production Research*. **43**(17), pp. 3703 – 3743.

[6.14] Stroud, I. and Xirouchakis, P., 2006, "Strategy features for communicating aesthetic shapes for manufacturing", In *International Journal of Computer Integrated Manufacturing*, **19**, pp. 639 – 649.

[6.15] Bernold, L.E. and Reinhart, D.B., 1990, "Process planning for automated stone cutting", In Journal of Computing in Civil Engineering, **4**(3), pp. 255 – 268.

[6.16] Heusinger, S., Rosso, R.S.U., Klemm, P,., Newman, S.T. and Rahimifard, S., 2006, "Integrating the Cax process chain for STEP-compliant NC manufacturing of asymmetric parts", In *International Journal of Computer Integrated Manufacturing*, **19**(6), pp. 533 – 545.

[6.17] Suh, S.H., Lee, B.E., Chung, D.H., and Cheon, S.U, 2003, "Architecture and implementation of a shop-floor programming system for STEP-compliant CNC", In *Comput-Aid Designe,***35**(12), pp. 1069 – 1083.

[6.18] Erdos G. and Xirouchakis,P., 2003, "STEP-NC Data Model Development for Wire-EDM Manufacturing", IFAC Conference, Budapest, Hungary.

[6.19] Zhang, C., Liu, R. and Hu., T., 2006, "On the futuristic machine control in a STEP-compliant manufacturing scenario". *In International Journal of Computer Integrated Manufacturing,* **19**(6), pp. 508 – 516.

[6.20] Fowler, J. 1995. *STEP for Data Management, Exchange, and Sharing,* Technology Appraisals, Ltd., Twickenham, UK.

[6.21] Xu, W.X. and He, Q., 2004, "Striving for a total integration of CAD, CAPP, CAM and CNC", In *Robot Computer-Integrated Manufacturing,* **20**, pp. 101 – 109.

[6.22] Ho, K.M., Newman, S.T., Allen, R.D., 2005, "STEP-NC compliant information modelling for wire electrical discharge machining component manufacture", In *Proc Inst Mech Eng Part B J Eng Manufacturing,* **219**(10), pp. 777 – 784.

[6.23] Xu, X.W. and Newman, S.T., 2006, "Making CNC machine tools more open, interoperable and intelligent-review of the technology", In Computers in Industry, **57**, pp 141-152.

[6.24] Brecher, C., Vitr, M. and Wolf, J., 2006, "Closed-loop CAPP/CAM/CNC process chain based on STEP and STEP-NC inspections tasks", In *International Journal of Computer Integrated Manufacturing,* **19**(6), pp. 570 – 580.

[6.25] Hardwick, M. and Loffredo, D., 2006, "Lessons learned implementing STEP-NC AP-238", In *International Journal of Computer Integrated Manufacturing,* **19**(6), pp. 523 – 532.

[6.26] Xu, X.W., Wang, L. and Rong, Y., 2006, "STEP-NC and Function blocks for Interoperable Manufacturing", *In IEEE transactions on Automation science and engineering,* **3**(3), pp. 297-308

7

Open Platform Development for STEP-compliant CNC

Tianliang Hu[1], Chengrui Zhang[2], Riliang Liu[3] and Lin Yang[4]

School of Mechanical Engineering, Shandong University, 27 Jingshi Road, Jinan 250061, People's Republic of China
Email: [1] tlhu@mail.sdu.edu.cn
[2] zhangchengrui@gmail.com
[3] liuriliang@sdu.edu.cn
[4] s3n@163.com

Abstract
To implement STEP-Compliant CNC, the implementing platform is very important. How to design open, STEP-Compliant CNC software and hardware platform becomes a hot research topic. A lot of research was focused on the STEP-NC data pre-processing of the STEP-Compliant CNC, but how to design flexible, open structure software and hardware platform for STEP-Compliant CNC implementing was not described in detail. In this chapter, the requirements of STEP-Compliant CNC system are analysed. Based on these requirements, an engine based platform is given as the software platform according to the reactive characteristic of the CNC system. In this platform, the software structure is designed as the Decision Unit (DU), Function Description Data (FDD) and System Engine (SE). The intelligent control rules and algorithms are designed in DU. The control rules of the NC functions are designed as the FDD, which works like a dictionary. FDD can use the resource of DU with specially designed Dynamic Link Library (DLL). SE is built upon this dictionary. When the system started, SE gives response to the stimuli according to the description of FDD. It can call the decision making functions through the description of FDD. Statechart method is used for the modelling of control rules, an FDD generator being designed to convert the modelling data of control rules to FDD. With this method, FDD can be easily derived. For this flexible software architecture, a corresponding flexible hardware platform is also designed. In this chapter, an Ethernet based industrial fieldbus is used for the hardware platform. The implementing details are given. With this design methodology, system upgrading can be easily done by simply regenerating of FDD and hardware reconfiguration.

7.1 Introduction

CNC system plays a very important role in modern industry. Since Numerical Control (NC) was introduced in the 1950s, the latest technology of electronics, control theory and programming techniques can be used in metal cutting. In the 1970s computer was incorporated into the NC system and thus NC evolved into Computer Numerical Control (CNC). The advent of CNC brought a massive improvement to the machine tools [7.1]. With the development of NC/CNC, the machining method of the working piece becomes more and more complex. Meanwhile, the NC programming task becomes very burdensome. It is evident that modern offline Computer Aided Manufacturing (CAM) technology has eased NC programming a lot. However, the programming interface, known as G & M code (ISO 6983), essentially stays the same. In recent years, how to add intelligence and autonomy into the CNC system has become a hot topic. But because ISO 6983 is just a description of tool path and switch signals, it causes great data loss from the CAD data to the CNC system. That limits CNC as an executor. In recent years STEP-Compliant CNC has been proposed to aim at making a seamless connection in the CAD/CAM/CNC data chain [7.2, 7.3]. In a lot of research, how to interpret STEP-NC data and how to use it to build a STEP-Compliant CNC system was discussed. R. D. Allen presents a STEP-NC compliant computational environment used to demonstrate agent-based process planning, resulting in the generation of STEP-NC code [7.4]. Xu et al. retrofitted a legacy CNC system and developed the STEPcNC as a STEP-Compliant NC converter [7.5]. Suh et al. gives an integrated STEP-Compliant NC development [7.6]. R. Liu et al. proposed a three-stage planning procedure in online STEP-NC data planning, and the basic theory and techniques for CNC based on STEP-NC for milling has been discussed [7.7, 7.8]. However, because STEP-Compliant NC is a very new concept, a lot of experiment and new method verification need to be done. To build a flexible implementing platform for the research of STEP-Compliant CNC system and also to test the advance of STEP-NC data model is very essential. In this chapter, an open platform for STEP-Compliant CNC development is proposed and the implementing scenario is given, with which the rapid development and verification of STEP-Compliant NC functions can easily be done.

7.2 Requirement Analysis

The requirement analysis has two levels: function level and implementing level.

7.2.1 Function Level Requirement Analysis

Till now, the STEP-Compliant CNC technology is still under research and there's no definite method. Also the implementing method is not fixed. A lot of experiments and prototype testing need to be done in order to use fully the enriched information that carried by ISO14649 data model. Currently, the STEP-Compliant CNC under research can be classified into three types [7.8].

Interface Type

As shown in Figure 7.1, the simplest way to interface the STEP-NC codes with CNC is probably to add an interpreting module to CNC. The interpreting module converts the STEP-NC based program into G&M codes and then feeds them into the NC kernel. In this case, the interpreter acts as a comprehensive post-processor transplanted from CAM to CNC, and a conventional CNC can be used with little modification. It should be noted that part program must be well organized in advance on a CAM system by a programmer who is familiar with the machine resource, for this kind of STEP-NC controller has no more intelligence than a traditional CNC except for the capacity for interpreting the STEP-NC codes. Since this kind of STEP-NC controllers depends greatly on more capable CAD/CAM systems, Newman et al. [7.9] presented a number of CAD/CAM solutions and implemented a system (AB – CAM) for the STEP-compliant manufacture in the light of this kind of STEP-NC controller.

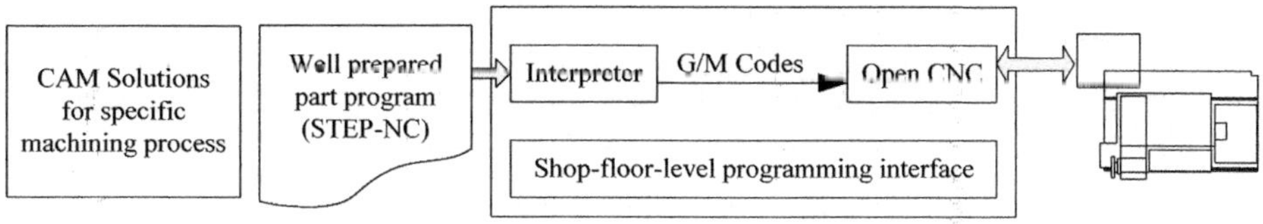

Figure 7.1. Interface Type STEP-Compliant CNC implementation

Embedded CAM Type

As shown in Figure 7.2, the CAM-embedded STEP-NC controller is basically a combination of a STEP-NC code interpreter, a basic CAM and a conventional CNC. The interpreter translates the physical file into the internal data format, the CAM makes decisions on the machine-specific details and generates G&M codes, and finally the CNC executes the codes. In this case, the STEP-NC controller can be implemented with certain intelligences; however the problems between CAM and CNC are not solved but hidden.

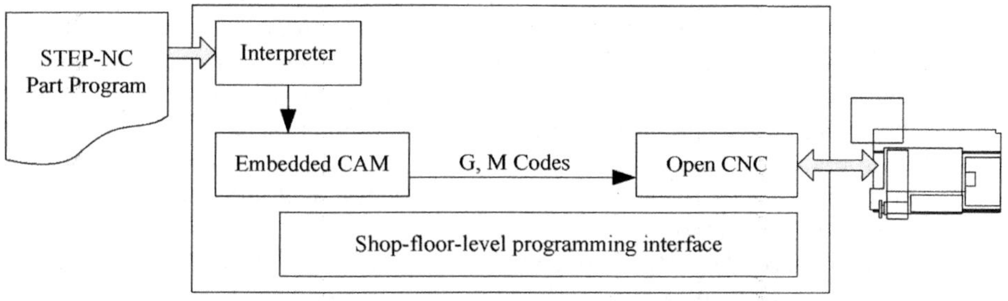

Figure 7.2. Embedded CAM Type STEP-Compliant CNC implementation

Intelligent Type

STEP-NC data interface gives a seamless data integration with CAD/CAM. How to take full advantage of STEP-NC data model is a hot topic. Much of research has been done on how to build the intelligent CNC with the help of STEP-NC data model. R. Liu and H. Lan discussed an agent-based STEP-Compliant CNC structure

[7.8, 7.10]. S. H. Suh et al. built an integrated STEP-Compliant CNC prototype, in which a CAD-CAM-CNC data chain was realized [7.6] and a conceptual framework for designing and implementing an intelligent CNC system was also presented [7.11]. Among this research, the generic structure of intelligent STEP-Compliant CNC is as shown in Figure 7.3.

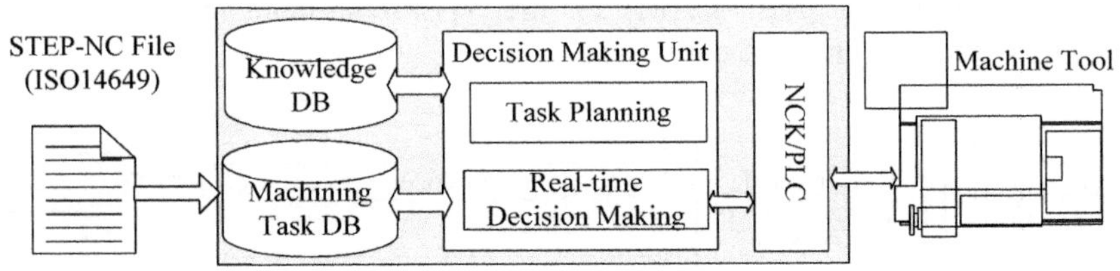

Figure 7.3. Intelligent STEP-Compliant CNC structure

As shown in Figure 7.3, the information of STEP-NC file are converted and saved in Machining Task DB based on STEP. Decision Making Unit of STEP-Compliant CNC makes task planning and real-time decision making. The control command then sends to the interface to convert to control signals.

As mentioned above, only the third type is the real intelligent STEP-NC controller. With this structure, the information of the STEP-NC file can be fully used by the Decision Making Unit and then can be converted to the control command directly. It is an ideal type of STEP-Compliant CNC. In this chapter, the development details will be under this structure.

To develop STEP-Compliant CNC under this structure, an implementing platform is urgently needed in the prototype development and function test. To build such kind of platform, the requirements below need to be fully considered:

- First, a well designed software structure is needed to ease the design burden. The CNC system is a very complex system. Programming and debugging such kind of system is very time consuming and needs well trained engineers who have enough knowledge of NC machining, computer software, control theory and electronics engineering. STEP-Compliant CNC system is not just an upgrading of the current systems. It is a totally new structure and nearly everything needs to be redesigned. A well designed software platform and programming method is urgently needed.
- Second, secondary development ability is needed for a STEP-Compliant CNC system. Till now STEP-Compliant CNC has not been a mature technology and usually it needs a lot of modification after the system was built. And also, even for mature CNC systems, it is very hard to integrate everything at the very beginning when the system is designed due to the rapid development of the machining method. Secondary development ability is very useful.
- Third, a well designed flexible hardware platform is needed. To implement the flexible software structure, the flexibility of hardware structure is essential. The hardware modules, such as motor control modules and I/O

modules, should be well designed to be easily removed or added to the system.

7.2.2 Implementing Level Requirement Analysis

The CNC control system is a typical reactive system. For this kind of systems, they are usually driven by stimuli [7.12]. These stimuli can be typically classified into two types. One type is external, like operator's requests and external driven signals from the controlled object. The other is internal stimuli, like periodic interrupt services. The handling of these stimuli can be free time handling and real-time handling. For the stimuli which are time critical, like external interrupts, periodic interrupt services, should be handled real-time. Normal stimuli, like normal request from control panel, can be processed at free time. The typical system work flow is as shown in Figure 7.4.

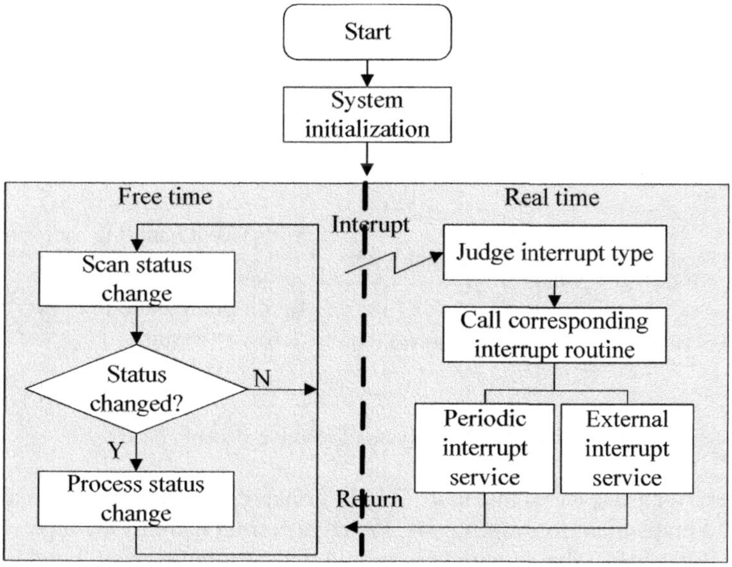

Figure 7.4. Typical workflow of embedded control system

As shown in Figure 7.4, when the system starts, after system initialization, the system will enter free time handling. It repeatedly scans status change and processes it. When an interrupt comes, the real-time handling will be activated and a corresponding interrupt service routine will be called.

Considering these characteristics, in the development of STEP-Compliant CNC system, some key functions which need to be realized are listed as below:

- A new system structure is needed to fulfil the reactive characteristics and also the compliance with STEP
- Second, to support the secondary development ability and provide a function test bench for STEP-Compliant CNC, system function design should be

separated from this structure to make the modification and upgrading of functions available
- Third, to make the secondary development easier, an easy-to-use modelling tool for complex functions is needed
- Fourth, to ensure software flexibility, a flexible hardware should be built to ensure the flexibility of the whole system at the hardware level

With the above considerations, an engine based structure is proposed as the software platform to fulfil the above demands. For the hardware in the traditional integrated hardware platform, the controlled axis and I/O amount is fixed. That makes a lack of flexibility. In the fieldbus based hardware platform, all controlled devices are connected with a fieldbus. Every controlled device, like motor driver and I/O modules, can be added to or removed from the system easily. That makes the hardware very flexible. Lately the Ethernet based fieldbus has become the main trend [7.13]. In this chapter, the EtherMC (Ethernet-based fieldbus for Motion Control) fieldbus platform is developed to be the hardware platform.

7.3 System Structure

The reactive system structure is usually considered to embrace three layers as illustrated in Figure 7.5 [7.15].

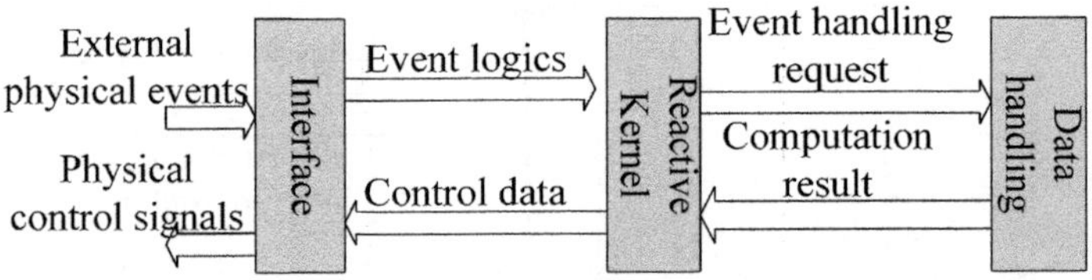

Figure 7.5. Three layers of reactive system [7.15]

As described in [7.15], in Figure 7.5, Interface deals with the external physical events and maps them to event logics. Reactive Kernel handles the logic inputs and outputs. It decides what computations and what outputs must be generated in reacting to input. The Data handling layer performs classical computations requested by the reactive kernel.

This structure gave a clear processing flow layer by layer. It can fulfil the processing demands of reactive systems. However, to the secondary development ability, it can't give a clear interface to the redesigning of system functions. In order to separate system function design from the system and give an interface for the system function redesigning as well, in this research, to realize the demands of the intelligent STEP-Compliant CNC, the CNC structure is considered as Figure 7.6, which is called Engine Based CNC Structure.

In this structure the software of CNC is composed of four parts: System Engine (SE), Engine Machine Interface (EMI), Decision Unit (DU), and Function Description Data (FDD). Mapping Data of EMI can be modified by Human Machine Interface (HMI), so that it can adapt to different controlled objects. In order to separate function design from complex system and gain secondary development ability as well, the Reactive Kernel and Data handling in Figure 7.6 are redesigned as System Engine (SE), Function Description Data (FDD), and Decision Unit (DU). DU is composed of Machining Database, Knowledge Database, and Dynamic Link Library (DLL). Below is a more specific description of these three parts:

- EMI. It is composed of Event Generator, Machine Driver, and Mapping Data. Event Generator converts external physical events to event, generates internal events, and then sends them to SE for future handling. Machine Driver converts control command data generated by SE to physical control signals. The mapping between internal data and external physical signals are stored in the Mapping Data and can be modified through HMI.
- SE. Engine Kernel of SE processes Event logics generated by EMI according to the handling method stored in FDD. Run-time Data is designed to record run time information and can be accessed at run-time by Engine kernel.
- FDD. FDD stores system function description, which represents event handling method. System Behaviour Logics is separated from Reactive Kernel of Figure 7.6 as a part of FDD. Also Action handling of Figure 7.6 is described in Action Function Library and can be called for computation by Engine Kernel of SE.
- DU. DU is in charge of the intelligent decision making. The DLL of DU is a function library. It is composed of functions which perform task planning and real-time decision making according to the machining task description of Machining DB and rule description in knowledge DB. These functions can be called by the action function library of FDD. New decision rules and algorithm can be reedited by replacing the Knowledge DB and DLL.

The EtherMC fieldbus is developed as the hardware platform. It can be driven by driver interface of SE. With this bus, all motion devices and I/O resources can be flexibly added or removed to the system conveniently.

Here we take the cutting tool wear to show how the system works. If too wear occurs, the system reacts as the following steps:

- Step 1: Tool wear was inspected by the online sensors.
- Step 2: Signal transfers to EtherMC hardware as I/O signal.
- Step 3: Machine Driver of EMI get the signal via EtherMC Driver Kernel. And then it looks up the Mapping Data of EMI to find what the signal stands for. Then the signal is mapped to events and posted to SE.
- Step 4: SE looks up FDD to get the event processing method. The it processes the event sent from EMI (tool wear event) as the method described in FDD. Special functions are called from DU (such as function for looking up the machine source to find new available cutting tools, function for re-planning machine tasks according to the new cutting tool parameters, etc.).
- Step 5: Tool wear process is done. The system starts the new tasks which are driven by periodic servo events generated by EMI.

Every time we need to change the tool wear processing algorithms, only tool wear related FDD and DLL of DU need to be changed.

Through this example we can see, with this structure, the NC function design can be done by editing FDD and DU even at shop-floor level. Then by reconfiguring EtherMC devices, the secondary development ability can be realized.

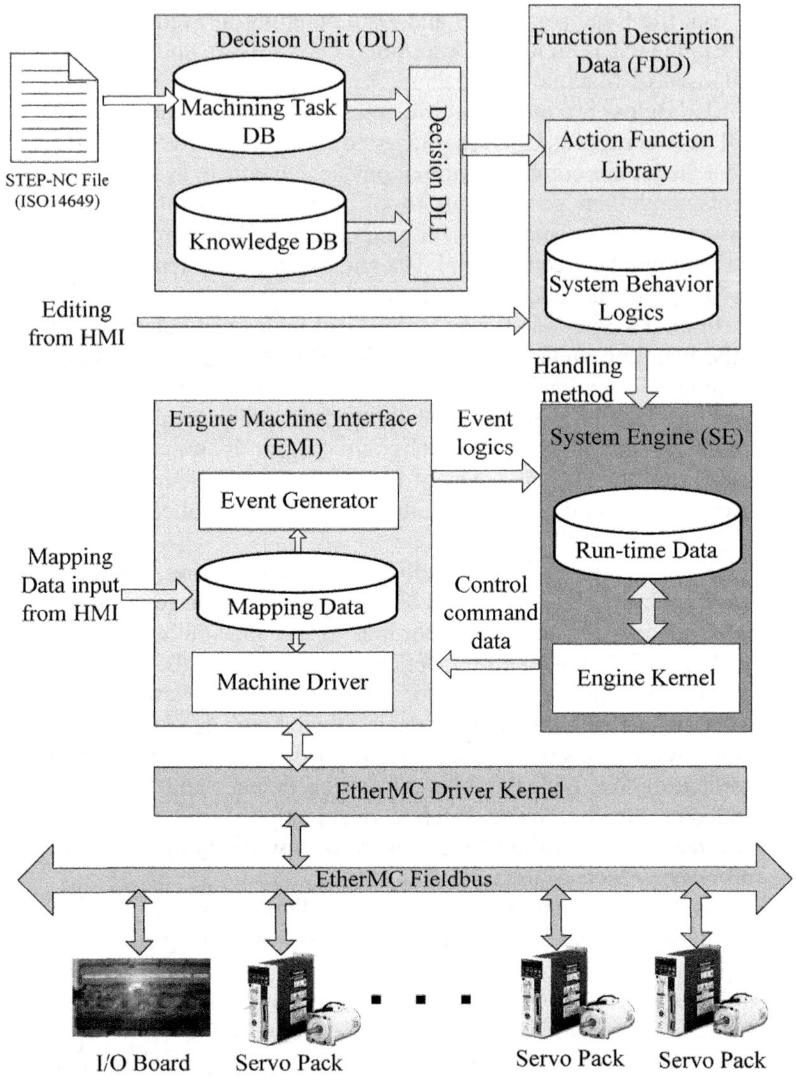

Figure 7.6. Engine based structure

7.4 Design Specification of Engine Based System

As described above, engine based system has five components: DU, FDD, EMI, SE, and hardware platform. This section will discuss design specification of engine based system.

7.4.1 Design of Decision Unit (DU)

For decision making of the system, there are many algorithms. Even for mature systems, these algorithms are not fixed. The upgrading of new algorithms is very important for the intelligent systems. In this chapter, DU is designed as a plug-in of the system. That makes the more efficient algorithms can be used or tested at the very time when they are developed.

As shown in Figure 7.6, DU is composed of Machining DB, Knowledge DB, and DLL. Machining DB saves machining information that converted from STEP-NC file. Knowledge DB saves processing methods and rules to support intelligent decision making. DLL of DU works as an interface of the DU. Developers can generate this DLL or replace Knowledge DB according to the specified format. All these functions can be called by the action functions of FDD.

7.4.2 Generation of Function Description Data (FDD)

To fulfil the demand of secondary development ability, FDD was separated from the system to record the processing rules of the system. How to design FDD format and give users secondary development ability as well becomes a key problem. Here Statechart [7.14, 15] is used as the modelling method, and then a tool is developed to convert Statechart model description file to FDD.

Statechart Modelling

Considering the reactive characteristic of the embedded control system processing rules, State Machine [7.16] is a traditional method. By using State Machine, reactive system can be divided into possible states. When an event comes, it will search available transitions among the states and active specific actions. Finite State Machine (FSM) [7.17] is a traditional method for modelling. But it is not very suitable for complex systems because of the limitation of its description method. Statechart method is an enhancement of FSM [7.16]. It provides a modelling method for complex event-driven reactive systems. Compared to conventional state machine methods, it extends conventional state-transition diagrams with hierarchy, concurrency and broadcast communication as D. Harel [7.16] described, which makes the description of complex reactive system with flat state machine possible.

There are several commercial statechart modeling tools. Stateflow of MATLAB® is one of the most popular ones. Figure 7.7 gives an example of Statecharts model which are built with stateflow of MATLAB® [7.18]. As seen in this figure, there are three kinds of commonly used components in Statecharts: States, Transitions, and Junctions. States can have parallel ("And") sub-states or exclusive ("Or") sub-states. For example, StateA and StateB are two parallel sub-states of MainState; StateA1 and StateA2 are two exclusive sub-states of StateA. Transitions can be clarified into two groups, default transitions and normal transitions. Default transition is executed as the default transfer action. All transitions may have trigger events, conditions, condition actions, and transition actions as shown in Figure 7.7. There are two kinds of junctions in this example: History Junction and Connective Junction. History junction records the previous active state of the ancestor state in which they reside in. Connective junctions are used as decision point to make decisions according to different conditions. Take the transition to StateA1 for example. When event ev_A1 happens, and if there's no history record in StateA, default transition to Connective Junction will taken place. Then if condition c_A1 is satisfied, the transition from Connective Junction to StateA1 will be triggered and Condition Action f_A1 will take place. Also Transition Action f_A2 will be called. Then StateA1 will become active and the enter action will be called simultaneously.

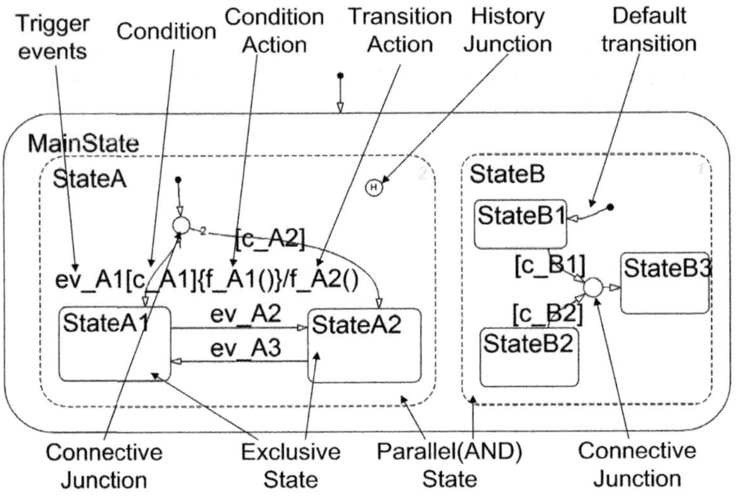

Figure 7.7. An example of Statechart Mode

Simulation

Most of Statechart modeling tools also have simulation functions. In MATLAB®, Simulink® can be used as the modelling tools to simulate statecharts which is built with Stateflow.

Generation of FDD

After simulation, the correct Statechart model can be derived. In most statechart modelling tools, this model is just a description of the system, and all the information are mixed together. In MATLAB®, this file is called Model Description Language (MDL) file. In this research, an FDD data format is designed and also an FDD generator is developed to sort the system information from the original MDL file to FDD. The mapping relationship between MDL and FDD is shown as Figure 7.8. In order to easily locate corresponding transition and the actions, in FDD, System Behaviour Logics and Action Function Library are separated. Take State for example, the descriptions in MDL file and corresponding information in FDD is described as below.

Description in MDL file

```
State {
 id 45
 labelString "JogYRapid\n"
 "en:spdCmdY = JogSpdY *2;\n"
 "spdCalculate();\n"
 "posCurY+= spdCurY;"
 ...
 treeNode  [62 0 63 0]
 type  OR_STATE
  ...
  }
```

System Behaviour Logics of FDD

```
id: 45
name: JogYRapid
descendent ID: 0
sibling ID: 63
sibling Type: 1
ancestor ID: 63
enter Function ID: 20
duration function ID: null
exit Function ID: null
```

System Action Functions of FDD

```
//variable definition
float spdCmdY; //speed command generated
float JogSpdY;  //Jog speed of Axis Y set through HMI
...
//function pointer definition
typedef void (*pSddVoidFunction)();
typedef bool (*pFSddBoolFunction)();
...
pSddVoidFunction thePSddVoidFunc[10000];    //void type function pointer
pSddBoolFunction thePSddBoolFunc[10000];    //bool type function pointer
...
//functions
void f0000()
{
...
}
void f0001()
{
...
}
...
```

```
void f0020()
{
        spdCmdY = JogSpdY *2;
        spdCalculate();
        posCurY+= spdCurY;
}
...
//Assign function pointer
thePSddVoidFunc[0] = &f0000
thePSddVoidFunc[1] = &f0001
...
thePSddVoidFunc[20] = &f0020;
...
```

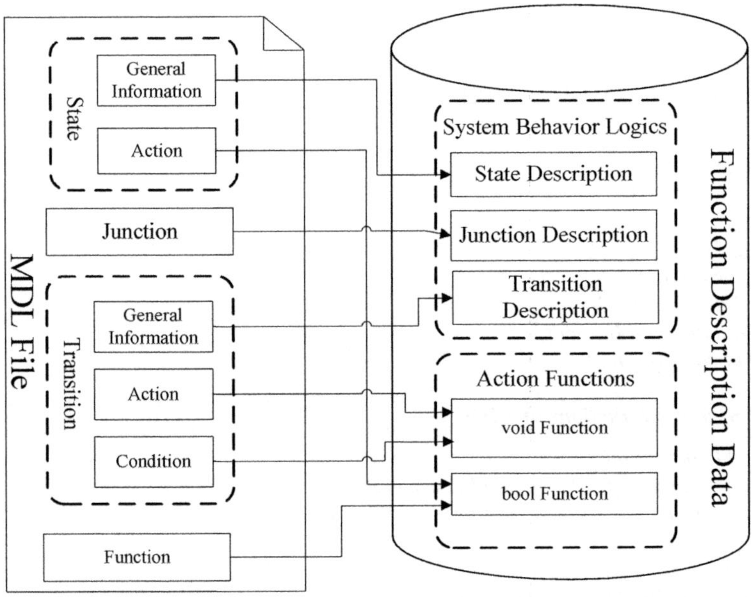

Figure 7.8. Mapping between MDL file and Function Description Data

For Action Function Library of FDD, they are formatted with standard C style functions in which the function pointer is used. The function ID in System Behaviour Logics is used as the function pointer offset to locate the specified function. With this design, System Behaviour Logics can be saved to a standard database, and Action Functions Library can be compiled to a Dynamic Link Library (DLL). Then both of them can be easily replaced even after the whole system is built. In this research, an FDD Generator is developed to convert MDL file to FDD. With Statechart modelling, simulation, and FDD Generator, FDD can be easily derived to ensure the shop-floor level development.

7.4.3 Design of EMI

EMI is designed as a bridge between embedded system and its controlled object. As shown in Figure 7.6, it consists of Mapping Data, Event Generator, and Machine Driver. Mapping Data can be modified through HMI and designed as a dictionary. Taking an Input port for example, its description in Mapping Data is in Table 7.1.

Table 7.1 An example of mapping data

Properties	Value
Source	Input terminal 1
Type	Digital input
Effective when	Low to high
Event ID	10
Priority	EMERGENCY

With this data format, every Input/Output port can be configured to trigger a corresponding event. The Event Generator monitors the signals according to the rules defined in Mapping Data. If event is monitored, it will post a message to SE with event ID and priority. For emergency event, Event Generator will first notify the Machine Driver to take emergency treatment and then post it to SE for further processing. For a real-time event, it will be posted to SE directly. If a free time event happened, it will wait in the free time event queue of SE and then be processed at free time when SE is available. After event processing of SE, command data is generated from SE and sent to Machine Driver of EMI. Machine Driver will convert these command data to physical signals according to the specification of EtherMC Driver.

7.4.4 Design of SE

SE, as its name, is the engine of the system. It consists two parts, Run-time Data and Engine Kernel.

Run-time Data

Run-time Data records the run-time activated states of the system. To fulfill the run-time access and modification demands, and also considering the characteristic of the States recorded in FDD, it is special designed as a multi-search tree-shape, as shown in Figure 7.9.

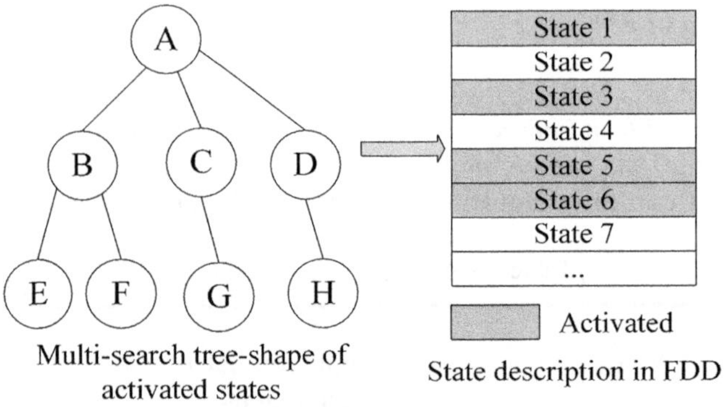

Figure 7.9. Multi-search tree-shape of state and state description in FDD

In this tree shape structure, each node stands for an activated state, and there's a corresponding description in FDD. It has four pointers: pointer to parent, pointer to left sibling, pointer to right sibling, and pointer to first child. It can dynamically grow and destroy both on width and depth according to the action of Engine Kernel and FDD.

Engine Kernel

Engine Kernel accepts stimuli sent from EMI and gives response to the Machine Driver of EMI. The workflow of Engine Kernel is as Figure 7.10.

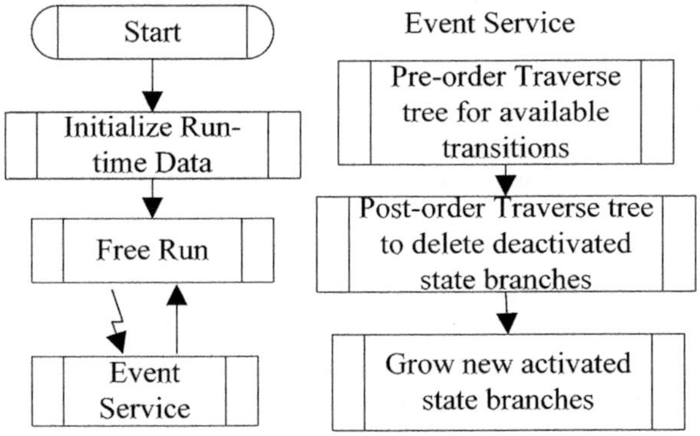

Figure 7.10. Workflow of Engine Kernel

After the system is started, Engine Kernel will first initialize the runtime data (Multi-search tree-shape of activated states). Then it will enter free run mode. When an event comes, event-service will be called to process the event. The Engine Kernel will do a pre-order traverse of the tree shape and look for the FDD to find available transitions. After the transition is found, corresponding tree branches will be deleted and new activated branches will be added in. In the transitions, related actions functions will be called from Action Function DLL of FDD as the description of System Behaviour Logics of FDD.

With this design, FDD and DU are separated from the whole design and work as the dictionary of SE. A secondary development implementation can be easily gained just through redesigning FDD and reconfiguring EMI.

7.4.5 Design of EtherMC Hardware Platform

The design of software shows good flexibility. To build such a kind of platform, an EtherMC based platform was developed for the implementation. EtherMC is an Ethernet based field bus. The design specification is shown as below.

EtherMC Topology

A tree-shape topology was used in the EtherMC fieldbus software. As shown in Figure 7.11, a chain or tree topology is used. It means that every slave node can be connected or removed from the system easily. This topology can cooperate well with the reconfiguration of the software.

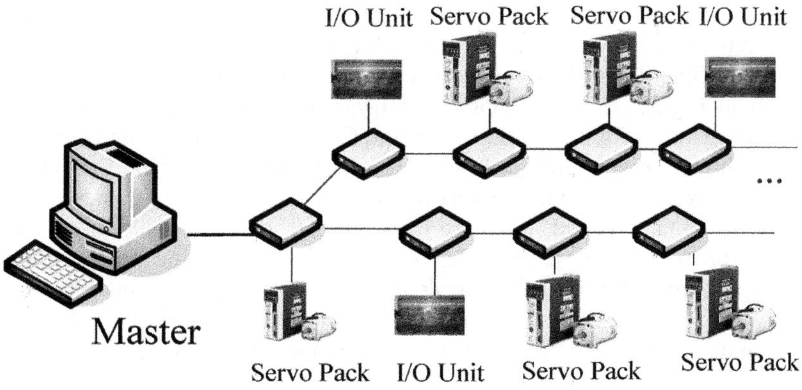

Figure 7.11. EtherMC topology

EtherMC Master Driver

The EtherMC Master Driver Kernel is developed with standard C Language under standard Windows NT operating system. As shown in Figure 7.12. An EtherMC Network Device Interface Specification (NDIS) protocol driver is developed under windows kernel layer. EtherMC API is developed to be used by the application layer applications. In this research, EtherMC API functions are called by the engine based software system.

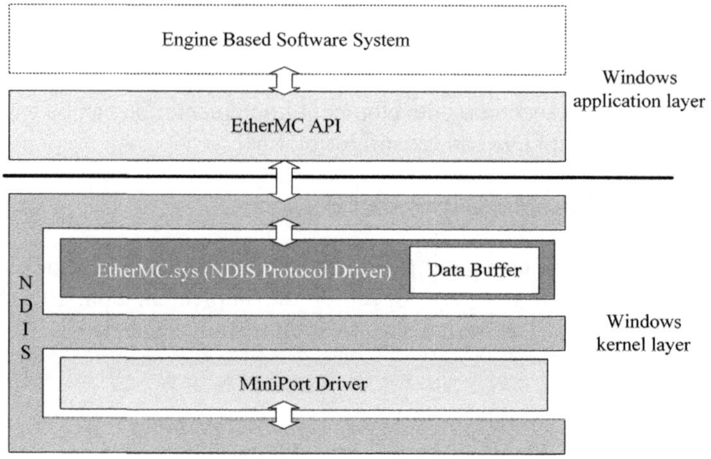

Figure 7.12. EtherMC Master Driver

EtherMC Slave

The communication functions of EtherMC Slave are implemented with an Field Programmable Gate Array (FPGA) based structure is built as Figure 7.13.

As seen in Figure 7.13, the data linker layer is composed of Mailbox Control Unit, Synchronization Control Unit, Command Control Unit, and Data Control Unit. All this data linker layer is built in FPGA. The Mailbox Control Unit is in charge of normal data exchange. Synchronization Control Unit is in charge of the synchronization management, such as data transfer delay measurement and the synchronization with other slave nodes. Command Control Unit parses command sub-datagram. Data Control Unit controls massive data exchange and periodic data exchange. With this design, the synchronization of all the slave nodes can be controlled and also flexible configuration can be done for each slave node.

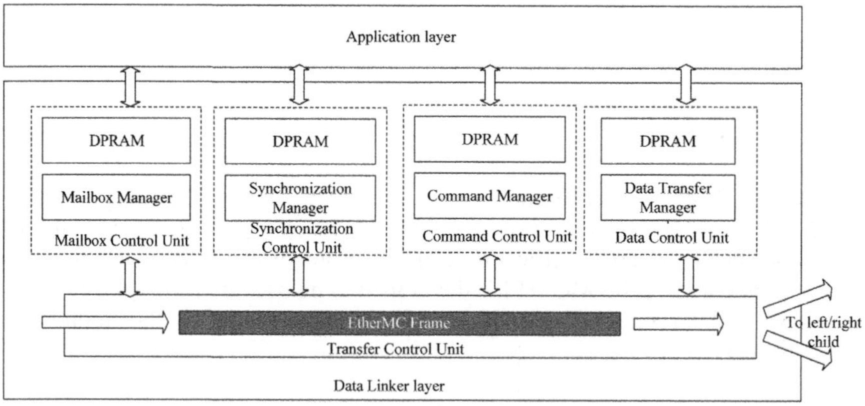

Figure 7.13. EtherMC Slave structure

With EtherMC, the flexibility of the hardware can be guaranteed. When the users need to do upgrading on this platform, the hardware can be reconfigured easily.

7.4.6 Secondary Development Scenario

With this system structure, system secondary development can be easily derived following the steps below:

1 Change Knowledge Database of DU or Decision DLL which contains the more efficient decision algorithm to make the system have up-to-date decision making ability.

2 Reedit FDD with the following steps if needed:

- Add or modify Statechart model to fulfil desired new features of the system.
- Do simulation with Statechart simulation tools. If requirements are not fulfilled, modify the Statechart model until all requirements are fulfilled.
- Convent the new Statechart model to FDD.
- Configure Mapping Data of EMI if there are any changes of system Input/Output related devices and rules of internal events.
- Replace the FDD (Behaviour Description Data and Action Function DLL) of the old system.

By reediting DU and FDD, the system with new decision making algorithm and NC functions can be realized.

7.5 Prototype Development

This section will show the development of STEP-Compliant CNC system with engine based structure.

7.5.1 Design of DU

In the design of DU, first, a knowledge database is needed. Then the Decision DLL is needed to provide an interface to other modules. In this case study, the Decision DLL provides functions which do task planning according to worksteps of the STEP-NC data which saved in Machining Task Database. Some C language style functions are designed as in Table 7.2.

Table 7.2. Function specification of DU Decision DLL

Function Name	Description
TaskPlan	Does task planning based on the Workingsteps described in STEP-NC file and the resources of the machine tool
ReplanTask	Replans the unfinished tasks. It can be called when the working condition changes in order to adapt the new working conditions
ResetCycle	Resets the task array. When the machining is restarted, all tasks will be executed
GenerateNewTask	Gets the next machining task from the planned task array
GenerateRoute	Generates the basic tool route using the given machining task, and save the route to route buffer
DoInterpolation	Does interpolation according to the route which saves in the route buffer

7.5.2 Design of FDD

As mentioned in Sections 7.2 and 7.3, because of the reactive characteristics, FDD can be modelled with stateflow of MATLAB®.

Statechart Modelling of FDD with Stateflow of MATLAB®

Figure 7.14 shows the Statechart model of the CNC prototype. When the system is started, the two parallel subchart, "Motion" and "logic" of statechart become activated. "Motion" takes care of the motion control of the system, both manual motion and automatic machining. "Logic" takes care of logic control of the system. "Motion" and "Logic" are synchronized by events and variables. In "Motion" subchart, the Statechart "auto" is in charge of auto machining. In this prototype, its main functions are: task planning, tool path generation, control command generation, and also single task function (for machining debugging). The more specific description of "auto" statechart is described as below:

- Top statecharts model

Figure 7.15 shows the "auto" statecharts top model. It consists of six subcharts: "RTProcess", "Initialization", "EnableNextSingleTask", "reSchedule", "CycleReset" and "Nop". "RTProcess" responses to "SERVO_PR" event and generates machining command. "Initialization" subchart gives response to "CYCLE_START" event and does initialization of task planning and related parameters. "EnableNextSingleTask" subchart take cares of executing next command request of users when the system runs under "single task" status. "reSchedule" subchart reschedules tasks according current machining status when it is requested. "CycleReset" subchart gives response to "CYCLE_RESET" event. It will reset all current running machine tasks when "CYCLE_RESET" event happens.

- "RTProcess" statecharts model

"RTProcess" generates motion command based on the machine tasks which are generated by "Initialization" state. As shown in Figure 7.16, it has four substates: "GetNewTask", "GetRoute", "CmdGeneration" and "Idle". "GetNewTask" state get tasks from the task queue. "GetRoute" generates basic route (line circle, curves etc.) according to the tasks generated by "GetNewTask". Then "CmdGeneration" does interpolation and generate motion command of every axis. In these substates, functions like "GenerateNewTask", "GenerateRoute" and "DoInterpolation" are called from the DU.

As shown in Figure 7.16, in this model, the substates process events as below:

- "CmdGeneration" gives response to servo request event"SERVO_PR". When the route buffer is not empty (the value of "bRouteBufEmpty" is false), the interpolation operation will be called. The "GET_ROUTE" event will be triggered when the route buffer reached a preset level. The "GetRoute" state will give response to this event and generate route to fill the route buffer. If the route buffer is empty, and the user gives a command of running the new task, the "GetNewTask" state will be activated and generate new task, or the it will turn to the "Idle" state.

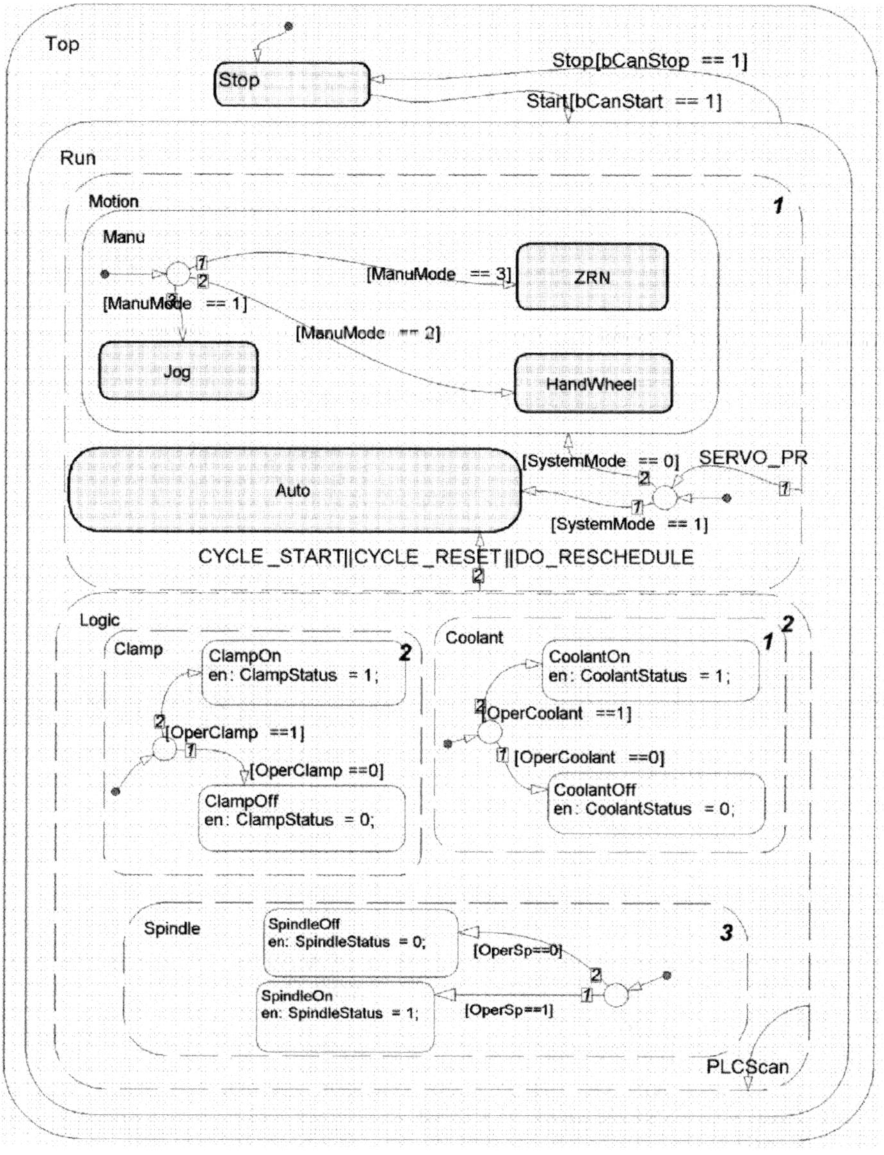

Figure 7.14. Statechart model of the CNC prototype

- "GetRoute" state gives response to the "GET_ROUTE" event. If there's still unfinished task ("bTaskConsumed" is false), the "RouteGeneration" state will be activated. When needed, it will request the "GetNewTask" state to generate a new task. If it is in "Single Task" status, it will not request for a new task unless the user requests; instead, it will enter "Idle" state to wait for the next task execution request.

- "GetNewTask" state will be activated when "GET_TASK" event happens. If all the tasks are done, it will mark internal variables and then enter "Idle" state; if not, it will get new task and fill the task buffer for the processing of "GetRoute" state.

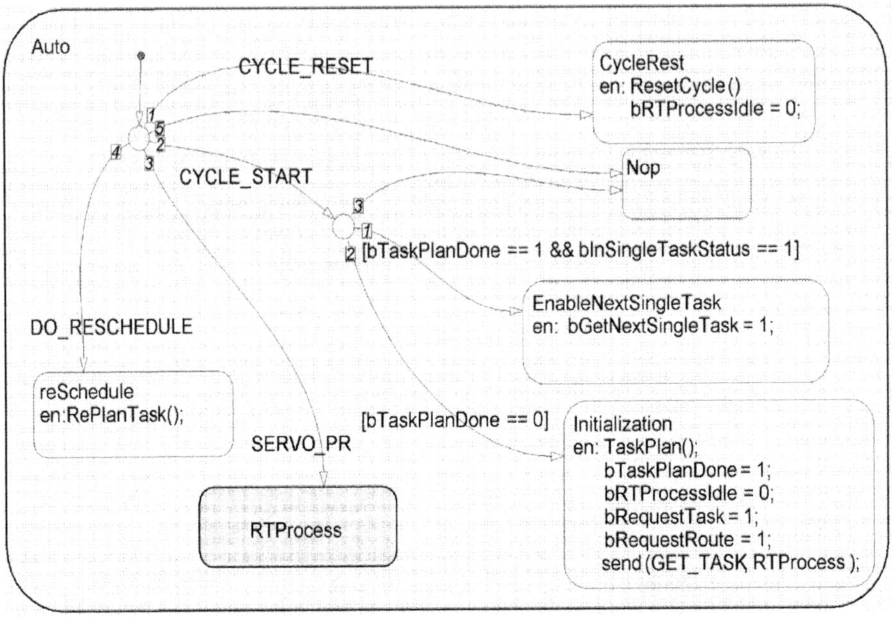

Figure 7.15. Statechart design of "Auto"

Simulation with Simulink of MATLAB®

After FDD statechart modelling with stateflow of MATLAB®, the model can be simulated with the simulation tools. In MATLAB®, it is Simulink®. Figure 7.16 shows the simulation environment of Simulink®. In the middle is Statechart model. With the parameters and events input from the left, the simulation results can be monitored in the right part. Also state transition behaviour can be monitored simultaneously in Stateflow of MATLAB®. Figure 7.17 shows the simulation environment.

Generation of FDD

After simulation, if all requirements are fulfilled, a correct CNC prototype model can be derived. Then with the FDD generator developed in this research, the FDD can be derived. After that, we can modify the Action Function Library of FDD to add the function calls of Decision Making DLL at the appropriate place. Then compile the FDD Action Functions to DLL, and the FDD with new control rules and decision making algorithms is derived.

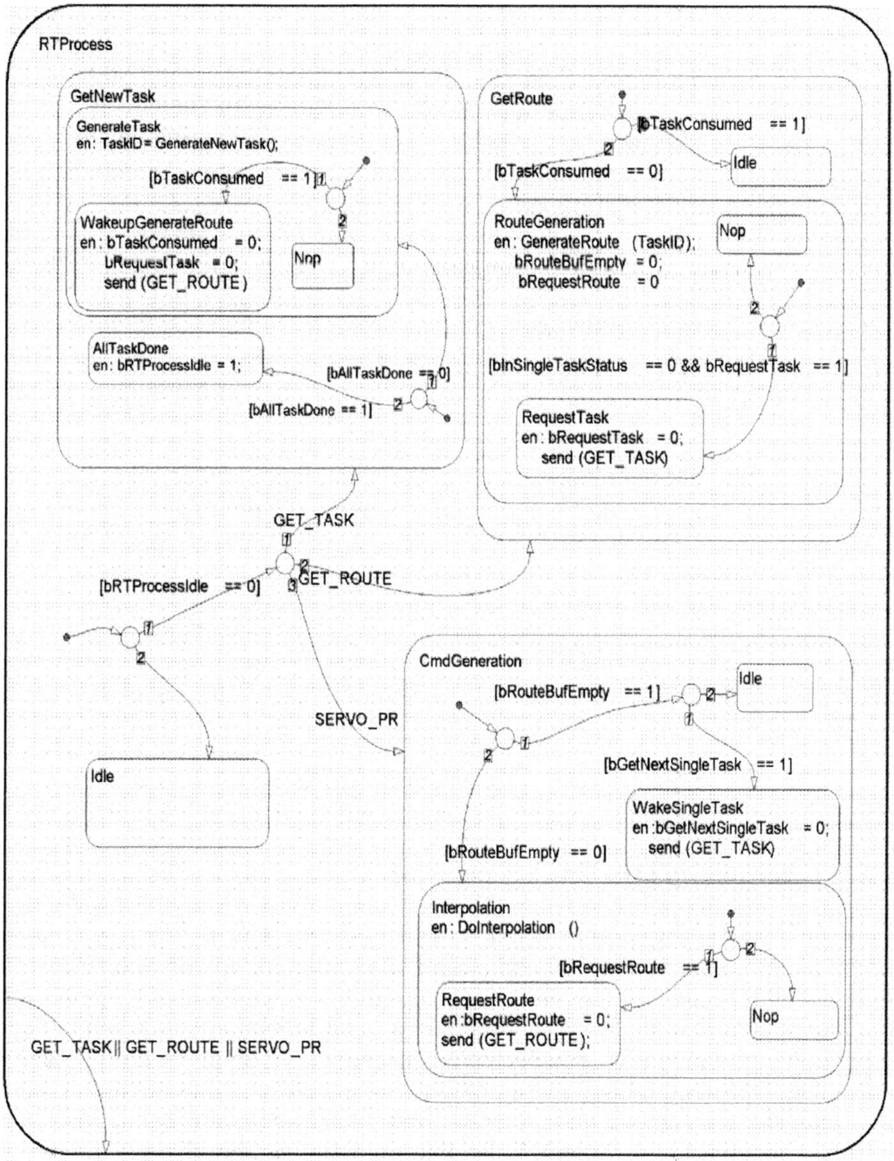

Figure 7.16. Statecharts model of "RTProcess"

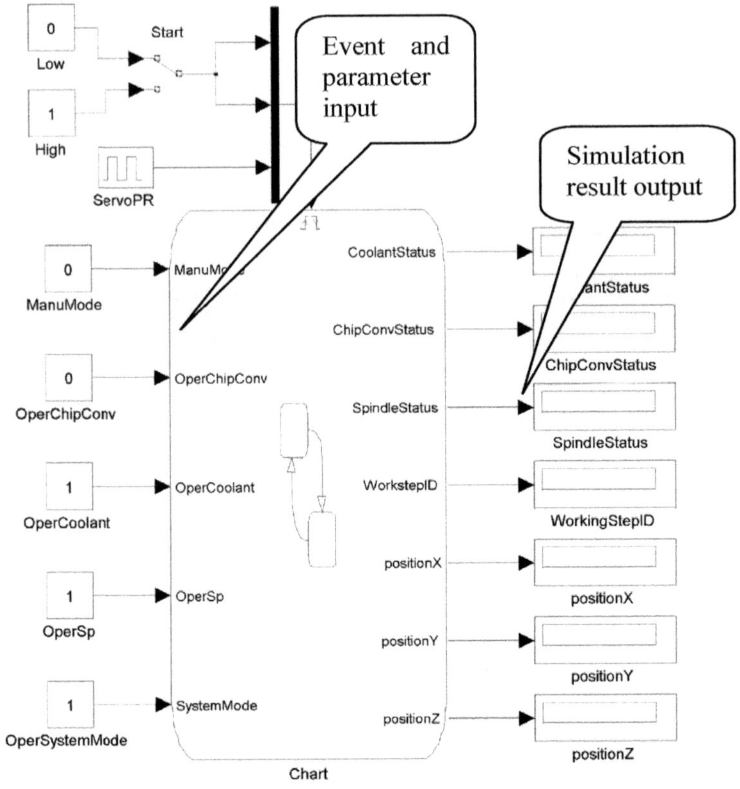

Figure 7.17. Simulation environment

7.5.3 Hooking up FDD with SE and HMI

New systems can be derived by copying FDD Behaviour data, FDD Action Function DLL, and Decision Unit to the installation directory. SE remains the same. The system with the modelled functions can be derived. Here we designed a virtual HMI to generate command to the system in order to test the functions, as shown in Figure 7.18. Every time secondary development is needed, for example to do NC function upgrading, Decision Making rules or algorithms upgrading, it can be easily done by the steps identified above with friendly design GUI and simulation environment. Finally, reconfigure the EtherMC fieldbus to adapting the changes of the NC functions.

With the steps above, the heavy burden of NC function programming is replaced by the graphical Statechart modelling and simulation.

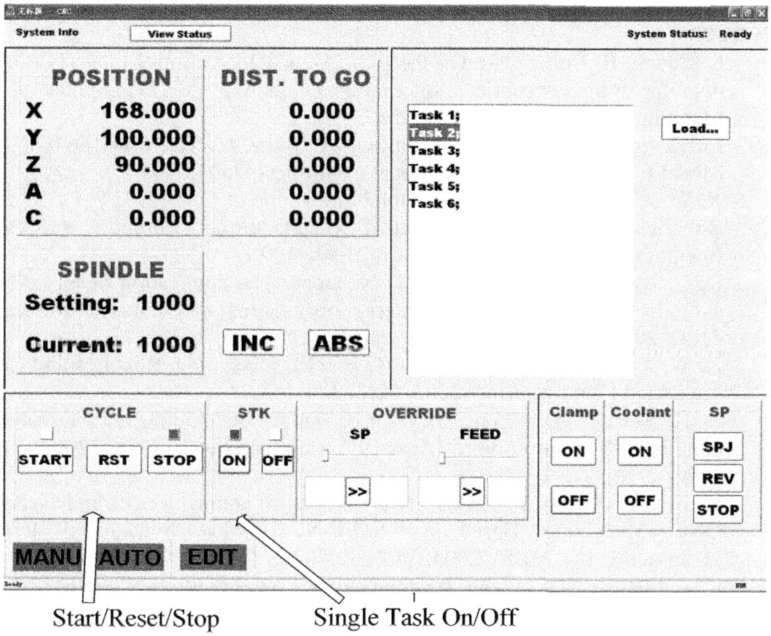

Figure 7.18. Prototype test

7.6 Conclusion

In this chapter, an engine based structure was proposed as the STEP-Compliant CNC implementing platform. With this structure, the flexibility of the software can be derived by reediting FDD with graphical statecharts modelling and simulation. Also the Ethernet based fieldbus-EtherMC, which is developed by our research group, can guarantee the flexibility of hardware platform via adding or removing devices to the fieldbus and hardware reconfiguration. With this design, the NC functions, including intelligent decision making algorithm, is separated from the flexible system. Every time when the NC function modification is needed, it can be easily done through graphical statecharts modelling and simulation. That makes the upgrading and secondary development very easy. With this method, the heavy design burden in the development or research of STEP-Compliant CNC system is greatly eased. Also the life cycle can be dramatically enlarged with continuous shop-floor upgrading.

Acknowledgment

This chapter is supported by National Natural Science Foundation of China (under contract 50675123) and the Research Fund for the Doctoral Program of Higher Education (under contract 20040422026).

References

[7.1] C. Zhang, R. Liu, T. Hu. On the futuristic machine control in a STEP-compliant manufacturing scenario. International Journal of Computer Integrated Manufacturing. 2006, 19(6):508-515.

[7.2] International Standards Organization. TC184/SC1/WG7, ISO 14649/FDIS, Data Model for Computerized Numerical Controllers. 2001.

[7.3] X. W. Xu, Q. He. Striving for a total integration of CAD, CAPP, CAM and CNC. Robotics and Computer-Integrated Manufacturing. Robotics and Computer-Integrated Manufacturing. 2004, 20(2):101-109.

[7.4] R. D. Allen, J. A. Harding, S. T. Newman. The application of STEP-NC using agent-based process planning. International Journal of Production Research, 2005 43(4). 2005, 43(4):655-670.

[7.5] X. W. Xu. Realization of STEP-NC enabled machining. Robotics and Computer-Integrated Manufacturing. 2006, 22(2):144-153.

[7.6] S. H. Suh, D. H. Chung, B. E. Lee, et al. Developing an integrated STEP-compliant CNC prototype. Journal of manufacturing systems. 2002, 21(5):350-362.

[7.7] R. Liu, C. Zhang, Y. Zhang, et al. Process Planning Model and Heuristics for CNC Machining Based on STEP-NC. China Mechanical Engineering (Transaction of CMES). 2004, 15(4):325-328.

[7.8] R. Liu. Basic Theory and Techniques for CNC based on STEP-NC for Milling. Vol. Ph. D Jinan: Shandogn Unviersity, 2004.

[7.9] S. T. Newman, R. D. Allen, J. R. S. U. Rosso. CAD/CAM solutions for STEP-compliant CNC manufacture. International Journal of Comupter Integrated Manufacturing. 2003, 16(7&8):508-515.

[7.10] H. Lan, R. Liu, C. Zhang. A multi-agent-based intelligent STEP-NC controller for CNC machine tools International Journal of Production Research. 2008, 46(14):3887-3907.

[7.11] S. H. Suh, S. U. Cheon. A Framework for an Intelligent CNC and Data Model The International Journal of Advanced Manufacturing Technology. 2002, 19(10):727-735.

[7.12] F. N. Lucena, H. K. E. Liesenberg. Transforming Statecharts into reactive system. In: XIX Conferencia Latinoamericana de Informatica Buenos Aires, AR, 1993, pp. 501-509.

[7.13] M. Felser, T. Sauter. Standardization of industrial ethernet : the next battlefield? In: 2004 IEEE International Workshop on Factory Communication Systems: IEEE, 2004, pp. 413- 420.

[7.14] D. Harel. Statecharts: A visual formalism for complex system. Science of Computer programming. 1987, 8(3).

[7.15] D. Harel, E. Gery. Executable object modeling with statecharts. Computer. 1997, 30(7):31-42.

[7.16] F. B. Schneider. The state machine approach: a tutorial. Technical Report 86-800. Cornel University. New York, 1986.

[7.17] A. Gill. Introduction to the theory of finite state machines.: McGraw-Hill, 1962.

[7.18] T. Hu, C. Zhang, R. Liu, et al. Function separated design method for real-time embedded system and its application. Computer Integrated Manufacturing Systems. 2008, 14(3):431-437,443

8

STEP-NC in Support of Machining Process Optimization

Liangji Xu

The Boeing Company
P.O. Box 3707, Seattle, Washington 98124-2207, USA
Email: liangji.xu@boeing.com

Abstract
Machining process optimization is the selection of machining parameters for a given process to achieve the maximum material removal rates within the process and machine limitations. Since the majority of those limitations directly relate to the cutting forces generated during the machining process, accurately calculating these cutting forces is not only critical to the optimization effort but also to the preservation of the equipment used in process. Calculation of the cutting forces requires knowing the cross-sectional geometry of each tool path over the course of the machining process. Although this geometrical information is available when three-dimensional (3D) modelling is applied in modern CAM systems, there has been no direct means to extract this information for use in the cutting force calculation and process optimization, until the recent work on ISO 10303 AP 238 (STEP-NC). This application protocol provides a new data model to transfer product data from CAM systems to computerized numerical controllers (CNC). It also contains the necessary data structure to implement the tool path geometry information into the process optimization. This protocol offers an unprecedented opportunity to control and manage the machining process based on explicit in-process information contained in the model that was previously unavailable. In this chapter, the fundamentals of cutting force calculation are explained, the tool path cross-sectional geometry in machining operations and its parameterization in ISO 10303 AP 238 are illustrated, the basic principles of force based optimization are described, along with a depiction of the optimization implementation plans. All of this demonstrates the vital role ISO 10303 AP 238 plays in machining process optimization.

8.1 Introduction

A machining process removes excessive material from a form of raw stock to generate a desired shape. The effectiveness of this process is generally evaluated by the volume of the material being removed in a given time, often denoted as the material removal rate. Machining process optimization is the selection of machining parameters for a given process to achieve the maximum material removal rates within the process and machine limitations.

Each machining process has an inherent set of limiting constraints that define the capability of a given machine system, which includes the major machine structural components, the spindle, tool holders, and cutting tools. The following is a list of constraints common to most machine systems:

- Machine structural stiffness
- Spindle torque, power and speed range
- Load capacity of the spindle bearings
- Axis drive capacity
- Bending moment and torque limitation of the tool holder
- The rigidity and wear-sustainability of the cutting tool
- Dimension error due to cutting tool deflection
- Dynamic characteristics of the cutting tool, holder and spindle

These constraints cover various aspects of the machine system, and they all have a direct relationship with the cutting forces generated during machining operations. Controlling these cutting forces within the machine system limits is key to the success of machining process optimization.

Although the forces can be measured with a dynamometer or similar measurement devices in a controlled environment, it is typically more desirable in industry to obtain the cutting force information during the planning stage of a machining process before the process is actually executed on a machine. Analytical modelling is the primary method used to calculate the cutting forces in machining processes.

The analytical study of cutting forces started in the mid twentieth century. Merchant [8.1–8.2] developed a cutting force model to calculate the force from the dimension of the uncut chip and the shear angle in chip formation. Further studies on the shear angle were done by Shaw et al. [8.3], Oxley [8.4], and Rowe and Spick [8.5] to improve the accuracy of the force calculation. Martellotti [8.6], Koenigsberger and Sabberwal [8.7], Kline et al. [8.8], and others [8.9–8.11] developed equations to calculate the forces based on cutting geometry in milling operations. While most force studies were concentrated in steady-state cutting condition, Merritt [8.12] developed the chatter loop concept to illustrate the dynamics in cutting processes, and Das and Tobias [8.13] explained the effects of the process dynamics to cutting forces. Andrew and Tobias [8.14], Tlusty et al. [8.15–8.18], and Altintas et al. [8.19–21] introduced the chatter theory in milling processes to study the system dynamics and its effects to cutting forces and process stability in milling operations. Based on the previous cutting force studies in 3-axis milling operations, Zhu et al. [8.22], Fussel et al. [8.23], and Ferry and Altintas [8.24–8.25] expanded the cutting force analysis into 5-axis milling processes.

Key to machining processes, cutting forces are used as a primary baseline to evaluate a machining process and optimize the machining parameters. Altintas and Spence [8.26–8.27] scheduled the feed rate in 2½-axis milling based on the predicted cutting forces. Yazar et al. [8.28], Fussel et al. [8.29], Tounsi and Elbestawi [8.30–8.31], and others [8.32–8.34] applied feed rate optimization to 3-axis machining of sculptured surfaces, where the arbitrary axial depth of cut significantly increases the complexity in modelling of the tool-stock engagement. The tool-stock engagement geometry becomes even more sophisticated in 5-axis machining with the introduction of angular motions of the cutter tool. Consequently, the difficulty of cutting force calculation and optimization also escalates. Ferry and Altintas [8.24–8.25] extended the optimization technology into 5-axis machining and applied it on machining of jet engine impellers. Most recently, Altintas and Merdol [8.35–8.37] developed the virtual milling concept and machining parameter optimization as an essential part of this concept.

In order to calculate the cutting forces and to optimize the cutting parameters, the geometry of the cutting edge engagement with the stock must be known. Solid modelling [8.26,8.27,8.38] and Z-buffer [8.22,8.28,8.29,8.32–8.34] are two primary modelling technologies used by researchers to investigate the tool-stock engagement geometry. Although successful research has been carried out in this area since the 1990s, the cutting force calculation and optimization have not been widely used in the manufacturing industry. A major obstacle is the difficulty of modelling the tool-stock engagement geometry during a production machining process so that the cutting forces in the process can be calculated.

Modern CAM systems use solid modelling or Z-buffer technologies to create 3D models of the stock, the cutting tools, and the product in various manufacturing stages. The geometrical information of the tool-stock engagement can often be derived from the established 3D models. However, it has been extremely difficult to extract this information from the CAM systems so that it can be used in the cutting force calculation and machining process optimization. This difficulty was resolved through the most recent work on ISO 10303 AP 238.

As introduced in Chapter 1, ISO 10303, the International Standard for the Exchange of Product Model Data (STEP), provides a neutral representation of product and process data so that it can be exchanged between computer systems through the entire product life cycle. Its application protocol AP 238, Application Interpreted Model for computerized numerical controllers, also referred to as STEP-NC, provides a new data model to transfer product data from CAM to computerized numerical controllers (CNC). The STEP-NC data model contains extensive machining process data such as part geometrical features, operation types, tool path trajectories, and the necessary data structure to implement the tool path geometry. STEP-NC provides a method to control and manage machining processes based on the explicit in-process information contained in its data model that was previously unavailable.

This chapter illustrates the basic principles of machining process optimization with an emphasis on the milling operation, and the vital role ISO 10303 AP 238 (STEP-NC) plays in such optimization. Section 8.2 explains the fundamentals of cutting force calculation and its requirement for tool path cross-section information. The variety of tool path cross-sections in machining operations and their parameterization in ISO 10303 AP 238 will be discussed in Sections 8.3 and 8.4. Section 8.5 describes the basic principles of force based feed optimization under multiple machine constraints. Various optimization methods such as tool life based optimization are reviewed in Section 8.6, along with a discussion on the influence of machine dynamics on optimization. Section 8.7 introduces optimization implementation plans in CAM, CNC, and independent optimization systems. Section 8.8 summarizes the discussions in STEP-NC based machining process optimization and illustrates its value to the manufacturing industry.

8.2 Cutting Force in Machining Processes

It is suggested by Merchant's study [8.1,8.2] that the cutting force can be calculated by the uncut cross-sectional chip area and a force coefficient:

$$F_c = K_{sc} A_0 \tag{8.1}$$

$$F_n = K_{sn} A_0 \tag{8.2}$$

where A_0 is the uncut chip cross-sectional area, F_c is the force in the cutting direction, F_n is the force perpendicular to the machined surface, K_{sc} and K_{sn} are cutting force coefficients for F_c and F_n. The cutting force coefficients capture the effects from the cutting edge geometry, material properties, cutting speed etc. These coefficients can be obtained either experimentally or through mathematical modelling, which is beyond the scope of this book.

The chip cross-sectional area A_0 in turning operations can be derived from the width of the cut w_c and the feed per revolution f_r, as shown in Figure 8.1a:

$$A_0 = w_c f_r \tag{8.3}$$

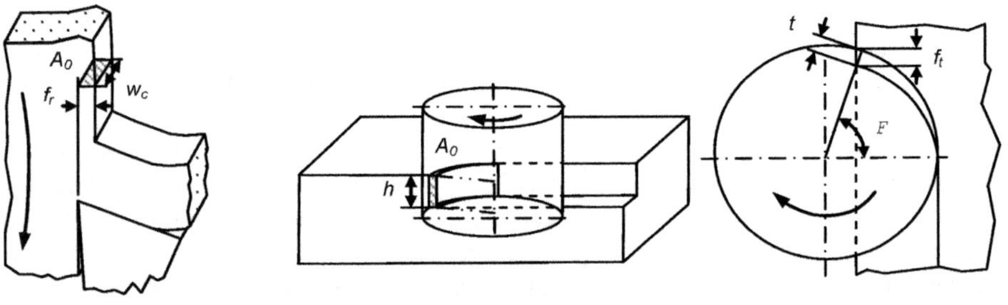

(a) (b)

Figure 8.1. Chip cross-sectional area in machining operations. a. Turning. b. Milling

In milling operations shown in Figure 8.1b, the chip thickness t is a function of the rotational angle ϕ of the milling cutter, which reaches 90° when the cutting edge is aligned with the feed direction:

$$t = f_t \sin\phi \tag{8.4}$$

where f_t is the feed per each flute of the milling cutter. It is a function of the feed f_d, spindle rational speed n_{sp} and the number of flutes of the cutter N_t:

$$f_t = \frac{f_d}{n_{sp} N_t} \tag{8.5}$$

Since the chip thickness t is a function of the rotational angle ϕ, the area A_0 in milling operations is also a function of ϕ as well as the axial depth of cut h. A_0 has the maximum value when ϕ is 90°:

$$A_0 = h f_t \sin\phi \tag{8.6}$$

The resulting force equations for a milling operation can be written as

$$F_{t\phi} = K_{st} h f_t \sin\phi \tag{8.7}$$

$$F_{r\phi} = K_{sr} h f_t \sin\phi \qquad (8.8)$$

where F_{tF} and F_{rF} are the tangential and radial force on an engaged cutter flute at cutter rotational angle ***F***; K_{st} and K_{sr} are the force coefficients for F_{tF} and F_{rF}.

It was found that the relation between the cutting forces and the chip thickness is non-linear, particularly when the chip thickness is thin (Figure 8.2a). In practice, piecewise linear equations are often used to simplify the calculation [8.32, 8.39] (Figure 8.2b, c).

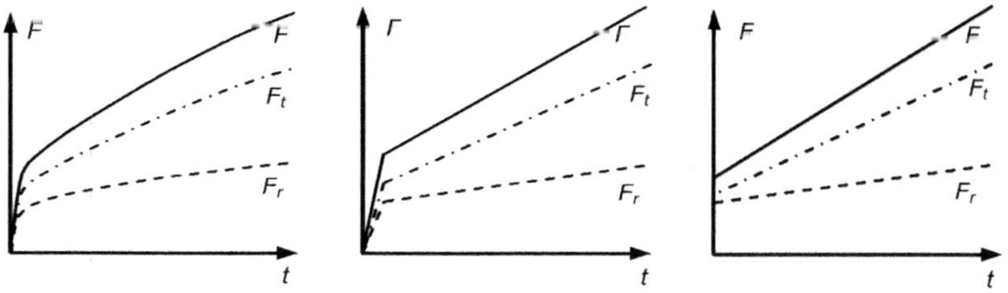

(a) (b) (c)

Figure 8.2. Cutting force and chip thickness (indicative). a. Non-linear relation between cutting force and chip thickness. b. and c. Simplified linear representations. F_t : tangential force; F_r : radial force; F: total force, vector sum of F_t and F_r

In order to calculate the cutting forces for a rotating cutter in milling operations, the instantaneous tool-stock engagement of each flute must be analyzed with the rotational angle. The majority of milling cutters have a nonzero helix angle. At a given rotational angle, the engagement condition of the cutting edge along the flute varies with the axial depth of cut (*AD*). To evaluate the cutting forces in such a circumstance, a milling cutter is divided into a number of thin discs along the axial depth of cut (Figure 8.3). A flute on a given disc starts its engagement into the stock at angle ***F***$_{st}$ and exits at angle ***F***$_{ex}$. The entry and exit angles of the engagement, ***F***$_{st}$ and ***F***$_{ex}$, are a function of the radial depth of cut (*RD*). The cutting forces generated by each disc at a given rotational angle can be calculated using Equations (8.7) and (8.8) if the angle falls between ***F***$_{st}$ and ***F***$_{ex}$. The total force generated by the cutting tool is the summation of the cutting force from each disc.

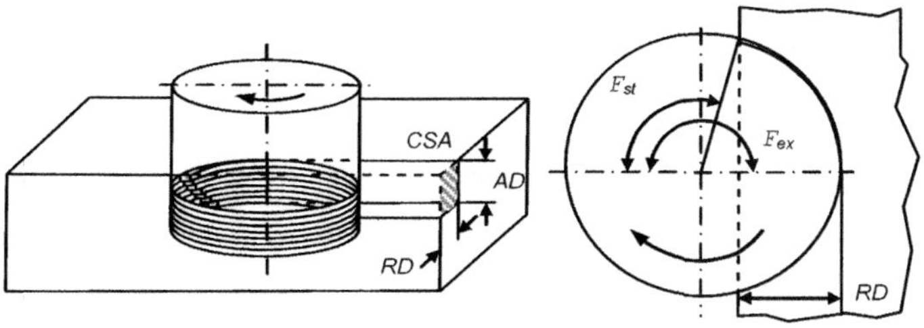

Figure 8.3. Cross-section, entry and exit angles in milling operations

Figure 8.3 illustrates the most basic milling operation case where the cross-sectional area *CSA* perpendicular to the feed direction is a rectangular area defined by *RD* and *AD*. Depending on the cutting tool geometry, the stock profile, and the position and orientation of the cutting tool, the cross-sectional area may not always be a convenient rectangular shape. Further discussions on the cross-section are held in Section 8.3.

In summary, the cutting forces in a machining operation are a function of the cutting parameters:

- Cross-section parameters (w_c in turning, *RD* and *AD* in milling)
- Operating parameters (spindle speed n_{sp}, feed f_r (turning) or f_d (milling))
- Geometrical parameters of the cutting tool
- Parameters of material cutting properties under given cutting conditions (cutting speed, cutting edge condition, etc.)

If the above parameters are known, the cutting forces can be calculated and force dependent measurements (power, bending moment, cutter deflection, etc.) can be derived and evaluated against the machine constraints. The operating parameters can then be adjusted to achieve an optimal material removal rate within the machine system limitations.

8.3 Tool Path Cross-section in Milling

A machining operation is usually comprised of multiple tool paths. A tool path is the cutter trajectory defined by the positioning parameters in a statement of a numerical controlled (NC) machine program. As discussed in Section 8.2, the cross-sections of tool paths is one of the most critical parameters in cutting force calculations and machining process optimization.

The path cross-section in a turning operation is simply the chip cross-sectional area A0 as illustrated in Figure 8.1. The path cross-section in a milling operation is defined as the area of the removed material along a tool path in the feed direction (Figure 8.3). An accurate cross-section of the tool path is needed to calculate the cutting force and optimize the machining process.

In production machining, the shape of the machined parts is often sophisticated. When the cutting tool moves along a pre-defined trajectory, the radial and axial depths often vary due to the positioning requirements of the tool and the variation of the stock shape. Consequently the cross-section continually varies along a tool path. Some simple examples are shown in Figure 8.4.

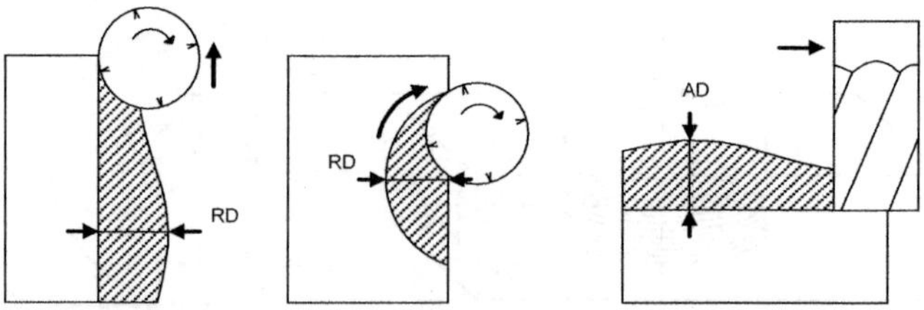

Figure 8.4. Examples of varying engagement in milling

Although it is possible to describe mathematically the relation of the cross-section to the cutting tool location and to optimize the operating parameters based on this relation, it is rarely done in practice due to the complexity of the required data structure. Instead, a constant cross-section, often the largest cross-section in the path, is normally used. Using the largest cross-section in each path for optimization purposes may not provide the most effective result for it places a machining operation towards a lower material removal rate so that the machine system limitations will not be exceeded at the largest cross-section. The accuracy and the effectiveness of this approach can be easily improved by reducing the tool path distance.

As discussed in Section 8.2, the chip thickness in a milling operation varies with the rotating angle of the cutting edge. Even with an identical radial depth value, having different entry and exit angles ($\boldsymbol{F}_{st}$ and $\boldsymbol{F}_{ex}$) will generate different cutting force profiles during the tool-stock engagement, and consequently impose different force impacts to the machine system. In order to calculate the cutting force correctly, both the radial depth and the tool-stock engagement angles $\boldsymbol{F}_{st}$ and $\boldsymbol{F}_{ex}$ have to be identified.

The following milling types are often applied in production: climb milling (down milling), conventional milling (up milling), slotting (channel cutting), and centre cutting, as shown in Figure 8.5 a–d, respectively.

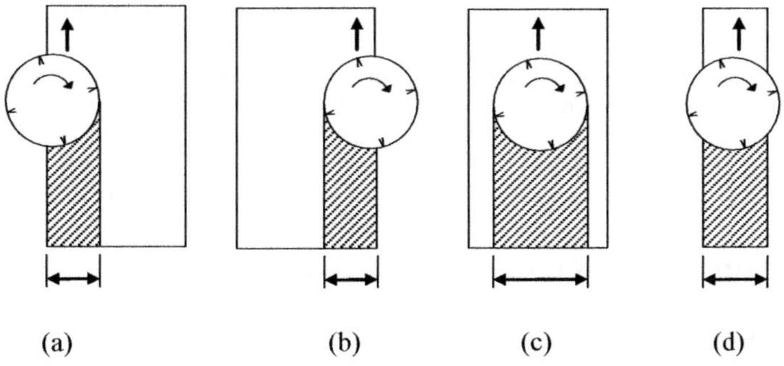

Figure 8.5. Milling types: a. Climb milling (down milling); b. Conventional milling (up milling); c. Slotting (channel cutting); d. Centre cutting

The chip thicknesses associated with these four milling types are shown in Figure 8.6. As discussed above, the chip thickness directly affects the cutting force, and consequently affects the impacts to the machine system.

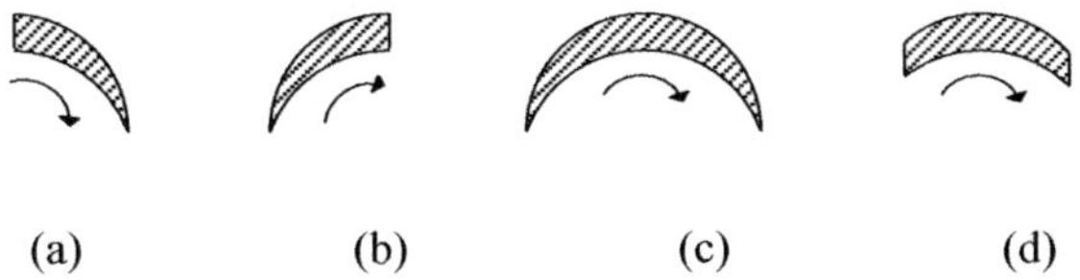

Figure 8.6. Chip thickness in different milling types: a. Climb milling (down milling); b. Conventional milling (up milling); c. Slotting (channel cutting); d. Centre cutting

In production machining, the axial engagement does not always start from the bottom of the cutter, and the radial engagement does not always create a side surface as shown in Figure 8.7a–b. The position of the cross-section relative to the cutting tool directly affects the cutting force and its impact to the machine system. For example, as the cross-section in Figure 8.7a moves towards to the tip of the cutting tool, the cutting force will create a larger bending moment against the machine system. Therefore, it is important to keep track of the cross-section location relative to the cutter origin. Also, the cross-sectional area in production milling operation is not always in rectangular shape due to various geometrical features of the stock and the milling cutter, as shown in Figure 8.7c. The variation of the location and shape of the cross-section adds a greater level of difficulty when parameterizing the cross-sectional area in a milling operation.

(a) (b) (c)

Figure 8.7. Examples of tool-stock engagements and cross-sections in milling operations: a. Nonzero-start axial engagement; b. Nonzero-start radial engagement; c. Nonrectangular cross-section

8.4 Parameterization of the Tool Path Cross-section

The need for accurate tool path cross-section information for cutting force calculations and machining process optimization has been established in Sections 8.2 and 8.3. The dimensions and the location of the cross-section can vary significantly in production machining. Methods to describe these arbitrary cross-sectional areas can be extremely sophisticated in some cases. Determining the number of parameters used to define these cross-sections requires a balance between the size of the STEP-NC data model and the accuracy in the representation of arbitary cross-sections. The number of parameters which describe the cross-section should be kept to a minimum to avoid overloading the STEP-NC data model. The following tool path cross-section parameters are defined in the latest corrigendum to ISO 10303 AP 238.

Dimension 0 shall describe the maximum axial depth of the tool contact cross-section, shown by *ADmax* in Figures 8.8 and 8.9. The maximum axial depth for milling shall be measured parallel to the tool axis, regardless of whether the direction of feed is perpendicular to the tool axis. The maximum axial depth for turning shall be measured parallel to the spindle axis:

- Dimension 1 shall describe the maximum radial depth of the tool contact cross-section, shown by *RDmax* in Figures 8.8 and 8.9. The maximum radial depth shall be measured perpendicular to both the tool axis and the feed direction
- Dimension 2 shall describe the location along the X axis where the maximum radial depth measure is located, shown by *Xmaxofs* in Figures 8.8 and 8.9
- Dimension 3 shall describe the location along the Y axis where the maximum axial depth measure is located, shown by *Ymaxofs* in Figures 8.8 and 8.9
- Dimension 4 shall describe the total area of the tool contact cross-section in the X−Y plane, shown by *CSA* in Figures 8.8 and 8.9
- Dimension 5 shall describe the location along the X axis of the centre of gravity (*CG*) of the tool contact cross-section, shown by *XCGofs* in Figures 8.8 and 8.9
- Dimension 6 shall describe the location along the Y axis of the centre of gravity of the tool contact cross-section, shown by *YCGofs* in Figures 8.8 and 8.9.

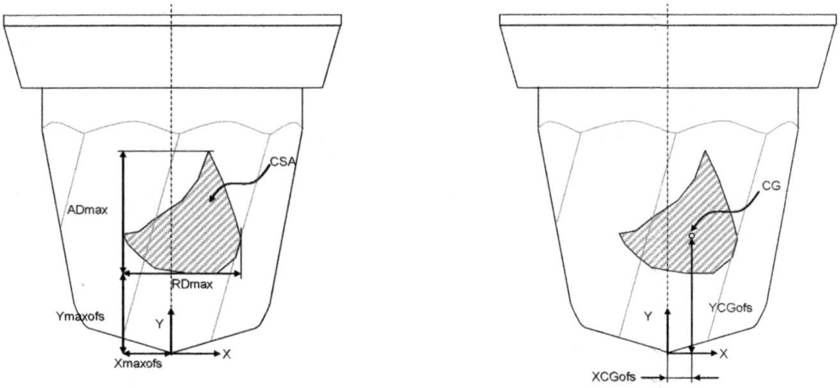

Figure 8.8. Cross-section parameters for milling

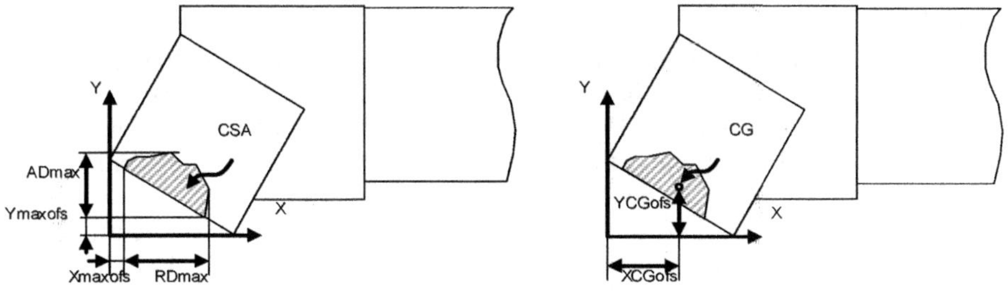

Figure 8.9. Cross-section parameters for turning

The maximum axial and radial depths *ADmax* and *RDmax* describe the dimensions of the tool-stock engagement in the axial and radial directions with their locations relative to the cutter coordinate origin defined by *Xmaxofs* and *Ymaxofs*. The cutter coordinate origin is the centre of the cutter tip for milling cutters and the point on the cutting edge profile closest to the spindle origin for turning cutters. The total area of the tool contact cross-section *CSA* quantifies the engagement area. The location of the engagement area was represented by distances of the area's centre of gravity *CG* to the cutter coordinate origin *XCGofs* and *YCGofs*. In production machining, the cross-section is often more complex than a simple rectangular shape. Hence the area of the cross-section and its centre of gravity are introduced in the parameterization to provide further information on the cross-section. Figure 8.10 shows some examples of the cross-section in milling operations.

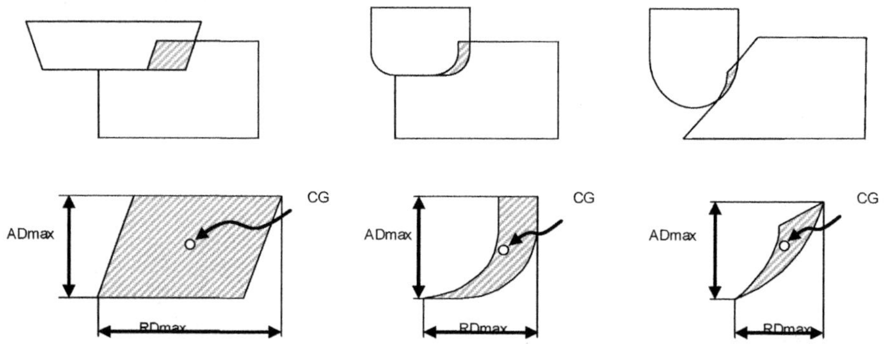

(a) (b) (c)

Figure 8.10. Examples of the cross-section with different milling cutters. a. Face mill; b. bull-nose end mill; c. ball-nose end mill

Plunge milling is a special milling operation where the feed direction is parallel to the milling cutter axial direction. The axial and radial depths described above do not apply to plunge milling. In this case, the cross-section can be parameterized as follows:

- Dimension 0 shall describe the maximum dimension of the tool contact cross-section in the radial direction in the polar coordinate system, shown by *Rmax* in Figure 8.11. The origin of the polar coordinate system is in the centre of the milling cutter. The X axis of the coordinate system is in the direction towards the next plunge operation location.
- Dimension 1 shall describe the maximum expansion angle of the tool contact cross-section in the polar coordinate system, shown by *Amax* in Figure 8.11.
- Dimension 2 shall describe the location in the radial direction where the maximum radial dimension of tool contact cross-section is located, shown by *Rmaxofs* in Figure 8.11.

- Dimension 3 shall describe the angular location from the X axis where the maximum expansion angle of the tool contact cross-section is located, shown by *Amaxofs* in Figure 8.11.
- Dimension 4 shall describe the total area of the tool contact cross-section in the X-Y plane, shown by *CSA* in Figure 8.11.
- Dimension 5 shall describe the location in the radial direction of the centre of gravity (*CG*) of the tool contact cross-section, shown by *RCGofs* in Figure 8.11.
- Dimension 6 shall describe the angular location from the X axis of the centre of gravity of the tool contact cross-section, shown by *ACGofs* in Figure 8.11.

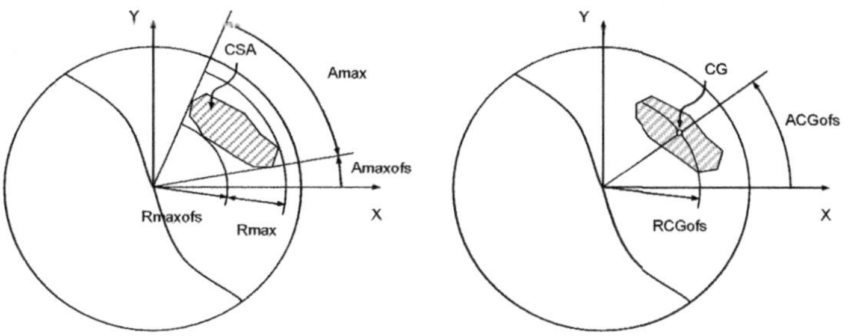

Figure 8.11. Cross-section parameters for plunge milling

These cross-sectional parameters in STEP-NC data model provide important information for quantification of tool-stock engagement, cutting condition studies, cutting force calculations and process optimization. In milling operations, the maximum radial depth *RDmax*, and the cross-sectional area *CSA*, can be used in tool wear studies and tool life-based optimization; the maximum axial depth *ADmax* and *CSA* can be used in calculating cutting force distribution along the cutter for process stability study; the location parameters *Xmaxofs*, *Xmaxofs*, *XCGofs* and *YCGofs* help determine the engagement position of the cutting force to the cutter and investigate the impact of the cutting force to the machine system. The same important roles are played by these parameters in turning operations as well.

8.5 Force-based Feed Optimization

As discussed in Section 8.2, the feed in machining processes directly relates to the chip thickness and consequently affects the cutting force. Adjusting the feed can effectively control the cutting force in order to achieve the highest productivity possible for a given machine system within its designed boundary. In force-based optimization, an increased feed will be suggested to achieve a higher material removal rate if the cutting force is lower than the machine system boundary.

Besides the cutting force limitation, there are also other restrictions within the machine system or the machining process itself which require the feed to be retained under a given limit. These restrictions include the speed limitation of machine axes, the constraints of the cutting tool, the requirements on the dimensional accuracy or surface roughness of the machined part, etc. Any feed change must be examined and restricted according to these limitations. The following sections discuss the derivation of the machining feed from the cutting force, the evaluation of multiple constraints in a machine system, and the commonly applied downward feed optimization method.

8.5.1 Feed Derivation

The majority of the force-dependent measurements, such as the bending moment and bearing load, have a linear relation with the cutting force. As discussed in Section 8.2, the cutting force is directly related to the chip thickness. The relation between the cutting force and the chip thickness shown in Figure 8.2 is redrawn in Figure 8.12 with the cutting force as the reference.

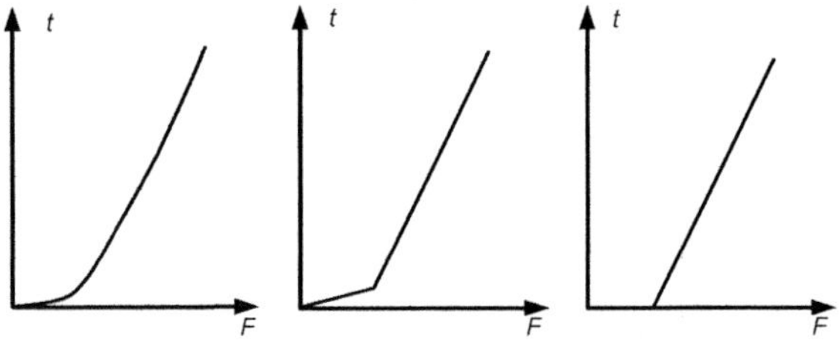

(a) (b) (c)

Figure 8.12. Relationship of cutting force and chip thickness (indicative). a. Exact form; b. and c. Simplified forms

With a known chip thickness, the feed can be easily derived. In turning operations, the feed is simply equal to the chip thickness. In a milling operation, the feed derivation is also simple:

$$f_d = \frac{f_t}{C_t} n_{sp} N_t \qquad (C_t \leq 1) \tag{8.9}$$

where n_{sp} is the spindle rotational speed, N_t is the number of flutes of the milling cutter, and C_t is the chip thinning factor, which is a function of the tool-stock engagement situation. The chip thinning effect will be discussed further in Section 8.6.3. To simplify the calculation, C_t can be set to 1, which places the derived feed to the conservative side.

The flowchart in Figure 8.13 shows the sequence of deriving the required feed from a known cutting force.

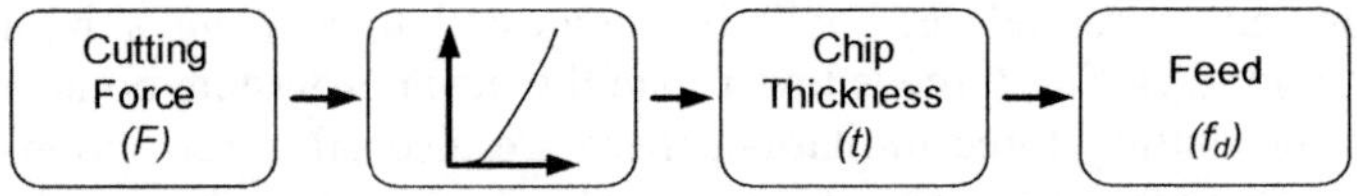

Figure 8.13. Derive the feed from a given cutting force

8.5.2 Multiple Machine System Constraints

As discussed in Section 8.2, there is a series of machine system constraints which limit the material removal rate of a machining process. Each of these constraints has to be evaluated during the optimization to ensure that they will not be violated. In order to evaluate these constraints with a common method, a load ratio, R_L, is used. This ratio is defined as the magnitude of a force-related measurement generated by the machining process over its corresponding system constraint. For example, the load ratio of the spindle power, R_{LP}, is defined as

$$R_{LP} = \frac{P_{act}}{P_{bound}} \tag{8.10}$$

where P_{act} is the power generated by the machining process, and P_{bound} is the machine spindle power boundary. Load ratios for other force-related measurements, spindle bearing force R_{LB}, bending moment R_{LH}, and cutter deflection R_{LD}, are defined similarly. The maximum load ratio among them, $R_{L\,max}$, is defined as

$$R_{L\max} = \max\,(R_{LP}, R_{LB}, R_{LH}, R_{LD}, \ldots) \tag{8.11}$$

In a machine system with multiple constraints, the maximum load ratio is used as a determining parameter in the calculation of the optimal feed for a machining process.

8.5.3 Downward Feed Optimization

Downward feed optimization is a common practice in force-based optimization. It sets the initial feed to the highest level within the restrictions and reduces the feed during the process whenever the initial feed causes a force-dependent measurement exceeding its machine constraint.

In downward feed optimization, the maximum load ratio $R_{L\,max}$ is calculated for each tool path. If all of the force-dependent measurements are within their respective limits, $R_{L\,max}$ will be less than 1 and the feed will remain unchanged. If one or more force-dependent measurements exceed their limits, $R_{L\,max}$ will be larger than 1. Then the highest allowable cutting force in this situation can be obtained from Equation (8.12) considering the linear relation between the cutting force and force related machine constraints:

$$F_{ca} = \frac{F_{cp}}{R_{L\max}} \tag{8.12}$$

where F_{cp} is the cutting force generated by the current programmed feed and F_{ca} is the highest allowable cutting force. With F_{ca} derived, the highest allowable feed can be calculated with the method described above in Section 8.5.2.

Figure 8.14 summarizes the process of downward optimization.

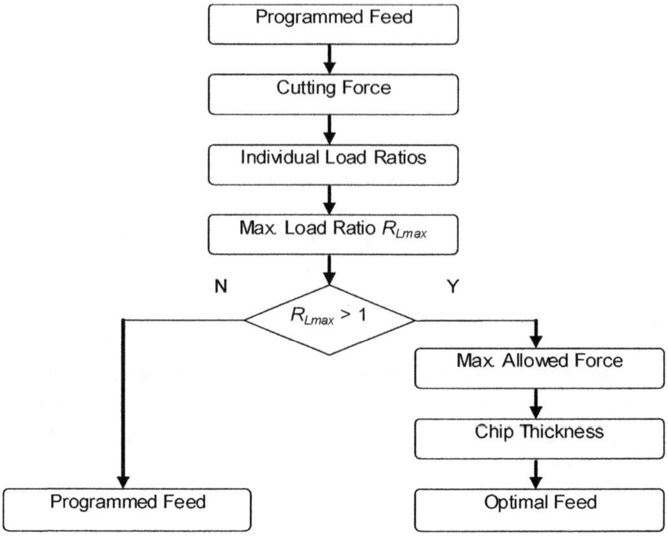

Figure 8.14. Derive the optimal feed from the cutting force

After the optimal feed is calculated, it will be sent to the STEP-NC data model to be assigned to the tool path. Instead of replacing the original programmed feed in the data model, feed override is often used in STEP-NC to form a new feed by multiplying the original feed with the feed override value. As the original programmed feed is known, the feed override value is simply the ratio of the optimal feed to the original programmed feed.

8.6 Other Optimization Methods

Besides the force-based optimization, other optimization methods are also used in machining processes for certain applications or to meet specific requirements. The following are a few such examples.

8.6.1 Tool Life-based Optimization

Cutting tool life plays an important role in machining processes planning and production cost reductions. Tool life is especially critical in the machining of heat-resistant materials, such as titanium and stainless steel, where the excessive temperature drastically expedites tool wear. It is often desirable to manage the cutting tool life to a specified time duration so that an optimal material removal rate can be achieved to minimize cutting tool costs and tool change time. However, due to the lack of the cross-section information in the machining process, it is extremely difficult for machine users to optimize the tool life. The STEP-NC data model closes this gap by providing the essential cross-section parameters for tool life optimization.

The cutting tool life can be expressed with expanded Taylor equation [8.40]:

$$TV_c^{\frac{1}{n}} t^{\frac{1}{m}} b^{\frac{1}{l}} = C \tag{8.13}$$

where T is the cutting tool life, V_c is the cutting velocity, t is the uncut chip thickness, b is the chip width, and l, m, n, and C are constants. This equation can be directly applied to turning operations. In milling operations, the uncut chip thickness is a function of the cutting tool rotational angle $\boldsymbol{F}$, rotational speed n_{sp} and the feed f_d. After neglecting the minor effect of the chip width b, the tool life equation for milling operations can be written as

$$T = C' n_{sp}^{\frac{1}{m}-\frac{1}{n}} f_d^{-\frac{1}{m}} \int_{\phi_{st}}^{\phi_{ex}} (\sin\phi)^{-\frac{1}{m}} d\phi \tag{8.14}$$

where C' is a constant representing the cutter and stock material properties. The cutter geometrical information such as the diameter and the number of flutes is also combined into C'. As discussed in Section 8.2, the entry and exit angles $\boldsymbol{F}_{st}$ and $\boldsymbol{F}_{ex}$ are a function of the radial depth of cut. With the cross-section parameters in the STEP-NC data model, $\boldsymbol{F}_{st}$ and $\boldsymbol{F}_{ex}$ can be easily derived.

Instead of using Equation (8.14), the relation between the tool life and the feed, speed, and the radial depth of cut is also often obtained through experiments in practice. Figure 8.15 shows the cutting parameters to maintain a selected cutting tool life of a cobalt milling cutter in titanium machining. With the spindle speed predetermined based on the cutting velocity, the optimal feed is a function of the radial depth of cut, shown as radial immersion (ratio of the radial depth of cut to the cutter diameter) in Figure 8.15.

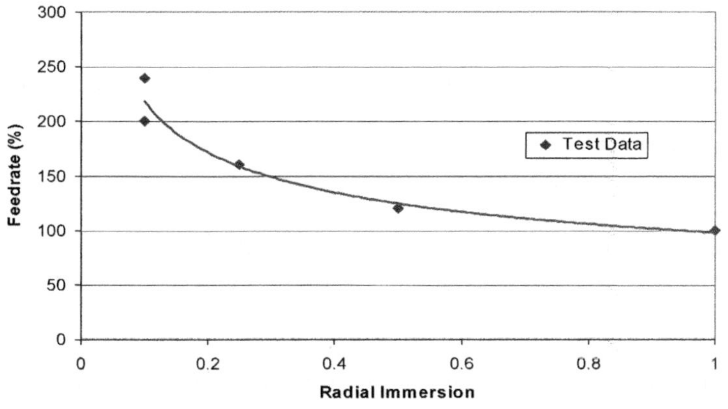

Figure 8.15. Cutting parameters of a selected tool life of a cobalt milling cutter used in titanium machining

After the relation between the feed and the radial depth of cut has been established, the mathematical expression of this relation can be brought into an optimization algorithm. The feed can then be adjusted according to the radial depth in each tool path in order to maintain a selected tool life. To ensure the optimized process is within the machine system limits, the adjusted feed and the subsequent cutting force need to be checked against the machine system constraints before the adjusted feed is issued to the machining process.

Optimizing the machining process to reduce the cutting tool related production costs has been a challenging task in the industry. With the cross-sectional parameters in the STEP-NC data model, the tool-stock engagement can be closely monitored and the corresponding feed adjustment can be made to manage effectively the tool life and reduce the production costs.

8.6.2 Volume-based Optimization

The goal of the volume-based optimization is to maintain a specific material removal rate over the course of a machining process.

Based on the energy equilibrium principle, a certain amount of energy, J_c, will be consumed when a given volume of solid metallic material, M_c, is converted to chips during a machining process:

$$J_c = K_m M_c \tag{8.15}$$

where K_m is a material property related factor. If divided by time, this equation can be rewritten as

$$P_c = K_m R_c \tag{8.16}$$

where R_c is the volume of solid metallic material being removed at a unit time, or material removal rate, and P_c is the power being consumed to achieve the given material removal rate.

In turning operations, R_c is a function of the cross-sectional area A_{cs} and the diameter of the stock D_s:

$$R_c = \pi A_{cs} D_s \tag{8.17}$$

The cross-sectional area in turning is the area of the uncut chip cross-section, which is a product of the feed per revolution f_r and the width of the cut w_c as described in Equation (8.3). Adjusting the feed in a turning operation can change the cross-sectional area and the material removal rate:

$$R_c = \pi w_c f_r D_s \tag{8.18}$$

Similarly, the material removal rate R_c is the product of the cross-sectional area A_{cs} and the feed f_d in milling operations. Adjusting the feed can result in a desired material removal rate:

$$R_c = A_{cs} f_d \tag{8.19}$$

Again, the key of volume-based optimization is to obtain the cross-section information of the tool paths. The cross-section parameters in the STEP-NC data model make the volume-based optimization easily achievable.

Equation (8.16) describes the relation between the material removal rate R_c and the cutting power P_c through a material factor K_m. However, using this equation to regulate the cutting power with the material removal rate often produces inaccurate results. One of the primary reasons for the inaccuracy is that the geometry and the condition of the cutting edge strongly affect the cutting force and consequently the cutting power. The material factor K_m in Equation (8.16) does not include the effects of the cutting edge geometry and its condition. With the same material removal rate from a given stock, the power may vary significantly as cutting edge condition changes over the course of a machining process. In addition, the relationship between the material removal rate and the cutting power is also a function of the chip thickness as discussed in Section 8.2. This is particularly important in milling operations where the chip thickness continuously varies with the rotation of the milling cutter.

Another cause for concern when using this approach is that the cutting power P_c obtained from Equation (8.16) is the average power. Although it is possible to correlate the power to the cutting force, the force derived from the average power will be the average force. The average force is useful in turning operations where the chip thickness is relatively stable, but it has limited applications in milling operations where the cutting force varies significantly with the radial depth of cut. As illustrated in Figure 8.16, the average forces in two different cutting scenarios are same, but the actual cutting forces are drastically different. As a result, although the average force may still remain within the machine system limits, the peak of the actual force may exceed the limits causing potential damage to the machine system. Therefore, the volume-based optimization has limited capability to protect the machine system from potential force-related damage.

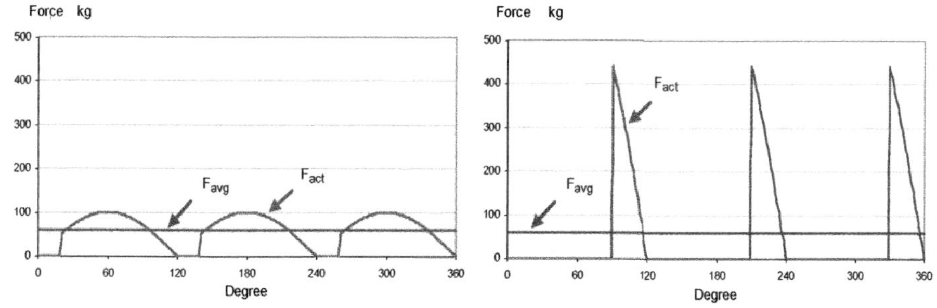

(a) (b)

Figure 8.16. Cutting force of a 3-flute milling cutter in one revolution. a. radial immersion 0.93; b. radial immersion 0.15. F_{avg}: average force; F_{act}: actual force

8.6.3 Constant-chip Optimization

The objective of the constant-chip optimization is to compensate the chip thinning effect by increasing the feed when the chip thickness decreases as the tool-stock engagement declines in either the radial or the axial direction.

As discussed in Section 8.2, the chip thickness in a milling operation varies with the cutter rotational angle within the tool-stock engagement defined by the entry and exit angles of the cutting edge, $\boldsymbol{F}_{st}$ and $\boldsymbol{F}_{ex}$, as described in Equation (8.20). Chip thickness is at its maximum value when the cutting edge is aligned with the feed direction, where $\boldsymbol{F} = 90°$ and the chip thickness t equals the feed per flute of the milling cutter f_t:

$$t = f_t \sin\phi \qquad (\phi_{st} \le \phi \le \phi_{ex}) \tag{8.20}$$

The programmed chip thickness t_p is defined as the feed per flute from the programmed feed, f_{tp}:

$$t_p = f_{tp} = \frac{f_d}{n_{sp} N_t} \tag{8.21}$$

When the tool-stock engagement includes the $\boldsymbol{F} = 90°$ case, the maximum thickness of the uncut chip t_m is equal to t_p. Otherwise, t_m is a function of $\boldsymbol{F}_{st}$ or $\boldsymbol{F}_{ex}$, whichever derives a larger chip thickness from Equation (8.20). In this case, t_m is always smaller than t_p. Such a phenomenon is often called "chip thinning" in the industry. The ratio of t_m over t_p is called the chip thinning factor C_t.

Constant-chip optimization is to increase the feed when the tool-stock engagement does not include the $\boldsymbol{F} = 90°$ case, so that the maximum chip thickness t_m remains equal to the programmed chip thickness t_p. Figure 8.17 illustrates this chip thinning compensation. Due to the small radial depth of cut, the maximum chip thickness t_m in Figure 8.17a is smaller than t_p from the original programmed feed f_d. To compensate the chip thinning, the feed is increased to f_d' and the feed per flute is increased to t_p' as shown in Figure 8.17b. As a result, the chip thickness t_m' remains equal to the programmed chip thickness t_p.

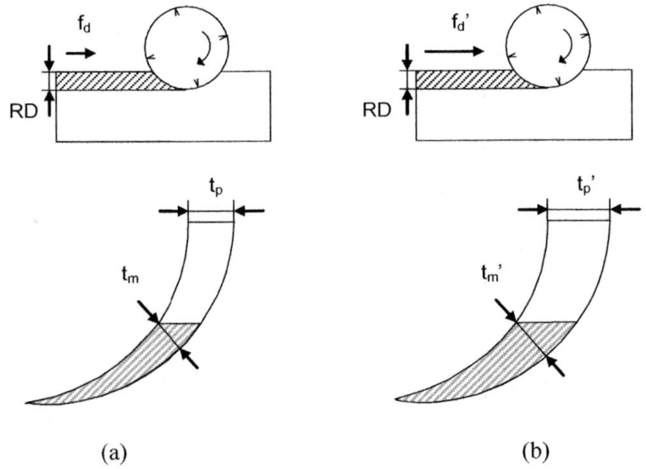

Figure 8.17. Compensate radial chip thinning by constant-chip optimization. a. original programmed feed; b. optimized feed

In addition to compensating for chip thinning in the radial direction, the constant-chip method can also compensate for chip thinning effects in the axial direction when the axial depth of cut is smaller than the cutter corner radius as illustrated in Figure 8.18.

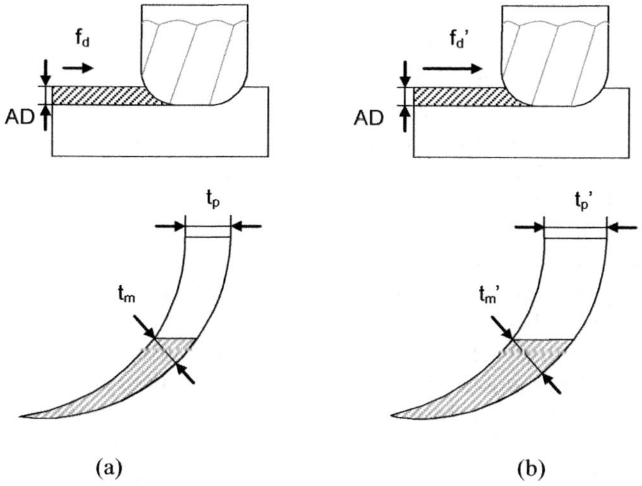

Figure 8.18. Compensate axial chip thinning by constant-chip optimization. a. original programmed feed; b. optimized feed

Similar to other optimization methods discussed above, the key for constant-chip compensation is to obtain the cross-section information of the tool-stock engagement. With the cross-section parameters in the STEP-NC data model, constant-chip optimization for both radial and axial chip thinning compensation is easily achievable.

8.6.4 Machine System Dynamics

In order to prevent excessive vibration occurring in machining operations, it is important to maintain a machining process in a stable condition prior to optimizing the process.

Self-generated vibration in a machining process, or “chatter”, can cause undesirable machined surfaces and potential damage to cutting tools and the machine spindle. Chatter often occurs at the cutting tools with a high L/D ratio (the ratio of the cutter length to its diameter or its cross-section dimension). Extensive studies on chatter and its prevention have been conducted by Tlusty, Altintas and others [8.13–8.21, 8.41,8.42]. Identifying the chatter frequencies through structural dynamic analyses of cutter-holder-spindle and adjusting the spindle speed accordingly is a common method in mitigating the chatter. Other approaches, such as reducing the axial depth of cut, using different cutter configurations or specially designed cutters are also useful. However, changing feed only has minimal impact on chatter due to its less dependence to the chip thickness [8.14, 8.41,8.42].

Besides the chatter occurring at the cutter-holder-spindle, excessive vibration may also take place if the stock, fixture, or machine components (spindle ram, column, table, etc.) are insufficiently supported. Therefore, system structural dynamic analyses should be carried out prior to the machining feed optimization and the structural compliance of the stock, fixture, or the machine components must be considered during the optimization.

8.6.5 Feed Lag

Each machine tool has an acceleration/deceleration limit based on its axis drive capability. Certain distance is needed for an axis to reach a specified feed. If the length of a tool path's trajectory is shorter than the required distance for acceleration/deceleration, the specified feed for this tool path may not be reached.

Modern CNC controllers have the capability to "look ahead" at the motion commands and adjust the actual feed to maintain the required positional accuracy during the transitions of the motion direction.

Due to the inertia of the machine structure, fixture devices and the stock, the actual feed may not follow the commanded feed closely during the transition from one feed level to another.

These factors contribute to the fact that the actual machine feed may "lag" from the feed specified in the NC program. Feed optimization may alter the feed at each tool path depending on the tool path geometry. The programmed feed may vary frequently through a machining process. However, due to the "lagging" of the actual feed, the accuracy and the effectiveness of the feed optimization may be compromised when the tool path distances are too small, as shown in Figure 8.19.

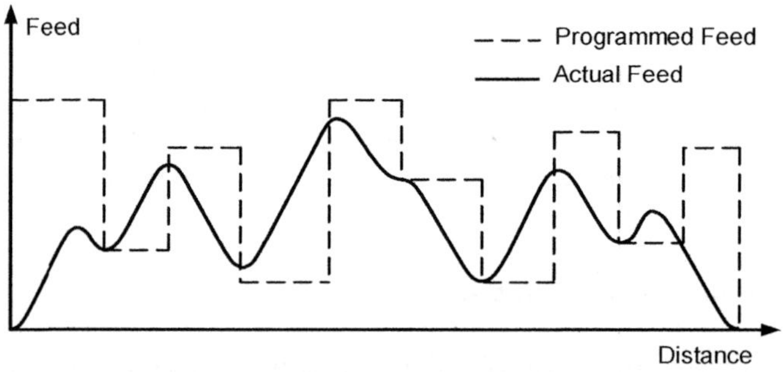

Figure 8.19. Feed lag (indicative)

Frequent feed change requires repeated acceleration/deceleration of machine axes. It raises the requirement to the machine axis drive system. To ease such a requirement, a low pass filter can be applied to the optimized feed to reduce the feed variation.

8.7 Optimization Implementation Plans

The importance of the geometrical information of the cross-section along the tool path to machining process optimization has been discussed above. In order to apply optimization to a manufacturing process, a suitable implementation plan needs to be developed to meet the manufacturing requirements. This section introduces potential implementation plans at various stages of manufacturing processes.

The STEP-NC data model provides a new data structure which permits manufacturing information to be maintained throughout a product's entire manufacturing process. The geometrical information of the tool path cross-section can be parameterized in a CAM system from the 3D modelling of the stock and the cutting tool, and exported to the STEP-NC data model along with other product information. With the cross-section information contained in the data model, the machining optimization can be carried out at the CAM system, on the CNC platform or as an independent process between the CAM and CNC.

8.7.1 Implementation at CAM

A Workplan can be optimized as soon as it is developed in a CAM system. The optimization can be integrated into a CAM system to become a standard procedure of the Workplan development (Figure 8.20).

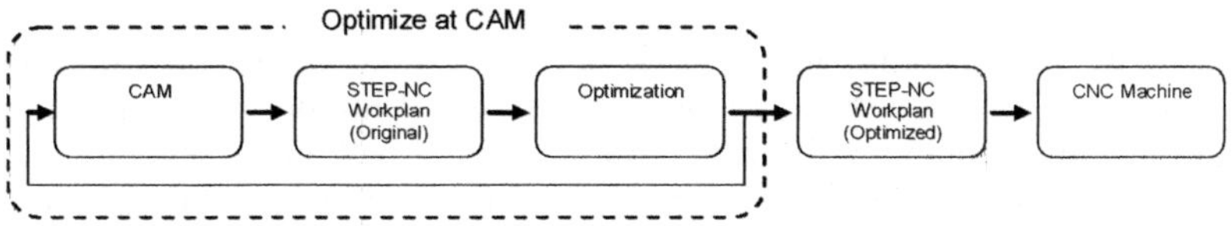

Figure 8.20. Machining process optimization at CAM

The primary advantage of optimizing at the CAM system level is the instant feedback of the optimization results for Workplan development. Based on the feedback from the optimization, improvement of the Workplan can be made to achieve further productivity enhancement. As an example, if the cross-sectional area in the tool paths of a machine operation in a Workplan is small, the optimization may increase the feed. However, the feed may reach one of its restrictions such as the machine axis speed limit, while the cutting force is still far below the force-related machine constraints. As a result, the machine's capability is not fully utilized. In this case, modifying the machine operation by increasing the cross-sectional area in each tool path and reducing the number of paths would be an effective approach to achieve a higher material removal rate and reduced machining time. The modified Workplan can go through the optimization again to obtain further improvements. This process can be iterated multiple times to achieve the maximum manufacturing efficiency. Moreover, the feedback of the optimization can be extended further upstream to the product design so that the product can be designed in a more "manufacturing friendly" and cost-effective manner.

The disadvantage of optimizing at the CAM level is that the detailed machine and cutting tool information may not be available at the development stage of the Workplan. Insufficient machine system information will affect the accuracy and the effectiveness of the optimization.

8.7.2 Implementation on CNC

With the fast evolution of computer technology, computational power of CNCs has been increased rapidly in recent years. Multiple CPUs and fast CPU-CPU communication are also common in modern CNCs. Today's CNCs are capable of handling additional tasks besides its core function of interpreting NC program commands and controlling machine axis motions. CNC is in the process of changing its role from a sole executing device to a system with "intelligence". With the implementation of STEP-NC into the CNC industry, optimization can be carried out in the CNC before the Workplan is executed (Figure 8.21).

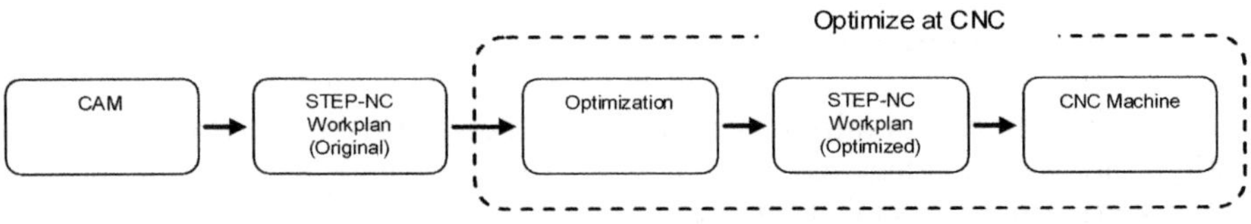

Figure 8.21. Machining process optimization at CNC

The advantage of optimizing a Workplan in the CNC is that the information of the machine system is readily available in the CNC. With the only missing piece of information – the geometrical information of the tool path cross-section – provided by STEP-NC, the optimization of a machining process can be achieved with minimum additional data inputs. Another advantage of optimizing a Workplan in the CNC is that changes made in the machine system during production will be registered at the CNC and instantly available for optimization. The machining process will always be optimized with the most recent machine system information.

Since this approach imbeds the optimization into the CNC platform, the optimized Workplan may only apply to the individual machines where the optimization enabled CNC resides. Machines without such capabilities may not be able to optimize their processes unless they are upgraded with optimization enabled CNCs.

8.7.3 Implementation with an Independent System

Machining process optimization can also be carried out by an independent system which imports a STEP-NC Workplan, optimizes it with machine system information, and exports the optimized STEP-NC Workplan to CNC machine tools (Figure 8.22).

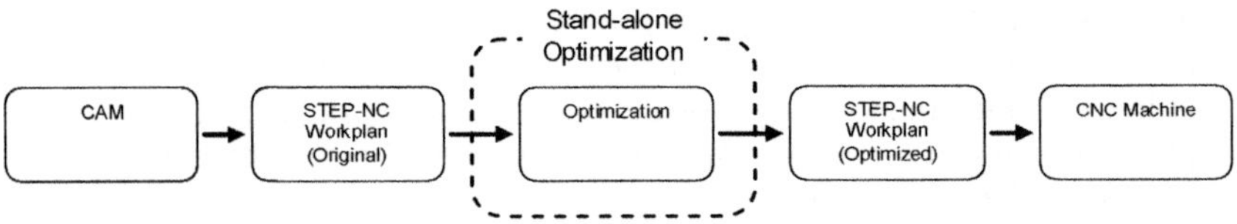

Figure 8.22. Machining process optimization with an independent system

This approach has the highest flexibility. With the geometrical information of the tool path cross-section contained in the STEP-NC Workplan, the optimization can be conducted at any preferred stage or location through the Workplan distribution from the CAM to the CNC. The optimization can be applied to a group of identical machines with same setup, or to each individual machine to compensate for the machine system differences. This approach is particularly useful in today's manufacturing industry where a product is often manufactured in multiple facilities on multiple machines around the world. The machining equipment, cutting tools, and machining process requirements may vary significantly between facilities. The optimization system needs to be highly capable and flexible so that it can meet the requirements of each individual facility.

The independent optimization system also has the highest portability. It can run on any computer platform and operating system, which greatly increases its applicability to the manufacturing facilities. This type of optimization system also has high interoperability. It can optimize Workplans generated by any CAM system and feed to any CNC platforms.

Unlike the optimization in CAM or CNC, this approach adds a new process in the Workplan distribution between the CAM and CNC. It also requires data entry of the machine system information, which can be transparent in the optimization at CAM or CNC.

While both the STEP-NC based CAM optimization software and STEP-NC optimization enabled CNC are not commercially available, independent STEP-NC based optimization using converters and adapters is currently being utilized.

8.7.4 Example of Optimization with an Independent System

A machining test was conducted to demonstrate the STEP-NC based machining process optimization with an independent system. A test part with typical airframe geometric features was designed and planned using a commercial CAM software package. The Workplan was output as a STEP-NC Workplan through STEP-NC plug-in software to the CAM system. Since existing CAM software is not capable of directly exporting the geometrical information of cross-sectional area along the tool path, special software was developed to generate the needed cross-sectional information and implement it into the STEP-NC Workplan.

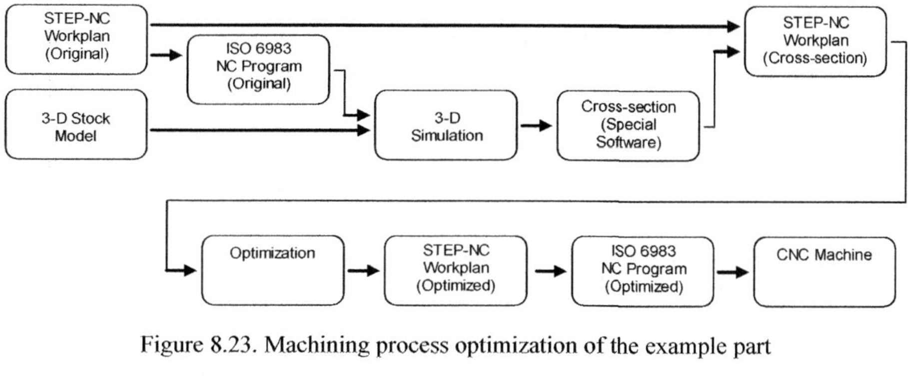

Figure 8.23. Machining process optimization of the example part

As shown in Figure 8.23, the original STEP-NC Workplan was converted to a conventional ISO 6983 NC program. The NC program and the 3D stock model were imported into a 3D simulation software package to simulate the tool-stock engagement. The special software extracts the geometrical information of the tool path cross-section from the 3D simulation software and associates it with the tool paths in the STEP-NC Workplan. The enhanced STEP-NC Workplan was sent to the optimization software to optimize the feed of the machining operations based on the selected CNC machine's capability. The optimized STEP-NC Workplan was again converted to an ISO 6983 NC program so that it could be executed on a regular CNC machine which was not STEP-NC compatible. The cross-section enhanced STEP-NC Workplan was optimized for two different machine tools. One is a horizontal and the other is a vertical machining centre. The machines are made by different manufacturers with different CNCs and located in different continents. Figure 8.24 shows the CAM model and the finished part produced by one of the machines with the optimized machining process.

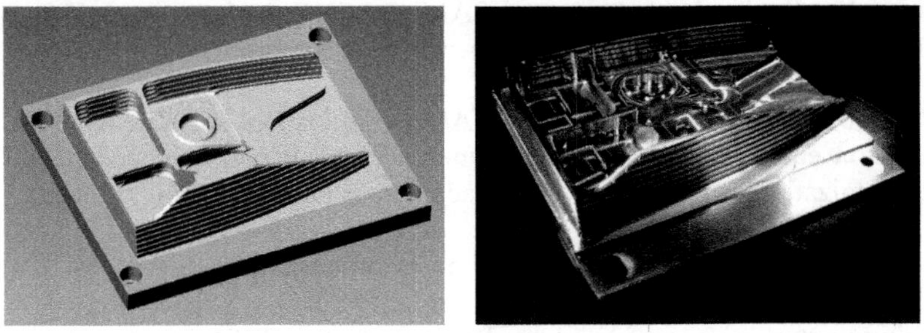

(a) (b)

Figure 8.24. Test part of STEP-NC optimization. a. CAM model; b. Finished part

8.8 Conclusions

Globalization intensifies competitiveness and stimulates reform of the manufacturing industry. Optimization of machining processes is an effective way to improve productivity and reduce the production costs.

Machining process optimization is the selection of machining parameters to achieve the maximum material removal rates within the process and machine limitations. Most machine limitations are directly related to the cutting forces generated during machining operations. In order to optimize a machining process, the cutting forces must be accurately calculated and controlled. An essential requirement of the cutting force calculation is the geometrical information of the tool paths over the course of the machining process. 3D modelling is a primary tool to obtain the tool path geometry. Due to the complexity of constructing 3D models, accurately and effectively modelling the tool-stock engagement for cutting force analysis is still being researched. 3D modelling is also a fundamental tool in modern CAM systems which have been widely used in the industry. However, there has been no direct means to extract the tool path geometry from the CAM systems for use in the cutting force calculation and process optimization. This situation is changed thanks to the recent work on ISO 10303 AP 238 (STEP-NC).

The STEP-NC data model contains extensive machining process data and offers an unprecedented opportunity to control and manage machining processes based on the explicit product information contained in the model. The STEP-NC model has the necessary data structure to extract tool path geometry information from a CAM system and bring this information into machining process optimization.

In today's industry-wide globalization, design and manufacturing are often carried out in different geological locations, sometimes in different continents. The manufacturing facilities and assigned machines may change frequently. With the tool path geometry available in the STEP-NC data model, the machining parameters of an NC program can be easily optimized according to each individual machine's constraints, so that the machine's capability can be fully utilized. Hence the maximum productivity and efficiency can be achieved.

Most existing optimization efforts are at the CAM level where the 3D models of the cutting tools and the stock are accessible. The STEP-NC data model with tool path geometry opens a new arena to optimize the machining process at the CNC level. So far, a CNC is primarily an execution device which interprets the motion commands in an NC program and controls the machine axes to accomplish programmed motions. With CNC level optimization, a CNC will be able to select intelligently the feed of a machine motion based on the predicted cutting force derived from the tool path geometry of that motion. Optimization at the CNC level has distinct advantages. One is the smooth transition of the machine feeds. Modern CNCs have the capability to "look ahead" at the future motion commands in an NC program and manage a smooth transition of the motion direction by manipulating the feed and the acceleration/deceleration of the machine axes. With CNC level feed optimization, the machine feed can smoothly transit from one desired magnitude to another depending on the predicted cutting force in each motion path. Another advantage is that the CNC contains all the machine setup parameters, and therefore the machining process optimization will be based on the latest machine condition. A further advantage is that the majority of the machine tool information and cutting tool dimensions are already stored in the CNC. A minimum number of data entries will be needed for the machining process optimization.

The STEP-NC data model also makes it possible to integrate the force based optimization algorithm into CAM systems. Existing optimization mainly controls the feed and the speed while maintaining axial and radial depths unchanged. Due to machine or cutting tool limitations, changing feed and speed alone may not achieve an optimal material removal rate in some situations. In these cases, the optimization results can be fed back to the CAM systems so that the axial and radial depths can also be adjusted. This significantly increases the scale of optimization so that machining processes can be fully optimized.

In addition to machining process optimization, the tool path geometry in the STEP-NC model also provides valuable information for process planning and machining equipment evaluation. With the tool path geometry over the course of a planned machining task available, the steady-state cutting force can be predicted and consequently the machining power and torque can be derived. With this information, the machine capability requirements can be estimated and potential candidate machines for the planned machining task can be identified. Combined with the material removal rate, machining time and the required production rate, the number of machines and the requirements of the manufacturing facility can also be outlined. As a result, the accuracy of the production planning can be greatly improved.

In conclusion, with the introduction of STEP and STEP-NC into the manufacturing industry, machining process optimization can be implemented into manufacturing facilities across the industry. With the subsequent improved productivity and reduced production costs, the efficiency of the manufacturing industry can be greatly improved.

References

[8.1] Merchant, M. E. 1944. Basic Mechanics of the Metal Cutting Process, *Journal of Applied Mechanics*, Vol. 11, pp. 168-175.

[8.2] Merchant, M. E. 1945. Mechanics of the Metal Cutting Process, I. Orthogonal Cutting and Type 2 Chip, *Journal of Applied Physics*, Vol. 16, pp. 267-275.

[8.3] Shaw, M. C.; Cook, N. H.; Finnie, I. 1953. The Shear-Angle Relationship in Metal Cutting, *Transaction of ASME*, Vol. 75, pp 273-288.

[8.4] Oxley, P. 1962. Shear Angle Solution in Orthogonal Machining, *International Journal of Machine Tool Design and Research,* Vol. 2, pp 219-229.

[8.5] Rowe, G. W. and Spick, P. T. 1967. A New Approach to Determination of the Shear-Plane Angle in Machining, *Journal of Engineering for Industry*; August, pp 530-538.

[8.6] Martellotti, M. E. 1941. An Analysis of the Milling Process, *Transactions of the ASME*, Vol. 63, pp. 677-700.

[8.7] Koenigsberger F. and Sabberwal A. J. P. 1961. An Investigation into the Cutting Force Pulsations during Milling Operations, *International Journal of Machine Tool Design and Research,* Vol. 1, pp 15-33.

[8.8] Kline, W. A., DeVor, R. E. and Lindberg, J. R. 1982. The Prediction of Cutting Forces in End Milling with Application to Cornering Cuts, Dynamics of High Speed Milling, *International Journal of Machine Tool Design and Research,* Vol. 22, pp 7-22.

[8.9] Bayoumi, A. E., Yucesan, G. and Kendall, L. A. 1994. An Analytic Mechanistic Cutting Force Model for Milling Operations: A Theory and Methodology, *Journal of Engineering for Industry,* Vol. 116, pp 324-330.

[8.10] Feng, H. Y. and Menq, C. H. 1994. The Prediction of Cutting Forces in the Ball End Milling Process, I. Model Formulation and Model Building Procedure, *International Journal of Machine Tools and Manufacture*, Vol. 34, pp 697-710.

[8.11] Feng, H. Y. and Menq, C. H. 1994. The Prediction of Cutting Forces in the Ball End Milling Process, II. Cut Geometry Analysis and Model Verification, *International Journal of Machine Tools and Manufacture*, Vol. 34, pp 711-719.

[8.12] Merritt, H. E. 1965. Theory of Self-Excited Machine-Tool Chatter, *Journal of Engineering for Industry*, November, pp 447-454.

[8.13] Das, M. K. and Tobias, S. A. 1967. The Relation between the Static and the Dynamic Cutting of Metals, *International Journal of Machine Tool Design and Research,* Vol. 7, pp 63-89.

[8.14] Andrew, C. and Tobias, S. A. 1962. Vibration in Horizontal Milling, *International Journal of Machine Tool Design and Research,* Vol. 2, pp 369-378.

[8.15] Tlusty, J. and MacNeil, P. 1975. Dynamic of Cutting Forces in End Milling, *Annals of the CIRP*, Vol. 24, pp 21-25.

[8.16] Tlusty, J. 1986. Dynamics of High Speed Milling, *Journal of Engineering for Industry,* Vol. 108, pp 59-67.

[8.17] Smith, S. and Tlusty, J. 1991. An Overview of Modeling and Simulation of the Milling Process, *Journal of Engineering for Industry*, Vol. 113, pp 169-175.

[8.18] Smith, S. and Tlusty, J. 1993. Efficient Simulation Programs for Chatter in Milling, *Annals of the CIRP*, Vol. 42/1, pp 463-466.

[8.19] Altintas, Y. and Budak, E. 1995. Analytical Prediction of Stability Lobes in Milling, *Annals of the CIRP,* Vol. 44/1, pp 357-362.

[8.20] Altintas, Y. and Lee P. 1996. A General Mechanics and Dynamics Model for Helical End Mills, *Annals of the CIRP,* Vol. 45/1, pp 59-64.

[8.21] Altintas, Y. and Weck, M. 2004. Chatter Stability of Metal Cutting and Grinding, *Annals of the CIRP,* Vol. 53/2, pp 619-642.

[8.22] Zhu, R., Kappor, S.G. and DeVor, R. E. 2001. Mechanistic Modeling of the Ball End Milling Process for Multi-Axis Machining of Free-Form Surfaces, *Journal of Manufacturing Science and Engineering*, Vol. 123, pp 369-379.

[8.23] Fussell, B. K., Jerard, R. B. and Hemmett, J. G. 2003. Modeling of cutting Geometry and Forces for 5 axis Sculptured Surface Machining, Computer-Aided Design, Vol. 35, pp 333-346.

[8.24] Ferry, W. B. and Altintas, Y. 2008. Virtual Five-Axis Flank Milling of Jet Engine Impellers – Part I: Mechanics of Five-Axis Flank Milling, *Journal of Manufacturing Science and Engineering*, Vol. 130, pp 011005-1-11.

[8.25] Ferry, W. B. and Altintas, Y. 2008. Virtual Five-Axis Flank Milling of Jet Engine Impellers – Part II: Feed Rate Optimization of Five-Axis Flank Milling, *Journal of Manufacturing Science and Engineering*, Vol. 130, pp 011013-1-13.

[8.26] Altintas, Y. and Spence, A. 1991. End Milling Force Algorithms for CAD Systems, *Annals of the CIRP,* Vol. 40/1, pp 31-34.

[8.27] Spence, A. and Altintas, Y. 1994. A Solid Modeller Based Milling Process Simulation and Planning System, *Journal of Engineering for Industry*, Vol. 116, pp 61-69.

[8.28] Yazar, Z., Koch, K., Merrick, T. and Altan T. 1994. Feed Rate Optimization based on Cutting Force Calculations in 3-Axis Milling of Dies and Molds with Sculptured Surfaces, *International Journal of Machine Tools and Manufacture*, Vol. 34, pp 365-377.

[8.29] Fussell, B. K., Jerard, R. B. and Hemmet, J. G. 2001. Robust Feedrate Selection for 3-Axis NC Machining Using Discrete Models, *Journal of Manufacturing Science and Engineering*, Vol. 123, pp 214-224.

[8.30] Tounsi, N. and Elbestawi, M. A. 2003. Optimized Feed Scheduling in Three Axes Machining. Part I: Fundamentals of the Optimized Feed Scheduling Strategy, *International Journal of Machine Tools and Manufacture*, Vol. 43, pp 253-267.

[8.31] Tounsi, N. and Elbestawi, M. A. 2003. Optimized Feed Scheduling in Three Axes Machining. Part II: Experimental Validation, *International Journal of Machine Tools and Manufacture*, Vol. 43, pp 269-282.

[8.32] Guzel, B. U. and Lazoglu, I. 2004. Increasing Productivity in Sculpture Surface Machining via Off-line Piecewise Variable Feedrate Scheduling based the Force System Model, *International Journal of Machine Tools and Manufacture*, Vol. 44, pp 21-28.

[8.33] Budak, E., Lazoglu, I. and Guzel, B. U. 2004. Improving Cycle Time in Sculptured Surface Machining Through Force Modeling, *Annals of the CIRP,* Vol. 53/1, pp 103-106.

[8.34] Erdim, H, Lazoglu, I. and Ozturk B. 2006. Feedrate Scheduling Strategies for Free-form Surfaces, *International Journal of Machine Tools and Manufacture*, Vol. 46, pp 747-757.

[8.35] Altintas Y. and Merdol, S. D. 2007. Virtual High Performance Milling, *Annals of the CIRP,* Vol. 56/1, pp 81-84.

[8.36] Merdol, S. D. and Altintas Y. 2008. Virtual Simulation and Optimization of Milling Operations – Part I: Process Simulation, *Journal of Manufacturing Science and Engineering*, Vol. 130, pp 051004-1-12.

[8.37] Merdol, S. D. and Altintas Y. 2008. Virtual Simulation and Optimization of Milling Operations – Part II: Optimization and Feedrate Scheduling, *Journal of Manufacturing Science and Engineering*, Vol. 130, pp 051005-1-10.

[8.38] El Mounayri, H. Spence, A. D. and Elbestawi, M. A. 1998. Milling Process Simulation – A Generic Solid Moduller Based Paradigm, *Journal of Manufacturing Science and Engineering*, Vol. 120, pp 213-221.

[8.39] Roth, D., Gary, P., Ismail, F. and Bedi, S. 2007. Mechanistic Modelling of 5-axis Milling Using an Adaptive and Local Depth Buffer, *Computer-Aided Design,* Vol. 39, pp 302-312.

[8.40] Shaw, M. C. 1984. *Metal Cutting Principles*, Oxford Science Publications.

[8.41] Smith, S. and Tlusty, J. 1990. Update on High-Speed Milling Dynamics, *Journal of Engineering for Industry*, Vol. 112, pp 142-149.

[8.42] Budak, E. and Tekali, A. 2005. Maximizing Chatter Free Material Removal Rate in Milling through Optimal Selection of Axial and Radial Depth of Cut Pairs, *Annals of the CIRP,* Vol. 54/1, pp 353-356.

9

Achieving a STEP-NC Enabled Advanced NC Programming Environment

Matthieu Rauch[1], Raphael Laguionie[2] and Jean-Yves Hascoet[3]

Modelisation Optimisation Process Production, Research Institute of Communications and Cybernetics of Nantes (IRCCyN) – UMR CNRS 6597, 1 rue de la Noe, BP92101, 44321 Nantes Cedex 03, France

Email: [1] matthieu.rauch@irccyn.ec-nantes.fr
[2] raphael.laguionie@irccyn.ec-nantes.fr
[3] jean-yves.hascoet@irccyn.ec-nantes.fr

Abstract

Modern manufacturing requires a flexible numerical chain of industrial products, in particular the relationships between CAD/CAM solutions and CNC. No longer can CNC controllers restrain their tasks at the execution of inflexible orders and choices made at earlier stages of the numerical chain. Thus, in a STEP-compliant environment, the CNC controller possesses a broad decision-making power to optimize the NC programming according to the machining equipment properties. The NC programming environment then has to face new challenges. In the first part of the chapter, an implementation method leading to STEP-NC advanced manufacturing is proposed. This approach is divided into three successive sceneries of STEP-NC deployment for progressive improvements. Industrial concerns can also use STEP-compliant applications with their current machining equipment. In the second part of the chapter, the STEP-NC Platform for Advanced and Intelligent Programming (SPAIM) developed at IRCCyN is presented. This platform controls current industrial machine tools directly from STEP-NC files, which benefits from this new data model. It also includes new tool-paths programming methods, such as pattern strategies for trochoidal milling and plunging tool-paths. The platform demonstrates the benefits of STEP-NC for industry and also forms a basis for future STEP-NC related research and validation.

9.1 Introduction

For the last 50 years, Computer Numerically Controlled (CNC) manufacturing has undergone numerous technological improvements. Modern machining centres with highly sophisticated CNC controllers are characterized as having high speed machining capacities and high accuracy. Computer Aided Manufacturing (CAM) software tools are responsible for preparing control commands for this powerful and complex manufacturing equipment. Meanwhile, new challenges were raised during the last century by the increasing demands of flexible, interoperable, and portable manufacturing systems. It is, therefore, essential that data communication between the components that make up the design and manufacturing chain be both standardised and integrated. In order to achieve such integration, bidirectional communication between each of these components is necessary.

The current NC programming language is the ISO 6983 [9.1], also known as G-codes, which was developed in the 1960s for NC machine tools. The low-level data conveyed by this old fashioned data standard not only breaks the numerical chain started at CAD/CAM level but also limits the flexibility and interoperability of manufacturing systems. Compared with G-codes, STEP-NC [9.2] provides a new vision to the numerical chain. Based on the STandard for the Exchange of Product model data (STEP), this new standard provides a mechanism for a bidirectional data flow from Computer Aided Design (CAD) to CNC. It also transfers a large part of the intelligence and decision-making jobs from CAM to CNC. The CNC controller, therefore, has a new role in the manufacturing process by being a close link to or directly generating the machining tool-paths and controlling the process data. In addition, this new standard also makes it possible for new NC programming methods to be employed to improve machining performance.

This chapter focuses on these new challenges that NC Controllers are facing. First, an implementation method leading to STEP-NC advanced manufacturing is proposed. This approach is divided into three consecutive scenarios of STEP-NC deployment so that the improvements can be made step by step. The ability of STEP-NC to generate and control high-level advanced tool-paths is discussed as well. Then, the STEP-NC Platform for Advanced and Intelligent Programming (SPAIM) developed at Institut Recherche Communication Cybernetique Nantes (IRCCyN) [9.3] is presented. It allows CNC systems to benefit from the advantages of STEP-NC numerical chain with current industrial equipment. The objectives are to make current industrial machining facilities compliant to STEP-NC numerical chains so that they can not only get familiar with this data model but also benefit from the advances conveyed by this data model. Indeed, STEP-NC will not be able to become widespread without endorsements from industry.

9.2 A New Role for the NC Controller into the Numerical Chain

The challenges imposed by modern industry lead to the necessity for redesigning the numerical communication chain between different components of manufacturing systems, especially between CAD/CAM software and CNC machines. CNC machines are no longer rigid machine tools for simply carrying out commands generated by CAM software. In contrast, the NC controllers in these machine tools are able to play an important role on the parameterization of machining operations because they can receive real-time process information and are privileged to adapt the programming parameters in a short period of time [9.4]. STEP-NC programming environments are consequently dedicated to advanced programming methods and high-level tool-paths generation.

9.2.1 Advanced CNC Programming and Machining

With the increased requirement for high volume, fast prototyping manufacturing at low cost in modern industry, it is found that the advanced programming method is highly necessary. In addition to the geometrical data and kinematics, it appears now that it is essential to take process-oriented parameters (tool loads, probing results, tool deflection, etc.) into account to program high-speed machine tools.

A method to improve NC machining processes was introduced in [9.5, 9.6] . This method named Intelligent Computer Aided Manufacturing approach (ICAM) was developed at IRCCyN. The main purpose of this method is to integrate process constraints in tool-path programming and control (Figure 9.1). Tool-path generation is shared by both CAM software and CNC. Basic tool-paths are generated by CAM software and sent to the CNC controller. Then, during the running of the program, these tool-paths are optimized by the CNC controller according to the real-time process data evaluation. As a result, tool-paths are adapted according to the real-time machine tool performance feedback.

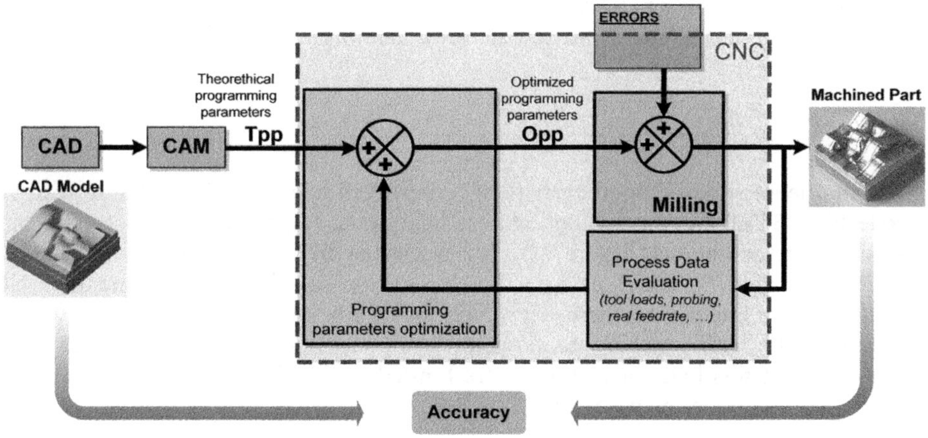

Figure 9.1. ICAM approach principles

This type of programming method, although not initially developed with STEP-NC format, shares the same vision of the relationships between CAM and CNC into a more interoperable environment. The implementation of such approach with G-codes is complicated and tedious because several subroutines and subprogram must be added to the machining program. This denies flexibility of a machine.

In contrast, STEP-NC data format proposes an ideal environment for the ICAM concept. From its beginning, STEP-NC standard is aimed at enabling advanced programming operations in order to propose a better control of tool-paths compared to G-Codes. Suh et al. in 2003 introduced the "New Intelligent Control" concept for STEP-NC programming [9.7]. This approach plainly announced the expectations with regard to advanced NC programming but no actual development was proposed to detail its implementation.

However, the implementation of STEP-NC Advanced NC Programming cannot be performed at once. Instead, it needs to spread step by step into the numerical chain. The reasons for this are as follows:

- *Cost-linked reasons*: investing in new NC machining equipment increases high cost for a company (material purchase, staff training, etc.). It is not possible to retrofit all the NC controllers of a company at once.
- *Technique-linked reasons*: current STEP-NC controllers are not mature. Very few of them are able to reach the same level of reliability and performance as the latest industrial G-code based CNC products, especially in 5-axis milling.
- *Pssychology-linked reasons*: industrial companies are not ready to get rid of G-codes that they know very well and to replace it with a new standard they do not know much. Time is needed before STEP-NC is trusted.

STEP-NC data format has to be spread gradually into the numerical chain, in order for the users to be familiar with it. It is also important for companies to work with a group of STEP-NC and G-code based CNC controllers in order to facilitate the transition from G-code to STEP-NC without affecting their business. It is proposed here to employ a three-level method to achieve advanced NC programming by using STEP-NC. The three levels are STEP-NC Interpreted Programming, STEP-NC Integrated Programming, and STEP-NC Advanced Programming.

STEP-NC Interpreted Programming

The first level is based upon current NC controllers which are only able to read G-code files. STEP-NC machining data are interpreted to generate the tool-paths and schedule the operations (Figure 9.2). It is necessary to convert these data into CNC machining code (G-code) using a post-processor so that the machine tool NC controller can understand them. However, G-code is only seen as a simple way to communicate with machine tool's axes using a G-code based industrial CNC controller. At this level, investments are limited to STEP-NC compliant CAM and STEP-NC interpretation software.

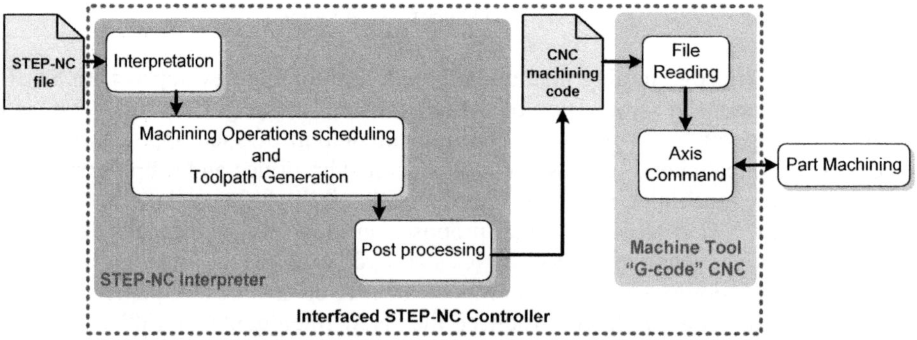

Figure 9.2. STEP-NC interpreted programming

Few significant contributions made to advanced programming can be observed at this level. However, the use of STEP-NC standard with current NC controller is now possible. This initial level enables the use of STEP-NC standard in an industrial environment. The user can benefit from the advantages of STEP-NC for CAD/CAM issues. For example, if some modifications of the programming parameters are made at the shop-floor level (i.e., into the STEP-NC controller), it is possible to send them to CAD/CAM using a STEP-NC file.

Eventually, generation of a G-code file and its implementation in a CNC machine tool is fully transparent to the user, because it is done by a STEP-NC interpreter. From the user's point of view, everything goes as if the NC controller of the machine tool is able to read and handle STEP-NC files.

STEP-NC Integrated Programming

In STEP-NC Integrated Programming, the CNC controller generates tool-paths and schedules manufacturing operations directly from a STEP-NC file (Figure 9.3). Machine tool characteristics (architecture, workspace, kinematics, capacities, etc.) are taken into account. It is also possible to integrate external data such as probing results into the tool-path generation process. Nevertheless, milling tool-paths cannot be modified during execution. STEP-NC numerical chain is now complete from a CAD model to the machined part. This offers interoperability and flexibility conveyed by the STEP data format. There is no stage between tool-path generation and axis control so that the integration of machine tool characteristics into the tool-paths generation process is made easier.

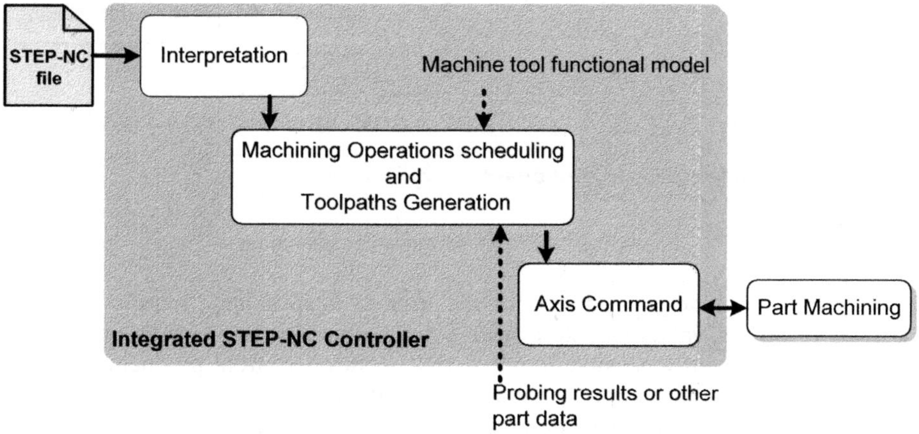

Figure 9.3. STEP-NC integrated programming

For a user, there is not much difference in comparison to the previous level. However, this level calls for a new generation of NC controllers that can handle high-level information to generate tool-paths and schedule the machining operations. This level of the approach is essential to prepare the development of advanced programming functions. There are high expectations associated with these NC controllers, so that the implementation of this level will be effective at a large scale into industrial concerns only in the medium term.

STEP-NC Advanced Programming

At this level, in addition to the capabilities developed at the previous level, machine tool NC controller evaluates process data online during the running of the process (Figure 9.4). The results obtained are employed for real-time manufacturing parameter optimization. To adapt to the requirements, the controller decides whether tool-path modifications are needed or whether further operations have to be added into the Workplan. For example, a semi-finishing operation can be inserted to the Workplan when the geometry obtained after roughing is not adapted to a sound execution of finishing operations.

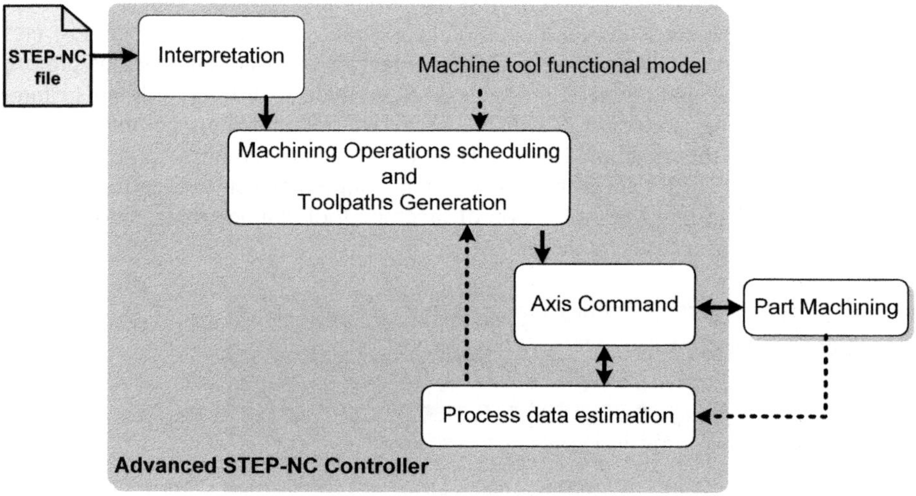

Figure 9.4. STEP-NC advanced programming

However, it is very important for the user to keep a large margin of action (display of results, validation, etc.), since a major objection to STEP-NC standard is that it would be hardly possible to prevent NC controllers from producing errors if everything was to be done automatically without any user intervention. The integration of the human-linked parameters should not been neglected.

This last level of the proposed method consists of the implementation of ICAM method using STEP-NC standard. It meets the requirements of the "New Intelligent Control" concept and stands consequently as the culmination of STEP-NC programming.

9.2.2 High-level Tool-path Generation

The purpose of the STEP-NC standard is to manage machining objects (features, Workingsteps, etc.) rather than machining explicit instructions. In a STEP-NC numerical chain, the CNC controller generates the explicit milling tool-path on its own, under the constraints of the STEP-NC file parameterization (tool-path type, tool diameter, retract plane position, etc.). This therefore enables new programming approaches for tool-path generation [9.8]. One of these approaches is developed at IRCCyN. It consists of using pattern strategy definitions for high-level machining strategies: plunge milling, trochoidal milling, drilling, etc. [9.9, 9.10]. This approach defines the tool-path as a combination of a guide curve and a milling pattern. The guide curve is built from geometrical data of machining features whereas milling pattern is only based on technological criteria. Figure 9.5 shows an example of pattern definition for a trochoidal tool-path [9.11].

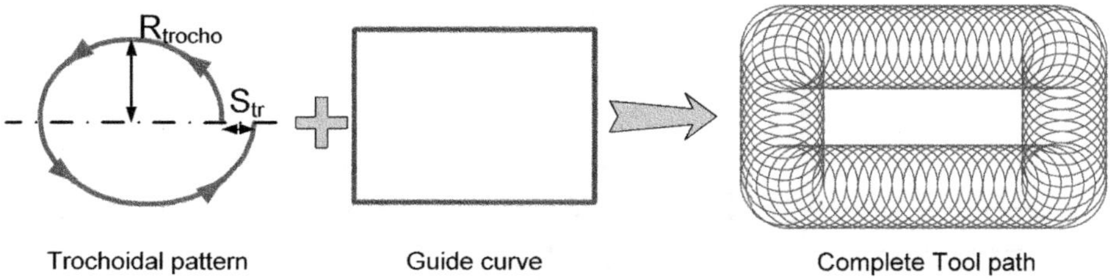

Figure 9.5. Example of pattern definition of a trochoidal tool-path

Geometrical modifications of the machining feature only affect the guide curve; the pattern remains independent. This approach provides a high flexibility of tool-paths generation and insures the technological constraints are met in spite of part geometry. Simplicity of construction, ease of modifications, and definition from a restricted number of parameters make pattern strategies a new way of programming tool-paths. Moreover, they are compliant with the STEP-NC standard. The integration of pattern strategies into ISO 14649 standard was proposed as well [9.10].

Experimental tests were performed using the IRCCyN STEP-NC platform on several CAD models and machining parameterization combinations. As shown in Figure 9.6, STEP-NC interpreter managed to read STEP-NC data and to generate appropriate tool-paths. These validation tests assess the effectiveness of both developed interpreter and programming approach using pattern strategies.

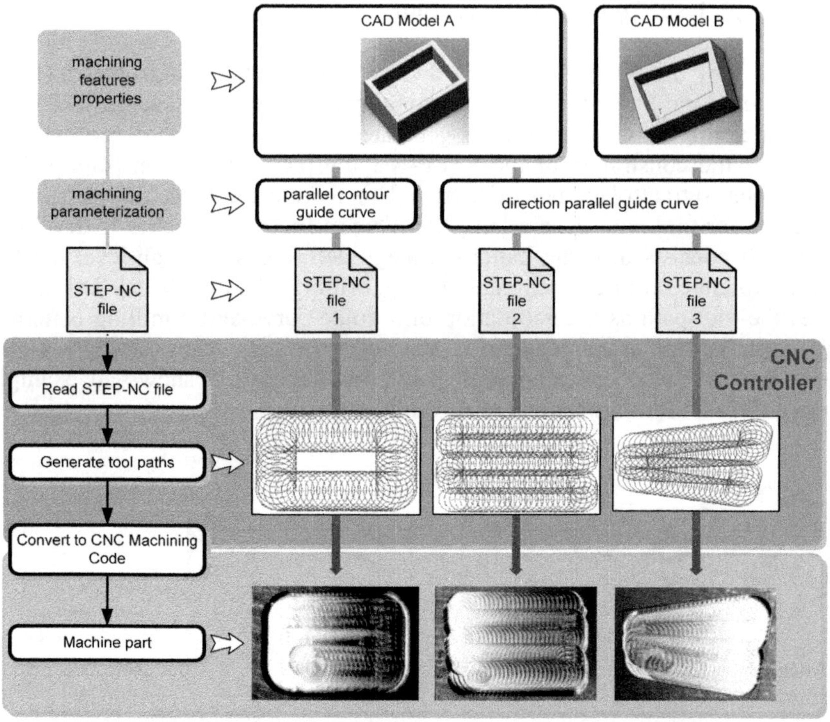

Figure 9.6. Experimental validation on pattern trochoidal tool-paths

9.3 STEP-NC Platform for Advanced and Intelligent Manufacturing (SPAIM)

The use of high-level information to communicate with CNC controllers calls for a new organization of the different CAM and CNC modules (i.e., tool-paths generation module, Workplan selection module, process parameterization module, etc.). In the case of milling, manufacturing a part consists of removing some material by the action of a cutting tool driven by an NC controller. All the motions and actions are controlled by low-level information generated at the CNC level, e.g. axis commands. All this information is totally transparent to the user. The electrical control and algorithms are not easily understood by humans. G-code programming can be seen as an upper-level whose details can be understood by humans but still remains quite difficult to interpret. The low-level G-code information is directly translated by the CNC controller to generate elementary motions and actions on the machine tool. In contrast, STEP-NC programming is based on high-level information such as feature geometry or high-level definition of the process data. STEP-NC format, which can be understood by humans, is an object oriented language.

However, it is still necessary to communicate with the CNC controller and control the different parts of the machine tool. This means that a translation from the high-level object oriented information to well-adapted and accurate low-level information is necessary. This job is carried out by the interpreter. Any STEP-NC interpreter is consequently machine tool specific as it makes the link between the STEP-NC file and the data required to control the machine tool's axes. The structure of a STEP-NC interpreter can be built in several ways by using different technologies to compute the high-level information in STEP-NC files to lower-level information for machining.

After a brief overview of the existing STEP-NC interpreters and CNC interfaces, this section details the STEP-NC Platform for Advanced and Intelligent Manufacturing developed at IRCCyN.

9.3.1 Machining a Part from a STEP-NC File

Much research work has been done for the development of STEP-NC CAD/CAPP/CAM/CNC chain. Software tools were developed based on ISO 14649 (STEP-NC Application Reference Model) [9.12, 9.13]. Most of these research projects are reviewed by Xu et al. in [9.14].

Among them, several STEP-compliant CNC prototypes have been proposed for milling and turning applications. They can be divided into two groups. The first category is made of STEP-NC interpreters which translate the STEP-NC machining data into G-code files for machining applications [9.15-9.17]. These software tools not only bring the benefits from the advances of STEP-NC CAD/CAM but also make this standard understandable by some current CNC controllers. The second category is made of CNC prototypes that are only based on the STEP-NC data model (it is not necessary to resort to G-code any more). Developed models have different objectives: setup of a STEP-NC environment for laboratory purposes [9.7, 9.18], proposal of new CNC architectures [9.19], STEP-NC enabled machining of a retrofitted turning centre [9.20, 9.21].

A STEP-compliant CNC interface is proposed by STEP Tools Inc. [9.22]. This software generates ISO 10303-238 files (STEP-NC Application Integrated Model) and works as a front application on a current CNC controller. Despite its performances, the current position of this software is more to unify the numerical chain rather than to enable shop-floor modifications and feedback to CAD/CAM.

Although various prototypes have been developed, there is still a lack of demonstration and implementation of STEP-NC on industrial NC machine tools in order to show its capabilities. Indeed, STEP-NC needs support and contributions from industry to spread on a broad scale in the manufacturing area. Development platforms are necessary to implement and validate new proposals and to optimize the tools for advanced programming. To achieve these goals, an integrated platform has been developed at IRCCyN Institute to control industrial machine tools directly from STEP-NC files [9.23].

9.3.2 A STEP-NC Platform for Industrial Machine Tools

SPAIM platform is based on STEP-NC Interpreted Programming approach, which allows an implementation on most industrial CNC controllers (Figure 9.7). Its implementation associates two main objectives. First, it stands as a demonstrator to demonstrate the benefits of STEP-NC; second, it serves as a development platform for future research and validations concerning the STEP-NC standard.

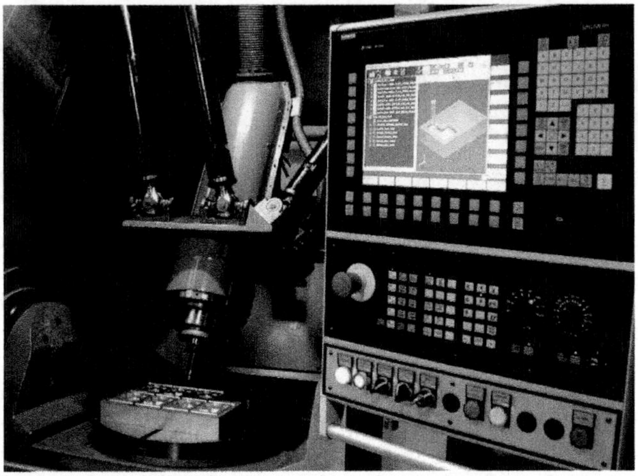

Figure 9.7. SPAIM platform at IRCCyN institute

The following sub-sections introduce the structure of SPAIM. Current and future research focusing in relationship to STEP-NC advanced manufacturing at IRCCyN are also introduced.

Presentation

STEP-NC object-oriented programming helps shift the tool-path generation to the shop-floor level. As a result, some intelligent and decision-making power can be transferred into the CNC controller. Self learning algorithms begin to be developed to produce better quality parts to compensate controlled errors [9.24]. In this case, the interpreter module is a key part of the STEP-NC controller because it translates STEP-NC manufacturing data into explicit tool-paths using manufacturing feature geometrical characteristics and programming parameterization of each manufacturing step.

Although it generates explicit machining tool-paths automatically, the interpreter still calls for user validation before sending the tool-paths to the machine tool. This stage is considered as compulsory because it informs the user about the movements that the machine tool will execute. Furthermore, the user can check whether proposed tool-path control points meet the constraints. If not, the user can modify the tool-path shape. After user validation, corresponding output file (G-code) is automatically executed by the controller.

The STEP-NC Platform for Advanced and Intelligent Manufacturing is composed of a Human Machine Interface (HMI) and several computation modules. HMI on the NC Controller displays a 3D visualization of the manufacturing data (CAD model, tool-paths, cutting tools, etc.), the machining parameters, and a tree-view of the STEP-NC file (Figure 9.8). The user can modify these data using the interface. After modification, the CAD model and explicit tool-paths are automatically regenerated and the STEP-NC file is updated. Then machining is executed directly from the STEP-NC file following a visual validation of the tool-path, machining parameters and modifications.

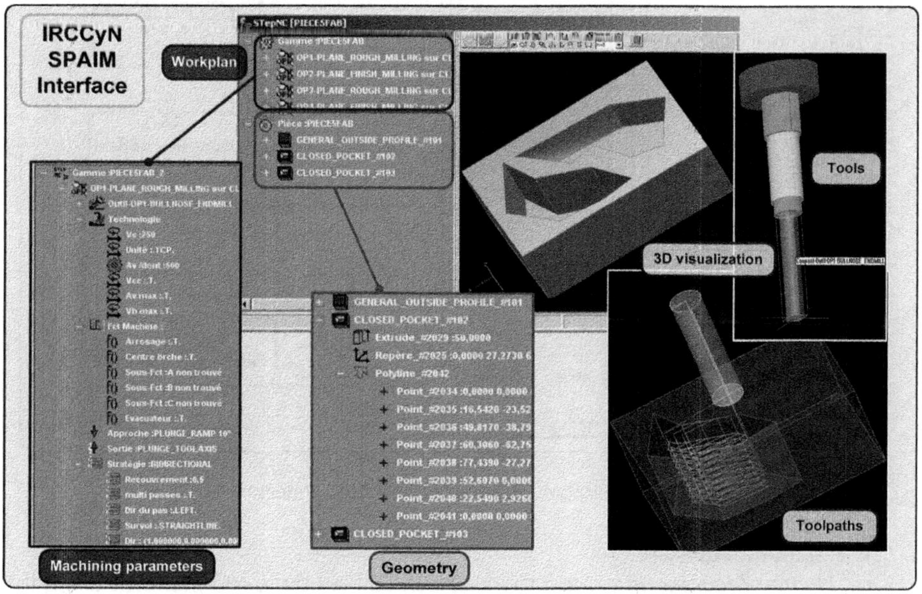

Figure 9.8. STEP-NC Human/Machine Interface on the NC controller

The SPAIM platform enables any implemented CNC to read and handle any STEP-NC files built in the format of ISO 14649 standard. It has already been implemented and validated on the high-speed manufacturing machine tool. This machine was designed by Fatronik company and named VERNE [9.25]. It has a parallel kinematics architecture and is equipped with a Siemens Sinumerik 840D NC controller (Figure 9.8). However, the geometrical transformation model of the machine tool is integrated into the NC controller. Another version of the IRCCyN platform has been developed for a Hermle C30U high speed machining centre equipped with a Heidenhain CNC controller.

Architecture of the Platform

The STEP-NC platform is composed of several modules controlled by Delphi applications (Figure 9.9). Some of them are installed on the CNC controller of the machine tool, others are on an external computer to save the computation power of the CNC hardware. However, all the modules running on the external PC can be implemented in the CNC computer if its computing capacity allows it. The adopted solution is totally transparent to the user. The external PC can be seen as an extension of the capacities of the CNC hardware.

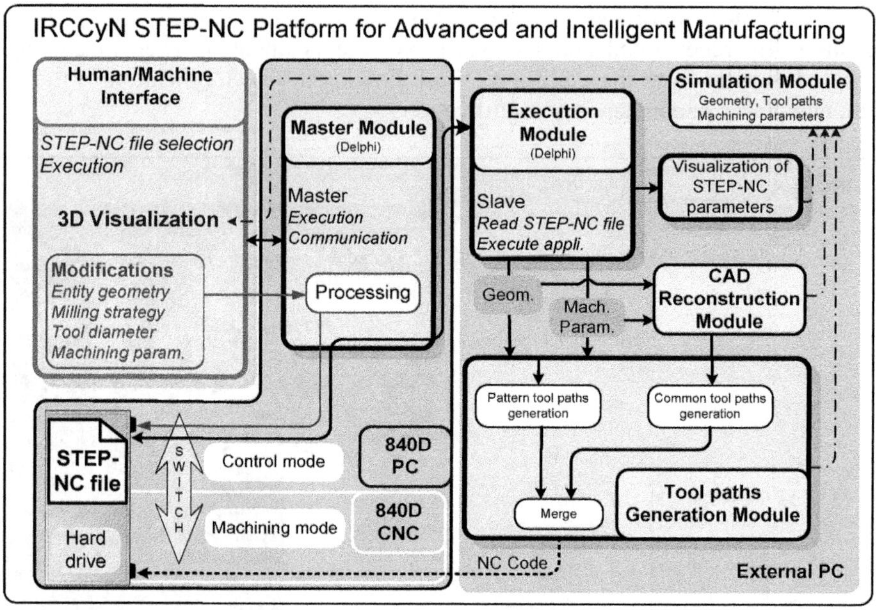

Figure 9.9. Architecture of SPAIM concept

The main modules are:

- *Human Machine Interface*: the user can control the CNC platform using this module. It displays the results of analysis and computations made by the interpreter using data from the simulation module on the screen.
- *Master module*: it belongs to the PC part of the NC controller. This Delphi program is directly linked with the HMI and sends the orders to other modules at the user's request. For parameter modification, it locates and replaces the corresponding elements in the STEP-NC file.
- *Execution module*: it distributes orders from the Master module via a local or an Internet network. This module reads and analyses the STEP-NC file through the master module and sends the requested information for processing to the tool-path generation module and to the simulation module.
- *CAD reconstruction module*: this module rebuilds the CAD geometry from the entity description in the STEP-NC file. This Delphi based automatic tool sends corresponding commands to Delcam PowerSHAPE CAD software [9.26] for geometry reconstruction. The CAD model is then used by the tool-path generation module to generate common strategies. This module provides feedback from the STEP-NC file to the CAD model as well.
- *Tool-path generation module*: it is divided into two components which are run in parallel. The first component handles every common strategy related

to the ISO 14649 standard (e.g. contour parallel, bidirectional, etc.). It uses the tool-path generation module of commercial software (i.e., Delcam PowerMILL [9.27]). The second component has been developed at IRCCyN. It handles pattern strategies in the case of trochoidal milling and plunge milling tool-paths. According to the manufacturing data, the execution module collects the corresponding tool-path generation module for each machining operation and merges the tool-paths results before sending the NC code to the controller for execution.

- *Simulation module*: it sends back the results of the different computations (e.g. STEP-NC file analysis, 3D geometry and explicit tool-paths as a VRML file, etc.) to the HMI for visualisation.

The tool-path generation module is based on a vendor component for the generation of common strategies. The IRCCyN SPAIM platform can benefit from the skills and performance of the CAM software. Moreover, it shows that, even if the numerical chain is redistributed, all the current knowledge is still needed. Many CAD/CAM companies are still reluctant to develop STEP-NC compliant applications because they tend to believe, incorrectly, that this new vision of the relationship between CAM and CNC brings about no benefits to them.

9.3.3 Benefits of a STEP-NC Enabled Controller

STEP-NC interpreted programming makes it possible to use the STEP-NC standard with the existing machine tools and NC controllers, which still understand G-code programming. The example of the IRCCyN SPAIM platform shows the feasibility of it. At this first level of the integration of STEP-NC, most of the advantages of STEP-NC at the CNC level are already evident (Figure 9.10):

1. A STEP-NC file can be directly read and executed on several machine tools without any modification. This compatibility is enabled by the high-level description of geometry and process data without any specificity to a single machine tool. All machine tool information and functional models are provided by the CNC platform. For example, the execution of the same STEP-NC file on different machine structures would use different spindle speeds, feed-rates, or tool-paths, because each NC controller would compute the most suitable and efficient parameterization, according to the target equipment.
2. Modifications of the geometry and the machining parameters can be achieved at shop-floor level directly on the CNC HMI. These modifications automatically lead to regeneration of the tool-paths, geometry displays and STEP-NC model tree update. The corresponding STEP-NC files and a CAD model are updated as well.
3. Feedback from CNC to CAD/CAM software is possible since the STEP-NC file is always up to date. Modifications can be done at shop-floor level during the first manufacturing phase of a part. This knowledge feedback enables process planning level to learn and improve the future manufacturing phases.

4. Optimizing the machining parameters and the tool-paths is easier at the CNC level. The SPAIM platform provides new ways of optimization based on the STEP-NC data model.
5. STEP-NC file has a small size because of its high-level data. This situation reduces the transfer time and is well suited for the Internet-based collaborative manufacturing.

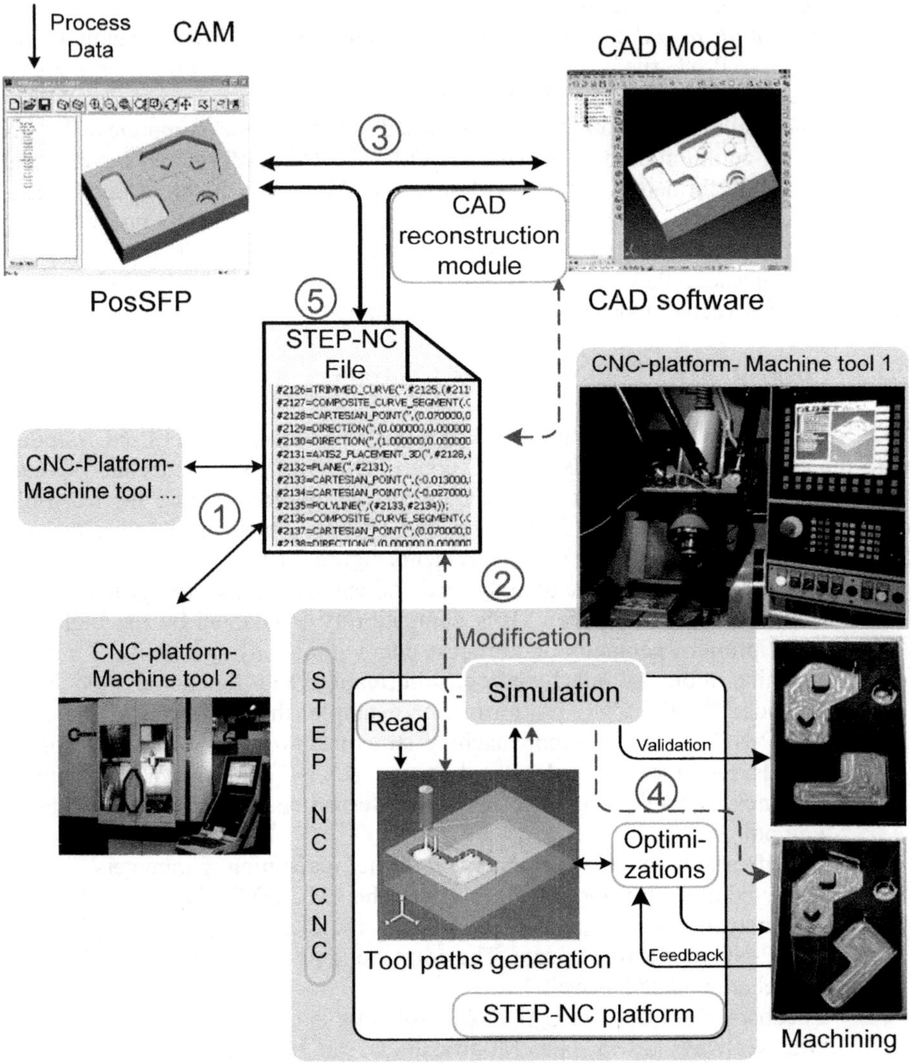

Figure 9.10. STEP-NC platform in the user numerical chain

9.3.4 Toward Advanced CNC Programming

The presented platform is an example of the improvements conveyed by the STEP-NC data model. It opens the door for future advanced programming methods. It is not only a demonstrator but also a development platform for future research and validations based on ISO 14649 standard.

Thus, new simulation and optimization approaches can be implemented because of the high-level data included in a STEP-NC file. For example, a real feed-rate simulation module and a tool deflection module will be added to the existing SPAIM platform. The implementation of these modules will be based on research works already done at IRCCyN on 3D solid simulation [9.28] and tool deflection compensation [9.29]. As shown in Figure 9.11, the SPAIM platform will consequently make each part of the manufacturing numerical chain interoperable. It presents a comprehensive environment dedicated to advanced programming.

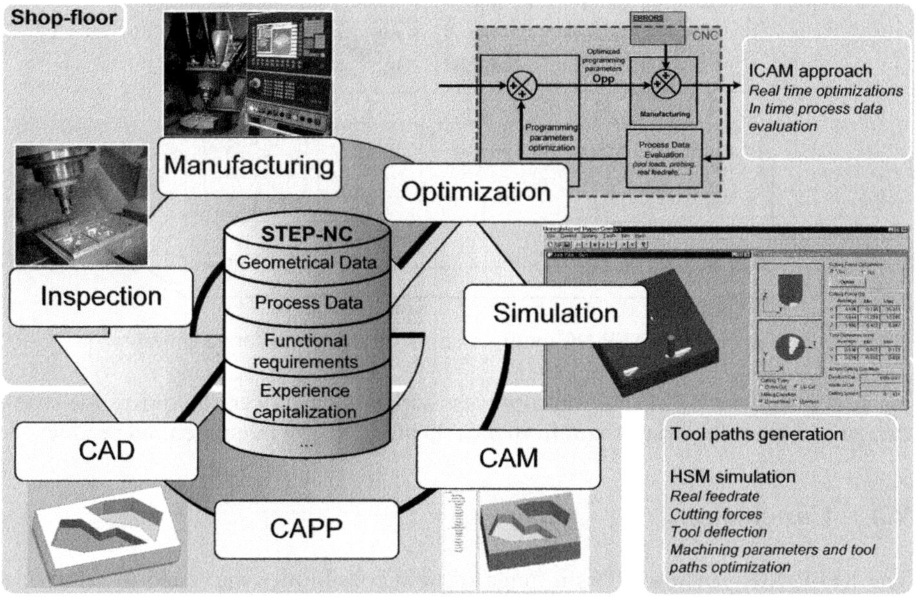

Figure 9.11. Advanced programming implementation in a comprehensive STEP-NC environment

Other work is also carried out for implementing the ICAM concept and process data feedback into the SPAIM platform. Tool-path regeneration based on on-machine inspection has been carried out with G-codes. It shows satisfying results. The real-time adaptations of machining parameters (i.e., feed-rate) and tool-paths are possible by using process data evaluation from the CNC controller (motor amperage, delivered power, real feed-rate, articular coordinates of the joints, etc.). The integration of a STEP-NC compliant ICAM approach is underway.

Future developments at IRCCyN on the STEP-NC standard and advanced NC programming methods will not be limited to machining. Instead, they will include turning and rapid manufacturing processes. The integration of other models such as wire electro discharge machining [9.30] should be considered as well. The development of the STEP-compliant process models and platforms is as essential to achieve the goal of combined processes interoperability. Such modules are under development and the feedback from the NC controller to a CAD model is already available for milling and rapid manufacturing.

The long-term purpose is to develop a comprehensive STEP-NC multi-process supervision platform as shown in Figure 9.12. In this environment, the digital model of a part can reflect the modifications and updates concerning each manufacturing process. Moreover, interactions between two processes will be available. Manufacturing of industrial parts which need a combination of several manufacturing processes will be optimized. The use of the STEP-NC standard will ensure interoperable and bidirectional data flow between all stages of the product development process.

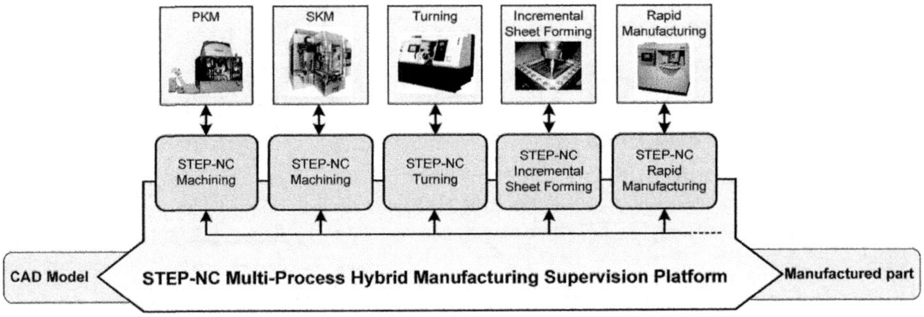

Figure 9.12. STEP-NC multi-process hybrid manufacturing concept

As a result, the STEP-NC multi-process supervision concept optimizes the whole design and manufacturing chain from a CAD model to the manufactured product.

9.4 Conclusions

The STEP-NC standard offers a range of new possibilities that need to reconsider the current practices of manufacturing processes. The G-code standard is out-of-date and is hardly able to support the future of NC programming. G-code though is still used in many of the current STEP-NC interpreters but in a way that is totally transparent for the user. This situation falls short of intelligent programming based on STEP-NC, but is considered a necessary step toward final implementation of the STEP-NC concept.

In this chapter, the proposed approach presents as a road map to achieve advanced programming applications in a STEP-compliant manufacturing environment. It is indeed important that we gradually deploy and integrate the STEP-NC data models into CNC. The architecture of the SPAIM concept is an example of it. This platform demonstrates the current advantages of the STEP-NC data model by showing its performances with industrial manufacturing equipment. It also proves to be a powerful piece of equipment to carry out further development. The bases of a STEP-NC multi-process manufacturing platform are introduced.

The architecture of the proposed platform suggests that all practicians involved in the present NC chain have a role to play in the development of STEP-NC. This new standard and data model aim to improve data exchange between all the modules involved in the NC chain. It would be wrong to state that all elements of the NC chain have to be reinvented (e.g. CAD modellers, tool-path generators, etc.). This is because the knowledge is already available. STEP-NC is a data model standard that extends this knowledge. It calls for fundamental and even cultural changes. Yet, it still relies on existing capabilities and know-how in the manufacturing industry.

References

[9.1] ISO_6983-1 Numerical control of machines -- Program format and definition of address words -- Part 1: Data format for positioning, line motion and contouring control systems.

[9.2] ISO_14649-1 Industrial automation systems and integration -- Physical device control -- Data model for computerized numerical controllers -- Part 1: Overview and fundamental principles.

[9.3] IRCCyN. Institut de Recherche en Communications et Cybernetique de Nantes. Available online at: www.irccyn.ec-nantes.fr (accessed 2008)

[9.4] R. Laguionie, M. Rauch, J. Y. Hascoet and S. Tichadou. 2008. The future for CNC machine tools programming. *Information day on High Speed Machining of aluminium alloys and composites*. Cetim, Senlis (France). 21 January.

[9.5] J. Y. Hascoet and M. Rauch. 2006. A new approach of the tool path generation in manufacturing operations using CNC data. *International Conference on High Speed milling*. Suzhou (China). 11-13 mai. 101-113

[9.6] M. Rauch. 2007. Optimization of CNC machine tool programming – Application to parallel kinematics machines. *PhD Thesis*. Ecole Centrale Nantes / University of Nantes

[9.7] S. H. Suh, B. E. Lee, D. H. Chung and S. U. Cheon 2003. Architecture and implementation of a shop-floor programming system for STEP-compliant CNC. *Computer-Aided Design*. 35 (12): 1069-1083

[9.8] M. Rauch and J. Y. Hascoet 2007. Rough pocket milling with trochoïdal and plunging strategies. *International Journal of Machining and Machinability of Materials*. 2 (2): 161-175

[9.9] J. Y. Hascoet, M. Rauch and S. H. Suh. 2007. Relevance of Step-Nc standard for high-level toolpaths generation. *International Conference on High Speed Milling*. San Sebastian (Spain). 21-22 mars.

[9.10] R. Laguionie, M. Rauch and J. Y. Hascoet 2008. Toolpaths programming in an intelligent STEP-NC manufacturing context. *Journal of Machine Engineering*. 8 (1): 33-43

[9.11] R. Laguionie, M. Rauch and J. Y. Hascoet. 2008. Integration of Pattern Strategies in STEP-NC standard : Application to Trochoidal milling. *Assises Machines et Usinage Grande Vitesse 2008*. Nantes (France). 8-9 June.

[9.12] ISO_14649-11 Industrial automation systems and integration -- Physical device control -- Data model for computerized numerical controllers -- Part 11: Process data for milling.

[9.13] ISO_14649-12 Industrial automation systems and integration -- Physical device control -- Data model for computerized numerical controllers -- Part 12: Process data for turning.

[9.14] X. W. Xu, H. Wang, J. Mao, S. T. Newman, T. R. Kramer, F. M. Proctor and J. L. Michaloski 2005. STEP-compliant NC research: The search for intelligent CAD/CAPP/CAM/CNC integration. *International Journal of Production Research.* 43 (17): 3703-3743

[9.15] H. Wang, X. Xu and J. D. Tedford 2007. An adaptable CNC system based on STEP-NC and function blocks. *International Journal of Production Research.* 45 (17): 3809-3829

[9.16] S. T. Newman, R. D. Allen and J. R. S. U. Rosso 2003. CAD/CAM solutions for STEP-compliant CNC manufacture. *International Journal of Computer Integrated Manufacturing.* 16 (7/8): 590

[9.17] S.-J. Shin, S.-H. Suh and I. Stroud 2007. Reincarnation of G-code based part programs into STEP-NC for turning applications. *Computer-Aided Design.* 39 (1): 1-16

[9.18] S. H. Suh and S. U. Cheon 2002. A Framework for an Intelligent CNC and Data Model. *The International Journal of Advanced Manufacturing Technology.* 19 (10): 727-735

[9.19] M. Minhat, V. Vyatkin, X. Xu, S. Wong and Z. Al-Bayaa A novel open CNC architecture based on STEP-NC data model and IEC 61499 function blocks. *Robotics and Computer-Integrated Manufacturing.* In Press, Corrected Proof

[9.20] X. W. Xu 2006. Realization of STEP-NC enabled machining. *Robotics and Computer-Integrated Manufacturing.* 22 (2): 144-153

[9.21] S. Habeeb and X. Xu 2008. A novel CNC system for turning operations based on a high-level data model. *The International Journal of Advanced Manufacturing Technology.*

[9.22] STEP Tools. ST-Machine STEP-NC for CAM/CNC. Available online at: http://www.steptools.com/products/stmachine (accessed November 2008)

[9.23] J. Y. Hascoet and R. Laguionie. 2008. STEP-NC research at the IRCCyN. *56th ISO TC 184/SC1 Plenary Meeting.* Busan (Korea). 29-31 October.

[9.24] S. Kumar, A. Nassehi, S. T. Newman, R. D. Allen and M. K. Tiwari 2007. Process control in CNC manufacturing for discrete components: A STEP-NC compliant framework. *Robotics and Computer-Integrated Manufacturing.* 23 (6): 667-676

[9.25] Y. S. Martin, M. Gimenez, M. Rauch and J. Y. Hascoet. 2006. VERNE – A New 5-Axes Hybrid Architecture Machining Centre. *5th Chemnitzer Parallelkinematik Seminar.* Chemnitz (Germany). April 25-26 657-676

[9.26] Delcam. PowerSHAPE - CAD Design and Modelling Software. Available online at: http://www.powershape.com (accessed 2008)

[9.27] Delcam. PowerMILL - Your Total Manufacturing Solution. Available online at: www.powermill.com (accessed 2008)

[9.28] A. Dugas, J. J. Lee, M. Terrier and J. Y. Hascoet 2003. Development of a machining simulator considering machine behaviour. *Proceedings of the Institution of Mechanical Engineers, Part B: Journal of Engineering Manufacture.* 217 (9): 1333-1339

[9.29] T. I. Seo, P. Dépincé and J. Y. Hascoet. 1997. Paths Compensation For Tool Deflection In End Milling. *Ifac-Ims Intelligent Manufacturing Systems.* Seoul (Korea). July 21-23.

[9.30] A. Sokolov, J. Richard, V. K. Nguyen, I. Stroud, W. Maeder and P. Xirouchakis 2006. Algorithms and an extended STEP-NC-compliant data model for wire electro discharge machining based on 3D representations. *International Journal of Computer Integrated Manufacturing.* 19 (6): 603-613

10

STEP-compliant CNC Systems, Present and Future Directions

Van Khai Nguyen[1] and John Stark[2]

CADCAMation SA, 103 route de Chancy, 1213 Onex, Geneva, Switzerland
Email: [1]vknguyen@cadcamation.ch, [2] jstark@cadcamation.ch

Abstract
In the early twenty-first century, manufacturing companies face increased productivity pressures and requirements for greater product variability. One of their biggest challenges is to make, while cooperating with multiple design and supply chain partners, the first part "correct and fast". This requirement leads to the need for a data exchange standard that allows disparate entities and their associated devices in a manufacturing system to share data seamlessly in a common format. This will enable, in the future digital factory foundation, the development and use of a "Smart and Ubiquitous NC machining workstation" that represents the brain and knowledge repository of the manufacturing system, based on high bandwidth information, real-time networking, and the ability to be adaptive (self-learning and flexible). The foundation of the Smart NC Machining Workstation is a STEP-compliant NC system connected to a Manufacturing Information Pipeline and distributed hubs for data acquisition from compliant devices as well as for agent-based communication with legacy data. This structure will support a self-learning decision controller providing intelligent science-based algorithms for automatic and optimised tool-path generation and on-line corrective software, preventive maintenance, hardware diagnostics based on statistical process control. Standardisation of the whole process will be achieved through the use of STEP and related standards to form an interoperable and adaptable solution across varied equipment and devices. Only such a standardised and digitised product-process-resource description will enable the transformation from resource-based to knowledge-based, networked and adaptive manufacturing.

10.1 The Traditional Numerical Control Environment

Over the last 50 years, machine tools have evolved from simple machines with controllers that had no memory, driven by punched tape, to today's highly sophisticated Computer Numerically Controlled (CNC) multi-process workstations. These workstations have capabilities such as multi-axis control, error compensation and multi-process manufacture (e.g. combined mill/turn/laser and grinding machines). These additional capabilities have made the programming task more and more difficult. However, conventional programming of Numerically Controlled

(NC) machine tools is still based on ISO 6983, a standard that dates back to the time of punched cards. ISO 6983 uses low-level codes to describe tool movements (such as G01) and switching instructions (such as M5), and does not support complex geometries such as spline interpolation. Programming with ISO 6983 results in large programs that are difficult to handle. As a result, offline software tools for CAD/CAM have become a necessity for controlled generation and verification of NC code. Another disadvantage is that the CAD description is not used directly on the machine, but has to go through a machine-specific post-processor (of which there are estimated to be more than 5,000 in existence). And ISO 6983 assumes information flow is from CAD to the shop-floor, so does not enable feedback of experience from the shop-floor to the designer. Another problem is that ISO 6983 does not support needs in the areas of five axis milling and high-speed machining.

In this environment, last-minute changes and corrections to complex programs are difficult to manage on the shop-floor, and control of program execution at the machine is very limited. Due to many different dialects and vendor-specific additions to the language, part programs are not interchangeable between different controls and machines.

Surveys carried out during the OPTIMAL project (see below) indicated the potential benefits that could be achieved with a more advanced standard. They showed major drawbacks of NC programming using ISO 6983. While 68% of all drawings surveyed were generated using CAD systems, only 10% of these were transferred to NC programming systems as electronic data. In 90% of the cases all contours were drawn again in the NC programming system.

ISO 6983 can be seen as a standard that holds back progress. Due to ISO 6983, although machine tools have changed in their intrinsic mechanical features, in the Internet age they have stayed far behind computers and other electronic devices in terms of communication, knowledge management and self-decision capacities:

- The programming language has basically remained the same with G codes and M codes as defined by ISO 6983 since 1982. It is based on a representation of the tool path digitised with respect to a tool size, and machine command status.
- Though there are many CAM tools to support NC manufacture, the problem of portability and interoperability from system to system remains an obstacle for cooperation among manufacturing partners. Although the part program is based on the above standard, each machine builder has extended the G-code to include proprietary functions and adaptations (e.g. cycles). As a result, for the same manufacturing technology, a specific post-processor is needed for each machine type and each numerical controller.
- The NC programming process is only *top-down*: there is no information feedback mechanism to allow adaptive or preventive correction at the top level for optimising the production line.
- Machining technologies and manufacturing processes are still based on disparate islands of automation unable to communicate with each other.
- There is a lack of end-to-end simulation and full verification because of, on the one hand, the shortage of reliable digital models (e.g. standardised representations of a machine-tool model and its performance), and on the other hand, the impossibility to connect easily different brand-name

machines and devices together as well as the difficulty to capture real-time data (e.g. temperature, vibration…) through sensors.

- Currently, for the manufacturing industry at large, the service and support for the use of a manufacturing machine is strictly dependent on the machine builder, whatever the nature of the user's needs (programming help, comprehensiveness of functionality, quality of results, knowledge acquisition, training and maintenance). Concerning simulation, analysis, part program creation and training and maintenance activities, the current situation is based mainly on equipment without real embedded functionalities. The majority of equipment manufacturers propose products with traditional support services, but they have not explored the association of powerful information technologies coupled to the equipment.
- Additionally, the factory floor is currently far from being a friendly environment compared to an office environment. A typical manufacturing facility has hundreds or thousands of machines and independent systems operating in "synergy" to ensure the fabrication of a product in a timely, cost-effective and quality manner. But each of these machines and production systems or devices that accumulate information on their specific operation is unable to communicate it to anyone or anything else. Dr David Patterson, inventor of the RISC processor, presented at the US AMT (Association for Manufacturing Technology) annual (2006) meeting his observations about the breakthrough progress being made by the computer industry. It has already put in place ways to connect sensors via wireless communication and merge all data with other services, thus realising the ubiquitous environment. If existing CNC systems, which are currently PC-based, were to embrace interoperability standards, a user could then monitor all brands of CNCs, thus leading to integrated solutions and services instead of a mixture of incompatible software and components.

10.2 The STEP-NC Standard

The ability to generate an NC tool path is now commonplace in CAD/CAM systems, but the technology used for programming and control of NC machines is still based on 1950s standards. STEP-NC takes advantage of advances in computing and controllers to overcome the limitations of ISO 6983. Using STEP-NC, an external system such as a process planning system or an integrated CAD/CAM system can deliver instructions for making a part on a machine tool and send those instructions to a CNC system containing an embedded CAM system.

The STEP-NC interface, which is based on an object-oriented data model, was developed in a series of research projects in which many industrial companies and universities participated. The first of these was the ESPRIT III project OPTIMAL (Optimised Preparation of Manufacturing Information with Multi-Level CAM-CNC Coupling) which ran from 1994 to 1997. Next was the European STEP-NC project (EP 29708) which started on January 1, 1999 and ran until December 31, 2001. For further developments and exploitation of the results, a global collaborative project involving the European Union (EU), Korea, Switzerland and the USA (with links to the 1999–2002 Super Model Project) was started within the scope of the Intelligent Manufacturing Systems (IMS) Project in 2002. Within the European Commission (EC)-funded STEP-NC project and the US Super Model Project, a large consortium validated and improved the existing data model for milling, and prepared models for additional technologies such as turning, wire-EDM, wood and glass cutting [10.1–10.4].

STEP-NC concentrates the standardisation effort on information content rather than on implementation technology. STEP-NC standardises how information about CNC machining can be added to parts represented in the STEP product model. The evolution of CNC manufacturing can be viewed from various perspectives, namely software, hardware, manufacturing processes and manufacturing paradigms.

Unlike traditional NC programming, STEP-NC provides an object-oriented data model for CNC with a detailed and structured data interface. It incorporates feature-based programming, with a range of information such as the feature to be machined, the type of tools used, the operations to perform and the Workplan. STEP supports 3D geometry data plus product information, assembly structure, configuration controlled assemblies, and manufacturing features. The information in the product model in this new interface created by a CAD/CAM system can be linked to the information required to control the machine. Figure 10.1 shows the major advances in NC manufacturing since its beginnings to the present.

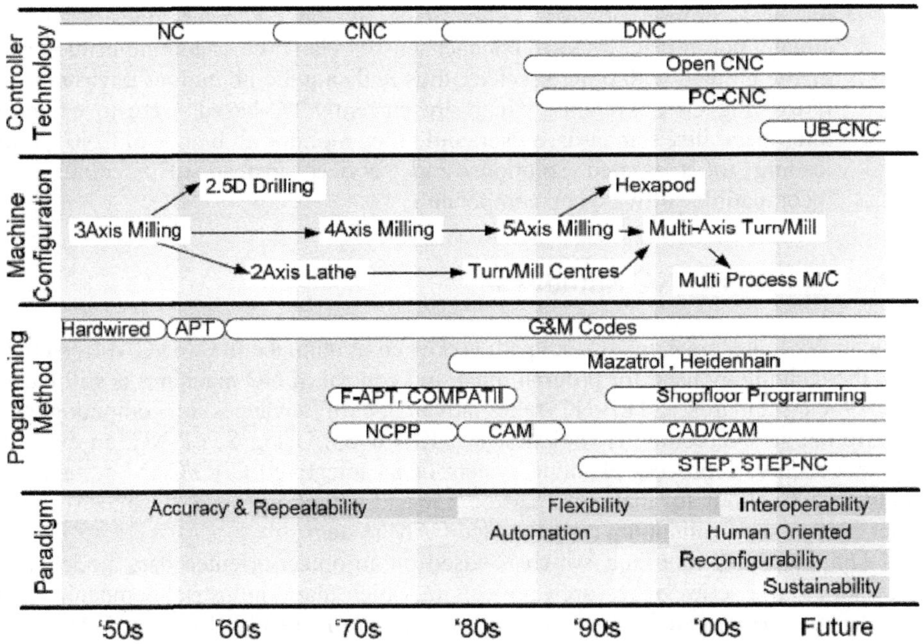

Figure 10.1. Evolution of CNC from the 1950s [10.27]

Basically, STEP-NC allows the seamless and comprehensive exchange of part information from conception to production.

10.2.1 Details of the STEP-NC Standard

The details of STEP-NC are included in the following documents:

- ISO 14649-1:2003 (Industrial automation systems and integration - Physical device control - Data model for computerised numerical controllers - Part 1: Overview and fundamental principles)
- ISO 14649-10:2004 (Industrial automation systems and integration - Physical device control - Data model for computerised numerical controllers - Part 10: General process data)
- ISO 14649-11:2004 (Industrial automation systems and integration - Physical device control - Data model for computerised numerical controllers - Part 11: Process data for milling)
- ISO 14649-12:2005 (Industrial automation systems and integration - Physical device control - Data model for computerised numerical controllers - Part 12: Process data for turning)

The "machining_schema" defined in ISO 14649-10:2004 contains the definition of data types that are relevant for milling, turning and grinding. In addition to this, an implementation needs at least one technology-specific part (e.g. ISO 14649-11 for milling, ISO 14649-12 for turning).

The process data is described using the EXPRESS language defined in ISO 10303-11. Data encoding is carried out using ISO 10303-21.

10.2.2 Characteristics of STEP-NC

One of the main characteristics of STEP-NC is the higher level of information that it holds. While a part program written according to ISO 6983 describes simple tool movements and switching instructions (G and M codes), the STEP-NC interface works at the level of manufacturing features (such as pockets and profiles), operations (such as drilling and roughing) and the sequence of workingsteps. Through this sequence of manufacturing operations on features, all activities necessary to produce the finished part from the raw piece can be described.

Since the data model enables the description of manufacturing operations linked to the original CAD geometry data, the resulting part program can bring a much higher quality of information to the shop-floor. By providing a complete and structured data model, no information is lost between the different stages of the process. Post-processors for machine-specific adaptations of NC programs are no longer needed. The rich information content results in higher flexibility, enabling last-minute changes or the correction of technological values within the part program, e.g. when a tool breaks and needs to be changed. In addition, the man-machine interface at the CNC controller can be more user-friendly.

Since the geometry of raw and finished parts is described using STEP, direct exchange of information between CAD, CAM and NC is possible, enabling an end-to-end CAD/CAM process chain. Geometric data can be imported directly from CAD systems. For example, when working with 2.5D geometry, manufacturing features can be imported from feature-based CAD systems. Then the technology information can be added to generate the part program. The new standard, which does not require machine-specific post-processors, enables Web-based strategies for distributed product development and manufacturing.

From a historical perspective it is clear why ISO 6983 evolved to ISO 14649 (STEP-NC). STEP-NC is the evolutionary response on the control side. Instead of being limited to a fixed sequence of instructions describing "how-to-do-it", the STEP-NC approach is to provide information about "what-is-to-be-made". This means that the machine or the operator has more flexibility to cope with missing tools or even to switch machines.

As well as the flexibility in coping with varying manufacturing conditions, the STEP-NC approach has significant benefits for innovative processes and novel architectures. For innovative manufacturing processes, such as Electrical Discharge Machining (EDM), the operator plays a key role in successful manufacture. A standalone CAM system may have little knowledge of the fine-tuning needed for production, nor of the possibilities of new machines. There is also a time-lag between the introduction of new developments into the process and their exploitation by traditional CAM systems. STEP-NC allows the process developer to be in control and to provide the necessary adaptation [10.5–10.8]. For instance, the STEP-NC Wire EDM Data Model is represented by Part 13 of ISO 14649 in conjunction with Part 10 of ISO 14649 (Figure 10.2). The geometrical features (that may be shared with other processes) are clearly separated from the technology specific to Wire EDM. The ruled surface is the main geometric feature that can be defined explicitly (B-Spline surface) or implicitly (defined on curves). It is an interoperable object thanks to its universal and mathematical formalism that exactly describes the "as-desired" part, thus bringing several additional advantages to calculate the Wire tool path:

- Conventional implicit ruled surfaces based on lines and arcs lead to geometrical inconsistencies
- Explicit ruled surfaces give a complete and precise geometry description (without any ambiguities)
- True 3D offsetting enabled
- Re-parameterisation possibilities of the explicit surfaces allow one to enhance continuity requirements for the surfaces transition

The STEP-NC approach can bring useful and intelligent solutions to the Wire EDM process in order to enhance seamlessly its integration with the product design (CAD) process.

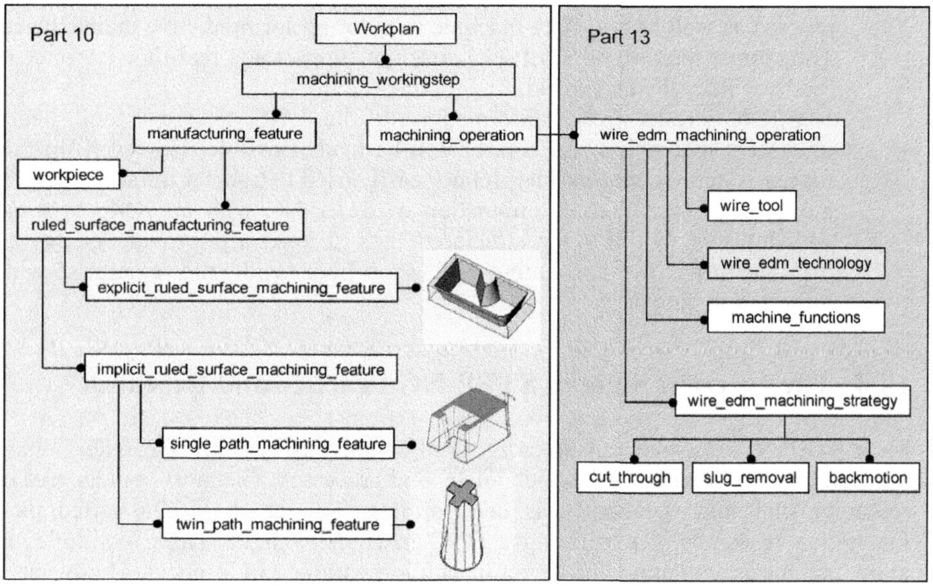

Figure 10.2. STEP-NC data model

10.3 Limitations of STEP-NC

Today, the STEP-NC concept is recognised as the main enabling "Information Technology" for a radical manufacturing industry transformation. However, there are many barriers preventing the acceptance of STEP-NC. These include the following:

- The STEP-NC protocol currently supports neither all manufacturing processes, nor the semantic data related to each process (e.g. tolerance treatment). Its future evolution towards semantic manufacturing enterprise integration will lead to a major overall improvement in STEP-based standards, thus facilitating the implementation of the missing elements in the enterprise information system to optimise the global manufacturing resources of an OEM and its suppliers
- Although STEP-NC has been proven to be the enabler to science-based NC applications, no added value and management services evolving towards collaborative and close-loop "engineering-to-manufacturing" services are being developed (e.g. there is a lack of high-level interpolation, a lack of integrated comparison of the "as-machined" with the "to-be-delivered" part with tolerance management and a lack of realistic simulation of the cutting process based on real machine behaviour)
- Manufacturing industry lacks a real "digital" process, "digital" being, in this context, the marriage of "virtual" simulation and "physical" information captured from sensors, thus enabling realistic validation of the machining

process as well as real-time machine monitoring for predictive maintenance. The current version of STEP-NC does not support any real-time parameters fed back from the real world.

- STEP-NC brings richer information to the CNC machine tool; hence intelligent machining and control will be made possible. However, the file format is neither easy to interpret nor easily used to transfer information over the World Wide Web. Combination of STEP-NC with an XML schema, which offers facilities for structured data access, will provide quality e-manufacturing by connecting the shop-floor with the enterprise-wide information network.

10.4 The Current State of STEP-NC Practice and Research

Most STEP-NC research can be categorised on the basis of the researcher's main focus in terms of manufacturing technology and processes. The most popular area of research from this viewpoint has been milling. Suh et al. [10.9] started their integration research by retrofitting a CNC machine with PC based controller to allow graphical simulation and direct machining without G&M codes. Hardwick et al. [10.10] provided one of the first overviews of STEP-NC compliant manufacturing with respect to milling technology. The research was supported by Hardwick's earlier work on virtual enterprises [10.11]. Based on Hardwick's research, Venkatesh et al. [10.12] used tool centre programming provided by AP238 in a joint effort between Boeing and NIST to illustrate an interoperable manufacturing scenario. Hardwick and Loffredo [10.13] published a paper on this implementation documenting the presentation with four CAM vendors and two CNC controls on two 5-axis machines with different axis configurations.

Some turning and turning/milling research has also been carried out, for example, by Rosso et al. [10.14] who investigated the use of STEP-NC for the manufacture of asymmetric rotational components. They concluded that ISO14649 Part 10's milling features can support the features these complex components require. Xu and Wang [10.15] developed a G-Code free lathe based on STEP-NC. Xu [10.16] described an approach in which the STEP-NC file was converted into a machine native format and passed on to the retrofitted lathe. This low-level language, while not G&M code, is still low-level and interpretation of STEP-NC code into this axis-movement language is not that different to translating STEP-NC into G&M codes. Chen et al. [10.17] proposed a RTCOBRA-based soft bus to realize a framework for turning based on STEP-NC. Eighteen functional modules were involved in the software-based framework.

Although application of ubiquitous, interoperability and knowledge-based technology to the manufacturing domain is only moving slowly, the number of projects in this area is steadily increasing. These efforts are being made in three main ways:

- Pro-active research projects, which are made up of industry-academic consortiums within R&D projects (PROMISE, ATHENA, KOBAS, ELIMA…). All of these projects address the issue of the Digital Factory of the Future, which is based on holistic concepts for the full product lifecycle with the aim to create knowledge-based and knowledge-driven networks, to integrate virtually concurrent enterprise activities and to integrate supply network planning and monitoring. However, none of the projects has created a foundation for vertical semantic integration in the Manufacturing space. Yet it can be argued that pragmatic and realistic horizontal integration can be deployed if, and only if, the vertical integration based on a standard and science-based formalism (such as STEP-NC with extension to semantics treatment) has previously been realised to ensure interoperability.

- Practical business cases (such as the STEP Tools suite, Siemens Archimedes project, FANUC STEP-NC controller, POSTECH controller patent...) can all be considered threats for standardisation and for SMEs. The vendor's strategy might be to impose a de facto standard. The acquisition of CAD/CAM provider UGS by control manufacturer Siemens is one kind of event that could be a catalyst for greater acceptance of standardisation that must propose extended standards encompassing the whole manufacturing/dismantling chain throughout the product lifecycle.
- Standardisation efforts (ISO TC 184 SC1 & SC4, NIST, ASME B5/TC56, OMAC, AMT project). Standards are required for the smooth and seamless exchange of part information from conception to production. The European initiatives (Esprit & IMS projects) have spread the STEP standard to the NC manufacturing process in the form of STEP-NC, and have encouraged many researchers and industries in the USA and Asia to carry out next generation projects. These address the subject of a more intelligent controller directly driven by an extended release of STEP-NC to sustain full interoperability through production networks. [10.18–10.20]

10.5 The Current Problem Statement

The remarkable opening of the world over the last 25 years has dramatically changed the landscape of both manufacturing industry competition and market vitality. Companies can now sell products worldwide. These "Global Products" provide huge opportunities. They allow billions of people to benefit from products to which they previously had no access. They allow companies to offer products to a global market of more than 6 billion customers and users. The opportunities of Global Products are not limited to just a few large manufacturing companies with thousands, or hundreds of thousands, of employees.

The opportunities are also there for small and medium enterprises (SMEs), with tens or hundreds of employees, of which there are millions throughout the world. The smaller company may sell its product direct to end users and consumers worldwide. Alternatively it may supply its product to a larger company, operating worldwide, that will include it in the products it offers to its customers. However, in addition to these opportunities, there are also many risks for manufacturing companies, one of which is a fast-changing environment. Among the changes that companies face are increased complexity, globalisation, geopolitical developments, social and health issues, changing business models, improved telecommunications, transport and travel, new technologies, new IS applications (such as Product

Lifecycle Management), new company structures, new customer requirements, changes to products, shareholder influences, financial market influences, regulation, deregulation, environmental concerns and sustainable development issues [10.21].

In the new globalised environment, manufacturing industry in countries with developed economies has to compete with manufacturing companies in low labour-cost countries. The survival of manufacturing companies will depend on their ability to change: large OEMs will be required to integrate and orchestrate manufacturing through the value chain by optimising their global supply chain. SMEs will need to create more value by adapting and reconfiguring their production to meet widely varying demands and intensified price competition. The new challenge for manufacturing industry in developed countries, mainly composed of SMEs (generating more than half of the total production output), is to reengineer the manufacturing process towards *mass customisation* (from yesterday's mass production) through the development and exploitation of new philosophies in industrial control engineering including process measurement and simulation. Currently, human-centred manufacturing processes, heavily dependent on highly skilled personnel and conventional machining equipment, typify the manufacturing SME. This drastically limits its agile performance and responsiveness, although the requirements for increased responsiveness to customers' demands have intensified in recent years. A major challenge for manufacturing companies under the productivity pressures (shorter time-to-market) and greater product variability is to make the *first part correct and fast*.

Practically their needs for improvement are twofold:

- To meet the highly individualistic customer desires for participation in the design and production procedures within the "configure-to-order" requirement
- To deal with the unpredictable pattern of demands in the "make-to-order" market

To meet these needs requires a "*Smart NC Machining Workstation*" that represents the brain and knowledge repository of the manufacturing system, based on high bandwidth information, real-time networking, and the ability to be adaptive (flexible and self-learning).

10.6 The Next Steps Beyond the State of the Art

During the timeframe of the STEP-NC projects, Product Lifecycle Management (PLM) emerged. PLM is the business activity of managing a company's products all the way across their lifecycles, from the very first idea for a product all the way through until it is retired and disposed of, in the most effective way. PLM can be seen as a horizontal cross-functional (across marketing, design, manufacturing, maintenance and disposal) counterpart to the vertical single-function approach of STEP-NC (CAD design to NC controller to machine tool to machined part).

The mid-1980s had seen the first Information Systems (IS) applications specifically developed to manage design engineering and manufacturing engineering data. These Engineering Data Management (EDM) systems, as they were called, matured into Product Data Management (PDM) systems in the 1990s. The twenty-first century then saw the emergence of Product Lifecycle Management (PLM) as a major pillar of corporate IS architecture. PLM manages a product throughout its lifecycle – "from cradle to grave" – making sure that everything works well with the product and making sure the product makes good money for the company. This requires full access to product data across the product lifecycle. [10.22]

This new activity of PLM emerged in about 2001. Before then, companies implicitly managed products across their lifecycles, but they did not do this, even conceptually, in an explicit, "joined-up", continuous way. Because it was not done explicitly, things fell through the cracks:

- Decisions were not co-ordinated
- Risks were not fully analysed
- Information got lost
- Customer requirements were misinterpreted
- Time was wasted
- Key relationships were ignored

As a result, although it appeared that everyone in the product development, manufacturing and support chain had done their work correctly, the product did not work properly in the field. By bringing together previously disparate and fragmented activities, systems and processes, PLM helps overcome the many problems that resulted from the old unconnected approach.

PLM also makes it possible to manage a company's projects to innovate and develop products, and their related services, all the way across the lifecycle. Without new products, company revenues will decline. Innovation activities are the source of growth and wealth generation in a company, and PLM makes them more effective.

PLM helps a company get control of its products and services, and enables it to take responsibility for them across the lifecycle. Mastering the activities in the lifecycle makes it easier to provide reliable products, sell services on them, and even sell services on competitors' products.

PLM is a holistic business activity addressing many components such as products, organisational structure, working methods, processes, people, information structures and information systems. Like STEP-NC, it is a new paradigm, a new way of looking at the world.

10.6.1 Data and Information in the PLM Environment

In the typical manufacturing company, with large numbers of people involved in the product lifecycle, there are many low-level problems in the PLM environment, in particular with product data. People often waste a lot of time looking for it. Sometimes they are unable to find the information they need. If they can find it, it may not correspond to the actual state of the product. For example, a facility drawing may not correspond to the actual physical facility layout. Developers may be unable to access rapidly a particular design among the mass of existing designs. To find specific information, they may have to search through several paper and electronic files. They lose valuable time. Some design engineers spend up to 80% of their time on administrative and information retrieval activities. They develop new designs which may be almost identical to existing designs with the result that unnecessary additional costs are generated as the new designs are taken through the various activities necessary for manufacture, and then supported during use.

As more and more data are generated on computer systems, it becomes more and more difficult, with control and management procedures introduced when data volumes were thousands of times lower, to track the location of data, to prevent unauthorised access and to maintain up-to-date product configurations. Organisations may have terabytes, petabytes, exabytes or even zettabytes (10^{21} bytes) of data. Companies hold thousands, or even millions, of drawings. 3D CAD part models may take up several gigabytes.

All types of product data and information are difficult to manage, even those as mundane as technical manuals. These can easily become outdated as products are upgraded, but the manual is often not updated. Similarly, logistics support data can get out of control. Inadequately documented configurations become difficult to maintain. Spares replenishment becomes inaccurate, and customers have to immobilise products while efforts are made to identify correct replacement parts. When the right part arrives, the right handling equipment and maintenance tools are not in place. Information about problems with product use is not fed back to developers, with the result that they design the same problems into the next generation of products.

For reasons such as these, product development and support in a whole range of manufacturing industries is delayed for a myriad of apparently random and minor, but cumulatively significant reasons. Product and service quality is erratic despite huge investments in technology and quality programs, and significant expenditure of management time.

In 2005, a new STEP standard, ISO 10303-239 Application Protocol: Product Life Cycle Support (PLCS), was published to address the information requirements of product lifecycle management. At the time of its publication, much of the product data addressed by PLCS was being created by independent processes which ignored, or failed to exploit, data already created. Some point-to-point exchanges had usually been established, but these did not address all the necessary configuration management issues across the product lifecycle. In particular, difficulties existed between the supplier and end-user communities, with neither having access to information created by the other. As a result, information which already existed was frequently re-generated or manually re-entered with a proliferation of errors over time. The feedback of data from the supply chain or in-service activities was patchy, or too vague to be useful [10.23].

The PLCS information model enables open information exchange and data consolidation of the large information set required for efficient and effective product support across the lifecycle. PLCS provides a means of achieving integration within and between organisations whilst, at the same time, reducing user dependency on specific software vendors through the use of a neutral and international standard for product data.

10.6.2 Next Generation Controller

The Smart NC Machining workstation is the next generation controller. It will offer new information-driven and science-based solutions enabled by the *synergic use of sensor, network and services (via software).*

Extensions will be made to the STEP-NC standard for traceability, feed/speed optimisation, and integrated machining and measurement. This will open up new perspectives for machining processes such as High Speed Machining and for other advanced manufacturing approaches that are currently very promising, but whose implementations are still limited due to the shortcomings of innovation in NC technology.

The foundation of the Smart NC Machining Workstation is a *Ubiquitous Manufacturing Platform* connected to a *Manufacturing Information Pipeline (MIP)* and distributed hubs for knowledge/data acquisition (from compliant devices) as well as agent-based communication (for legacy data) to support multi-sensor feedback. Semantic application interfaces provide bi-directional exchanges from sensors (inspection results, thermal readings, force monitoring, tool monitors…), machine models, CAx processes and business process software. This structure forms the kernel of the system and supports a *self-learning decision controller* providing corrective software, preventive maintenance and hardware diagnosis based on statistical process control. Standardisation of process control will be achieved through the use of STEP and related standards to form an interoperable and adaptable solution across varied equipment and devices. Only a really standardised and totally *digitised product-process-plant-resource description* will enable the radical transformation from traditional resource-based manufacturing to a knowledge-based, networked and adaptive manufacturing system that provides new and competitive added value.

This focus on the vertical integration dimension of the manufacturing process will allow industrial companies (mainly SMEs) to gain competitive advantages within the global economy, namely:

- To reduce "time to part" due to new capabilities such as geometry and process planning knowledge and availability at the factory floor level, thus offering equivalent *"wysiwyg" features* (what you see is what you get) and *"plug-and-produce" ability* equivalent to business software solutions.
- To preserve the design intent through science-based description of the "as desired" part. In this approach, the responsibilities of calculating tool end-positions (tasks previously performed by the host) will now be shifted to the new controller. This provides a safe environment for designing surfaces within the controller and eliminates the transmission of large data files. The creation of these files can be avoided by performing the tool-positioning function in real-time. The computational power needed for real-time operation is provided by a parallel-processing network. The network implements the B-spline or Bezier surface-representation and interpolation technique and the projection tool-positioning method. These algorithms will be executed through on a network of grid processors.

- To provide a *self-learning NC machining workstation* to analyse machine health based on intelligent interpretation of feedback from sensors and part measurement with respect to tolerance.

SMEs will then be able to run their businesses at an international level as customer-centric within a cooperation network, able to comprehend the "to-be-delivered" demand without ambiguity and to provide accurately the *"as-delivered"* documentation to end-users at the point of use, wherever the user might be.

In addition, one can envisage the integration of the manufacturing Information and Knowledge into the overall Product, Process and Resource management throughout the lifecycle of a product, thus delivering full traceability from design to individual product for all the different domains. In the manufacturing domain, (human) knowledge management based on closed-loop PLM applications will have an important impact on manufacturing enterprises. Whatever the type of company, OEM or SME contract manufacturer, the companies which come out on top in the global manufacturing environment, will depend crucially on the integration of supply-chain lifecycle information models and their ability to turn best the information into knowledge-based best practices. Identifying the value stream and creating a continuous and seamless process flow is a crucial challenge for manufacturing industry to gain competitiveness in emerging markets. With the application of the closed-loop PLM concept, the customer's voice as well as the machinist's know-how will be better and systematically considered in product development, manufacturing and servicing, thus enriching the information base. Additionally, information on the efficient and interoperable use of machines and systems will sustain green manufacturing. [10.24–10.26]

10.7 Conclusions

Under the impact of the enhanced information technology and competitive pressures that face developed countries, the actual situation as well as the next trends and challenges can be summarised as follows:

- The percentage share of manufacturing in employment and value added has decreased in developed countries for decades reflecting structural changes in the global economy.
- Global manufacturing is expected to grow, fuelled by emerging economic development.
- Manufacturing remains the basis for a high quality of life, as it provides us with manufactured goods for our comfort and our quality of life.
- Today's manufacturing is, and has to be, a service-oriented and high-tech industry, despite its inherited image of a dirty manual sector. Increasingly in manufacturing companies more emphasis is put on knowledge as a source of competitive advantage. This knowledge appears in many forms, such as new technologies, or how to manufacture complex products, or understanding of changing customer needs derived from customer data.
- Manufacturing industry contributes significantly towards environmental problems, not just during manufacture, but also during use and final disposal of products.

- New methods and approaches of combining manufacturing activities are generating new opportunities. The resulting manufacturing networks can take various different shapes, ranging from integrated supply chains around a key player to collaborative SME networks, strategic alliances, virtual organisations, extended enterprises, etc.

Although co-operation across company boundaries and the sharing of knowledge are seen as a panacea for different types of organisations, the emulation of the existing best practice by all players in the market seems more difficult than theory suggests. One reason for this is certainly insufficient interoperability and standardisation.

The new factors – globalisation, service orientation, knowledge intensity, and responding to environmental concerns – highlight the importance of undertaking research in the area of manufacturing on a global level.

The game is not over in developed countries. It continues on a global scale but the standards for interoperability and the rules for equity are still dramatically missing. As shown in Figure 10.3, STEP-NC and its further extension can be seen as a stepping stone towards a "brave new world" of manufacturing.

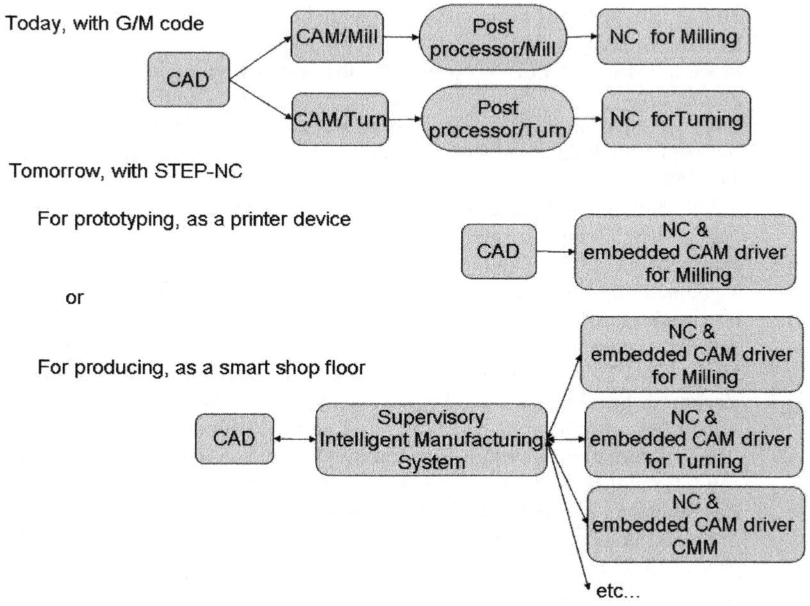

Figure 10.3. The change from G and M codes to STEP-NC

Acknowledgment

The authors would like to thank the European Commission for funding a series of research projects into this subject.

References

[10.1] Maeder, W., Nguyen, V., Richard, J., Stark, J., 2002, "Standardisation of the Manufacturing Process: the IMS STEP-NC project," in *Proceedings of the IPLnet Workshop,Saas-Fee, Sept.9-11 2002.*

[10.2] Ali I., Bedi S., 1992, *NC Controller for Surface Machining,* The International Journal of Advanced Manufacturing Technology, Volume 7, Number 3

[10.3] Hardwick M., Lofredo D., 2001, *STEP into NC*, Manufacturing Engineering - January 2001, pages 38-50

[10.4] Suh S.-H., Cheon S.-U., 2002, *A Framework for an Intelligent CNC and Data Model,* The International Journal of Advanced Manufacturing Technology, Volume 19, Number 10, 2002

[10.5] Suh S.-H., Cho J.-H., Hong H-D, 2002, *On the architecture of intelligent STEP-compliant CNC,* Int. J. Computer Integrated Manufacturing, Vol. 15, No. 2, 168–177

[10.6] Richard J., Nguyen V.K., 2004, *Interface intelligente pour intégrer CAO-FAO et MOCN*, Revue MSM informatique industrielle, 12/2004

[10.7] Richard J., Nguyen V.K., Stroud I., 2004, Standardisation of the Manufacturing Process: IMS/EU STEP-NC project on the Wire EDM process, International IMS Forum 2004

[10.8] Balic J., 2004, *Intelligent Computer Numerical Control Unit for Machine Tools*, Journal of Intelligent and Robotic System, Volume 40, Issue 4, Aug 2004; Pages: 343-358

[10.9] Suh S., Noh S. and Cho Y. 1995, *A PC-based retrofitting toward CAD/CAM/CNC intergration*, Computers and Industrial Engineering, Vol 28, No 1, pages 133-146

[10.10] Hardwick M., 2002, *Digital manufacturing using STEP-NC*, Tech, Paper-Soc of Manufacturing Engineers, MS02-242

[10.11] Hardwick M., Morris K., Spooner D., Rando T., Denno P., 2000, *Lessons learned developing protocols for the industrial virtual enterprise*, CAD Computer Aided Design, Vol 32, No. 2, pages 159-166

[10.12] Venkatesh S., Odendahi D., Xu X., Michaloski J., Proctor F., Kramer T., 2005, *Validating portability of STEP-NC tool center information*, Proceedings of the ASME International Design Engineering Technical Conferences and Computers and Information in Engineering Conference – DETC2005, pages 285-290

[10.13] Hardwick M., Loffredo D., 2006, *Lessons learned implementing STEP-NC AP-238*, International Journal of Computer Integrated Manufacturing, Vol 19, No 6, pages 523-532

[10.14] Rosso R.S.U., Newman S. T., Rahimfard S., 2004, *The Adoption of STEP-NC for the Manufacture of Asymmetric Rotational Components*, Proc of the ImechE, Part B:Journal of Engineering Manufacture, Vol, 218, No 11, pages 1639-1644

[10.15] Xu X.W., Wang J., 2004, *Development of a G-Code free, STEP-compliant CNC Lathe*, American Society of Mechanical Engineers, Computers and Information in Engineering Division, pages 75-82

[10.16] Xu X.W., 2006, *Realization of STEP-NC enabled machining*, Robotics and Computer Integrated Manufacturing, Vol.22, No.2, pages 144-153

[10.17] Chen X., Zhang C., Lan H, Zhai P., Wu H., 2005, *A Framework for CNC turning system based on STEP-NC*, Proceedings of SPIE - The International Society for Optical Engineering.

[10.18] Sääski, J., SalonenT., Paro J., 2005, *Integration of CAD, CAM and NC with Step-NC*, VTT Industrial Systems White Paper. VTT Working Papers 28

[10.19] Xu X., Wang H., Mao J., Newman S., 2005, *STEP-compliant NC research: the search for intelligent CAD/CAPP/CAM/CNC integration*, International Journal of Production Research, Volume 43, Issue 17 September 2005, pages 3703 – 3743

[10.20] Sokolov A., Richard J, Nguyen V.K., Stroud I., Maeder W., 2006 *Algorithms and an extended STEP-NC compliant data model for wire electro-discharge machining based on 3D representations,* International Journal of Computer Integrated Manufacturing, 19 (6) : 603-613

[10.21] Stark, J., 2007, *Global Product: Strategy, Product Lifecycle Management and the Billion Customer Question*, Springer, Berlin.

[10.22] Stark, J., 2004, *Product Lifecycle Management: Paradigm for 21st Century Product Realisation*, Springer, Berlin.

[10.23] Dunford, J., Bergström, P., Stark, J. 2007, *Standards-based PLM: Re-engineering the Aftermarket with PLCS*, Eurostep White Paper. http://www.eurostep.com/

[10.24] Meo, F., 2007, *Algorithms for the optimized generation of trajectories through splines, and their integration into open architecture control systems,* IPROMS, 2007

[10.25] Newman S., Nassehi A., 2007, *Universal Manufacturing Platform for CNC Machining*, CIRP Annals - Manufacturing Technology, Volume 56, Issue 1, 2007, Pages 459-462

[10.26] Li W., Red E., Jensen G., Evans M., 2007 *Reconfigurable Mechanisms for Application Control (RMAC): Applications, Computer-Aided Design and Applications,* 2007

[10.27] S.T. Newman, L. Ali, A. Brail, C. Brecher, P. Klemm, R. Liu, A. Nassehi, V.K. Nguyen, F. Proctor, R.S.U. Rosso Jr., I. Stroud, S-H. Suh, M. Vitr, L. Wang, X.W. Xu, *The Evolution of CNC Technology from Automated Manufacture to Global Interoperable Manufacturing.* The 2nd International Conference on Changeable, Agile, Reconfigurable and Virtual Production (CARV 2007), Toronto, Canada. 22-24 July 2007.

11

Standardised Process Control System for CNC Manufacturing

Sanjeev Kumar[1] and Stephen T. Newman[2]

Innovative Design and Manufacturing Research Centre, Department of Mechanical Engineering, University of Bath, Claverton Down, Bath, BA2 7AY, United Kingdom
Email: [1]s.kumar@bath.ac.uk, [2] s.t.newman@bath.ac.uk

Abstract
Manufacturing firms continuously strive to improve existing methods or develop new ideas to reduce production costs and lead times in order to provide quality assured parts rapidly to customers. Maintaining control over the processes involved in manufacturing is therefore vital. The manufacturing process chain includes product design, machining and measurement processes as well as the interpretation of measurement results. However, these process chain elements are currently regarded as separate islands of information. In the large majority of current manufacturing companies the machining and inspection processes are not integrated. This limits the process control capability in terms of modifying the machining process parameters. In-process modification of machining parameters such as the work coordinate system (WCS) offsets, tool diameter and length, etc. can lead to improvements in the quality of manufactured parts and also control the rejection rates for parts produced in batches. In this chapter, literature related to statistical process control and manufacturing data analysis are presented together with a commercial process control solution, namely the Renishaw Productivity+. The authors envisage the use of STEP-NC standards as one of the ways to integrate machining and inspection processes for CNC machine tools. Based on the standards a process control information model for CNC manufacturing has been specified. The final part of the chapter describes a standardised process control system together with a computational prototype based on this system and its application with a simple piece.

11.1 Introduction

CNC machining is at the final stage of the CAx process of a highly complex chain that starts from the design of a part through to its manufacture and inspection. In recent years, CNC machines have provided greater reliability, increased capabilities coupled with more advanced attributes of servomechanism control, increased processor speeds and user-friendly programming tools. Today the measurement accuracy using either On Machine Measurement (OMM) or Coordinate Measuring Machines (CMMs) has reached an acceptable limit in real-time production environments [11.1]. However, until now CNC machine tools have had limited process control capability. The majority of the current state-of-the-art CNC production systems use statistically process control techniques for the manual modifications of process plans at manufacturing shop-floor. In the current manufacturing setup, skilled engineers are required in addition to the machine operators for interpreting the inspection results obtained from either OMM or CMMs. The conventional method is to plot the control charts for the measurement results and investigate whether the process is under control or not, with subsequent corrective measures being taken for adjusting the process deviations [11.2]. However, this involves a large amount of scrap, especially if the process is out of control after the measurement of a machined batch of parts.

One of the primary reasons for the lack of process control systems is the use of independent machining programs and inspection programs for components on the machine tool. The need for integrating the machining programs along with inspection programs to achieve process control has been realised in industry [11.3] as well as by academic researchers [11.4]. Using this concept of integration for machining and measurement programs, two types of process control systems have recently been developed. A prototype process control system based on STEP-NC standards has been realised at the University of Bath [11.5]. This system enables the automatic modifications of the process parameters for feature placement locations for prismatic components. One industrial solution from Renishaw, a leading manufacturing company of precision measurement products, is a process control system which enables machine tool users to integrate measurement cycles into the machining part program. This system automatically provides compensation for tool offsets, fixturing error, tool wear, etc. into the CNC machine registry settings.

This chapter describes process control in CNC manufacturing and presents the advantages of such systems for CNC machining, with an example of the Renishaw process control system. A brief review of contemporary Statistical Process Control (SPC) techniques and manufacturing data analysis is also provided. The main goal of the chapter is to present a standardised approach for achieving process control in CNC manufacturing. A framework for standardised process control is presented together with information models which underpin this system. The final part of the chapter provides a general discussion of the use of process control systems and how they provide benefits through increasing the productivity of manufacturing systems.

11.2 Process Control

Process control has evolved as a term from a number of different industries such as the chemical process industries, semiconductor manufacturing industries, machining industries, etc. The function of a process control system in any industry is to monitor the behaviour of the process and provide appropriate feedback to adjust for undesired deviations. In this regard, the authors envisage a generic architecture for a process control system as presented in Figure 11.1. A similar architecture for process control has also been presented by Siu [11.6].

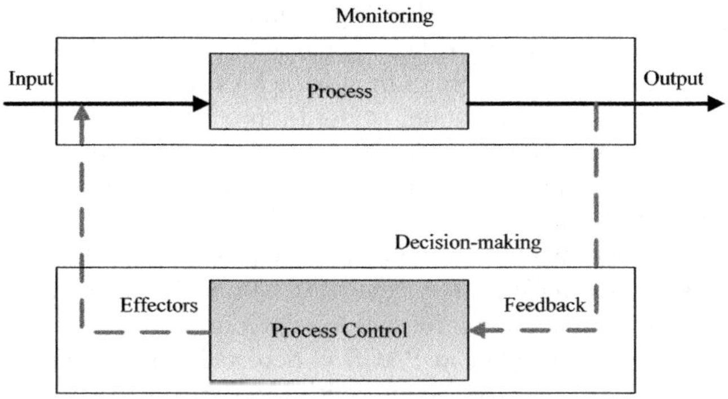

Figure 11.1. An architecture of a generic process control system

Process control systems primarily include two components, namely process monitoring and decision-making. Process monitoring is related to observing the acquired input and output parameters of the process involved in a particular system. The role of a decision-making system is to analyse the output results and input parameters associated with the involved processes within a particular system. Based on this analysis, a decision-making system further provides the compensation parameters to adjust the deviation in the output results from their desired values. This helps in controlling the individual process behaviour as well as overall outcome of the processes within a particular system. This system can be CNC machining, chemical process, wafer fabrication process, etc.

Process Industries

The importance of process control has been realised initially in the process industries to monitor the running of the manufacturing facilities [11.7]. According to this, process control is generally related to attaining control over mechanized processes, where materials undergo chemical or physical transformations. Process control has also been recognised at different levels inside the factory (plant) such as a cell or workstation. In the chemical process industries, Marlin et al. [11.8] observed process control from a viewpoint of plant economics, production requirements and capacity, and equipment capacity.

IBM Austin faced a challenge in 1989 related to controlling the manufacturing process involved in Electronic and Assembly and Test (ECAT) facilities [11.9]. To resolve this issue, a methodology was developed that included the reduction of raw paste ingredients to a minimum, standardisation and control of manufacturing process. A Dynamic Manufacturing Process Control (DMPC) system for ECAT facilities was implemented by Pearsall and Raines [11.10] that incorporated changes in customer feedback into the manufacturing process.

CNC Manufacturing

CNC machine tools commercial utilisation has increased radically over the last few decades. Manufacturing requirements along with quality requirements have become progressively more stringent for achieving high-quality products [11.11]. In this regard, Mou observed that a major U.S. manufacturer Caterpillar started a Process Variability Reduction (PVR) program for achieving consistently higher part accuracy by thermal compensation, closed-loop process control and chatter avoidance. Traditionally, process plans were checked for feasibility using charting techniques to verify control over the manufacturing processes. Due to intrinsic variations in manufacturing processes, the feasible process plan alone did not assure the process control and manufacturing of quality products. In this regard, a closed-loop Process Analysis and Control System (PACS) was developed by Cheraghi et al. [11.12]. This system utilised the information related to dimensions of manufactured parts. The methodology adopted in PACS is depicted using a flowchart in Figure 11.2. Figure 11.2 shows a closed-loop control system in terms of selection of a process plan, on-line dimensional checking of parts, process adjustment routines and rerouting of machining processes. The methodology shows an iterative process where the modifications in machining processes are being provided until the feasible dimensions of part are obtained, or else the part is declared as scrap. Although a closed-loop control methodology has been demonstrated, it contains several deficiencies, such as not identifying what is being fed back in the system at its preceding levels, how it is being fed back and what types of variations in machining and measurement are being adjusted.

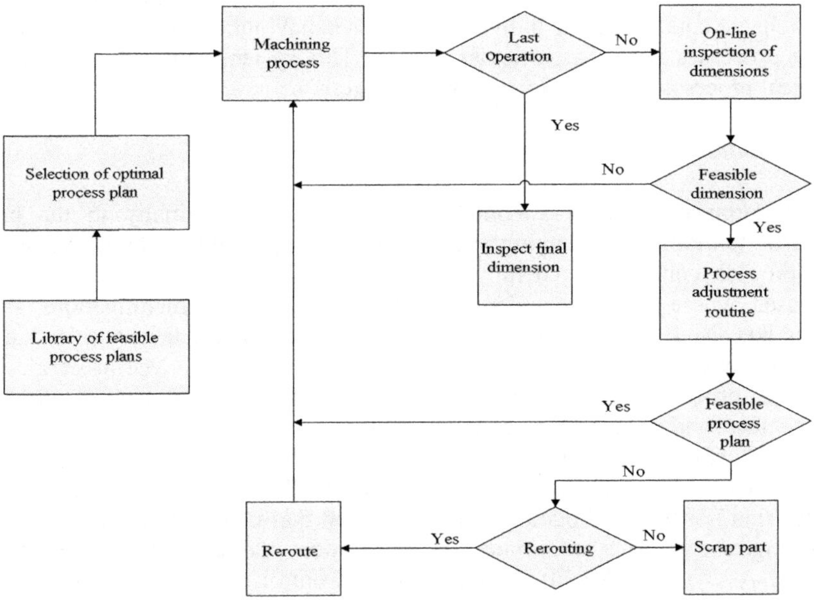

Figure 11.2. Closed-loop PACS methodology [11.12]

Ribeiro et al. [11.13] defined process control as integrated process control having two attributes. The first is the radical reduction of the effort necessary to analyse and interpret quality data and the second to provide total access to the information relating to both processes and product quality. Liang et al. [11.14] reviewed machining process monitoring and control and concluded that automation in machining processes is useful for both large batch production environments and small batch jobs. It has also been recognised that process automation can autonomously tune machining parameters such as feed, speed, and depth of cut, etc. for increasing the performance of the machine tool in terms of part tolerances, surface finish, operation cycle time, etc. In the context of CNC manufacturing, the monitoring parameters, i.e. inputs and outputs, are essentially related to the measurements for the part. The data gathering techniques, i.e. OMM and CMM, are being used in current CNC manufacturing systems for acquiring these input parameters.

11.2.1 Definitions

The authors define process control in CNC manufacturing as the "ability to monitor machining parameters and apply corrective measures where appropriate, in order to provide confidence in the machine tools to produce consistently parts within the desired tolerance limits". The authors believe that the process control systems developed based on this definition are able to monitor and control the processes on-line and, as an effect, the productivity of the manufacturing systems is increased. This process controlled system is intended to provide the corrective measures (in-process) for:

- Fixturing errors
- Tool wear (which relates to surface roughness)
- Thermal drift in part dimensions during machining
- Feature placement location deviations

The advantages of such system are outlined as follows:

- Consistent production of parts in the desired tolerances
- Increased productivity of the system in terms of number of parts being manufactured
- Maintaining the desired process capability throughout production cycle by automatically compensating for the disturbances associated with the involved machining processes fixturing, etc.
- Improvement in overall process cycle time
- Reduced machine downtime
- Reduced sampling size for CMM inspection (more confidence in machining process on a M/T to produce parts within desired tolerance limits)
- Better tool handling
- Reduced number of manual operators on the shop-floor
- Reduced scrap and rework

11.2.2 Requirements for Developing Process Control Systems

One of the primary reasons for the lack of process control systems is the use of independent machining programs and inspection programs for components on the machine tool. Measuring manufactured parts and storing the results in an appropriate format is an important requirement of process control in machining. The authors' vision for achieving process control is to utilise standards for integrating high-level information and knowledge across the process chain of product design, process planning, machining, measurement and data feedback. This enables the specification of the machining know-how at the abstract level, i.e. what and how to manufacture and inspect, and also what and how to feedback data by monitoring the integrity of the data throughout the process control chain. The requirements for development of a process control system are availability of data models, feedback requirements and data transfer mechanism. The data models are required for the product information, process information and feedback information. The product information consists of geometry and representation techniques for features and tolerances. The process information is related to generating a process plan which includes both machining and inspection information together. The feedback information is related to inspection items and their results format. The feedback requirements are used to construct the decision-making engine for the process control system. These feedback requirements are in terms of what parameters are needed to be feedback to modify the control of the processes. The data transfer mechanism means in what format and how the information between different stages of the CAx chain (i.e. design, process planning, measurement, machining) is to be exchanged. This can be either in the ASCII text file format (ISO 10303-Part 21) [11.15] or XML format (ISO 10303-Part 28) [11.16] for transferring the data to CNC machine tools.

11.2.3 Process Control Solutions for CNC Machine Tools

The work in development of process control for CNC manufacturing has been in progress for the last two decades; however there are only a few solutions available for closed-loop machining. In academics, STEP-NC standards based solution has been proposed. As an industrial solution for CNC machine tools, a process control system for closed-loop machining is provided by Renishaw.

STEP-NC Based Closed-loop Machining

Over the last 5 years, a number of research activities have been carried out towards achieving Closed-loop Machining (CLM) using STEP-NC. In this area, there are four leading research centres: the Laboratory for Machine Tools and Production Engineering (WZL) at Aachen University, Germany led; the National Institute for Standards and Technology (NIST) in US; the University of Auckland led; and Advanced Machining Processes and Systems (AMPS) at the University of Bath, UK.

The first prototype for CLM was demonstrated in WZL at Aachen University [11.17]. WZL showed the verification and integration of inspection standards using a STEP-NC based CAx chain. The demonstration was carried out on two STEP-NC based controllers, namely a Sinumerik 840D with STEP-NC enabled ShopMill controlling a Chiron machine tool and a WZL-NC control connected to a Maho 600E machining centre. The ISO 14649 Part 16 standard was used by Brecher et al. [11.4] to create a closed-loop inspection system. Fred Proctor at NIST, USA in association with Boeing, General Electric and Unigraphics demonstrated their initial work on CLM in May 2005 [11.18]. The demonstration revealed the use of probing results gathered on a CNC by modifying AP 238 data. The proposed system utilizes three Conformance Classes (CC), namely CC1, CC2 and CC3 of AP238. The advantages of this CLM system are claimed to be the inclusion of a few operational errors, faster setup times and the ability to perform flexible manufacturing.

The original research in CLM by Newman et al. [11.19] was carried out at Loughborough University where a STEP-NC compliant inspection framework [11.20] was developed using ISO 14649 Part 16 standards. This framework was able to generate a generic STEP-NC measurement process plan which would be read by different types of feature-based inspection systems and controllers. This system enables the storing of measurement results for part features in a generic process plan file but did not present any methodology to utilise these results and provide feedback in the CNC machining process loop. Xu's research on CLM is captured in the framework of a STEP-NC enabled real-time system described in Zhao [11.21]. This system specified an OMM methodology and how it can be incorporated during various stages in machining operations. However, this system did not provide the methodology for closing the machining process loop.

KTH, the Royal Institute of Technology (Stockholm), Eurostep, Sandvik Cornomant, Scania and STEP Tools demonstrated the recent advances in closed-loop CNC machining [11.22] in March 2008. This event exhibited how machining information such as tolerances, wireless measuring equipment and ISO 13399 [11.23] tool descriptions can be integrated with STEP-NC to compensate for increased precision machining of parts. A prototype of cylinder head block containing 2½ D hole features was machined with hand-held wireless callipers being used to capture the tolerance data of the part.

An Industrial Process Control System

Renishaw plc provides an industrial solution for process control on CNC machine tools generally known as Productivity+. This has gained wider acceptability in manufacturing practices worldwide. This system provides the following capabilities which leads towards achieving process control in current state-of-the-art in CNC machining environment:

- Integrating machining and inspection cycles together in one program which provides the ability to perform inspection wherever required. This enables in-process automatic modifications such as WCS updates, tool length and diameter updates, etc. to attain control over the machining processes and consistently produce parts within desired tolerance limits.

Example: In manufacturing of the parts (e.g. mould parts, etc.) which involve the roughing and finishing operations, process control may prove to be advantageous for increasing productivity of such systems. In machining of such parts, the process control system makes use of measured data (in-process) for a feature after roughing operations and provides automatic modifications for carrying out the finishing operation. These actions could be either to continue machining or update the tool length, re-cut features then perform finishing or stop machining. In the absence of a process control system, these decisions are carried out manually or require time consuming off-line measurement operations for required features which essentially increases the machine down-time and decreases the production efficiency of such a system.

- In the current state-of-the-art CNC manufacturing systems, low level programming codes, i.e. G codes (ISO 6983) are used to perform machining instructions. These commands only describe the tool path approach direction and do not posses any associative information related to part design and machining process knowledge. This essentially means that manufacturing knowledge is not preserved, maintained and used throughout the process chain and thus control over the processes is not achieved. However, the process control system can be used at the CAM programming level of the process chain by which the information related to machining of parts can be preserved and, together with the measurement information, process control information can be provided.

The framework of the Renishaw process control system has been presented in Figure 11.3. Figure 11.3 shows the utility of the Productivity+ system in current manufacturing process chain comprising of a CAD system, CAM system, post-processors and a machine tool to provide the process control.

The process control system is able to read ISO programs converted from CAM systems, which contain the machining instructions, and then add the inspection routines for the required features of the component into the same machining program. This machining and inspection program is post-processed into the respective format of a range of machine tool controllers. Using this system, an internal closed-loop control on the machine tool is realised. The feedback obtained for the process control system is in the form of modified machine parameters such as tool offsets or NC variable updates inside the machine tool controllers. In Figure 11.3, the process control methodology has been highlighted using a dashed rectangle. A CNC report is obtained in the end which lists the nominal, actual and deviated measurements of desired features which can be further used to carry out the statistical investigation such as evaluating process capability indexes, etc. Another functionality of this process control system is that it also supports its CAM plug-in (Figure 11.4). This provides parametric programming advantages such as using this plug-in, selected cutting processes and its specified parameters can be modified.

The Renishaw process control system provides a collision detection visualisation which enables the users to simulate the probing operations and detect errors during simulation such as ill defined inspection operations, collisions, etc. This system compensates for fixturing errors, thermal drift, etc.

With the use of this process control system, modifications are obtained in-process for a number of errors involved during manufacturing of components which are compensated. These errors can be categorised as operator errors, fixturing errors, tool wear errors and part deformation errors. With the help of the Renishaw process control system, CNC manufacturers can reduce the waste by minimising scrap parts production. The throughput in terms of batch sizes at reduced machining cycle times can be increased. Since this process control software provides in-process checking and modifies for dynamic sources of errors, there is also no need to check each and every part on the dedicated measurement resource such as a CMM. Instead of checking each part, sampled inspection can be carried out for different batch sizes.

This reduces machine down-time and considerably enhances the total productivity in the manufacturing system.

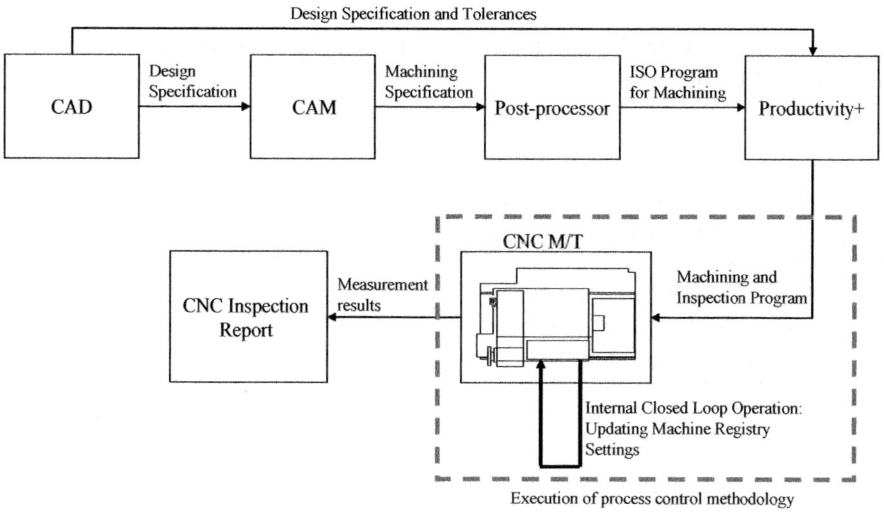

Figure 11.3. Architecture of Renishaw process control system

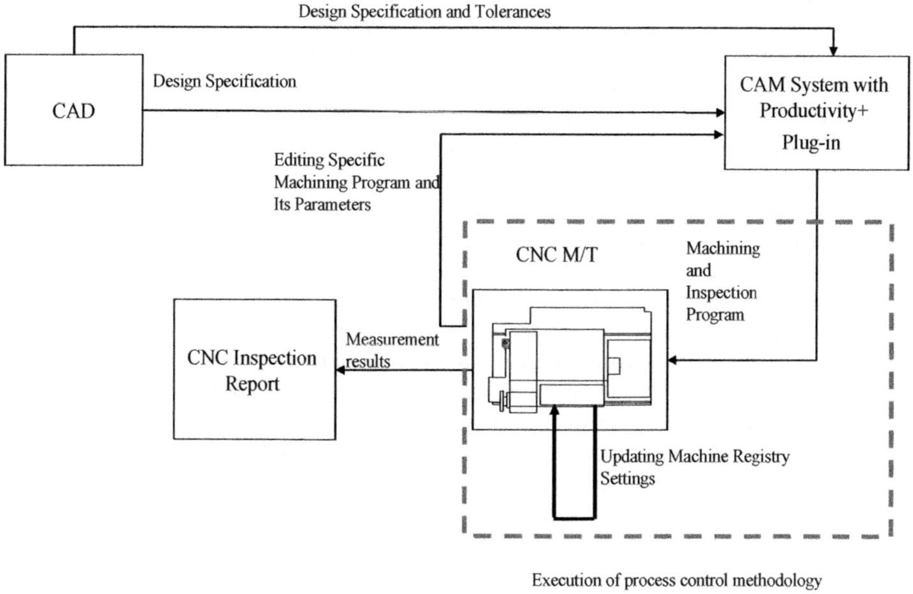

Figure 11.4. Architecture of Renishaw process control system as a plug-in to CAM system

11.3 Review of Process Control Systems

Statistical Process Control has traditionally been used in manufacturing systems to analyse and interpret measurement results and provide compensation parameters. This section provides a brief review of these developments together with Manufacturing Data Analysis.

Statistical Process Control(SPC)

A survey [11.24] conducted 30 years ago reported that measurement errors are not compensated appropriately using SPC tools. This survey also demonstrated that measurement errors and their occurrence should be modelled in the design stage of these available control techniques for effectively providing compensated parameters. This widely adopted technique used in quality control today is SPC, where the gathered process output and data are plotted on a graph. This graph which is also known as a run chart shows the sample points based on the historical records that lie outside the process limits [11.25]. Based on this, modifications in the desired processes are made by operators on the shop-floor. Harmol [11.26] described statistical process quality control techniques that help in reducing process disturbances and result in production of higher quality products. These measures are capable of detecting the uncertain states of machining processes at an early stage and subsequently rectify them to obtain better machining performance. A more recent survey [11.27] of SPC tools demonstrated that these tools should be used for identifying problems and should only be applied to problems for which the investigation into error occurrences has been completed.

Ben-Gal and Singer [11.28] carried out a study for feedback controlled processes and suggested that conventional SPC schemes fail to monitor nonlinear and finite-state processes. A new methodology termed as Context Statistical Process Control (CSPC) had been proposed for monitoring a state-dependent process of varying dependence order. This methodology ensured the monitoring of the interdependent processes which can vary according to the fluctuations in the process state. In another study conducted by Albazzaz and Wang [11.29], two other types of SPC tools, namely Principal Component Analysis (PCA) and Independent Component Analysis (ICA), have been discussed. Motorcu and Güllü [11.30] applied statistical process control in machining of spherodical cast iron parts. In this approach, quality problems such as out of tolerance limits and out of circularity errors during machining are reduced. Machining experiments were conducted and, based on these data sets, X-R control charts were constructed. The machine capability (C_p) and process capability (C_{pk}) were also determined to show the disturbances in the machining process behaviour. However, this approach doesn't automatically provide the compensation parameters for the deviations in process behaviour.

Manufacturing Data Analysis (MDA)

Lee [11.31] proposed manufacturing data analysis for determination of production error causes and provided feedback data based on geometric deviations obtained from the part measurements. This analysis was supported by a suite of product data models entitled Model Oriented Simultaneous Engineering System (MOSES). A feature-based analysis was undertaken using decision trees or networks associated with features. A measurement graph of the product model was constructed which analysed the deviation of the measured features from their nominal values and categorised them into one of three SPC classes, namely upper-fault, satisfactory and lower-fault. This system only provides feedback for the errors associated with individual features.

Menq et al. [11.32] proposed a number of integrated knowledge and automation technologies for dimensional inspection in manufacturing. Automation technology was developed for eliminating the inconsistency in manual operation for obtaining reliable inspection results. Dimensional inspection knowledge was developed to enable intelligent inspection planning, automatic inspection execution and consistency of results for the comparative analysis.

Rentoul et al. [11.33] also presented a methodology for manufacturing data analysis for interpreting production errors from measurement data. This methodology was integrated with Rules and a System of Rules (RASOR) constraint modeller where tolerances extracted from the CAD system are modelled as constraint rules. This system was primarily developed to ascertain the relationship between inspection points and a CAD model.

Mou [11.11] presented a compensation methodology by integrating pre-process, real-time, process-intermittent and post-process measurement results into manufacturing equipment. This approach is based on the fast characterisation of multi-axis machines, on-machine sensing and analysis, post-process inspection, analysis and an adaptive error correction system. A mathematical model for describing resultant positioning errors was developed. This model is based on the relationship between the actual position and the ideal position of the tool tip and inspection sensor. The information obtained from the post-process inspection analysis and positional model was used to improve the machine performance and SPC as well. Bagshaw and Newman [11.34] carried out manufacturing error diagnosis based on manufacturing data analysis of the results obtained from the part measurements on a CMM. A Product data Analysis Distributed Diagnostic Expert Systems (PADDES) was developed which provided feedback for elimination of manufacturing errors for the typical component features machined on a 3-axis vertical machining centre.

11.4 A Standardised Process Control Framework

A framework for process control using STEP-NC standards in CNC manufacturing is presented in Figure 11.5. This uses a standardised machining and measurement process plan for describing the part design, machining and measurement processes. This same plan also includes the tolerancing information of the part along with a provision to store measurement results for respective inspection items. A feature-based CNC controller is used for machining and measurement. After manufacturing of the parts, the standardised process plan is updated with actual measurement results that are later analysed using standardised compensation methodologies and corrective actions are fed back. Based on these corrective measures, the process plan is modified and used to manufacture future parts.

Recent research efforts related to standards have revolutioniszed the conventional method of CNC programming by introducing new standards such as STEP-NC which claim to improve radically the information interface with the CNC machine tool. These new standards aim to provide a new level of information for the manufacturing of discrete components on the CNC machine. This information consists of the parts geometric features together with the manufacturing methods which are required to manufacture the part. The authors have recognised STEP-NC as an enabler for achieving process control due to its ability to provide seamless integration between product and manufacturing process information models, thus maintaining the integrity of data across the process chain.

With the development of STEP-NC standards for measurement combined with the application of touch trigger probes, there is enormous potential to realisze process control in a standardised manufacturing environment. These standards provide the flexibility to store the inspection results in the same file containing the geometrical, manufacturing process, tooling and measurement information.

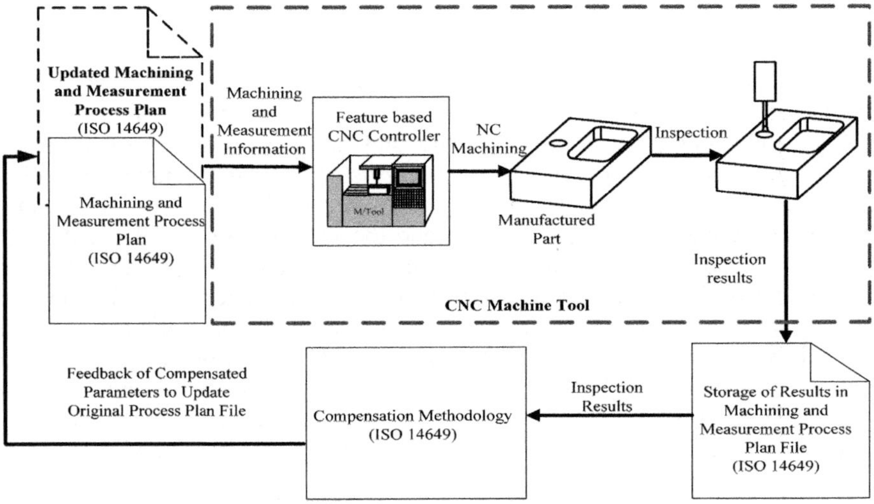

Figure 11.5. A framework for standardised process control

Compensation methods have been used for analysing these inspection results that enable the modifications of a feature's geometrical parameters, machining process and tooling parameters to be carried out at the preceding levels of the manufacturing CAx chain. These methods use optimisation techniques for analysing the measured results and provide feedback parameters to satisfy the desired criteria such as minimising the dimensional errors for a component and its features.

The machining of the parts is carried out on the basis of a STEP-NC process planning file which provides the information related to part design, manufacturing and inspection. This activity is controlled by the STEP-NC standards, i.e. ISO 14649 Part 10 [11.35], ISO 14649 Part 11 [11.36] and ISO 14649 Part 16 [11.37]. The mechanism to carry out this activity is a feature-based controller attached to a CNC vertical machining centre. The manufactured parts are then measured using a touch trigger probe, based on the STEP-NC inspection standard ISO 14649 Part 16. This activity provides the measurement results which are further analysed in the next activity, i.e. feedback. The feedback activity is automated by optimisation techniques to provide the compensation parameters in an updated STEP-NC file. The functional view of the standardised process control framework based on STEP-NC standards is provided in Figure11.6.

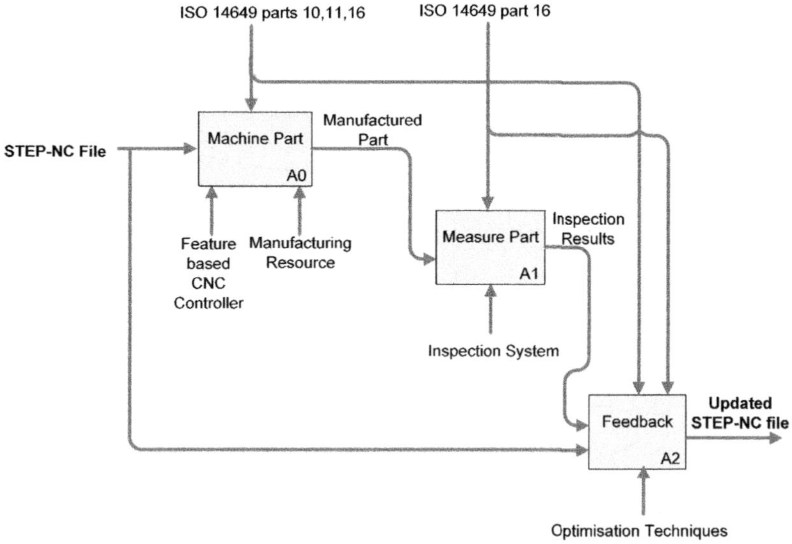

Figure11.6. Functional view of the standardised process control framework

11.5 Process Control Information Model

The authors process control information models for CNC manufacturing have been developed on the ISO 14649 suite of STEP-NC standards. This model and its constituents are detailed below.

11.5.1 STEP-NC Compliant Product and Manufacturing Information Model

The STEP-NC compliant product and manufacturing information model offers a standardised structure for information related to the component and its features, material, tolerances, machining and inspection specification. In addition, it is further used to generate a machining and inspection process plan for a component. This product and manufacturing information model is depicted in Figure 11.7.

This model provides information related to inspection activities, inspection items, and tolerances along with the provision to store the inspection results for these items of a component. The model also provides the process information in terms of Workplan, machining and probing Workingsteps. A linkage of the inspection results for parts with the associated design data, process data and manufacturing data is established based on these information models and enables standardised feedback of the corrective actions. This product information model defines the following information:

- Part design, GD&T specifications and its features
- Machining specifications
- Inspection specifications
- Inspection results

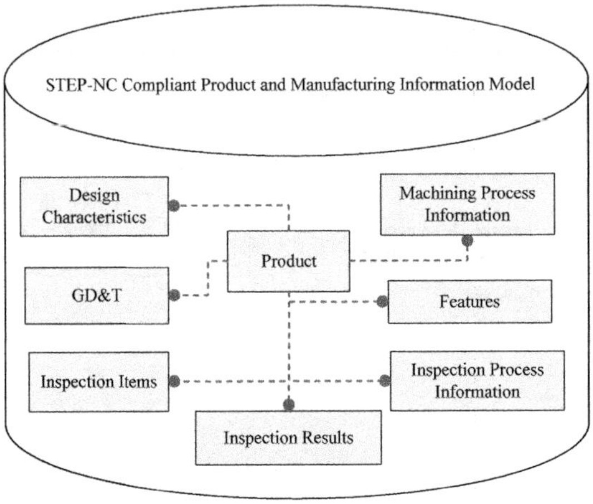

Figure 11.7. STEP-NC compliant product and manufacturing information model

Information Model for Part Design, GD&T Specifications and its Features

The product model based on ISO 14649 Part 10 defines the component as a workpiece and associates this with its shape tolerance, its geometry and its boundary geometry. The shape tolerance is the global tolerance for the component and is defined in the absence of any other tolerances. The component geometry is defined based on ISO 10303-514 [11.38]. The boundary geometry helps in defining the component either as a box, cylinder or geometry based on an advanced_ B-rep_shape_representation attribute referenced from ISO 10303 Part 514. The tolerances for the component and its features are characterised based on definitions provided by ISO 14649 Part 16. These tolerances have been categorised as dimension tolerances, applied shape tolerances and pose tolerances. The dimensional tolerances include toleranced_length_measure and toleranced plane angle measure. The upper and lower limits for the length and angle measurement are specified. The applied shape tolerances contain straightness, flatness, circularity, cylindricity, line shape and surface shape. The pose tolerances comprise of angularity tolerance, circular runout tolerance, concentricity tolerance, parallelism tolerance, perpendicularity tolerance, position tolerance, symmetry tolerance and total runout tolerances.

Information Model for Machining Specifications

This product information model defines machining features based on ISO 14649 Part 10 and is a subtype of the 2½D manufacturing features. The tool movements for 2½D machining occur mostly in the XY plane and the Z axis is set to a certain depth for removing a layer of material. This depth is provided by the elementary_surface attribute that denotes actual location of the feature in the co-ordinate system of the component. This is defined by a plane that includes the lowest points of the feature as measured in its local co-ordinate system.

Information Model for Inspection Specifications

The product information model uses ISO 14649 Part 16 to represent the inspection items of a part. Inspection items are defined by attaching tolerances to the geometrical elements which could be either by the attributes of the manufacturing feature or relations within one feature and relations between two different items. These attributes can be length of a rectangular pocket, distance between two sides of a pocket or parallelism for a hole towards a plane. These inspection items represent toleranced dimension items, toleranced spanning dimension items, toleranced pose items and toleranced shape items. A toleranced dimension item specifies an inspection item with a toleranced dimension which is generally one geometrical attribute. A toleranced spanning dimension item is defined for more than one feature. A toleranced pose item represents the transition as well as rotation whereas a toleranced shape item describes a tolerance for a shape. These inspection items are related to the underlying geometry that may be a surface or a manufacturing feature. This information is represented using a toleranced_shape attribute of these items, together with a tolerance attribute signifying the tolerance which is applied for the specified geometric item.

Information Model for Inspection Results

The STEP-NC compliant product information model provides the provision for storing the measurement results for inspection items of a component. This model uses ISO 14649 Part 16 that defines inspection result entity as a container to store the result of the inspection activity. The inspection result as a selected attribute signifies the category by which the measured result is stored. The locations to store these measured results are identified based on which attributes, i.e. length_measure, plane_angle_measure, plus_minus_value, offset_vector, pose have been measured. One such attribute pose is defined as follows:

```
ENTITY pose;
    name:      STRING;
    Its_pose: axis2_placement_3d;
END_ENTITY;
```

The attribute its_pose uses axis2_placement_3d referenced from ISO 10303 Part 42 [11.39] and is used to store the measured results in the form of its placement coordinates. A link has been established among the measured results, inspection items, manufacturing features, Workingsteps and executables entities of ISO 14649 suite of standards by this product information model and presented in Figure 11. 8. These measured results do not modify the existing geometry elements or their nominal attributes, and are stored as additional elements in the machining and inspection plan of a component. The relationship between the measured results and other entities of the product information model provide the opportunity in terms of developing the methods using the STEP-NC data structure. These methods aim to provide the compensation data and actions within the nominal machining and inspection file.

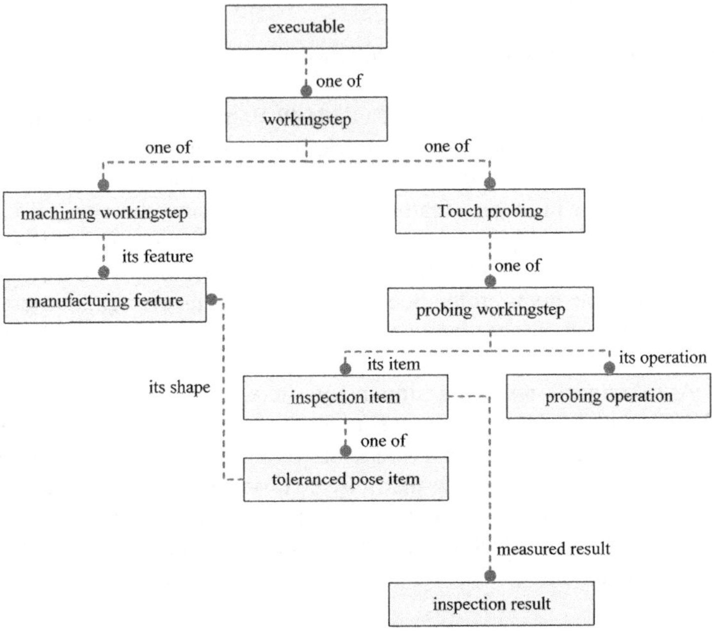

Figure 11.8. Association of measured results with various entities in the STEP-NC compliant product model

11.6 A Computational Prototype of Standardised Process Control System (SProCS)

SProCS has been developed using the standardised process control framework and information model, (discussed in Sections 11.4 and 11.5. This system is based on the STEP-NC suite of standards that utilises ISO 14649 objects and creates various methods using a Java-based object-oriented platform. The two major elements of this platform are a STEP-NC compliant probing simulator and STEP-NC compliant compensator. The input to this platform is a nominal STEP-NC process plan of a part. This process plan contains information related to the part geometry, features and their associated geometric and shape tolerances, machining operations required for the features and various machining process parameters.

The role of the STEP-NC compliant probing simulator is to read this process plan and identify various inspection items. Based on these inspection items, the probing operations are simulated. This simulation is based on the mean and standard deviations in the measurement of inspection items for manufactured parts. The STEP-NC compliant probing simulator generates various STEP-NC part files updated with the measured results for these inspection items. A Java platform which supports ISO 14649 objects has been created for the simulator, which automatically generates text encoded STEP-NC process plans updated with probing results. This standardised process plan consists of nominal part design requirements, machining specifications and measurement results. This ensures that the manufacturing context in terms of machining and tolerance information for parts is not lost.

Another vital component of SProCS is a STEP-NC compliant compensator which reads and stores probing results and associates them with their respective inspection items and manufacturing features. The decision-making engine of SProCS is based on the Standardised Positional Deviation (SPaDe) model that analyses the nominal measurements and probing results. The feedback is provided in a standardised format using Java objects and interoperable STEP-NC standards. The output of this compensator is a STEP-NC process plan updated with the compensation parameters. This process plan is provided to a feature-based CNC controller to carry out the production of parts and achieve standardised process control.

The architecture of computational platform of SProCS is presented in Figure 11.9. Figure 11.9 exhibits the input, output, Java-based platform and decision-making engine of SProCS.

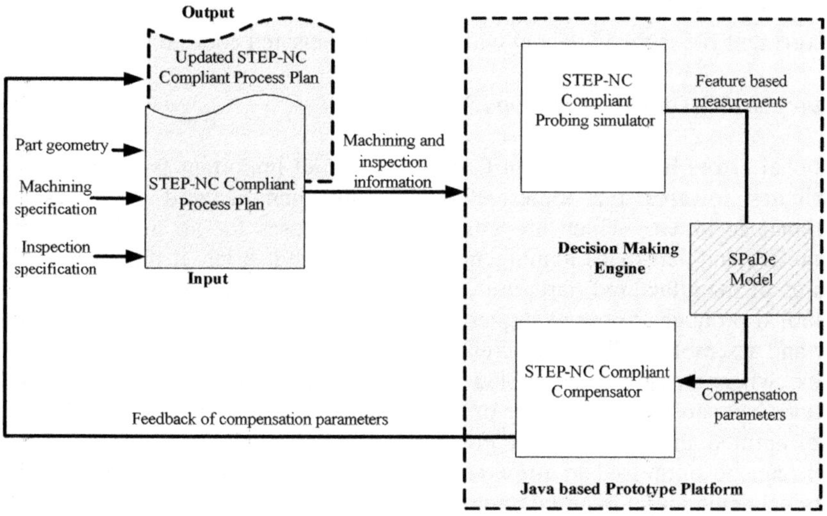

Figure 11.9. Representation of SProCS platform

STEP-NC Compliant Probing Simulator

A STEP-NC compliant probing simulator has been designed for reading a nominal STEP-NC machining and measurement process plan file and generating the updated STEP-NC compliant machining and measurement process plan with the measurement results for the required inspection items. The probing results in the nominal process plan are simulated based on the mean and standard deviation for the identified inspection items and updated process plans with probing results generated. This simulator outputs STEP-NC Part 21 files updated with inspection results which are stored in user defined location and are used further for providing compensation actions.

STEP-NC Compliant Compensator

The role of the STEP-NC compliant compensator software platform is to generate a compensated machining and measurement process plan. This software platform has been conceptualised in a Java environment, and utilises objects based primarily on the ISO 14649 schema. The compensated process plan comprises of the modified parameters based on the feedback from the SPaDe model. The operational structure of the STEP-NC compliant compensator platform is represented using a flow chart in Figure 11.10. This methodology starts [11.40] with reading the STEP-NC compliant machining and measurement files and subsequently the inspection results are stored. These results are then associated with the respective manufactured features and their nominal feature placement coordinates. The SPaDe model is used to analyse these measurement results and provide compensated feature location placement coordinates. The nominal STEP-NC compliant machining and measurement file is then updated with these compensated coordinates.

Standardised Positional Deviation (SPaDe) Model

Positional errors have been identified [11.11] as an important type of error which contributes towards the imperfections in the manufactured parts, especially, positional accuracies which are vital for part features for assembled products in automotive and aerospace manufacturing. In these industries, if the positions of the features of manufactured parts are out of alignment, there would be failure in the assembled products. Some examples are die and mould manufacturing, chassis and body and assembly, cylinder head engine assembly, etc. In the case of cylinder head blocks, when these are assembled onto engines, the imperfection in positional locations can cause engine failure or an increase in fuel consumption. In this study, a model entitled the Standardised Positional Deviation Model has been developed which can be optimised to provide compensation parameters for minimising the positional deviation of manufactured part features.

This model has been developed by considering the machine tool as a "black box" for evaluating its performance over the positional accuracies of the manufactured part features. The term black box means that the information related to machine tool axis linkages and different geometric error components such as roll, yaw and pitch is unknown. The known factors are the design requirements and measured positional locations of part features. The input parameters for this model are the nominal and measured positional locations of part features as shown in Figure 11.8. This model then evaluates the machine tool static error components based on the nominal and measured positional locations of part features, and provides the compensated positional locations of part features. The compensation parameters for this deviation can be directly fed back to manufacturing process planning. This is in accordance with the tolerancing method of manufacturing process control. These machine tool static error components are modelled in terms of linear translation and angular rotation values of machine tool for a part feature.

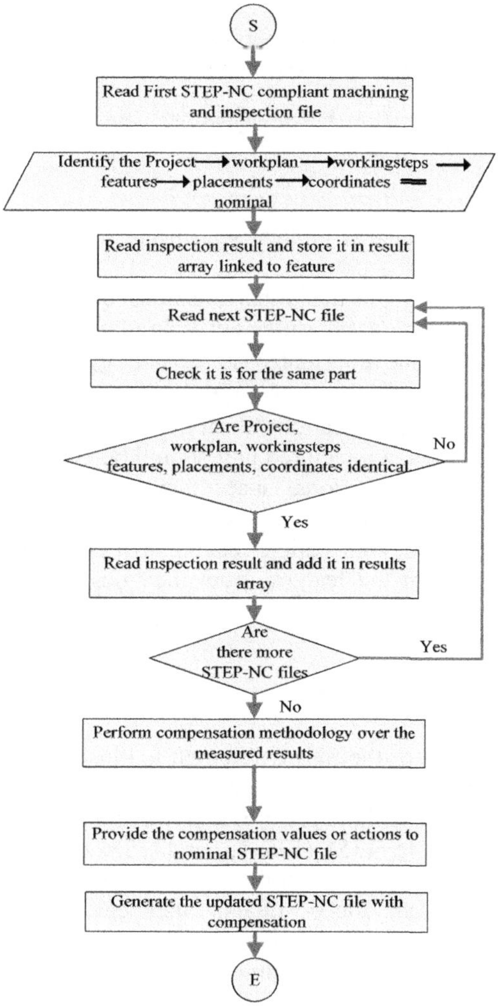

Figure 11.10. A methodology for generating a compensated STEP-NC compliant machining and measurement file

The part features and machine tool are represented in their Local Coordinate System (LCS). A Homogeneous Transformation Matrix (HTM) is used as a mathematical tool to study the positional deviations of the part features in terms of translation and rotation. Positional inaccuracies for part features are the deviations of the actual positional locations from its measured positional locations (Equation 11.1):

$$Positional\ Inaccuracies\ (PI)_i = [P_{measured\ i}]-[P_{actua\ li}] \quad (11.1)$$

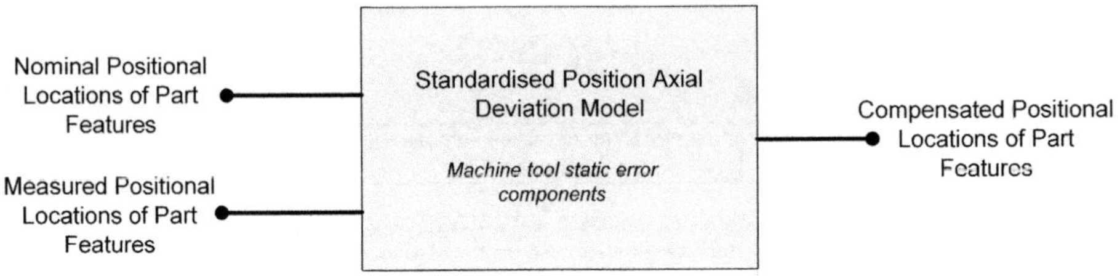

Figure 11.11. Representation of SPaDe

In (11.1), $P_{\text{measured }i}$ is the positional locations of the *i*-th machining feature obtained after the probing or measurement operations. The coordinates for this measured feature in 2D are represented as x_{mi} and y_{mi}. If there are a number of measurements for the same feature location, then x_{mi} and y_{mi} signify the means of the measurements for that feature location. Ideally the dimensions and locations of the machining features of the manufactured part should be equal to the nominal values. However, in reality, this positional inaccuracy is not zero which means the positional locations of these machining features have deviated from its nominal locations. Since the machine tool static errors are always present and repeatable, this results in the deviations of the features from their desired locations and has been signified by P_{actual_i}.

11.7 Realisation of SProCS

A simulated environment of the SProCS system is illustrated using a prismatic test part. The CAD design of this prismatic part in 2D is shown in Figure 11.12. This part design comprises of four hole features and their nominal placement coordinates in X and Y are (15, 15), (15, 135), (135, 15) and (135, 15). The required machining processes for this test piece are face milling and drilling operations. The inspection items in this case study are the positional tolerances for the four hole features. SProCS is realised for this test part and an updated process plan is obtained which is able to manufacture parts with minimised positional deviations of features placement locations. The detailed descriptions of the machining operations, processing parameters, tooling, fixturing, and probing operations are not presented in this section. This section only focuses on presenting the working of SProCS towards generating a compensated machining and measurement file.

A STEP-NC compliant process plan in Part 21 file format for the test part is created. This includes a Workplan, machining Workingsteps, probing Workingsteps, nominal design requirements of tolerances and placement locations for the manufacturing features, tooling, machining process parameters, security planes, probing strategies, probing tools and inspection items. SProCS reads this nominal process plan using the STEP-NC compliant probing simulator. The probing simulator reads and stores the nominal feature placement coordinates and simulates the probing results for these features based on the mean and standard deviation for the production profile. The generation of these files using SProCS is shown in Figures 11.13 and 11.14.

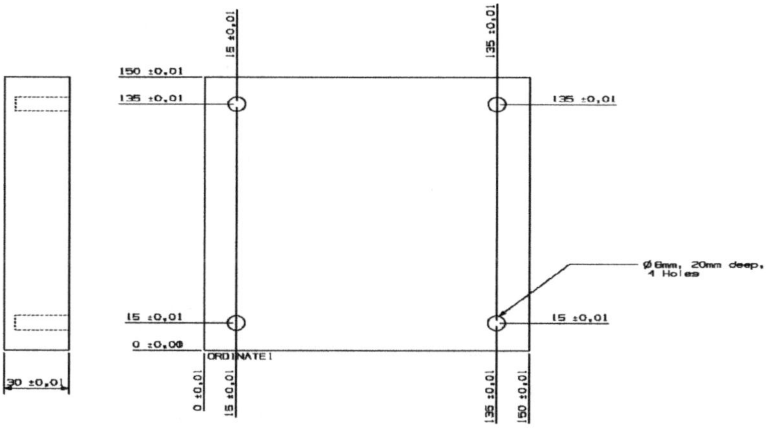

Figure 11.12. Design of a test part

The feedback module (Figure 11.10) represents the STEP-NC compliant probing simulator platform. In this screenshot, a user interface of the simulator has been shown. In the user interface of the simulator are two options, namely reading the nominal STEP-NC compliant machining and measurement file and generating updated STEP-NC compliant machining and measurement file. The values of mean and standard deviation are provided as inputs which are shown by *m* and *s*. The number of measurements to be simulated is shown by *n*. In the right hand box of the simulator user interface, an excerpt of a nominal STEP-NC machining and measurement file is shown, and in another box the simulated measurements are presented. This simulator simulates the measurement results using the mean as 0.1 and standard deviation as 0.001 for the positional axial deviations for manufactured parts. There are 100 measurement results obtained for the probing operation of the 4 hole features. This simulator generates 100 updated STEP-NC process plan files with inspection results which are stored in another location on the computer. One of these updated files is shown in Figure 11.14.

These 100 simulated STEP-NC files embedded with inspection results are read by the STEP-NC compliant compensator. For the undertaken case study, the nominal (denoted by *N* followed by a suffix for the feature number) and mean of the measured values (denoted by *M*) are obtained as: [*N1* (15.0, 15.0) and *M1* (15.009, 14.008)], [*N2* (15.0, 135.0) and *M2* (14.003, 134.098)], [*N3* (135.0, 135.0) and *M3* (135.092, 135.094)] and [*N4* (135.0, 15.0) and *M4* (135.096, 14.099)].

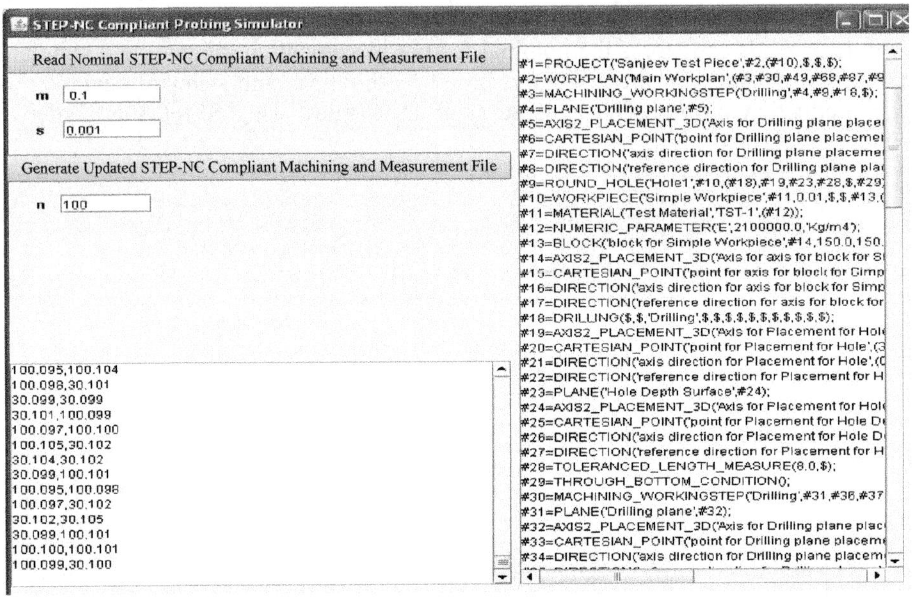

Figure 11.13. Reading of machining and measurement file and simulating probing results using STEP-NC compliant probing simulator

Based on the nominal values and mean values for the four hole features, the SPaDe model of the SProCS platform is optimiszed using the quasi-Newton algorithm for the dimensional control of the features positional locations. The compensated parameters (denoted by *C* followed by a suffix for feature number) for the four hole features are obtained as follows: [*C1* (14.97, 14.89)], [*C2* (15.07, 134.98)], [*C3* (135.02, 134.93)] and [*C4* (134.09, 14.92)].

Standardised feedback in the form of compensated feature placement coordinates are fed back into the manufacturing CAx chain. The output of SProCS is a standardised manufacturing information file which essentially is a compensated STEP-NC compliant machining and measurement file. The compensated parameters are highlighted using a dashed rectangular block in the output obtained from SProCS, and is shown in Figure 11.12. In Figure 11.12 it is shown that SProCS provides an output by using ISO 14649 classes developed for providing standardised feedback. In the updated Part 21 file (Figure 11.15), the feature placement coordinates are fed back into the container for Cartesian coordinates for the respective hole features (shown in #20, #39, #58, #77).

```
Updated STEP-NC Compliant Machining and Measurment Process Plan File 97 - Notepad
File Edit Format View Help
#80=PLANE('Hole Depth Surface',#81);
#81=AXIS2_PLACEMENT_3D('Axis for Placement for Hole Depth',#82,#83,#84);
#82=CARTESIAN_POINT('point for Placement for Hole Depth',(0.0,0.0,-20.0));
#83=DIRECTION('axis direction for Placement for Hole Depth',(0.0,0.0,1.0));
#84=DIRECTION('reference direction for Placement for Hole Depth',(1.0,0.0,0.0));
#85=TOLERANCED_LENGTH_MEASURE(8.0,$);
#86=THROUGH_BOTTOM_CONDITION();
#87=PROBING_WORKINGSTEP('Hole1',#88,$,#89,(#91));
#88=PLANE('Security plane',$);
#89=PROBING_OPERATION('probing round hole axial position',$,#90,$);
#90=PROBING_STRATEGY('null');
#91=TOLERANCED_POSE_ITEM('axial position',#92,#9,#98);
#92=INSPECTION_RESULT(#93,$,($));
#93=POSE('measured posiiton',#94);
#94=AXIS2_PLACEMENT_3D('Axis for Placement for Hole',#95,#96,#97);
#95=CARTESIAN_POINT('point for Placement for Hole',(15.092,15.099,0.0));
#96=DIRECTION('axis direction for Placement for Hole',(0.0,0.0,1.0));
#97=DIRECTION('reference direction for Placement for Hole',(1.0,0.0,0.0));
#98=POSITION_TOLERANCE(0.1,());
#99=PROBING_WORKINGSTEP('Hole2',#88,$,#100,(#102));
#100=PROBING_OPERATION('probing round hole axial position',$,#101,$);
#101=PROBING_STRATEGY('null');
#102=TOLERANCED_POSE_ITEM('axial position',#103,#36,#109);
#103=INSPECTION_RESULT(#104,$,($));
#104=POSE('measured posiiton',#105);
#105=AXIS2_PLACEMENT_3D('Axis for Placement for Hole',#106,#107,#108);
#106=CARTESIAN_POINT('point for Placement for Hole',(15.107,135.106,0.0));
#107=DIRECTION('axis direction for Placement for Hole',(0.0,0.0,1.0));
#108=DIRECTION('reference direction for Placement for Hole',(1.0,0.0,0.0));
#109=POSITION_TOLERANCE(0.2,());
#110=PROBING_WORKINGSTEP('Hole3',#88,$,#111,(#113));
#111=PROBING_OPERATION('probing round hole axial position',$,#112,$);
#112=PROBING_STRATEGY('null');
#113=TOLERANCED_POSE_ITEM('axial position',#114,#55,#120);
#114=INSPECTION_RESULT(#115,$,($));
#115=POSE('measured posiiton',#116);
#116=AXIS2_PLACEMENT_3D('Axis for Placement for Hole',#117,#118,#119);
#117=CARTESIAN_POINT('point for Placement for Hole',(135.11,135.101,0.0));
#118=DIRECTION('axis direction for Placement for Hole',(0.0,0.0,1.0));
#119=DIRECTION('reference direction for Placement for Hole',(1.0,0.0,0.0));
#120=POSITION_TOLERANCE(0.3,());
#121=PROBING_WORKINGSTEP('Hole4',#88,$,#122,(#124));
#122=PROBING_OPERATION('probing round hole axial position',$,#123,$);
#123=PROBING_STRATEGY('null');
#124=TOLERANCED_POSE_ITEM('axial position',#125,#74,#131);
#125=INSPECTION_RESULT(#126,$,($));
#126=POSE('measured posiiton',#127);
#127=AXIS2_PLACEMENT_3D('Axis for Placement for Hole',#128,#129,#130);
#128=CARTESIAN_POINT('point for Placement for Hole',(135.103,15.083,0.0));
#129=DIRECTION('axis direction for Placement for Hole',(0.0,0.0,1.0));
#130=DIRECTION('reference direction for Placement for Hole',(1.0,0.0,0.0));
#131=POSITION_TOLERANCE(0.4,());
```

Figure 11.14. STEP-NC compliant process plan file updated with measurement results generated by SProCS

11.8 Conclusions

This chapter describes recent research developments together with state of the art systems for process control in CNC manufacturing environments. It identifies the need to integrate machining and inspection for achieving a process controlled system for CNC manufacturing. Investigation of STEP-NC standards for the design, machining and inspection showed the richness of the information model in terms of providing machining, measurement and results data in the same process plan file. These standards combined within an object-oriented environment of machining and measurement information objects enables feedback into a Standardised Process Control System, and subsequently results in generation of a modified process plan to close the machining and measurement loop. An analysis of the information flow in the existing process control loop and standardised process control loop has been conducted.

Compensated STEP-NC Compliant Machining and Measurment Process Plan File 1 - Notepad

File Edit Format View Help

```
#1=PROJECT('Sanjeev Test Piece',#2,(#10),$,$,$);
#2=WORKPLAN('Main Workplan',(#3,#30,#49,#68,#87,#99,#110,#121),$,$,$);
#3=MACHINING_WORKINGSTEP('Drilling',#4,#9,#18,$);
#4=PLANE('Drilling plane',#5);
#5=AXIS2_PLACEMENT_3D('Axis for Drilling plane placement',#6,#7,#8);
#6=CARTESIAN_POINT('point for Drilling plane placement',(0.0,0.0,10.0));
...
#18=DRILLING($,$,'Drilling',$,$,$,$,$,$,$,$,$,$,$);
#19=AXIS2_PLACEMENT_3D('Axis for Placement for Hole',#20,#21,#22);
#20=CARTESIAN_POINT('point for Placement for Hole',(14.97,14.89,0.0));
#21=DIRECTION('axis direction for Placement for Hole',(0.0,0.0,1.0));
#22=DIRECTION('reference direction for Placement for Hole',(1.0,0.0,0.0));
#23=PLANE('Hole Depth Surface',#24);
...
#35=DIRECTION('reference direction for Drilling plane placement',(1.0,0.0,0.0));
#36=ROUND_HOLE('Hole2',#10,(#37),#38,#42,#47,$,#48);
#37=DRILLING($,$,'Drilling',$,$,$,$,$,$,$,$,$,$,$);
#38=AXIS2_PLACEMENT_3D('Axis for Placement for Hole',#39,#40,#41);
#39=CARTESIAN_POINT('point for Placement for Hole',(15.07,134.98,0.0));
#40=DIRECTION('axis direction for Placement for Hole',(0.0,0.0,1.0));
#41=DIRECTION('reference direction for Placement for Hole',(1.0,0.0,0.0));
...
#55=ROUND_HOLE('Hole3',#10,(#56),#57,#61,#66,$,#67);
#56=DRILLING($,$,'Drilling',$,$,$,$,$,$,$,$,$,$,$);
#57=AXIS2_PLACEMENT_3D('Axis for Placement for Hole',#58,#59,#60);
#58=CARTESIAN_POINT('point for Placement for Hole',(135.02,134.93,0.0));
#59=DIRECTION('axis direction for Placement for Hole',(0.0,0.0,1.0));
#60=DIRECTION('reference direction for Placement for Hole',(1.0,0.0,0.0));
#61=PLANE('Hole Depth Surface',#62);
#62=AXIS2_PLACEMENT_3D('Axis for Placement for Hole Depth',#63,#64,#65);
...
#75=DRILLING($,$,'Drilling',$,$,$,$,$,$,$,$,$,$,$);
#76=AXIS2_PLACEMENT_3D('Axis for Placement for Hole',#77,#78,#79);
#77=CARTESIAN_POINT('point for Placement for Hole',(134.096,14.92,0.0))
#78=DIRECTION('axis direction for Placement for Hole',(0.0,0.0,1.0));
#79=DIRECTION('reference direction for Placement for Hole',(1.0,0.0,0.0));
#80=PLANE('Hole Depth Surface',#81);
#81=AXIS2_PLACEMENT_3D('Axis for Placement for Hole Depth',#82,#83,#84);
#82=CARTESIAN_POINT('point for Placement for Hole Depth',(0.0,0.0,-20.0));
#83=DIRECTION('axis direction for Placement for Hole Depth',(0.0,0.0,1.0));
#84=DIRECTION('reference direction for Placement for Hole Depth',(1.0,0.0,0.0));
#85=TOLERANCED_LENGTH_MEASURE(8.0,$);
#86=THROUGH_BOTTOM_CONDITION();
#87=PROBING_WORKINGSTEP('Hole1',#88,$,#89,(#91));
#88=PLANE('Security plane',$);
#89=PROBING_OPERATION('probing round hole axial position',$,#90,$);
#90=PROBING_STRATEGY('null');
#91=TOLERANCED_POSE_ITEM('axial position',#92,#9,#98);
#92=INSPECTION_RESULT(#93,$,($));
#93=POSE('measured posiiton',#94);
#94=AXIS2_PLACEMENT_3D('Axis for Placement for Hole',#95,#96,#97);
#95=CARTESIAN_POINT('point for Placement for Hole',(15.11,15.087,0.0));
#96=DIRECTION('axis direction for Placement for Hole',(0.0,0.0,1.0));
```

Figure 11.15. Generation of compensated STEP-NC compliant machining and measurement process plan

It is shown that current process control loop includes a number of irreversible data conversion systems and fails to maintain the homogeneity in information at various stages of manufacturing. Whereas the standardised process control framework comprises of STEP-NC based interoperable standards which provide the powerful advantage of bi-directional information exchange at different stages of the manufacturing CAx chain. The computational prototype of the developed system has been shown to close the existing process control loop by modifying the placement coordinates of the hole features in the modified STEP-NC program. These modifications are based on the feedback obtained in the machining loop and compensate for the deviations in the locations of the measured holes from its nominal positions.

Acknowledgment

The authors acknowledge the support of University of Bath, especially Department of Mechanical Engineering for supporting this research in a form of PhD studentship as well as resource availability for Sanjeev Kumar in pursuing this research. In addition to this, the work reported in this chapter has been supported by a number of grants for Engineering and Physical Sciences Research Council (EPSRC), involving a large number of industrial collaborators. In particular, current research is being undertaken as part of the EPSRC Innovative design and Manufacturing Research Centre at the University of Bath (reference GR/R67507/01). The authors gratefully express their thanks for the advice and support of all concerned.

References

[11.1] Kwon, Y., Ertekin, Y.M. and Tseng, B., 2005. In-process and post-process quantification of machining accuracy in circular CNC milling. *Machining Science and Technology,* 9(1), pp.27-38.

[11.2] Hamrol, A., 2000. Process diagnostics as a means of improving the efficiency of quality control. *Production Planning & Control,* 11(8), pp.797-805.

[11.3] Renishaw, 2007a. *Evolution of the Renishaw productivity system^{tm}*. Renishaw Plc (White Paper). pp.1-12

[11.4] Brecher, C., Vitr, M. and Wolf, J., 2006. Closed-loop CAPP/CAM/CNC process chain based on Stepand STEP-NC inspection tasks. *International Journal of Computer Integrated Manufacturing,* 19(6), pp.570-580.

[11.5] Kumar, S., Nassehi, A., Newman, S.T., Allen, R.D. and Tiwari, M.K., 2007. Process control in CNC manufacturing for discrete components: A STEP-NC compliant framework. *Robotics and Computer-Integrated Manufacturing,* 23(6), pp.667-676.

[11.6] Siu, T.Z., 2001. Cycle to cycle feedback control of manufacturing processes. *Thesis (Master of Science).* MIT, Massachusetts.

[11.7] Edgar, T., 1987. Current problems in process control. *Control Systems Magazine, IEEE,* 7(2), pp.13-15.

[11.8] Marlin, T.E., Perkins, J.D., Barton, G.W. and Brisk, M.L., 1991. Benefits from process control: Results of a joint industry-university study. *Journal of Process Control,* 1(2), pp.68-83.

[11.9] Jayaraman, R. and Toole, P., Jr., 1993. Manufacturing systems simulated. *Spectrum, IEEE,* 30(9), pp.60-62.

[11.10] Pearsall, K. and Raines, B., 1994. Dynamic manufacturing process control. Components, *Packaging, and Manufacturing Technology, Part A, IEEE Transactions on,* 17(1), pp.153-158.

[11.11] Mou, J., 1997. A systematic approach to enhance machine tool accuracy for precision manufacturing. *International Journal of Machine Tools and Manufacture,* 37(5), pp.669-685.

[11.12] Cheraghi, S.H., Chen, X., Twomey, J.M. and Arupathi, R., 1999. A closed-loop process analysis and control system for machining parts. *International Journal of Production Research,* 37(6), pp.1353.

[11.13] Ribeiro, J.L.D., Caten, C.S.T. and C., F., 2001. Integrated process control. *The International journal of quality & reliability management,* 18(4), pp.444.

[11.14] Liang, S.Y., Hecker, R.L. and Landers, R.G., 2004. Machining process monitoring and control: The state-of-the-art. *Journal of manufacturing science and engineering,* 126(2), pp.297.

[11.15] ISO 10303-21: 1994. Industrial automation systems and integration — product data representation and exchange, part 21: Implementation methods: STEP-file clear text encoding of the exchange structure. ISO.

[11.16] ISO 10303-28: 2002. Industrial automation systems and integration — product data representation and exchange, part 28: Implementation methods: Xml representations of express schemas and data. ISO.

[11.17] IMS, 2004. *Description of the prototype CAD-CAM-CNC chain for inspection.* IMS, STEP-NC consortium, (Project Deliverable ID6.5.2).

[11.18] Kramer, T.R., Proctor, F., Xu, X. and Michaloski, J., 2006. Run-time interpretation of STEP-NC: Implementation and performance. *International Journal of Computer Integrated Manufacturing,* 19(6), pp.495-507.

[11.19] Newman, S.T., Allen, R.D. and Rosso Jr, R.S.U., 2003. CAD/CAM solutions for STEP-compliant CNC manufacture. *International Journal of Computer Integrated Manufacturing,* 16(7-8), pp.590-597.

[11.20] Ali, L., 2006. Development of a STEP-NC compliant feature-based inspection framework for prismatic parts. Thesis (PhD Thesis). Loughborough University.

[11.21] Zhao, F., Xu, X. and Xie, S., 2008. STEP-NC enabled on-line inspection in support of closed-loop machining. *Robotics and Computer-Integrated Manufacturing,* 24(2), pp.200-216.

[11.22] STEP Tools, 2008. International STEP-NC demonstration of feed optimisation, high speed machining, tolerance-driven tool compensation, and traceability http://www.Steptools.Com/library/stepnc/2008_sweden.

[11.23] ISO 13399-1: 2006. Cutting tool data representation and exchange — part 1: Overview, fundamental principles and general information model. ISO.

[11.24] Dorris, A.L. and Foote, B.L., 1978. Inspection errors and statistical quality control: A survey. *IIE Transactions,* 10(2), pp.184 - 192.

[11.25] Taguchi, G., 1986. Introduction to quality engineering. *Asian Productivity Organization.*

[11.26] Hamrol, A., 2000. Process diagnostics as a means of improving the efficiency of quality control. *Production Planning & Control,* 11(8), pp.797-805.

[11.27] Ahmed, S. and Hassan, M., 2003. Survey and case investigations on application of quality management tools and techniques in SMIs. *International Journal of Quality and Reliability Management,* 20(6), pp.795-826.

[11.28] Ben-Gal, I. and Singer, G., 2004. Statistical process control via context modeling of finite-state processes: An application to production monitoring. *IIE Transactions, 36(5),* pp.401-415.

[11.29] Albazzaz, H. and Wang, X.Z., 2004. Statistical process control charts for batch operations based on independent component analysis. *Industrial & Engineering Chemistry Research,* 43(21), pp.6731-6741.

[11.30] Motorcu, A.R. and Güllü, A., 2006. Statistical process control in machining, a case study for machine tool capability and process capability. *Materials & Design,* 27(5), pp.364-372.

[11.31] Lee, M.K., 1990. Data feedback in an integrated design to manufacturing systems. *Thesis (PhD Thesis).* Loughborough University, Loughborough, UK.

[11.32] Menq, C.H., Yau, H.T. and Lai, G.Y., 1992. Automated precision measurement of surface profile in CAD-directed inspection. *IEEE transactions on Robotics and Automation,* 8(2), pp.268-278.

[11.33] Rentoul, A.H., Mullineux, G. and Medland, A.J., 1994. Interpretation of errors from inspection results. *Computer Integrated Manufacturing Systems,* 7 (3), pp.173-178.

[11.34] Bagshaw, R.W. and Newman, S.T., 2002. Manufacturing data analysis of machine tool errors within a contemporary small manufacturing enterprise. *International Journal of Machine Tools and Manufacture,* 42(9), pp.1065-1080.

[11.35] ISO 14649-10: 2002. Industrial automation systems and integration — physical device control — data model for computerized numerical controllers, part 10: General process data. ISO.

[11.36] ISO 14649-11: 2002. Industrial automation systems and integration — physical device control — data model for computerized numerical controllers, part 11: Process data for milling. ISO.

[11.37] ISO 14649-16 WD: 2004. Industrial automation systems and integration — physical device control —data model for computerized numerical controllers — part 16: Data for touch probing based inspection. ISO.

[11.38] ISO 10303-514: 1999. Industrial automation systems and integration -- product data representation and exchange -- part 514: Application interpreted construct: Advanced boundary representation. ISO.

[11.39] ISO 10303-42: 1994. Industrial automation systems and integration — product data representation and exchange, part 42: Integrated generic resources: Geometric and topological representation. ISO.

[11.40] Kumar S., Nassehi A., Vichare P. and Newman S.T., 2008, The standardisation of process control for CNC manufacturing, *Proceedings of 18th International Conference on Flexible Automation (FAIM) in Skövde,* Sweden, June

12

A STEP-NC Compliant Methodology for Modelling Manufacturing Resources

Aydin Nassehi[1] and Parag Vichare[2]

Innovative Design and Manufacturing Research Centre, Department of Mechanical Engineering, University of Bath, Claverton Down, Bath, BA2 7AY, United Kingdom
Email: [1] a.nassehi@bath.ac.uk, [2] p.vichare@bath.ac.uk

Abstract
Manufacturing enterprises of all sizes, from small subcontractors and job shops to trans-national aerospace and automotive giants, rely on a vast array of resources to generate added value and accomplish their business goals. These resources are comprised of human resource, knowledge resources and technological resources. Achieving business goals requires relevant, timely and well-calculated decision-making. A decision making process requires a model of the decision domain to be constructed with the necessary fidelity to achieve acceptable results and to determine the best course of action in the problem context. A reliable representation of manufacturing resources is therefore necessary for making correct decisions along the manufacturing chain. In this chapter a STEP-NC compliant methodology for representing technological manufacturing resources is presented. This modelling approach is based on mechanical elements that constitute machine tools and other manufacturing hardware together with their kinematic links and is developed with a focus on supporting process planning decisions. Models for various types of machines are presented at the end of the chapter to highlight the flexibility of this approach in modelling manufacturing resources.

12.1 Introduction

In rational decision making it is necessary to develop an understanding of the problem domain. To make consistently correct decisions, it is necessary for this representation of the context to be accurate, timely and reflect the aspects of the problems that affect the possible decisions. From a manufacturing point of view, making effective decisions requires a precise, relevant and valid model of the decision space. In order to automate the decision making process or to enable human operators of manufacturing systems to make more informed decisions it is imperative to have access to such comprehensive models for products, processes and resources.

This chapter focuses on the development of a methodology for representation of manufacturing resources with the ability of easy integration with the process and product models of STEP-NC. The chapter is organised so that first the requirements and various perspectives for manufacturing resource models are investigated. A theoretical framework for materialisation of a universal approach to modelling manufacturing resources is then presented followed by the EXPRESS schema for representation of resources. The models for a number of example CNC machine tools are then utilised to demonstrate the applicability of the modelling methodology to a varied range of machining resources. The chapter is concluded by an enumeration of future developments envisioned for the modelling approach.

12.2 Manufacturing Resource Modelling

STEP-NC aims to provide a comprehensive methodology for representation of manufacturing data [12.1, 12.2]. Figure 12.1 presents a classification of manufacturing information required in CNC-based production. The current versions of STEP-NC provide information models for representation of Workplans and through integration with other parts of STEP (i.e. AP224) allow part geometry and tolerances to be represented as well.

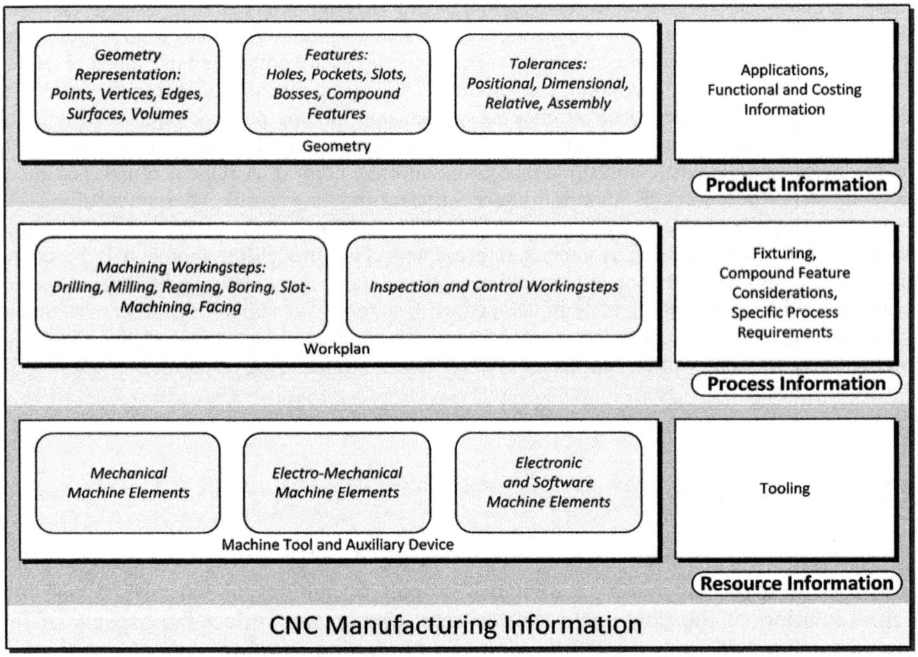

Figure 12.1. Manufacturing information classification (adapted from [12.1])

12.2.1 Manufacturing Resource Representation Methodologies

Research into manufacturing resource modelling has been conducted by many teams around the world and a number of approaches have been explored by researchers to represent resource information.

Graphical Simulation and Virtual Manufacturing Systems

These approaches focus on tool path simulation through accurate graphical representation of physical machine elements. Such simulations would minimise trial errors in design [12.3]. Models produced based on this approach contain accurate graphical representations of the physical elements of a machine tool. Software packages such as Machine Tool Builder in the Siemens NX suite [12.4], Vericut from CGTech [12.5] and VNC from Delmia [12.6] are capable of creating machine representations using the virtual modelling approaches. The software packages often allow kinematic relationships between the various physical elements of the machines to be defined and associated with the geometry. These models have been used in conjunction with Web-enabled technology to develop Internet-based virtual machine tools [12.7–12.10]. While most of the modellers based on this approach focus on machine tools, some allow robots and auxiliary manufacturing resources such as gantries, pallet changers, etc. to be modelled as well. Machine tool builder Mazak in particular has a module-based simulations system capable of simulating entire flexible manufacturing cells [12.11]. Altintas et al. provide a thorough report of virtual machine tools [12.12].

Process Capability Representation

This approach represents the manufacturing resource as a collection of process capabilities. For example, Heusinger et al. proposed manufacturing models in two parts: resource availability and process capabilities of available workstations [12.13]. This approach closely relates process data with resource data. In creation of data models, such links would result in tight coupling of entities [12.14] which is highly undesirable in model design. In addition, process capability is a collection of derived attributes from a manufacturing resource's mechanical, electro-mechanical and software/electronic elements. The integrity of the data contained within a model with a high number of derived attributes is difficult to maintain. Furthermore, combination of this approach with other approaches creates a considerable amount of data redundancy.

Hierarchical Classification

This approach is based on creating a hierarchy of machine tools. It relies on classifying machine tools according to some guidelines into "lathes", "milling centres", "EDM machines", "parallel kinematic machines", etc. Shinno et al. proposed a series of methods to generate a structural model of machine tools utilising modular components [12.15–12.17]. These components would form a conceptual machine tool model in the preliminary product development stage. The research was mainly geared towards the needs of the machine tool builders and utilised the hierarchical approach. Yang and Xu [12.18] have presented a machine tool data model to support STEP-NC using the hierarchical approach. ASME B5.59-2 standard also uses the same approach for modelling machine tools.

Kinematic Representations

Approaches based on kinematics representations allow valid mechanical movements of the machine to be determined based on graphs representing the links between individual mechanical machine elements. Suh et al. [12.8] employed kinematic representations to allow tool-path simulation on the virtual machine tool. Figure 12.2 shows the kinematic chain for a 5-axis machining centre using a similar convention.

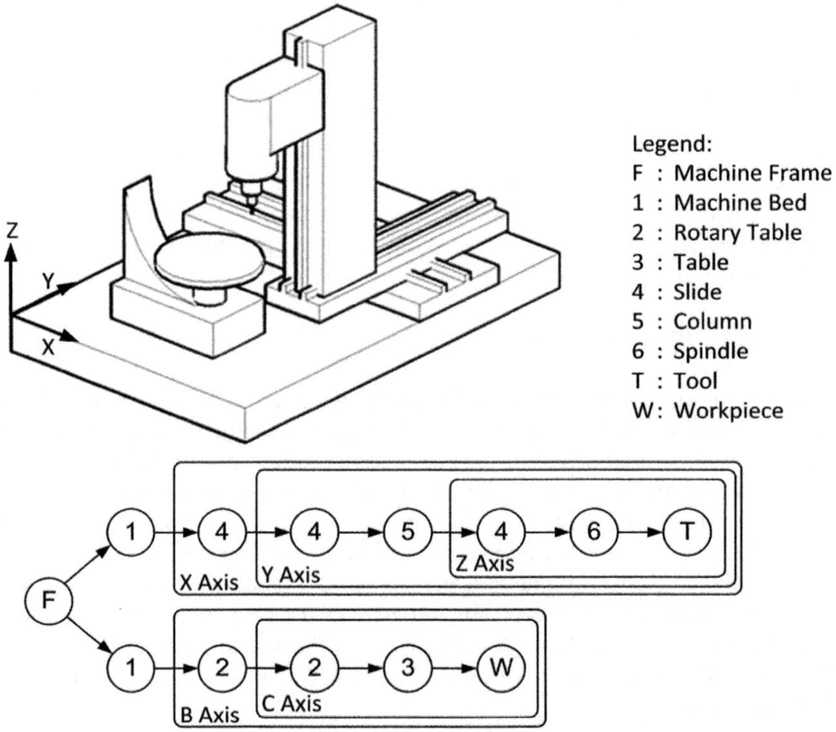

Figure 12.2. Kinematic chain representation of 5-axis machining centre (adapted from [12.8])

By following the kinematic chain from workpiece to tool (or end-effecter in the case of other manufacturing resources) it is possible to ascertain the position and orientation of the tool if the positions of axes are known. Calculating the axes positions based on a required position and orientation is possible too, albeit with complex calculations. These calculations entitled "inverse-kinematics" are commonly seen in robotics and several algorithms for providing numerical answers to the ensuing equations have been developed.

This approach to modelling manufacturing resources is very flexible but the current implementations face difficulties when machine kinematics is not serial. For example, most implementations are unable to represent Stewart Platforms [12.19] and other parallel kinematic machine tools while maintaining kinematic integrity.

Other Approaches

Depending on the purpose and requirements of the resource model, other approaches have been chosen. Costing software systems such as SEER for Manufacturing by Galorath [12.20] represents machine tools as a set of simple equations that convert part complexity into a time estimate. Preventative maintenance systems represent machine tools as statistical entities to provide aid in decision making [12.21]. Rule-based machine tool models [12.22] and knowledge representation methodologies such as "LOOM" and its first order logic query language [12.23] have also been used to represent manufacturing resources. These representations are usually geared towards specific purposes and are not directly invoked during manufacturing resources' primary function of production.

A study of the above approaches shows that the purpose of modelling has shifted from machine tool modelling for tool path verification towards advanced process planning [12.24].

To develop a STEP-NC integrated model for resources it is therefore necessary to ascertain the main focus and purpose of the modelling. In addition, for the model to be STEP-NC compliant, it needs to be created using the STEP modelling language EXPRESS. The necessity of development of a manufacturing resource model to augment STEP-NC was formally identified in a 2007 joint meeting of sub-committees one and four of the ISO Technical Committee 184, the developers of STEP-NC, and a new work item proposal (NWIP) was put forward. This NWIP aims to formalise the manufacturing resource model as Part 110 of ISO14649.

12.2.2 Perspectives for Resource Modelling in the Context of Manufacturing

An important issue in developing models and modelling approaches is to specify the focus and the purpose of the models. In the context of CNC manufacturing a number of different perspectives can be identified.

Process Planning

Models utilised for process planning should be able to represent the functional capabilities of the modelled resources. These models are used in conjunction with product models to generate process plans or refine the fidelity of existing process plans from the macro level to the micro level.

Resource Specific Tool-path Generation

A variation of the models used for process planning are those that are utilised to generate machine specific tool-path instructions. These models contain a representation of the mechanical configuration of the machine in addition to the software model of the controller. CAM post-processors use such models to convert part programmes from the product coordinate space to machine coordinate space. Most of the implementations of these models are hard coded into proprietary software systems. CNC-controller-based shop-floor programming systems use a specific type of these models to convert generic graphical input into low-level axis movement instructions.

Multi-route Process Planning

At the other end of the process planning spectrum are multi-route process plans where, depending on the availability and efficiency of resources, products can be manufactured using alternative procedures. Figure 12.3 shows a simple multi-route process plan where the process capability for each step is identified.

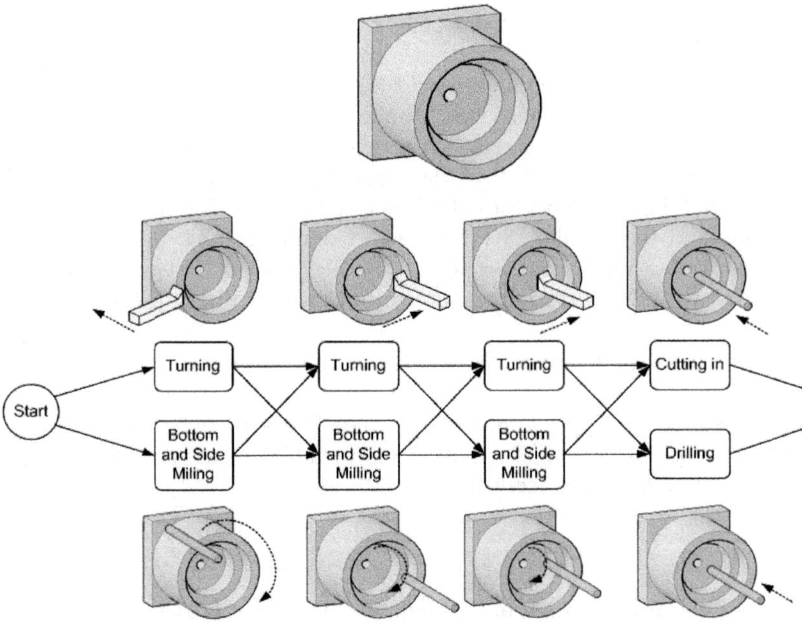

Figure 12.3. A simple multi-route process plan

To create multi-route process plans effectively it is necessary to have access to resource process capabilities; however some modelling approaches (i.e. hierarchical) would limit the choice of resources for a specific step. For example, in Figure 12.3 it is possible to take any path in the graph from start to finish on a current generation of turning centre with C-Axis and driven tools but most hierarchical models will limit the process to turning type operations on the top row due to early ability assessment.

Intelligent CNC Controllers

Every CNC controller has an internal model of the machine tool. These models can range from simple axis configurations as used on older generation controllers to complete graphical models used on the latest generation of CNC controls with collision detection capabilities. In order to develop an intelligent CNC controller, capable of handling a variety of machine configurations, a generic modelling approach is necessary.

STEP-NC Compliant Bi-directional Data Flow

With one of the promises of STEP-NC being enabling of bi-directional information exchange, resource models are necessary to retain the semantics of the transferred information. A modelling approach based on this perspective can enable the seamless information exchange in global manufacturing scenarios and thus enable interoperability and resource fluidity.

Tool-path Verification and Collision Detection

This perspective is the original reason for the development of virtual manufacturing systems. To reduce costs of running part programmes on production equipment at reduced speeds or with soft materials to verify them, virtual simulations can be utilised to confirm that a particular part program is error free. Furthermore, it is possible to optimise cutting forces (by modelling the interaction surface between the material and tool) and determine collisions. In modern systems using this methodology, the virtual part resulting from the virtual machining process can be compared to the original CAD model to determine the extent of the variations.

Business and Management Perspective

In the business and management perspective, manufacturing resources are seen as generators of added value that perform a valuable transformation at a cost. Modelling approaches based on this perspective therefore create representations of resources in terms of time, cost, personnel, and generated added value. Costing systems, enterprise resource planning applications and production scheduling systems use models developed in this perspective.

12.2.3 Modelling Approaches in the Context of Modelling Perspectives

Table 12.1 cross tabulates the various modelling approaches and modelling perspectives and gives an overview of the suitability of each method in the context of each perspective. Kinematic representations meet the requirements of most perspectives with the notable exceptions of tool-path verification systems and business and management perspectives.

A modelling approach based on the kinematic approach with extensibility to accommodate other approaches such as geometrical representations of virtual manufacturing can meet the requirements of a wide range of perspectives. Consequently the modelling perspective of this chapter is constructed based on the kinematic approach with the capacity to relate entities from other approaches to the data elements.

Table 12.1. Suitability of various modelling approaches to modelling perspectives

Suitability of modelling approaches for various modelling requirements *0 = Least Suitable* *5 = Most Suitable*	Process planning	Resource specific tool-path Generation	Multi-route process planning	Intelligent CNC controllers	STEP-NC compliant bi-directional data flow	Tool-path verification and collision detection	Business and management perspective
Graphical simulation and virtual manufacturing systems	1	0	0	2	1	**5**	1
Process capability representation	**5**	0	4	0	1	0	2
Hierarchical classification	4	1	3	0	1	1	1
Kinematic representations	**5**	**5**	**5**	**5**	**5**	2	1
Other approaches	1	0	0	0	0	0	**5**

12.3 A Modelling Framework for Technological Manufacturing Resources

In order to construct the modelling framework, technological manufacturing resources have to be formally defined. For the purpose of this chapter, a technological manufacturing resource is any physical device that manipulates matter using physical tools or end-effectors to generate added value. Machine tools, auxiliary devices in flexible manufacturing cells, robots, pallet loading and unloading systems, and automatic warehousing systems are examples of these resources. Each type of these resources has mechanical elements, electro mechanical elements and electronic and software elements that function in coordination to produce the added value.

The focus here is specifically on CNC machine tools and their auxiliary devices. In this section a formal definition of this type of technological resources is provided together with an enumeration of their various types of elements and their kinematic representations.

12.3.1 CNC Machine Tools and Auxiliary Devices

A CNC machine tool can be defined as a physical device that manipulates a raw or blank workpiece to produce a desired geometrical shape within specific tolerances. This is achieved by controlled interaction of tools mounted on the machine and the raw workpiece, removing (or depending on the type of the CNC machine tool) adding material.

Auxiliary devices are physical devices that are in close proximity to the CNC machine tool and perform mechanical functions to assist in the value addition process of the machine tool. Examples for such devices are pallet loaders and unloaders, tool magazines and bar feeders.

12.3.2 Mechanical Elements, Electro-mechanical Elements and Electronic Elements

A CNC machine tool is constructed of mechanical elements such as frames, tables, slides, tool holders, spindles, chucks, turrets, etc. Elements are connected to others via mechanical joints. Sliding joints, spherical joints and rigid joints are just a few examples of the different types of joints connecting the mechanical elements.

A number of mechanical elements are driven by electro-motors. These servos are part of the electro-mechanical elements of CNC machine tools. The main function of servos is to convert electric signals into controlled motion of the mechanical elements in the machine. Encoders and sensors are utilised in conjunction with servos to achieve the desired level of control. These devices convert mechanical motion of machine elements into electrical signals, to be processed by the CNC.

The controller is the electronic element in CNC machine tools and is responsible for running the control software. This piece of software interprets part programmes written in a pre-defined language and generates electric signals based on the instructions that are sent to the servos. The feedback from the servos is also analysed by the controller to ensure that the process is being executed within specified parameters. Figure 12.4 shows the interactions of these elements.

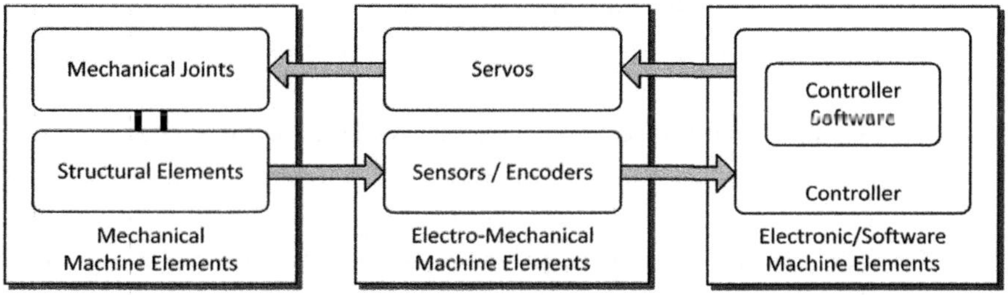

Figure 12.4. Interaction of various elements in a CNC machine tool

A comprehensive modelling approach should consider all three types of elements. In this chapter, however, the modelling approach is mainly concerned with the mechanical elements of the machine. The connection between the electromechanical and mechanical elements are, nevertheless, addressed in the approach to allow future developed models for electro-mechanical and electronic elements to interface with the generated mechanical models.

12.3.3 Kinematic Chains

As mentioned in Section 12.3.1, a machine tool can be seen as a mechanical chain between a tool and a workpiece. This chain can be modelled as a kinematic chain. The classic approaches for mechanical chains including D-H [12.25] representations and directed graphs [12.8] are incapable of representing parallel kinematics in an adequate fashion. This is because in the case of parallel links, references to parent nodes are required as well as the forward kinematic links. One possible solution is to store a backward reference to parent nodes in each kinematic node. Figure 12.5 shows a parallel kinematic chain with backward references. When expressed as equations, serial kinematic machines result in a series of equations with axis positions as variables and position and orientation of the tool in relation to the workpiece as parameters. In the case of parallel kinematics, additional conditions are added to the equation system. The machine is in a valid state if acceptable values can be found for all variables while meeting the conditions.

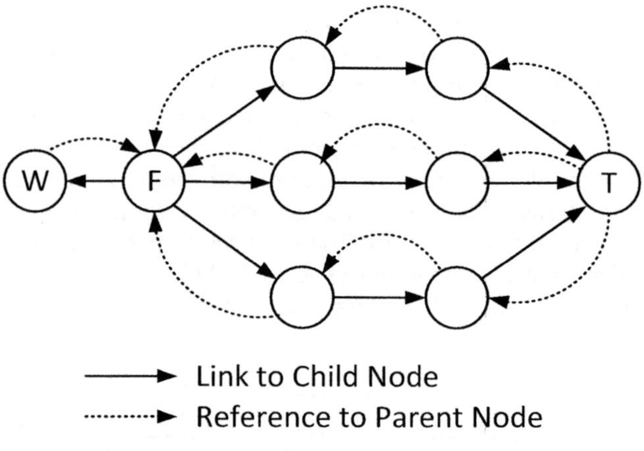

Figure 12.5. A parallel kinematic chain with backward references

12.4 The STEP-NC Compliant Schema for Representation of Machine Tools and Auxiliary Devices

As indicated in Section 12.2.1, a STEP-NC compliant methodology will require modelling entities to be defined using the EXPRESS language. Entities representing mechanical machine elements, kinematic joints and axes of movement are therefore defined with their EXPRESS-G representations depicted.

12.4.1 Mechanical Machine Element

The mechanical machine element entity allows data about physical elements of the machine to be stored. Figure 12.6 shows the EXPRESS-G diagram for the mechanical machine element entity. As seen in the figure, each mechanical machine element is attached to a kinematic joint. It also has a shape representation that can be utilised for realisation of virtual manufacturing systems. A mechanical machine element can also hold a workpiece or a tool.

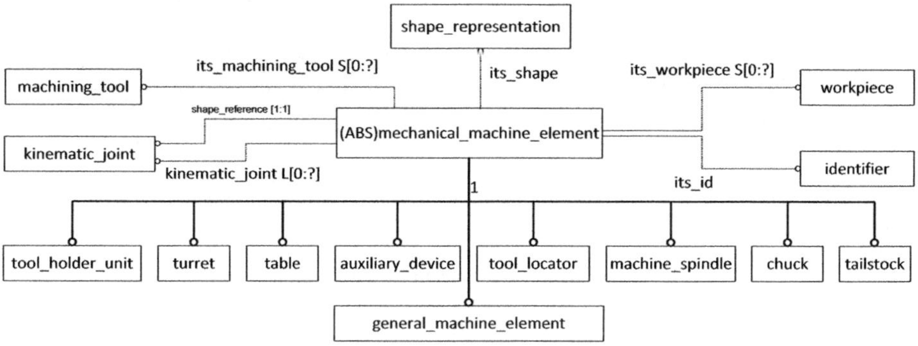

Figure 12.6. EXPRESS-G diagram of the mechanical machine element entity

The entity is formally defined using EXPRESS as follows:

```
ENTITY mechanical_machine_element
   ABSTRACT SUPERTYPE OF (ONEOF(tool_holder_unit, turret,
   table, auxiliary_device, general_machine_element,
   tool_locator, machine_spindle, chuck, tailstock));
   its_id: OPTIONAL identifier;
   its_workpiece: OPTIONAL SET[0:?] OF workpiece;
   its_machining_tools: OPTIONAL SET[0:?] OF machining_tool;
   its_kinematic_joint:LIST[0:?] OF kinematic_joint;
END_ENTITY;
```

12.4.2 Kinematic Joint

The entity kinematic joint is defined to represent the kinematic relationship of the mechanical machine elements with each other. Figure 12.7 shows the EXPRESS-G diagram for the kinematic joint entity. In addition to the kinematic joint, an entity called *transfer_placement* is also defined to specify the backward reference to the parent kinematic joint as well as the placement of the child kinematic node in the parent's coordinate system. This is achieved by using an *axis2_placement_3D* entity.

In addition to forward links to the mechanical machine elements and backward references to previous joints, a kinematic joint can also have linear and rotational axes of movement. Mechanical properties can also be defined for the mechanical joint to specify additional attributes.

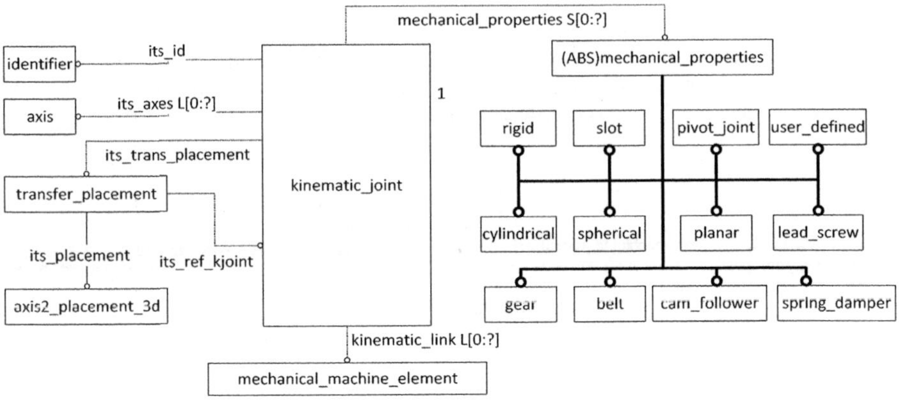

Figure 12.7. EXPRESS-G diagram of the kinematic joint entity

The formal EXPRESS definitions for kinematic joint, transfer placement and mechanical properties entities are as follows:

```
ENTITY kinematic_joint;
  its_id:OPTIONAL identifier;
  its_axes:OPTIONAL LIST[0:?] OF axis;
  its_trans_placement: OPTIONAL transfer_placement;
  kinematic_link:OPTIONAL LIST[0:?] OF
    mechanical_machine_element;
  mechanical_properties:OPTIONAL SET[0:?] OF
    mechanical_properties;
END_ENTITY;

ENTITY transfer_placement;
  its_placemenet:axis2_placement_3D;
  its_ref_kjoint:OPTIONAL kinematic_joint;
END_ENTITY;

ENTITY mechanical_properties
  ABSTRACT SUPERTYPE OF (ONEOF (rigid, slot, pivot_joint,
    user_defined, cylindrical, spherical, planar, lead_screw,
    gear, belt, cam_follower, spring_damper));
END_ENTITY;
```

By defining the above entities it is possible to represent both serial kinematic and parallel kinematic machines in the same manner. It is also notable that additional mechanical properties can be defined in the future as the approach has high cohesion and low coupling with these entities. The properties can later be utilised to model machines from other (i.e. environmental impact) perspectives.

12.4.3 Axes of Movement

To create dynamic models of machine tools it is imperative to represent accurately axes of movement. However, to be able to create snapshots of the machine at a given point in time, it is also necessary for the representation to provide a mechanism to store the nominal value of the axis according to its coordinate system. Figure 12.8 shows the EXPRESS-G diagram for the axis entity. The entity utilises the direction entity to specify the axis direction within its coordinate system. For rotary axes this is the normal vector while for linear axis it is a vector pointing along the axis. The direction of the normal vector is determined according to the right hand rule. Ranges and snapshot values for linear axes and rotary axes are stored as lengths and angles respectively.

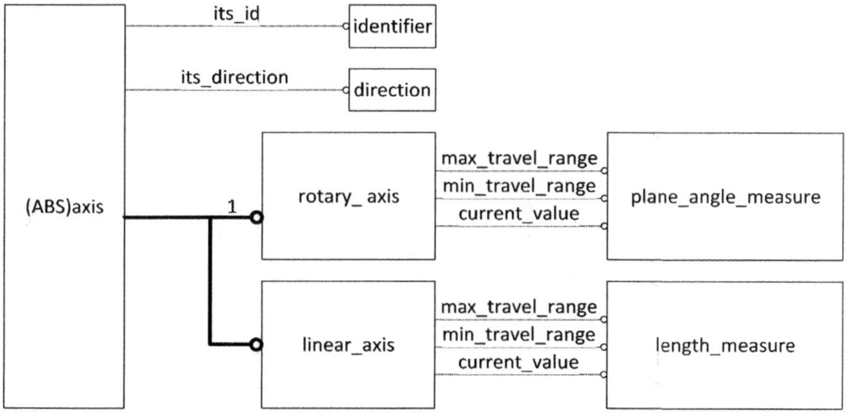

Figure 12.8. EXPRESS-G diagram for the axis entity

The formal EXPRESS representation of the axis entity is as follows:

```
ENTITY axis
   ABSTRACT SUPERTYPE OF (ONEOF (rotary_axis, linear_axis));
   its_id:OPTIONAL identifier;
   its_direction:OPTIONAL direction;
END_ENTITY;
```

12.4.4 Additional Entities Required for Resource Representation

The entities defined in this section enable the modelling of the kinematic relationship of the mechanical elements in a machine tool or an auxiliary device. Geometrical representation of the elements can also be attached to these entities. To have a complete model of the manufacturing resource, however, it is necessary to supplement this data with more information. This information should include an entity defining the name, location, make, model and vendor of the resource in addition to people responsible for the maintenance and use of the resource.

The attributes required for these representations can vary from one organisation to another and could thus be modified in accordance with the requirements.

Models of cutting tools and workpieces conforming to ISO13399, ISO14649 Parts 111 and 121 and ISO10303-224 can be combined with the manufacturing resource model to provide a complete view of the manufacturing scenario.

12.5 Example Models

In this section the entities defined in Section 12.4 are used to create models of an increasingly complex typology of machine tools. For each case the instances of the entities are shown together with a geometric representation of the machine tool.

Figure 12.9 provides a key for the representation of the entities introduced in Section 12.4 in the figures in this section.

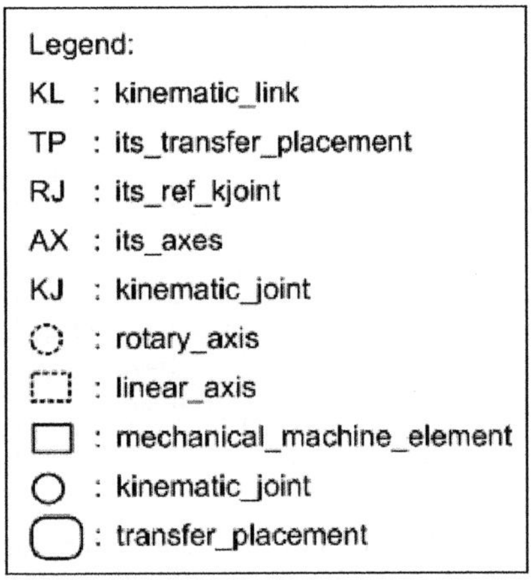

Figure 12.9. Representation of EXPRESS entities in machine models

12.5.1 2-Axis Lathe

The ubiquitous 2-axis lathe is one of the oldest configurations of metal cutting machine tools. The part is fixed in a chuck attached to the spindle. Cutting tools, usually static, are mounted on a moving arm with two linear axes of movement. Figure 12.10 shows a graphical representation of typical 2-axis lathe.

The mechanical entities and the kinematic joints are indicated on the figure. The relationship of these entities, the kinematic joints and placement references are illustrated in Figure 12.11 as an instance diagrams based on the EXPRESS entities defined within the schema.

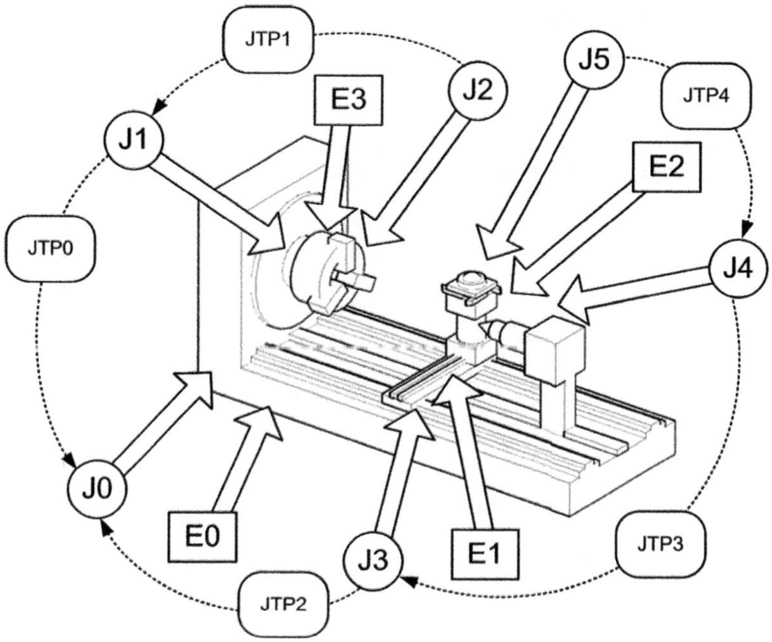

Figure 12.10. Geometric model of a 2-axis lathe

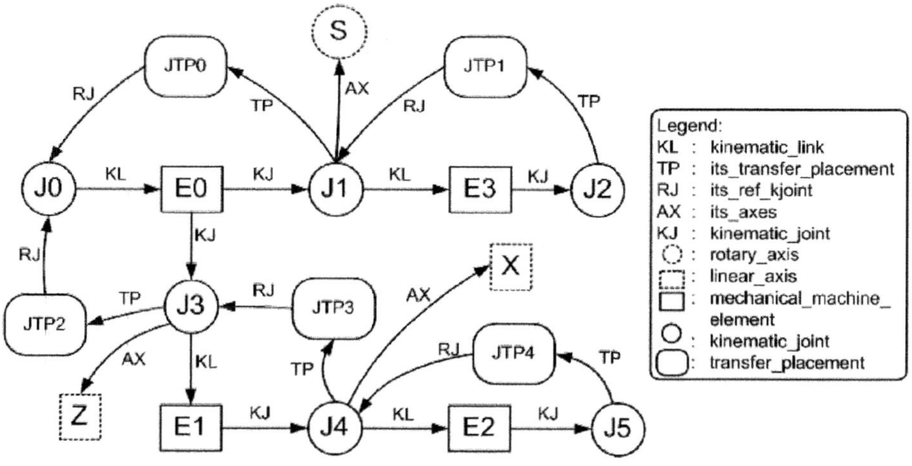

Figure 12.11. Data model for a 2-axis lathe

12.5.2 3-Axis Milling Centre

The first CNC machines were 3-Axis milling machines. They are still one of the most prominent configurations of metal-cutting machine tools in the industry. The tool and the table are connected with three linear axes of movement as shown in Figure 12.12. Figure 12.13 shows the instances of the entities for the model.

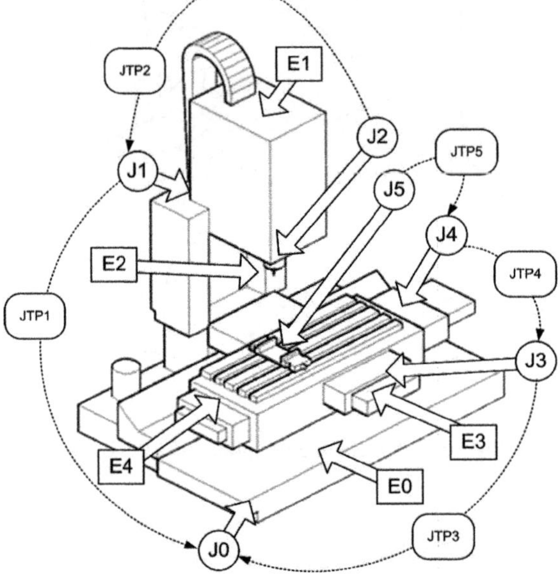

Figure 12.12. Geometric model of a 3-axis vertical machining centre

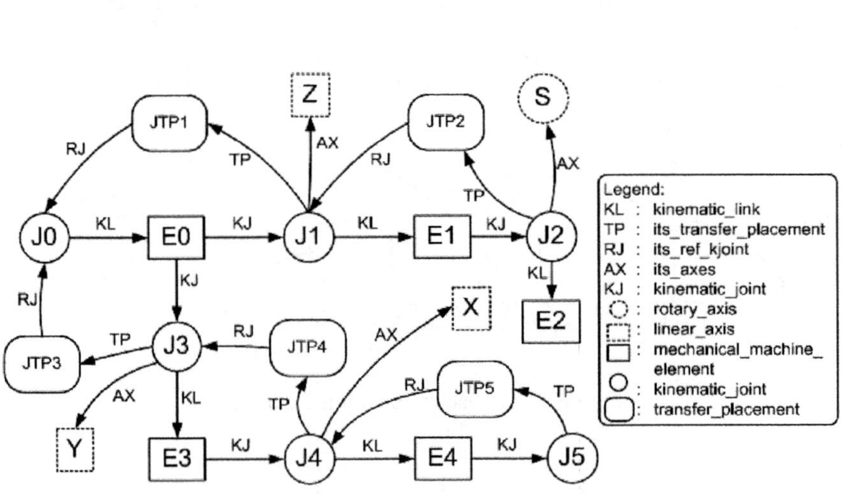

Figure 12.13. Data model for a 3-axis vertical machining centre

12.5.3 5-Axis Milling Centre

Manufacturing sculptured surfaces requires machines with 5 axes of movement, three linear and two rotational. These metal-cutting machines can be found in numerous configurations. Figure 12.14 shows a typical configuration of a 5-axis machining centre. Figure 12.15 illustrates the instances of entities forming the machine model.

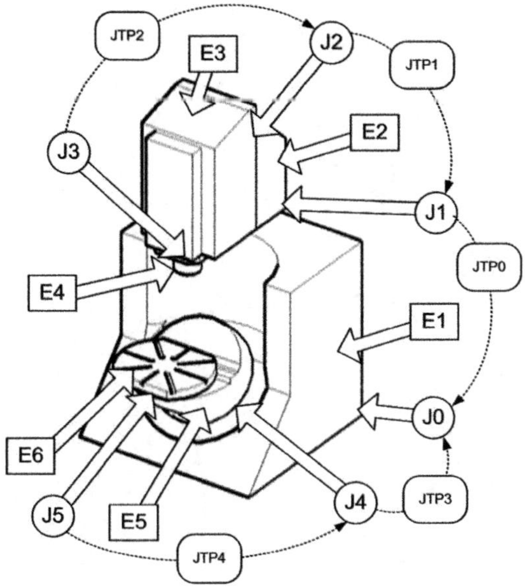

Figure 12.14. Geometric model of a 5-axis machining centre

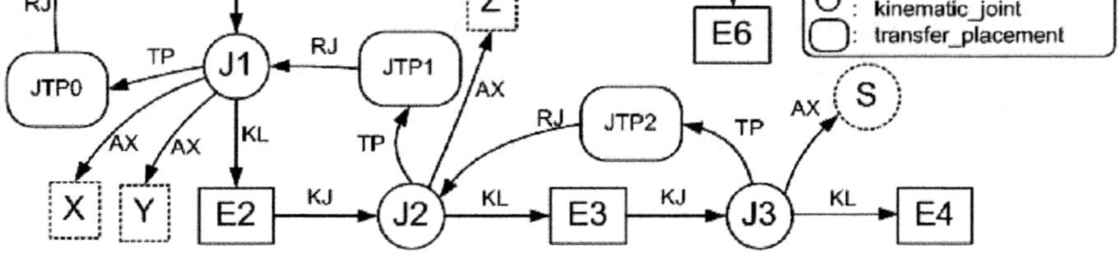

Figure 12.15. Data model for a 5-axis machining centre

12.5.4 Parallel Kinematics Machine

A particular challenge for most modelling methodologies arises when representing parallel kinematic machines. In these machines, movements result from parallel motion of more than one axis. Figure 12.16 shows an example of parallel kinematic machines.

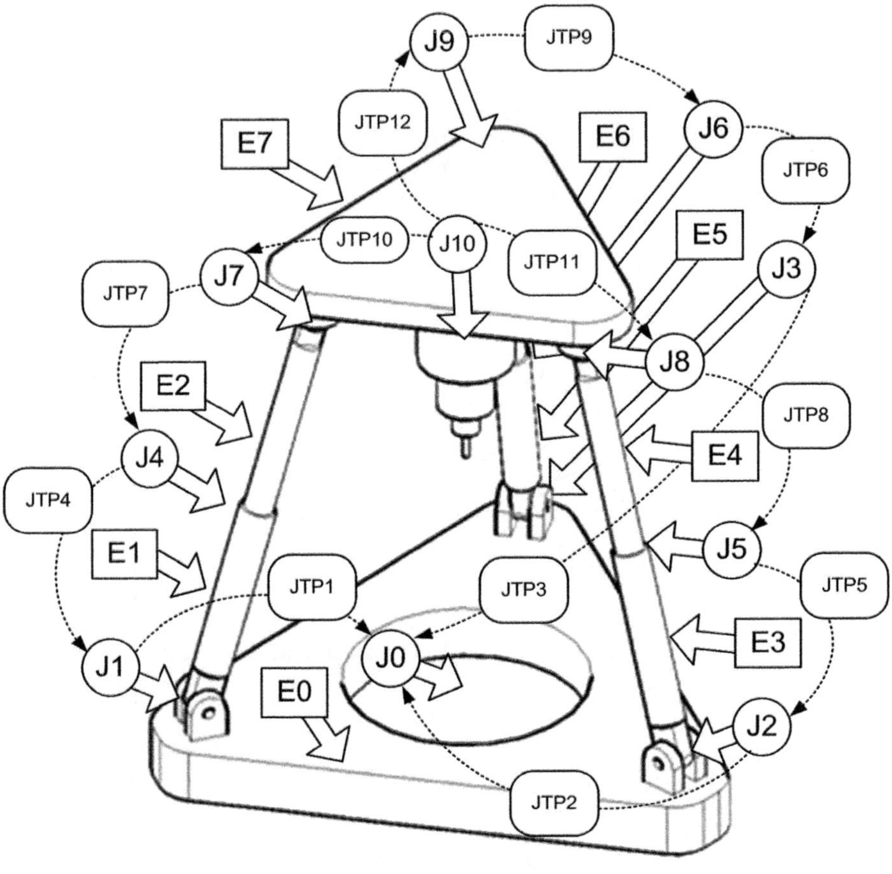

Figure 12.16. Geometric model of a parallel kinematic machine

Figure 12.17 shows the instances of entities representing the parallel kinematics machine. In joint 10, the multiple transfer placements ensure that the machine is in a feasible state by ensuring that the position of joint 10 when calculated from any one of the links is consistent with the position as calculated from the other links.

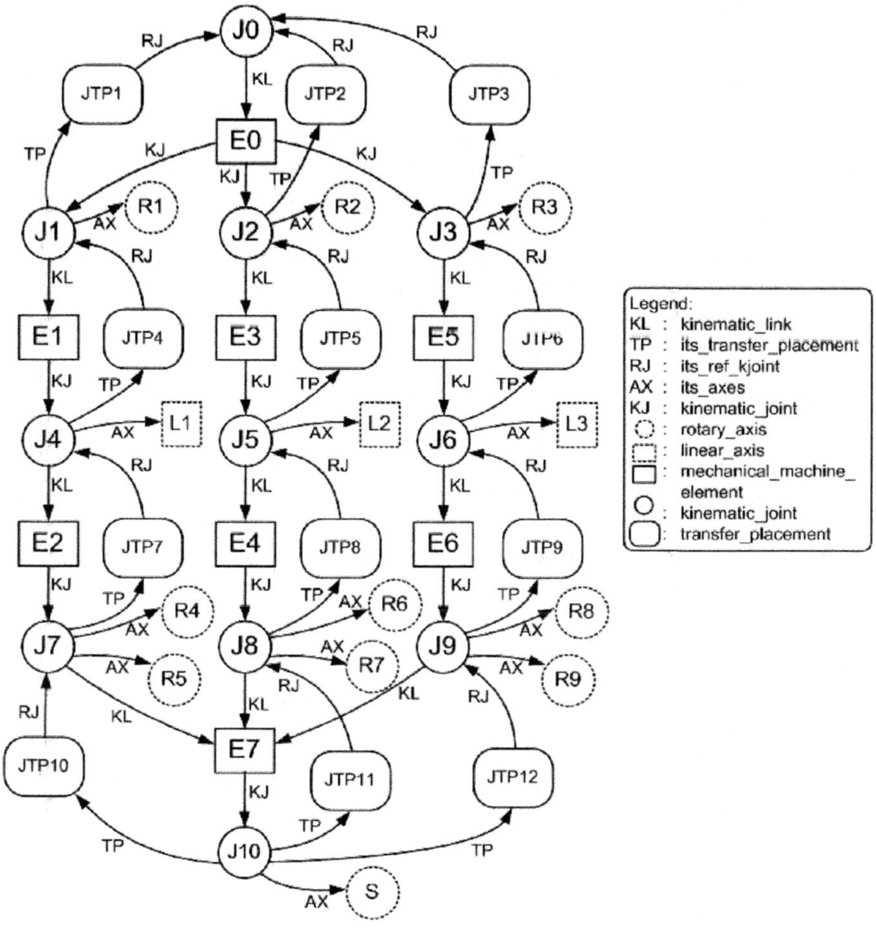

Figure 12.17. Data model of a parallel kinematics machine

12.6 Future Developments

In this chapter a unified approach for representing the information for the mechanical aspects of machine tools, regardless of their configuration was presented. However, in order, to enable intelligent STEP-NC manufacturing, extensions to this data model to encompass the electro-mechanical and electronic elements of manufacturing resources are necessary.

Furthermore, the proposed model in this chapter can be extended to represent information about production resources that perform processes that are not significant in terms of kinematics. Such processes include heat treatment, layer laying in composite manufacturing and laser sintering in rapid manufacturing.

12.7 Conclusions

STEP-NC presents an integrated representation of the information within the domain of manufacturing. While the current parts of the standard are capable of representing data about the desired product and the processes that are utilised to manufacture the product from raw material, the information about resources is not currently captured. This chapter presented an approach for modelling metal cutting machines and their associated devices to be used in conjunction with other Parts of STEP-NC to realise the next generation of manufacturing. This resource model can be utilised for process planning, process verification, machinability determination and visualisation of the manufacturing process.

Acknowledgment

The work reported in this chapter has been undertaken as part of the EPSRC Innovative Design and Manufacturing Research Centre at the University of Bath (Grant reference EP/E00184X/1) and supported by a number of industrial companies. The authors gratefully acknowledge this support and express their thanks for the advice and support of all concerned.

References

[12.1] Nassehi, A., Newman, S.T., Xu, X.W. and Rosso Jr., R.S.U. 2008. Toward interoperable CNC manufacturing. *International Journal of Computer Integrated Manufacturing*. 21(2):222-230.

[12.2] Xu, X.W. and Newman, S.T. 2006. Making CNC machine tools more open, interoperable and intelligent – a review of the technologies. *Computers in Industry*. 57(2):141-152.

[12.3] Ehmann, F.K., Kapoor, S.G., Devor, R.E. and Lazoglu, I. 1997. Machining Process Modelling: A Review (Invited paper) *The 75th Anniversary Issue of the ASME Journal of Manufacturing Science and Engineering*. 119:655-663.

[12.4] Siemens. 2008. *NX Machining: A complete solution for machine tool programming*. [Online]. Available: http://www.plm.automation.siemens.com/en_us/Images/nx%20machining%20brochure%20W%203_tcm53-4561.pdf [20 October 2008].

[12.5] CGTech. 2008. *About VERICUT Software* [Online]. Available: http://www.cgtech.com/usa/products/about-vericut/ [20 October 2008].

[12.6] Delmia. 2008. *DELMIA – PLM Solutions – Digital Manufacturing* [Online]. Available: www.delmia.com [20 October 2008].

[12.7] Ong, S.K., Jiang, L. and Nee, A.Y.C. 2002. An Internet-Based Virtual CNC Milling System. *International Journal of Advanced Manufacturing Technology*. 20:20-30.

[12.8] Suh, S.-K., Seo, T., Lee, S.-M., Choi, T.-H., Jeong, G.-S. and Kim, D.-Y. 2003. Modelling and Implementation of Internet-Based Virtual Machine Tools. *International Journal of Advanced Manufacturing Technology*. 21:516-522.

[12.9] Yao, Y., Liu, C. and Li, J. 2005. Web-based Virtual Machining and Measuring Cell, In *Proceedings of the 9th International Conference on Computer Supported Work in Design*. 2:673-678.

[12.10] Seo, Y., Kim, D. and Suh, S. 2006. Development of Web-based CAM System. *International Journal of Advanced Manufacturing Technology.* 28:101-108.

[12.11] Mazak. 2008. *Mazak Europe* [Online]. Available: www.mazakeurope.com [20 October 2008]

[12.12] Altintas, Y., Brecher, C., Weck, M. and Witt, S. 2005. Virtual Machine Tool. *CIRP Annals – Manufacturing Technology.* 54(2):115-138.

[12.13] Heusinger, S., Rosso, R., Klemm, P., Newman, S.T. and Rahimifard, S. 2006. Integrating the CAx Process Chain for STEP-Compliant NC Manufacturing of Asymmetric Parts. *International Journal of Computer Integrated Manufacturing.* 19(6):533-545.

[12.14] Stevens, W.P., Myers, G.J. and Constantine, L.L. 1974. Structured Design. *IBM Systems Journal.* 13(2):115-139.

[12.15] Shinno, H. and Ito, Y. 1987. Computer Aided Concept Design for Structural Configuration of Machine Tools: Variant Design Using Directed Graph. *Transactions of ASME, Journal of Mechanisms Transmission Automation Design.* 109(3):372-376.

[12.16] Shinno, H., Ito, Y. and Hashizume, H. 1991. A Decision Making Methodology for Basic Layout Design of Machine Tools. *JSME International Journal, Series III.* 34(2):290-294.

[12.17] Shinno, H., Yoshioka, H. and Marpaung, S. 2006. A Structured Method for Analysing Product Specification in Product Planning for Machine Tools. *Journal of Engineering Design.* 17(4):347-356.

[12.18] Yang, W. and Xu, X. W. 2008. Modelling Machine Tool Data in Support of STEP-NC Based Manufacturing. *International Journal of Computer Integrated Manufacturing.* 21(7): 745–763.

[12.19] Steward, D. 1966. A Platform with Six Degrees of Freedom. In *Proceedings of the Institute of Mechanical Engineers (Part I).* 180(15):371-386.

[12.20] Galorath, 2008. *SEER for Manufacturing: Project Management Software* [Online]. Available: http://www.galorath.com/index.php/products/manufacturing/C6/ [20 October 2008]

[12.21] Mobley, R. K. 2002. *An Introduction to Predictive Maintenance.* Butterworth-Heinemann.

[12.22] Seo, Y., Hong, D., Kim, I., Kim, T., Sheen, D. and Lee, G. 2005. Structure Modelling of Machine Tools and Internet-based Implementation. In *Proceedings of the 2005 Winter Simulation Conference.* 1699-1704.

[12.23] University of Southern California, 2007. *Loom Project Home Page* [Online]. Available: http://www.isi.edu/isd/LOOM/ [20 October 2008]

[12.24] Vichare, P., Nassehi, A., Kumar, S., Newman, S., Zheng, L. and Dhokia, V. 2007. Towards a STEP-NC Compliant Model for Representation of Machine Tools. In *Proceedings of DET2007, 4th International Conference on Digital Enterprise Technology.* 721-733.

[12.25] Desai, J. P., 2005. D-H Convention, in *Robotics and Automation Handbook,* edited by Thomas R. Kurfess. CRC Press

13

Development of Digital Semantic Machining Models for STEP-NC Based on STEP Technology

Fumiki Tanaka[1], Makoto Yamada[2], Satoshi Mitsui[3], Takeshi Kishinami[4], Kiyoshi Akama[5], Tsukasa Kondo[6] and Masahiko Onosato[7]

[1, 5, 7]Graduate School of Information Science and Technology, Hokkaido University, Kita-14 Nishi-9, Sapporo, Hokkaido, Japan
Email: [1] ftanaka@ssi.ist.hokudai.ac.jp
[5] akama@cims.hokudai.ac.jp
[7] onosato@ssi.ist.hokudai.ac.jp

[2, 6]Department of Mechanical Engineering, Hakodate National College of Technology, 14-1, Tokura-Cho, Hakodate, Hokkaido, Japan
Email: [2] myamada@hakodate-ct.ac.jp, [6] kondo@hakodate-ct.ac.jp

[3]Department of Information System Engineering, Asahikawa National College of Technology, 2-2-1-6, Shunkoudai, Asahikawa, Hokkaido, Japan
Email: mitui@asahikawa-nct.ac.jp

[4]Kushiro National College of Technology, 2-32-1, Otanoshike, Kushiro, Hokkaido, Japan
Email: kisinami@office.kushiro-ct.ac.jp

Abstract
To realize the articulate link between design and manufacturing, it is necessary to realize interoperable digital tools which represent the semantics of information independently of any implementations. In manufacturing, the interoperable CNC machining systems can be characterized as being capable of (1) seamless information flow without any errors related to the quality of the shape information, (2) feature-based machining, and (3) autonomous CNC. In this research, modelling and implementation of the Digital Semantic Machining Model that fulfils the above requirements are proposed. The Digital Semantic Machining Model is based on STEP-technology, which includes representation and quality inspection of shape data, feature model and feature recognition process model, and feature-based CNC machining process model.

13.1 Introduction

The movement toward concurrent engineering drives the need for frequent transformation of the product data between different digital tools such as CAD, CAM, CAPP, CAE, etc. To realize the articulate link between design and manufacturing, it is necessary to realize interoperable digital tools which represent the semantics of information independent of any implementations. However, some fundamental problems which cause manufacturing inefficiency have been pointed out and are as follows [13.1]:

- Interoperability among heterogeneous software systems and tools is generally not available. The design, manufacturing, and supply chain software systems are not interoperable across software ownership boundaries.
- The use of sub-optimal process parameters increases the manufacturing cost because of increased tool wear or under-utilized tools and machines.

Today, all product data, such as geometry representation data, tolerance data, etc., are not used throughout product development processes (Figure 13.1).

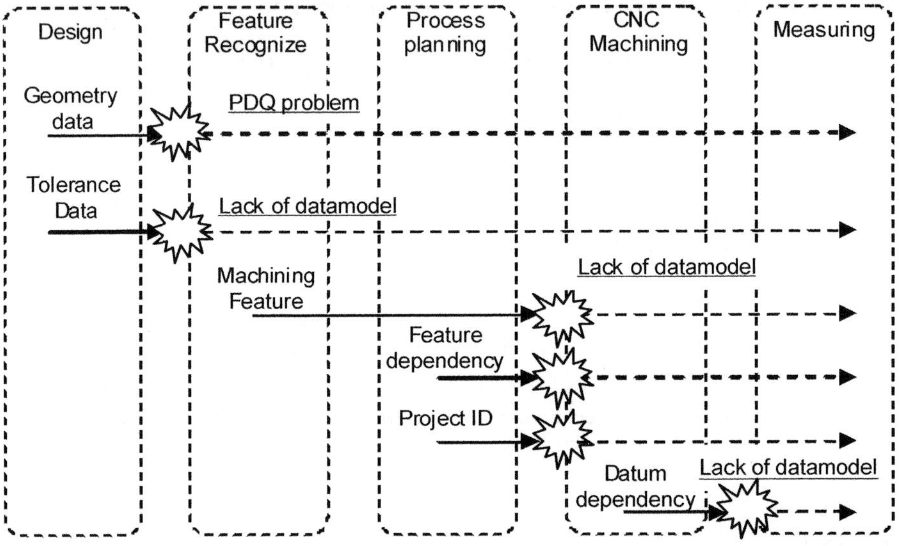

Figure 13.1. The problem of the communication in product development processes

An interoperability error makes product model data unusable when applying digital tools. If an interoperability error arises in collaborative engineering applications, tracking the error back to its source is expensive (if not impossible) and requires a significant amount of time. The National Institute of Standards and Technology (NIST) report titled 'Interoperability Cost Analysis of the US Automotive Supply Chain' estimates that the economic cost of bad interoperability in the US automotive industry is 1 billion dollars per year [13.2]. Thus, it is necessary to ensure that the quality of the shape information overcomes the limitation of interoperability of the shape information [13.3].

Engineering departments of manufacturing companies face increasingly frequent and unpredictable customer demands, and interoperable CNC machining systems have become key systems to meeting these customer demands. This type of system can be characterized as being capable of: (1) seamless information flow; (2) feature-based machining; (3) autonomous CNC, etc. [13.4]. However, current CNC systems and NC data do not have these capabilities.

In order to realize the articulate link between design and manufacturing and the traceable manufacturing system, it is necessary to represent the semantics of information independently of any implementations. Moreover, the quality of data that represent information should be ensured during the product life cycle. In this research, the Digital Semantic Machining Model that fulfils the above requirements is proposed. In this research, modelling and implementation of the Digital Semantic Machining Model is described.

13.2 Digital Semantic Machining Model

The digital semantic model consists of the quality assurance of the product model data, the semantic modelling for humans (such as designer and worker), understandable and computer readable data, and the functional modelling for supporting the product life cycle [13.5]. The digital semantic machining model is an application specific model of a digital semantic model in order to support products for machining. This section describes the concept of the digital semantic machining model and overview of our research projects.

13.2.1 Basic Concepts of the Digital Semantic Machining Model

In order to ensure the product model data, the Product Data Quality Checking method for the ISO Standard product model (ISO 10303-214 [13.6]) is adopted. In order to support all the product lifecycle activities, the functional model for each activity is developed based on the related ISO Standard, such as ISO 10303-224 [13.7], 238 [13.8], 240 [13.9], and ISO 14649 [13.10–13.12]. In order to represent the semantics of the data, the EXPRESS language [13.13] for data modelling and XML for representing the data [13.14] are adopted. The EXPRESS language was proposed and standardized in ISO TC184/SC4. The elements of the digital semantic machining model such as PDM data, geometric data, application attribute (machining feature), and application management (project ID), shown in Figure 13.2, are modelled based on the ISO Standards, In this research, three research themes shown in Figure 13.2 were chosen to solve typical problems in current product lifecycle. T-1 is the research about the representation and quality inspection of shape data. T-2 is the research about the feature recognition process model for 5-axis machining. T-3 is the research about the feature-base CNC machining process model. The detail of the T-1 is described in Section 13.3, the T-2 is in Section 13.4, and the T-3 is in Section 13.5, respectively.

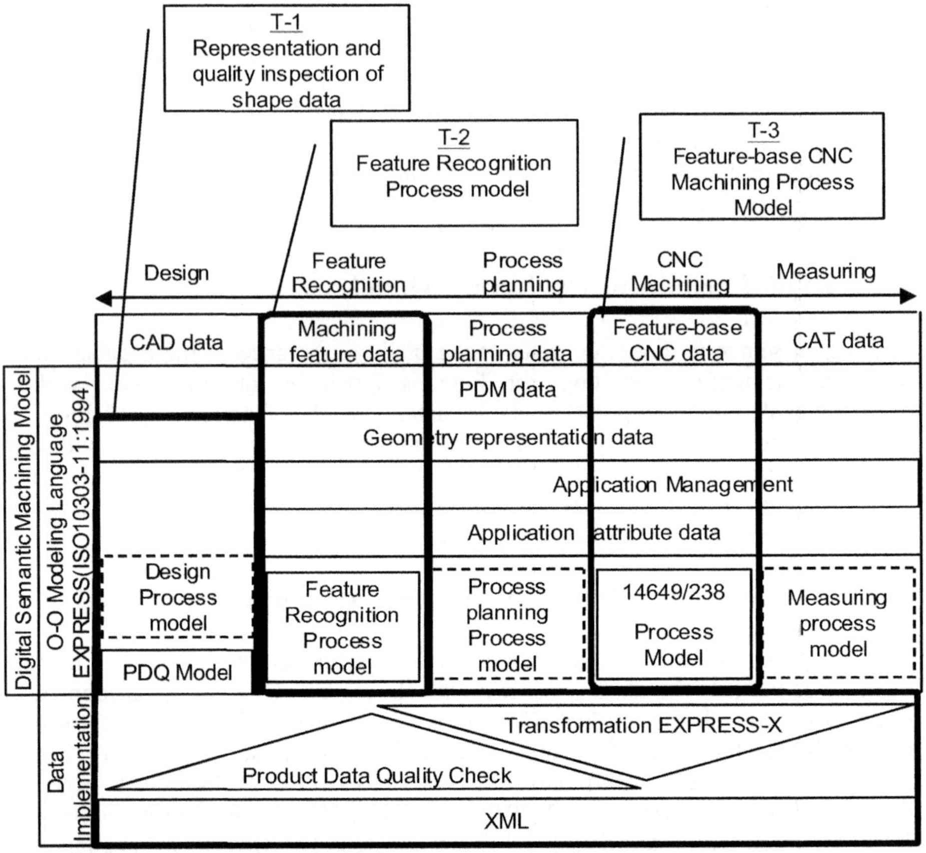

Figure 13.2. The elements of the digital semantic machining model and research theme

13.2.2 Modelling and Implementation of Digital Semantic Machining Models

Figure 13.3 shows the modeling and implementation of digital semantic machining models for 5-axis machining. The manufacturing procedures of the product using proposed digital semantic machining models are described below. First, the designed product model (Application Protocol (AP) -214 data) is checked in terms of its data quality based on the product data quality assurance method. Second, the machining features for 5-axis machining are recognized and ISO 14649 CNC data are generated. The machining features for 5-axis machining are also proposed as elements of the digital semantic machining model. Finally, the product is manufactured using the machine tools for ISO 14649 CNC data model.

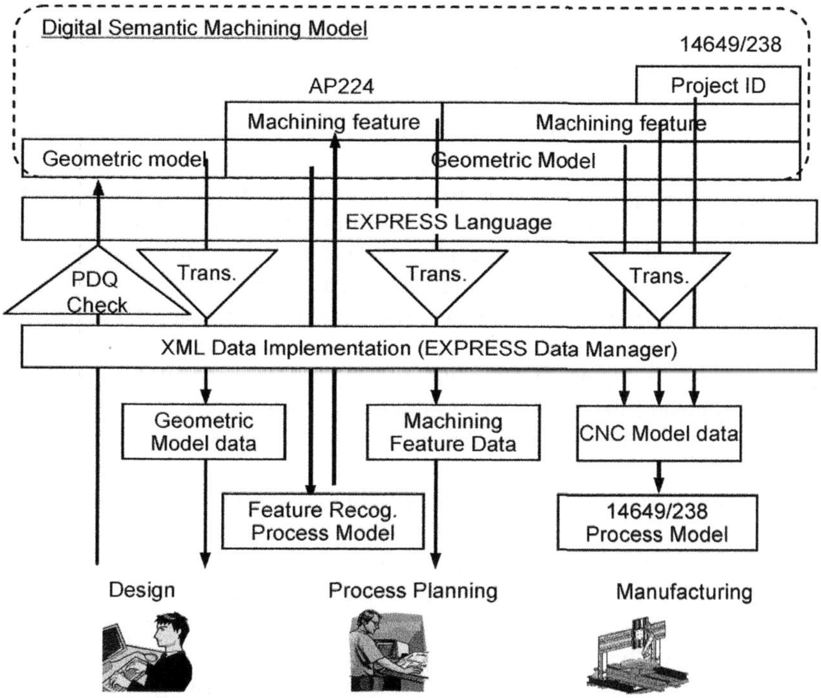

Figure 13.3. Modelling and implementation of digital semantic machining models

13.3 Product Data Quality Assurance Method

In order to avoid interoperability errors, it is necessary to develop quality checking software for the shape information of the product model data and to verify the shape data using the checking software whenever the shape data are transformed. Some research efforts have been undertaken in this area [13.15–13.21]. However, many previous researchers have addressed the identification and repair of shape errors based on the specific CAD model. Communication is usually performed between different CAD systems, or different application systems such as CAD, CAM, CAE, etc. Therefore, an approach based on a neutral file is necessary to achieve the real data quality assurance of the shape data of the product model. A classification for model quality errors was developed by the Strategic Automotive Product Data Standards Industry Group (SASIG) task force on Product Data Quality as SASIG-PDQ [13.22] for product data quality in the automotive industry. SASIG-PDQ has unified the emergent national recommendations for product data quality which were developed in JAMA [13.23] and VDA [13.24], etc. The criterion has its own code. The examples of the criteria and code are large segment gap (G-CU-LG), self-intersecting curve (G-CU-IS), large edge gap (G-LO-LG), and self-intersecting loop (G-LO-IS) [13.23]. However, these criteria are developed for automobile industries and described in natural language (English) with a certain amount of ambiguity in the measurement methods of the product data quality. The product data quality issue is mostly common to all manufacturing industries, but product data quality criteria and acceptable values are slightly different from industry to industry. Therefore, the product data quality program should be developed in the specific industry's community.

To solve these problems, it is recommended that the criteria be described in formal language or programming language. In this research, a software platform for checking quality of shape data of STEP product model [13.6] is proposed [13.25]. The proposed method is based on the Equivalent Transform (ET) language [13.26] which is a kind of rule-based language. The advantages of using ET include ease of developing the checking programs for the shape data as the constraint rules. In this chapter, current problems of checking the quality of product data and proposed method of checking the quality of product data are described. An example is also shown in order to demonstrate the effectiveness of the proposed system.

13.3.1 Current Problems of Checking the Quality of Product Data

At present, interoperability errors are one of the critical problems in concurrent engineering. In order to avoid an interoperability error, it is necessary to develop a quality checking mechanism for the shape information of the product model data and to verify the shape data whenever they are transformed. Some commercial software systems have been developed for implementing PDQ guidelines. However, their algorithms are not open, and their accuracy cannot be verified. Even if their algorithms are presented in natural language, there is no assurance for equivalence between algorithms and implemented software.

Moreover, the PDQ criteria defined for one industry need to be extended easily to other industries. In order to specify the algorithms for the user, the data format of the product data also needs to be open and described in an unambiguous way.

In brief, current problems for PDQ diagnostics are as follows (Figure 13.4):

- In order to avoid interoperability errors in collaborative engineering environments it is necessary to verify the shape data using various PDQ-tools before the shape data are used in different CAx systems. Therefore, many PDQ-tools (or criteria) for specific combinations of CAx systems are needed.
- PDQ criteria such as SASIG-PDQ guidelines are described in natural language, and a certain amount of ambiguity exists in the measurement methods of the product data quality. Therefore, it is difficult to implement a PQQ-tool.
- A PDQ-tool uses closed (secret) algorithms in which the consistency between tools cannot be ensured. Even if a PDQ tool presents its algorithms in natural language, there is no assurance for equivalence between algorithms and implemented software, and the PDQ criteria need to be extended easily.

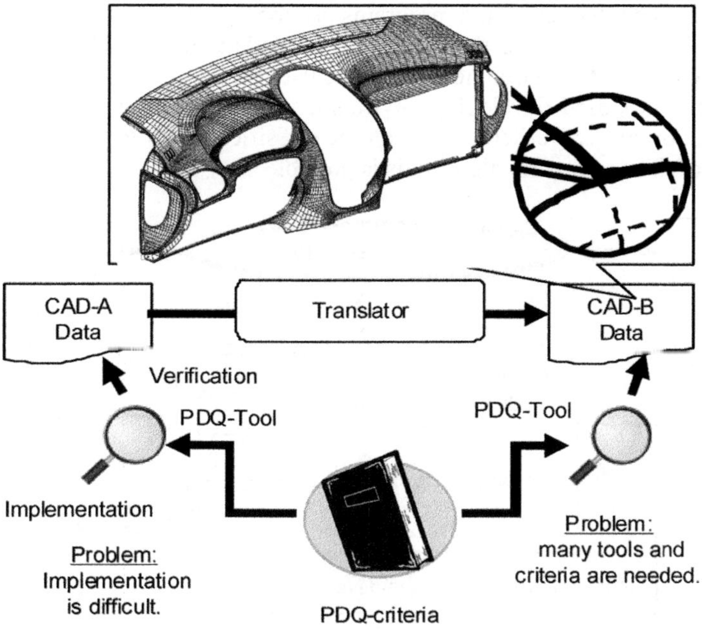

Figure 13.4. Current problems of checking the quality of product data

13.3.2 A Software for Checking the Quality of Product Data

To solve the problems mentioned above, a software platform for checking the quality of shape data of the STEP product model for CAD/CAM environment is proposed in this research as shown in Figure 13.5. The constituents of the proposed method are as follows:

- The target shape data of the product model should be a neutral file which is independent of any specific CAx systems. Moreover, the data structure of the product model data should be known by all users in order to specify the PDQ criteria of the shape data. Then the target product model data uses STEP AP 214 [13.6] for the purposes of the automotive industry, and it is widely supported by many CAD systems. The data structure of AP 214 data is explicitly described in formal language to avoid any ambiguity of the PDQ criteria.
- PDQ criteria should be described in formal language. Describing the criteria in formal language means representing the diagnostics algorithms in some kind of programming language. In this research, the PDQ criteria are described in the ET language. The described quality inspection algorithms are referred to as quality criterion gauges.
- The description language (or formal language) should have the ability of extending the existing criteria easily, and anyone should be able to use it. As mentioned above, the description language of PDQ criteria is the ET language in this research. The advantages of using ET include ease of

developing the checking programs as the constraint rules. The software platform for checking the quality of shape data of the STEP product model for CAD/CAM environment is also developed.

Using the proposed developing environment of the PDQ inspection algorithm, systematization of PDQ criteria based on the classification of PDQ criteria from the viewpoint of their geometric algorithms is possible.

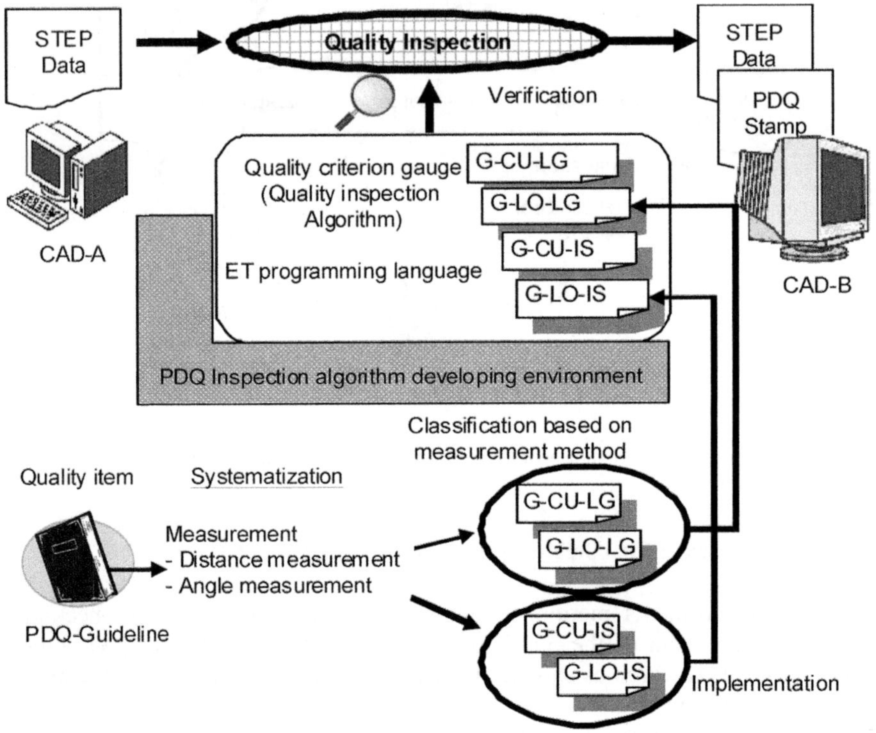

Figure 13.5. Software platform for checking the quality of shape data

13.3.3 Constituents of Proposed Method

As mentioned above, the PDQ criteria are described in the ET language which is a kind of rule-based language. A program to solve a problem in ET is a set of rules, and the program is made by accumulating the rules as shown in Figure 13.6. The equivalent transformation rules are written as shown in this figure. The leftmost atom, which is the atom to be replaced, is called "Head." Each atom consists of predicate and arguments. The left atoms enclosed with curly brackets are the condition-part of the rule. The atoms on the right of the arrow are called "Body." The right atoms enclosed with curly brackets are the execution part of the rule. The example of the adding the two values is also shown in this figure.

Using ET language, a PDQ criterion gauge can be written as shown in Figure 13.7. "Head" consists of the quality_criterion_name and arguments. The arguments are variables which represent a product model, measurement_accuracy, measurement_threshold, and result. The condition-part of the rule consists of an atom of extracting the inspected elements from a product model. "Body" consists of two atoms. First atom calculates the measured_data of inspected_element according to the measurement_accuracy. Second atom evaluates the measured data according to the evaluation_operator and measurement_threshold. The measurement rules can be created easily in the proposed software platform, and the measurement rules are used in many quality criteria. Moreover, if PDQ inspection procedure needs to be used for many quality criteria, the root rule can be created from these quality criteria.

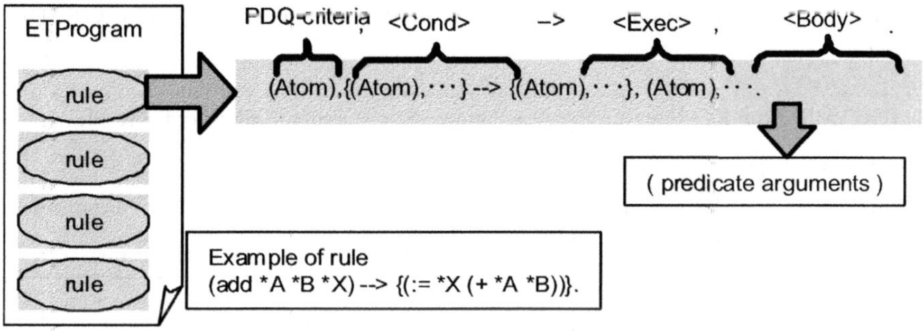

Figure 13.6. Equivalent transformation program language

```
(quality_criterion_name  *model
                         *measurement_accuracy  *measurement_threshold  *result),
{(extract_inspected_element_name *model *inspected_element)}
-->
    (quality_measurement_name *inspected_element
                              *measurement_accuracy *measured_value),
    (evaluate *criterion_name *evaluation_operator *inspected_element
                              *measured_value  *measurement_threshold *result).
```

Figure 13.7. The problem of the communication in product development processes

13.3.4 Example of Checking Quality of Product Data

An example is shown to demonstrate the ability of the representation of ET, the correctness of the PDQ criteria example for a test data, and the availability of practical use of the PDQ gauge. The test data sets including the known error set are provided by JAMA/JAPIA for PDQ diagnostics experimentations. The target quality item is G-FA-IS [13.23]. G-FA-IS is the criterion that the pair of loops are in the same face that intersect each other. The gauge example is shown in Figure 13.8. This gauge calculates the maximum distance between each point on edge of the loops using the rule named "distance_between_loops", and evaluates whether measured value is larger than the specified measurement_threshold or not.

Test data and the inspection result of the G-FA-IS are shown in Figure 13.9. This data contains the error edge_loop elements of which the maximum distances between each point on them are 0.1, 0.012, 0.01, 0.008, and 0.001 respectively. From the result the number of errors and error sources and reasons are easily found. The error data show that the distance between each point on edges of the loops is 0.001288 and smaller than the specified recommended threshold value (0.01). Using these data, the designer can correct his own data. Because the result is correct according to the specification of the test data, the specified quality criterion is correct.

```
(G-FA-IS *model *measurement_accuracy *measurement_threshold *result),
{(extract_inspected_element_FALPs *model (*loop1 *loop2 *face))}
--> (distance_between_loops (*loop1 *loop2) *measurement_accuracy *distance),
    (evaluate G-FA-EG > (*face) *distance *measurement_threshold *result).
```

Figure 13.8. G-FA-IS gauge example

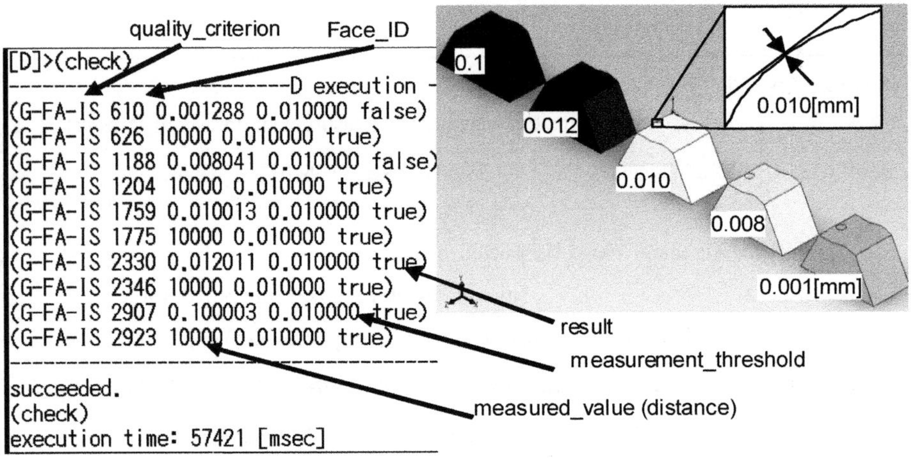

Figure 13.9. Test data and Inspection result of the G-FA-IS using proposed gauge

13.4 Machining Features for 3+2 Axis Machining

Finishing processes for injection moulding moulds and die casting dies are not typical machining processes because they consist of many thin and deep cavities. To solve this problem, mould and die machining using indexing tilted angles of the tool axis on a 5-axis machining centre (referred to as *3+2-axis machining* in this research) was proposed [13.27]. In this chapter, the concepts of 3+2-axis machining, the definition of the machining feature for 3+2-axis machining, and the extracting method of the machining feature are described. Extracted features are represented in Region entity of STEP-NC. The Operation and Strategy are represented in Five_axes_const_tilt_yaw entity and other STEP-NC entities.

13.4.1 Concepts of 3+2 Axis Machining

Today nearly all die and mould makers use high speed cutting or machining in cavity and core manufacturing. Optimized tool-path generation, to maintain a constant chip load in machining complex sculptured surfaces, is offered by research centers as well as some software suppliers [13.28]. However, finishing processes for injection moulding moulds and die casting dies are done by EDM because these applications require the machining of many thin and deep cavities. Otherwise, long solid carbide milling cutters are necessary for machining these moulds and dies. One of the candidates for solutions is using 5-axis machining. To make good use of 5-axis machine tools, it is necessary to solve more complicated interference problems as well as optimal tool orientation determination problems. The research on 5-axis machining has been mainly concerned with tool position and orientation computations [13.29–13.34]. However, two additional rotation axes cannot move more rapidly than three translation axes, and efficient high speed machining cannot be realized for a 5-axis simultaneous control. There have been few researches dealing with determination of tilted angle using 5-axis machine tools [13.35, 13.36].

To solve these problems, mould and die machining using an indexing tilted angle of tool axis on a 5-axis machining center is proposed in this research. In the proposed machining process, high speed machining is done on each indexed tool axis. Typical machining operations with a milling tool are classified into 3-axis machining and 5-axis machining as shown in Figure 13.10. As mentioned above, we propose mould and die machining which is carried out on an indexing tilted tool axis using a 5-axis machining center shown in this figure. In the proposed machining process, the orientation of the cutting tool remains fixed during high speed machining. This kind of machining is referred to as 3+2-axis machining in this research. To guarantee the surface integrity, the transition region should be character-line such as edge or fillet.

13.4.2 Machining Feature in 3+2 Axis Machining

In this machining process, the machinable area for each indexing tilted angle is defined as a machining feature for 3+2 axis machining as shown in Figure 13.11. In this research, the cutting tool is a ball end mill for mould and die machining. The machinable areas, where cutter and holder interference does not occur, are the machining portion using effective cutting edge to avoid machining around the center of the rotation of the ball end mill tool. In Figure 13.11, two machining features derived from one tilted angle (+20) and two machining features derived from another tilted angle (–20).

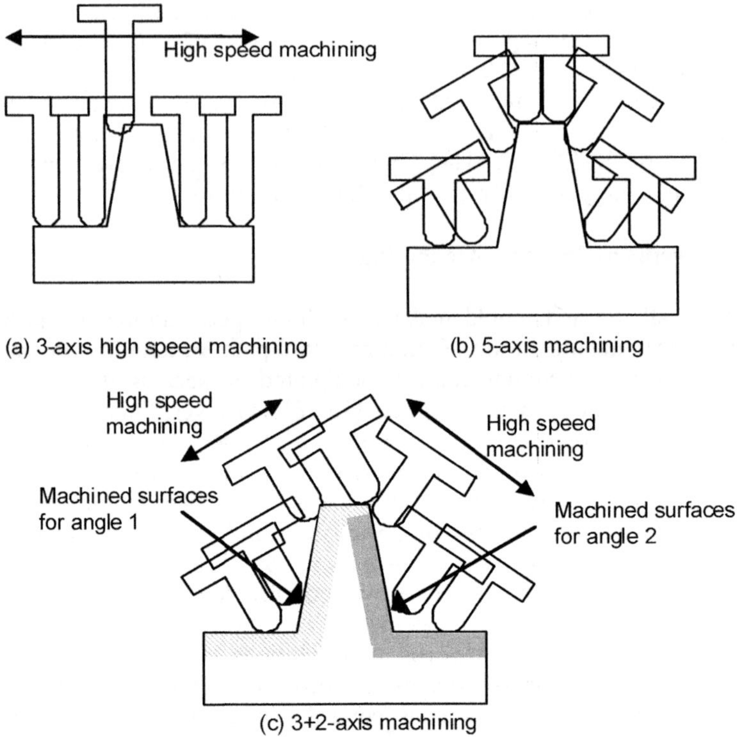

Figure 13.10. 3-axis high speed machining, 5-axis machining, 3+2-axis machining

To extract the machining feature, the indexing angle should be determined automatically. In this research, the required machining shape is given as a facet of a solid model. The procedure for calculating the indexing angles and obtaining the features are as follows:

- The candidates of the indexing angles in consideration of the contact position and orientation on the cutting edge to the required shape are calculated.
- For each calculated candidate, the machining features (machinable areas), where cutter and holder interference does not occur, are derived from the information on the cutting edge of the tool, the shank and holder shape using the inverse offset method with the state flag.
- The indexing angle for which the machinable area is maximum is selected. Then, unmachinable facets are derived based on the machinable area in the selected indexing angle.

If unmachinable areas exist, they are fed back to the computing process of the candidates of the indexing angles in order to derive other indexing angles. Consequently, the list of the indexing angles and the machinable areas are derived from the process of 1 to 3.

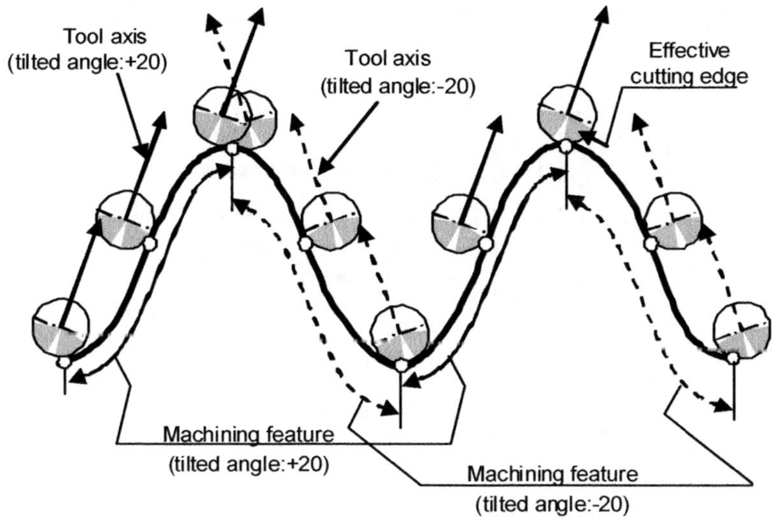

Figure 13.11. Software platform for checking the quality of shape data

13.4.3 Extraction Method of 3+2 Axis Machining Features

To calculate the machinable area (feature) for each candidate tool indexing angle, the offset surface of the required surface should be obtained, and machining simulation should be done on the obtained offset surfaces. To do so simultaneously, the inverse offset method [13.37] is adopted in this research. An offset surface can be generated from the enveloped surface of the cutting edge, which is generated by moving the reverse cutting tool on the required shape.

The state flag is added to the inverse offset point to distinguish between the offset surface derived from a cutting edge part and the offset surface derived from a tool holder part. To obtain machinable facets, a reference flag to the required machining facet from the offset point derived by inverse offset operation is also added. The deriving method consists of the following procedures:

- A grid space area in the model space is set in the computer, and Z-Queue is prepared to memorize the Z coordinate value to each grid point. On that occasion, we set the maximum Z coordinate value as the intersection point of facets and Z-Queue, as shown in Figure 13.12.
- An inverse offset operation is applied to the previously obtained Z-Queue for the tool model with a tool shank part and a holder as shown in Figure 13.13. By this operation, the offset surface which can avoid holder interference, and the required machining surface can be obtained.

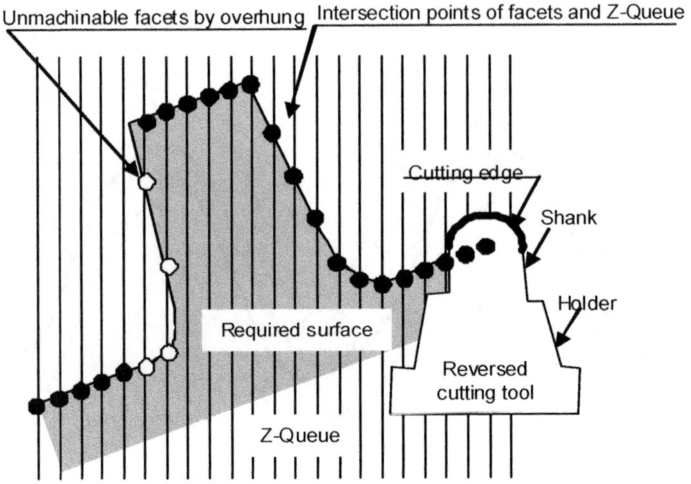

Figure 13.12. Generation of machining points on Z-Queue, and model of cutting tool

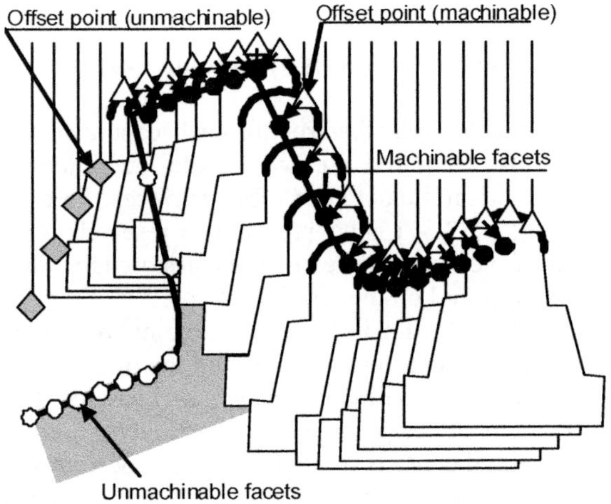

Figure 13.13. Deriving of machinable facets

13.4.4 Example

To demonstrate the effectiveness of the proposed methods, we applied the proposed method to a mould and die shape as shown in Figure 13.14. Figure 13.14 also shows the size of cutting tool.

Figure 13.15 shows the two machining features and indexing angles derived from the inverse offset method with the state flag. We determine the indexing angle in which the machinable area becomes the largest among the candidate indexing angles. The area in bright was judged to be the machinable facet of an indexing angle in each figure. The lines in Figure 13.15 also show the machinable offset boundary. In this sample shape, all of the required machining areas became machinable with the two indexing angles.

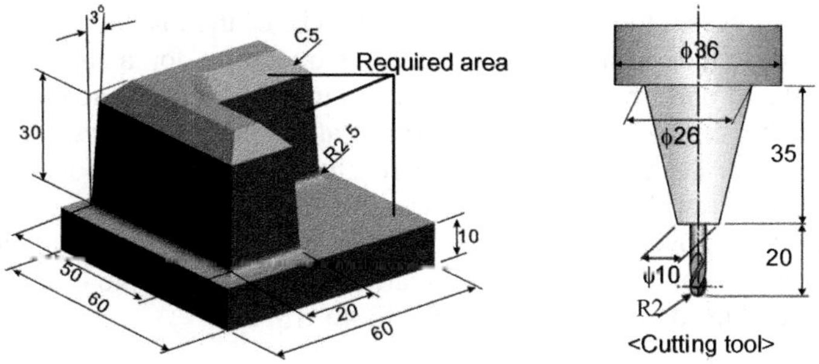

Figure 13.14. Target required shape and cutting tool

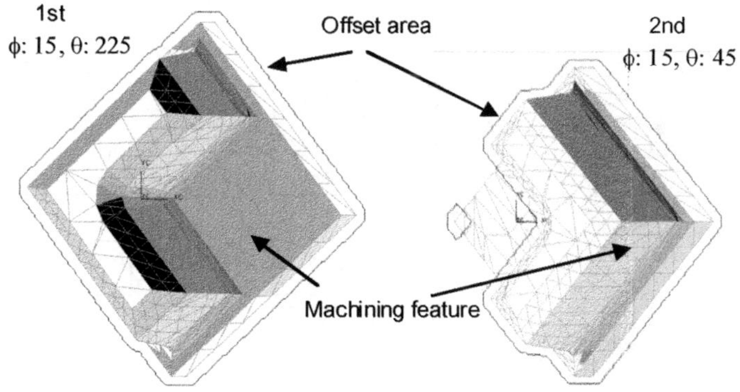

Figure 13.15. Case study for deriving machining features on each indexing angle

13.5 Machine Tools for ISO 14649 CNC Data Model

In this research, a prototype system of 5-axis machine tools for ISO 14649 CNC data model is developed [13.38]. In this chapter, the background and objective of this research, the machine tool model based on STEP kinematic model [13.39], and the prototype machine tool are described.

13.5.1 Background of Developing ISO 14649 Machine Tools

The fundamental principle of the ISO 14649 NC data model is the object-oriented view of programming in terms of manufacturing features [13.10], instead of direct coding of sequences of axis motions and tool functions defined in ISO 6983.

Together with the general process data described in ISO 14649-10 [13.11], ISO 14649-11:2003 [13.12] specifies the technology-specific data elements needed as process data for milling.

The overview of the data structure of ISO 14649 is shown in Figure 13.16. Workingsteps represent the essential building blocks of the machining process. Each Workingstep describes a single manufacturing operation for a machining feature using one tool and one strategy. The tool movement within the machining operation is determined by the technology-dependent strategy and additional parameters. In order to determine the tool movement to reach the first cut, optional information about the approach (plunge) strategy can be specified. Moreover, in order to determine the tool movement after finishing the last cut, optional information about the retract strategy can be specified. Research works of ISO 14649 and STEP-NC were investigated in some research groups [13.40–13.44].

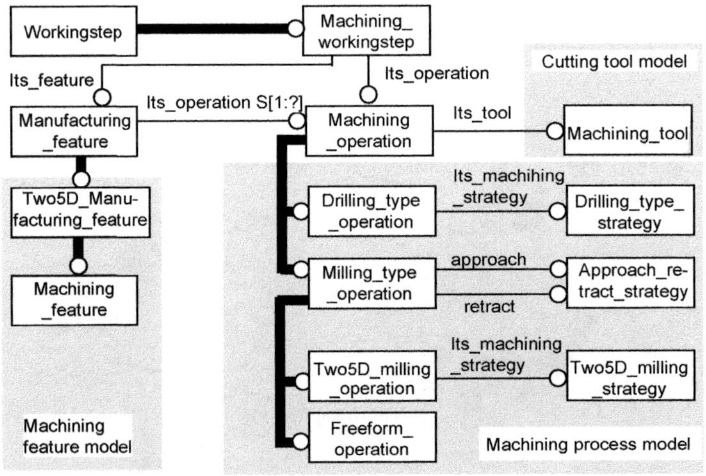

Figure 13.16. Case study for deriving machining features on each indexing angle

With a 5-axis CNC machine tool that is built according to the new CNC data model, a user only considers information about the motion of tools, such as the machining shape, the machining operation, and the cutting condition. The user can work without being influenced by the structure of the machine. In 2005, the OMAC STEP-NC Working Group tested on 5-axis parts using AP-238 CC1 machine-independent tool-paths.

Our studies aimed at two goals: making a CNC machine tool that directly inputs CNC model data and realizes 5-axis machining, and verifying the CNC model. Our CNC machine tool had a CNC-model data interpreter, built-in feature driven tool-path generator, and driving-axes controller.

13.5.2 Machine Tool Model Based on STEP Kinematic Model

The machine tool model consists of a kinematic model which describes a kinematic structure of a machine tool and a shape model which describes all the shapes of a machine tool. In order to describe a kinematic structure of a machine tool, particular entities are added to an information model standardized in ISO10303-105 [13.39]. A data structure of the machine tool model is shown in Figure 13.17. The 'mechanism' is described as a kinematic structure with a single link fixed to the ground using entity 'kinematic_structure' in ISO10303. The 'kinematic_structure' is represented by means of 'kinematic_joint' which describes a relationship between a pair of links. Each link is represented by a 'kinematic_link.' The 'kinematic_pair' defines the kinematic constraints between two adjacent links coinciding at a joint. ISO10303-105 covers 17 types of 'kinematic_pairs.' In this study, types of 'kinematic_pairs' for a machine tool are the only 'revolute_pair' which constrains the motion between two adjacent links to a rotation about a common axis, and the 'prismatic_pair' which constrains the motion between two adjacent links to a translation along a common axis. The 'su_parameters' describe the position between adjacent pairs and consists of six parameters [13.39]. Generally, the kinematic mechanism of a machine tool is considered to be two mechanisms from a base of a machine tool to a tool and from the base of a machine tool to a workpiece.

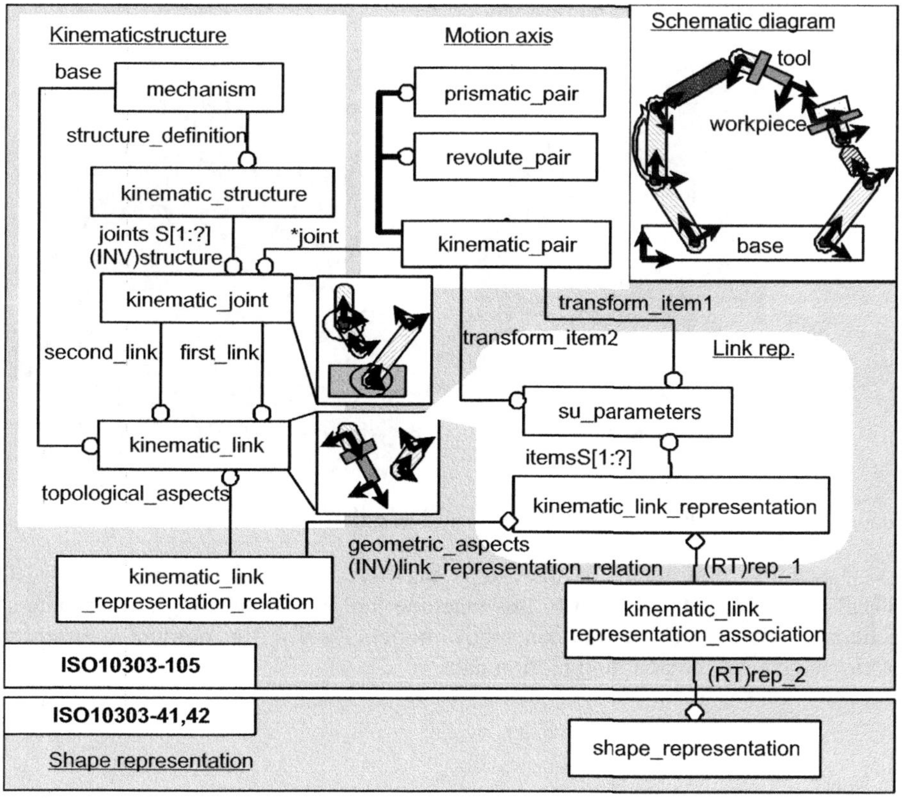

Figure 13.17. Kinematic model of machine tool

13.5.3 Prototype of 5-Axis Machine Tool for ISO 14649 CNC Data Model

We developed a 5-axis CNC machine tool that can process the machining feature model based on ISO 14649. The structure of this machine tool combines a tool (A-axis) and table (C-axis) tilt as shown in Figure 13.18. The machine tool contains a CNC-model data interpreter; the CNC kernel, with built-in tool-path and tool-orientation generation capabilities, executes the programmed Working_step. Furthermore, the CNC machine tool computes the position and angle of the driving axes by coordinate transformation processing, and it controls the driving axes. These functions are built into a personal computer.

The prototype machine tool is shown to the right of Figure 13.18. The driving axis for X, Y, and Z uses an AC linear motor in which high speed and high positioning are possible. An AC rotary motor is used for driving the A and C axis. This machine is equipped with a brushless motor with a maximum speed of 25,000 rpm. The speed is manually set. Although the cutting position changes depending on the orientation of the tool, and the cutting speed changes along with it, the spindle speed control cannot be adjusted.

Figure 13.18. Machine tool for ISO 14649 CNC data model

The software structure of the 5-axis machine tool is shown in Figure 13.19. STEP-NC XML data are read to this machine tool, and machining motion data are generated using machine tool kinematic model. Finally, the product is machined from the generated machining motion data.

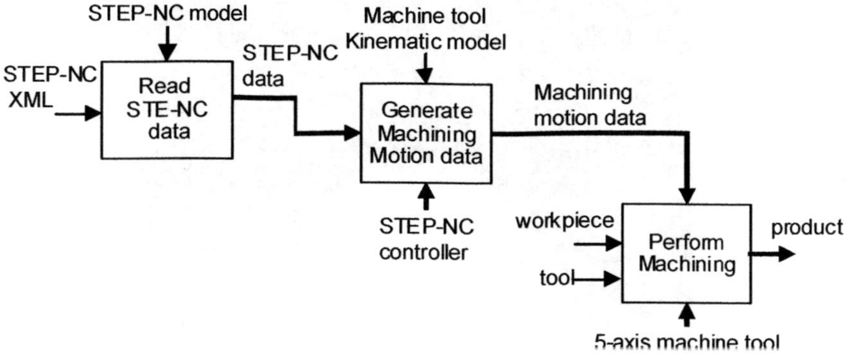

Figure 13.19. Software flow of machine tools

13.5.4 Practical Results

To verify the CNC data model, the machining data of a hemisphere was created, and the hemisphere was machined with a 5-axis machine tool using the data. A hemisphere was created by machining a free-form surface, and Cutter_Location_trajectory described the instance of a tool-path. The center position and ordinate of the tool were specified by Polyline, and Cartesian_point described each point. The orientation of the tool was a unit normal vector. Figures 13.20 shows the processing of the CNC-model data of a hemisphere that has a diameter of 80 mm in the XML format. The 5-axis CNC machine tool input this data and machined urethane foam using a ball end mill with radius of 3 mm as shown in Figure 13.21. The cutting conditions include a feed rate of 500 mm/min and a spindle speed of 25,000 rpm.

```
<machining_workingstep id=" i944">
<its_operation>
<freeform_operation
   xsi:nil=" true" ref=" i946" />
</its_operation>
</machining_workingstep>
<freeform_operation id=" i946">
<its_toolpath>
<toolpath_list xsi:nil=" true" ref=" i947" />
</its_toolpath>
</freeform_operation>
<toolpath_list id=" i947">
<its_list ex:cType=" list">
<cutter_location_trajectory
  xsi:nil=" true" ref=" i948" />
</its_list>
</toolpath_list>
<cutter_location_trajectory id=" i948">
<its_technology>
<milling_technology xsi:nil=" true" ref=" i950" />
</its_technology>
<basiccurve>
<polyline xsi:nil=" true" ref=" i953" />
</basiccurve>
<its_toolaxis>
<polyline xsi:nil=" true" ref=" i4712" />
</its_toolaxis>
</cutter_location_trajectory>
<milling_technology id=" i950">
<cutspeed>0.833</cutspeed>
</milling_technology>
<polyline id=" i953">
<points ex:cType=" list">
<cartesian_point xsi:nil=" true" ref=" i955" />
```

Figure 13.20. Description of CNC model data using XML (a part of data)

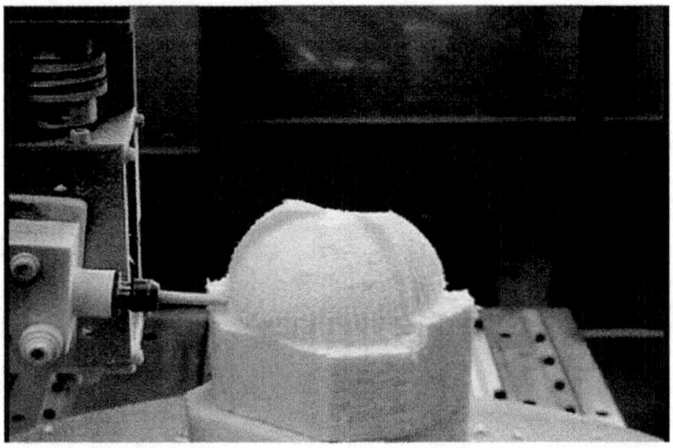

Figure 13.21. Machining using CNC model data

13.6 Conclusions

In order to realize the articulate link between design and manufacturing and the traceable manufacturing system, it is necessary to represent the semantics of information independently of any implementations. Moreover, the quality of data that represent information should be ensured during the product life cycle. In this research, the Digital Semantic Machining Model that fulfils the above requirements is proposed. Conclusions of this research are as follows:

- The problems which are faced in current manufacturing systems are pointed out. To solve these problems, the basic concepts of the digital semantic machining model are proposed. The overview of the Digital Semantic Machining Models is described. In this research, three research themes, the research about the representation and quality inspection of shape data, the research about the feature recognition process model for 5-axis machining, and the research about the feature-base CNC machining process model, were chosen to solve typical problems in current product lifecycle.
- Current problems of checking the quality of product data are pointed out and product data quality assurance method is proposed. The target shape data is the ISO standard 10303-214 and PDQ criteria are described in the Equivalent Transform (ET) language.
- Machining feature for 3+2 axis machining is proposed and the deriving method from product data is also proposed based on inverse offset method. The 3+2-axis machining is the machining process with indexing tilted angles of the tool axis on a 5-axis machining center. Extracted features are represented in Region entity of STEP-NC. The Operation and Strategy are represented in Five_axes_const_tilt_yaw entity and other STEP-NC entities. In order to convey semantics about 3+2 axis machining, however, important future work is to construct new features, operations, and strategies.
- Overview of the prototype 5-axis machine tool is described. The machine tool model based on ISO10303-105 is also proposed. Improving 5-axis machining accuracy by determining the optimal tool-path control method,

deviation compensation based on STEP-NC data will be future works of our research.

Acknowledgment

This research was financially supported by Ministry of Science and Education as a grand in aid for Science Research (14102032).

References

[13.1] Feng, SC., Stouffer, KA. and Jurrens, KK. 2005. Manufacturing planning and predictive process model integration using software agents. *Advanced Engineering Informatics*. 19(2): 135–142.

[13.2] National Institute of Standards and Technology. 1996. Interoperability Cost Analysis of the US Automotive Supply Chain, Planning Report #99-1. [Online] NIST Strategic Planning and Economic Assessment Office, Gaithersburg, Md. Available: http://www.nist.gov/ director/prog-ofc/report99-1.pdf.

[13.3] Tanaka, F., Kishinami, T. 2006. STEP-based quality diagnosis of shape data of product models for collaborative e-engineering. *Computers in Industry*. 57(3): 245–260.

[13.4] Xu, X. W., Wang, L., Rong, Y. 2006. STEP-NC and Function Blocks for Interoperable Manufacturing. *IEEE Trans. Automation science and engineering*. 3 (3):297 – 308.

[13.5] Tanaka, F., et al. 2008. Modeling and implementation of Digital Semantic Machining Models for 5-axis machining application. In *Proceedings of the 41st CIRP Conference on Manufacturing Systems,* University of Tokyo, May 26-28, 2008,. pp. 177–182.

[13.6] ISO 10303-214. 2003. Product data representation and exchange Part 214: Application Protocols: core data for automotive mechanical design processes.

[13.7] ISO 10303-224. 2006. Product data representation and exchange Part 224: Application Protocols: mechanical product definition for process plans using machining features.

[13.8] ISO 10303-238. 2007. Product data representation and exchange Part 238: Application Protocols: Application Interpreted Model for Computerized Numerical Controllers.

[13.9] ISO 10303-240. 2005. Product data representation and exchange Part 240: Application Protocols: process plans for machined products.

[13.10] ISO14649-1. 2003. Data Model for Computerized Numerical Controllers: Part 1 Overview and Fundamental Principles.

[13.11] ISO14649-10. 2004. Data Model for Computerized Numerical Controllers: Part 10 General Process Data.

[13.12] ISO14649-11. 2004. Data Model for Computerized Numerical Controllers: Part 11 Process Data for Milling.

[13.13] ISO 10303-11. 2004. Product data representation and exchange Part 11: Description methods: The EXPRESS language reference manual.

[13.14] ISO 10303-28. 2007. Product data representation and exchange Part 28: Implementation methods: XML representations of EXPRESS schemas and data, using XML schemas.

[13.15] Bohn, J.H. 1995. Removing Zero Volume Parts from CAD Models for Layered Manufacturing. *IEEE Computer Graphics and Application*, 15 (6): 27–34.

[13.16] Barequet, G. and Kumar, S. 1997. Repairing CAD Models. In *Proceedings of IEEE Visualization'97.*

[13.17] Krause, F.-L., Stiel, C., and Lueddemann, J. 1997. Processing of CAD-Data: Conversion, Verification and Repair. In *Proceedings of International Symposium on Solid Modeling and Applications*, pp.248–254.

[13.18] Ficco, M.M., Mandorli, F., and Otto, H.E. 1999. Error classification and recovery within CAD models reconstruction. In *Proceedings of the fifth symposium on Solid modeling and applications*, pp. 316–317.

[13.19] Gu, H., Chase, T.R., Cheney, D.C., Bailey, T., Johnson, D. 2001. Identifying, Correcting, and Avoiding Errors in Computer-Aided Design Models Which Affect Interoperability. *Journal of Computing and Information Science in Engineering.* 1 (2): 156–166.

[13.20] Contero, M., Company, P., Vila, C., Aleixos, N. 2002. Product Data Quality and Collaborative Engineering. *IEEE Computer Graphics and Applications.* 22 (3): 32–42.

[13.21] Bischoff, S., Kobbelt, L. 2005, Structure preserving CAD model repair. *COMPUTER GRAPHICS FORUM.* 24 (3):527-536.

[13.22] SASIG. 2005. Strategic Automotive Product Data Standards Industry Group– Product Data Quality.

[13.23] JAMA/JAPIA. 2003. *JAMA/JAPIA PDQ guideline V3.0 – PDQ Criteria part –*, [Online] Available: http://www.jama.or.jp/it/pdq.

[13.24] VDA Recommendation 4955/2. 1999. *Scope and Quality of CAD/CAM Data*, German Assoc. Automotive Industry (VDA), Frankfurt, Germany.

[13.25] Tanaka, F., Iwata, T., Akama, K., Kishinami, T., Onosato, M. 2006. Developing software platform for Checking Quality of shape data of STEP product model for CAD/CAM environment, In *Proceedings of FAIM2006*, pp. 823 – 830.

[13.26] Akama, K., Shimitsu, T. and Miyamoto E. 1998. Solving Problems by Equivalent Transformation of Declarative Programs. *Journal of Japanese Society of Artificial Intelligence (JSAI).* 13 (6):944– 952

[13.27] Tanaka, F., Yamada, M., Kondo, T., Kishinami, T., Kohmura, A. 2004. Software system for sculptured surface machining based on 3+2-axis high speed machining on a 5-axis machining center. In *Proceedings of JUSFA symposium 2004*, JL-006.

[13.28] Altan, T., Lilly, B., Yen, Y. C. 2001. Manufacturing of Dies and Molds. *Annals of the CIRP.* 50 (2): 405–423.

[13.29] Liu, X. W. 1995. Five-axis NC cylindrical milling of sculptured surfaces. *Computer-Aided Design.* 27 (12): 887–894.

[13.30] Morisige, K., Kase, K., Takeuchi, Y. 1996. The Method of Collision Avoidance for 5-Axis Control Machining Using 2-Dimensional Configuration Space. *Journal of the Japan Society for Precision Engineering.* 62 (1): 80–84.

[13.31] Lo, C. C. 1999. Efficient cutter-path planning for five-axis surface machining with a flat-end cutter. *Computer-Aided Design.* 31 (9): 557–566.

[13.32] Jun, C. S., Cha, K., Lee, Y. S. 2003. Optimizing tool orientation for 5-axis machining by configuration-space search method. *Computer-Aided Design.* 35 (6):549–566.

[13.33] B. Lauwers, P. Dejonghe, J. P. Kruth, 2003, "Optimal and collision free tool posture in five-axis machining through the tight integration of tool-path generation and machine simulation," *Computer-Aided Design*, 35 (5), pp. 421–432.

[13.34] Chiou, C. J., Lee, Y. S. 2002. A machining potential field approach to tool-path generation for multi-axis sculptured surface machining. *Computer-Aided Design.* 34 (5): 57–371.

[13.35] Suh, S. H., Kang, J. K. 1995. Process planning for multi-axis NC machining of free surfaces. *International Journal of Production Research*. 33 (10): 2723–2738.

[13.36] Gupta, P., Janardan, R., Majhi, J. Woo, T. 1996. Efficient geometric algorithms for workpiece orientation in 4- and 5-axis NC machining. *Computer-Aided Design*. 28 (8): 577–587.

[13.37] Kondo, T., Kishinami, T., Saito, K. 1988. Machining System based on Inverse Offset Method. *Journal of the Japan Society for Precision Engineering*. 54 (5): 167 – 172.

[13.38] Mitsui, S., Tanaka, F., and Kishinami, T. 2006. Machining Feature-Driven 5-Axis CNC Machine Tools. In *Proceedings of 11th International Conference on Precision Engineering*, pp. 169 – 174

[13.39] ISO 10303-105. 1996, Product data representation and exchange Part 105: Integrated application resource: Kinematics.

[13.40] Weck, M., Wolf, J., Kinitsts D. 2001. STEP-NC - The STEP compliant NC Programming Interface. In *IMS Forum*

[13.41] Newman, S. T., Allen R. D., and Rosso Jr, R. S. U. 2002. CAD/CAM solutions for STEP Compliant CNC Manufacture. In *Proceedings of the 1st CIRP(UK) Seminar on Digital Enterprise Technology*, pp. 123– 128.

[13.42] Suh, S. H., Lee, B. E., Chung, D. H., Cheon, S. U. 2003. Architecture and implementation of a shop-floor programming system for STEP-compliant CNC. *Computer-Aided Design,* 35: 1069– 1083.

[13.43] Xu, X. W., He, Q. 2004. Striving for a total integration of CAD, CAPP, CAM and CNC. *Robotics and Computer-Integrated Manufacturing*. 20: 101– 109

[13.44] Nassehi, A., Newman, S. T., Allen, R. D. 2006. The application of multi-agent systems for STEP-NC computer aided process planning of prismatic components. *International Journal of Machine Tools & Manufacture*. 46:559– 57

14

Development of a STEP-NC Network Management Protocol for Decentralized Manufacturing

Francesco Calabrese[1] and Amedeo Buonanno[2]

Senseable City Laboratory, Massachusetts Institute of Technology, 77 Massachusetts Avenue, Cambridge 02139, MA
Email: [1] fcalabre@mit.edu, [2] buonanno@mit.edu

Abstract
STEP-NC is a new model for data transfer and exchange between CAD/CAM and CNC that allows specifying machining processes rather than tool motions with respect to the machine axes. This chapter presents a novel procedure for the design of decentralized STEP-NC compliant manufacturing solutions. Considering the problem of dispatching and managing a network of STEP-NC compliant machines, a three-tier architecture is proposed to perform remote machining requests. The architecture is composed of a dispatcher, a series of managers dislocated in manufacturing units, and agents interfacing with the STEP-NC compliant intelligent CNC machines. An ad hoc STEP-NC Network Management Protocol describes the exchange of messages between tiers. To prove the effectiveness of the decentralized manufacturing architecture, this chapter shows its application in a test-bed and describes how to evaluate the performance of the designed solution.

14.1 Introduction

Today's CNC machines have enormous capabilities such as multi-axis control, multi-process manufacture, etc. which have complicated the programming tasks and made machine tools less adaptable. CNC machines use a 50-year-old language called ISO6983 (also called G-code) [14.1]. Some limitations of using these standards include: 1) manufacturing a part designed with a CAD/CAM software using an actual CNC machine requires converting, the CAM project into a ISO6983 program through the use of a post-processing module 2) ISO6983 defines the statements' syntax but, in some cases, leaves ambiguities in the related semantics [14.2]; 3) ISO6983 allows describing the tool centre path with respect to the machine axes rather than the features to be manufactured; 4) vendors usually supplement the language with further commands to provide new features and creating machine-specific languages.

To avoid such limitations, the manufacturing systems research community is defining the new standard ISO14649, also called STEP-NC [14.2, 14.3], which establishes a unified format for data transfer between CAD, CAM, SFP, and CNC, thus avoiding conversions and post-processing mechanisms.

STEP-NC allows for specifying machining process rather that tool motion with respect to the machine axes, using the object-oriented concept of Workingstep. In fact, the generation of the tool-path is performed at the latter stage of the manufacturing chain by a new breed of intelligent controllers called STEP-NC controllers. These controllers are usually classified in three categories [14.4]:

- *Conventional CNC control using STEP-NC*: a translation module, implemented on a PC, converts a STEP-NC program into ISO6983 commands, executable on a traditional CNC machine (see, e.g., [14.5] and [14.6]).
- *STEP-NC enabled control*: such a controller allows processing a STEP-NC file directly into the CNC machine (see, e.g., [14.7–14.9]). However, the controller utilizes only low-level machining information (already available in an ISO6983 program).
- *STEP-NC enabled intelligent control*: such a controller differentiates from the previous ones because it also allows performing high-level activities like: automatic part setup, automatic and optimal tool-path generation, accurate machining status and result feedback, adaptive control, etc. (e.g., [14.10, 14.17]).

Prototypes of controllers utilize both PCs and microcontrollers (see, e.g., [14.16]) to interpret the STEP-NC programs and generate the tool-paths. Some implementations have also used multi-agents systems to deal with the different features that an intelligent STEP-NC controller must implement [14.21].

The availability of the new STEP-NC standard and of intelligent CNC controllers allows for a manufacturing industry made of highly decentralized units, composed of different small industries, which seamlessly interoperate to provide variegated services to clients. One current challenge to creating a decentralized manufacturing industry is the dispatching and management of remote requests for machining (described by means of STEP-NC files) to different CNC machines in different manufacturing units. The problem involves the evaluation of machine availability, the feasibility of the Workingsteps using those machines, and the required machining times. Moreover, load balancing techniques must be considered to prevent bottlenecks and deadlocks.

In the past few years, the concept of decentralized production has been applied in the electronic photographic service market, and has driven to the development of a new environment called Common Picture eXchange (CPXe) [14.19]. In the manufacturing framework, that concept has been partially tackled in [14.23], where the authors described a distributed manufacturing environment in which factories possessing various machines are geographically distributed. When jobs requests need to be processed, the available factories produce feasible process plans, and a Genetic Algorithm is used to generate optimal or near-optimal process plans based on the chosen criterion. In [14.22] the problem has been particularized for the manufacturing with CNC machines, and issues such as how distributable, interoperable and reconfigurable CNC machine tools have come into focus. This chapter deals with the definition of a protocol for the exchange of dispatching and management information between the entities involved in the decentralized manufacturing solution. An architecture composed of three logical tiers is proposed in order to deal with this problem, and to define a modular solution that can be adapted to the manufacturing units scale. A STEP-NC Network Management Protocol has been defined to describe all the communications among the tiers for the purposes of supervision, dispatch, feasibility, and availability evaluation. This protocol draws on the Simple Network Management Protocol [14.18] used on the Internet to deal with the monitoring of networked resources such as routers.

To test the effectiveness of the architecture, a test-bed consisting of a simplified version of the dispatching/management system has been implemented, and the associated performance evaluated.

The chapter is structured as follows: Section 14.2 gives an overview of the STEP-NC standard; Section 14.3 provides a description of the dispatching/management problem, and proposes the STEP-NC Network Management Protocol solution; Section 14.4 describes an implemented dispatching/management solution, and highlights the advantages of the proposed approach proposing measures for evaluating the derived performance; Section 14.5 draws final conclusions.

14.2 Overview of STEP-NC

The ISO10303 standard (also called STEP, Standard for the Exchange of Product model data) [14.11–14.13] provides a mechanism for describing production data along the development chain, for CAD to 3D CAM design, allowing the implementation and sharing of product databases.

The STEP-NC standard is an extension of STEP and allows for the connecting CAD/CAM design to CNC. The information contained in a STEP-NC file is divided into three sub-sections (Figure 14.1):

- Workplan executables
- Technology description
- Geometry description

The Workplan is characterized by a series of executables whose order is pre-established, or dependent on the machining conditions if conditional controls are used. The executables can be of three types: Workingsteps, NC functions, and program structures, where the most important executables are the Workingsteps, which represent 2.5D (two5D_manufacturing_feature) and 3D (region) machining features. Each Workingstep also includes further subfeatures (such as planar_face, pocket, step, slot, round_hole, genereal_outside_profile, etc.) together with cutting condition information.

The technology description contains a detailed and complete description of all the Workingsteps to be executed in the Workplan. In particular, this description includes data regarding tools, machining strategies, definitions of the workpieces, etc. A complete technology description includes, for example, depth of hole to be machined, feed rate, rotational speed, and diameter of the tool.

As for the geometry description, all geometry data used in the various components are described using the ISO 10303 format.

In the proposed decentralized manufacturing solution, the STEP-NC description of machining is the main part of a client request, which is then processed by the Dispatcher in order to find the "best" manufacturing solution. The details of this process are described in the following section.

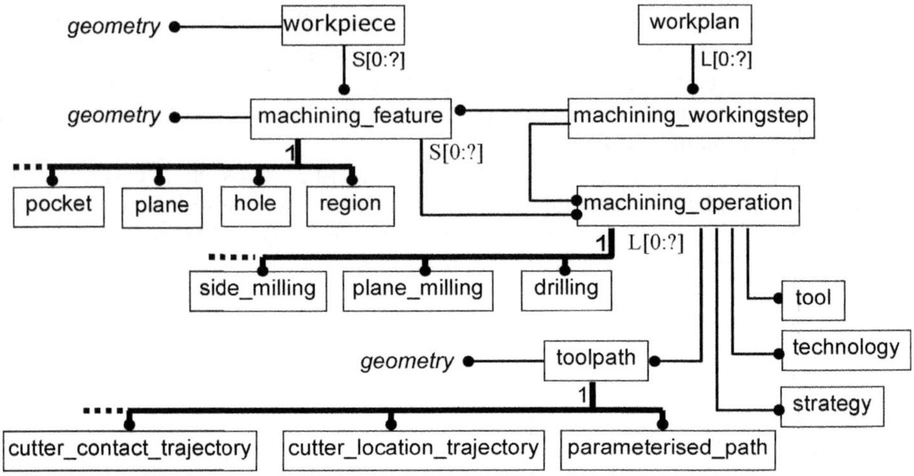

Figure 14.1. Feature-based STEP-NC data structure [14.14]

14.3 Decentralized Manufacturing Solution

In this section we consider the problem of dispatching remote requests for machining (described by means of STEP-NC files) to different CNC machines, dislocated in manufacturing units. A classic application scenario is the following: a remote client requests a particular machining by submitting a STEP-NC file. The dispatcher forwards it to the "most appropriate" CNC machine, and the resulting machined object is returned to the client.

This problem involves the evaluation of the availabilities of the CNC machines, the feasibility of the Workingsteps using those machines, and the required machining times. To solve this problem, a software architecture composed of three tiers is proposed (Figure 14.2):

- *Agents*, which represent the interface with the STEP-NC-compliant Intelligent CNC machines that execute the machining operations.
- *Managers*, which represent the entities that manage a set of CNC machines in a manufacturing unit and perform availability and feasibility checks. Each Manager keeps a database of available Agents, containing their characteristics (e.g., available tools, working plan) and their status (e.g., queue of requests, current tool). Moreover, it evaluates the best Agent to which it can allocate the request. The Manager manages the queues of allocated works per Agent, and uses this information to evaluate when the Agent will be available for the new machining.

- *Dispatcher*, which represents the entity that handles the machining requests made by remote Clients. It keeps a database of available Managers and performs the choice of the best Manager to which it can allocate the request.

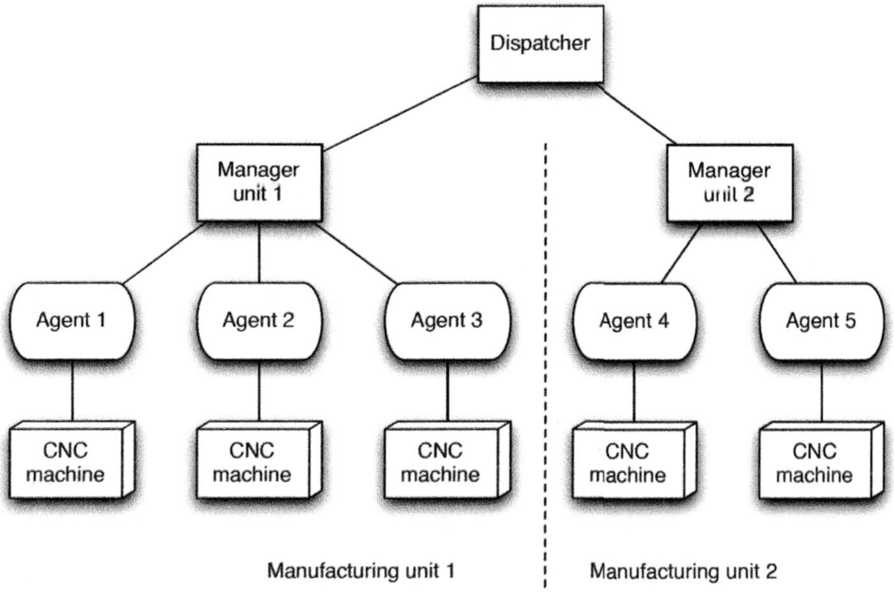

Figure 14.2. Proposed architecture

Observe that the concept of the machine-independent *Generic STEP-NC file*, which describes only "what-to-do" but not "how-to-do", as introduced in [14.22], is used in this architecture. It is sent by the Client during the machining request, and is forwarded to Managers by the Dispatcher. Once the machining has been allocated to an Agent, the Manager converts the *Generic STEP-NC file* to a *Native STEP-NC file*, specific for the selected CNC machine.

Next, we describe how the network management protocol is used to make the three components interact, analyze the details of each component and discuss different simplified and hybrid architectures.

14.3.1 STEP-NC Network Management Protocol

The communication protocol designed for the decentralized manufacturing solution draws on the Simple Network Management Protocol [14.18] used on the Internet for the monitoring of networked resources. The protocol has been extended and tailored to account for the specific requirements of the manufacturing problem through various messages (Table 14.1).

Table 14.1 Protocol messages

Type of message	Message	From	To
get	checkAvailability	Dispatcher	Manager
	checkAvailability	Manager	Agent
	checkWorkability	Dispatcher	Manager
	checkStatus	Manager	Agent
set	machiningRequest	Client	Dispatcher
	allocateWork	Dispatcher	Manager
	startWork	Manager	Agent
getBulk	getCharacteristics	Manager	Agent
inform	agentFree	Agent	Manager
	workAllocated	Dispatcher	Client
trap	toolChange	Agent	Manager
	error	Agent	Manager
	error	Manager	Dispatcher

Messages are formatted as XML documents and sent over HTTP, standard de facto for Web services.

The interactions between the components during the dispatching of a Client's request are shown in Figure 14.3.

The sequence begins with a machining request, which is composed of the following parameters:

- STEP-NC file
- Destination of the manufactured object
- Timing / price for the service

The Dispatcher then forwards the request to a series of Managers, which are available at the current time.

Each Manager knows the information about the connected Agents (tools, working area, etc.), and evaluates the feasibility of the machining using the connected Agents (*workability*). If at least one Agent meets the requirements provided by the Client (e.g., in terms of feasibility and time) then the Manager locks this resource for a fixed time, and a positive response message sends back to the Dispatcher.

The Dispatcher collects all the positive and negative response messages, selects the "optimal" solution, and contacts the Manager, wherein the machining request is allocated to the Agent's queue. Bearing in mind that the term "optimal" can depend on the Client's request, a subsequent explanation that better describes this parameter will be provided.

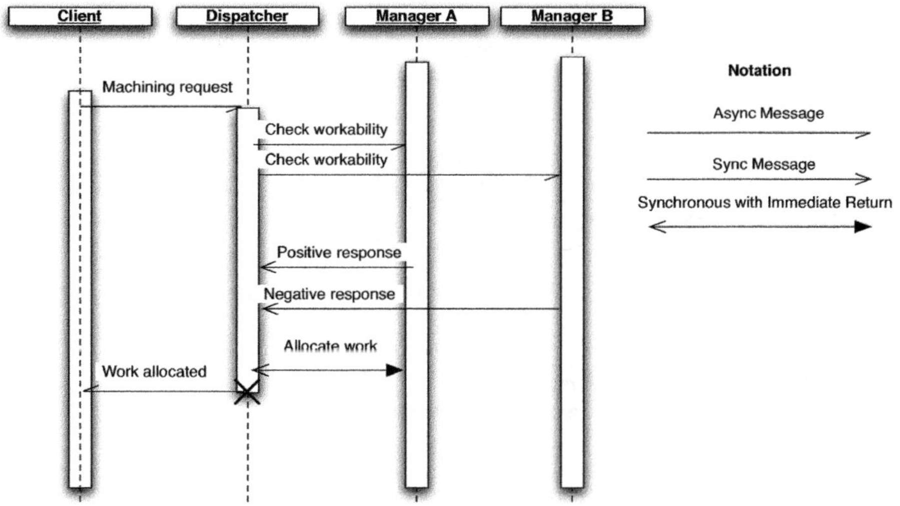

Figure 14.3. Sequence diagram of the interaction among tiers while performing a machining request

14.3.2 Details of the Components

In this section the behaviours of the components involved in the decentralized manufacturing solution are described.

Dispatcher

Figure 14.4 shows a high-level state diagram of the Dispatcher. When the Dispatcher begins, it pauses until one Manager is available, in order to be able to receive machining requests from the Clients (service up). When a Client sends a request, the Dispatcher evaluates the workability of the task by sending a *checkWorkability* message to all available Managers. Once all the Managers reply (or a timeout occurs), the Dispatcher evaluates the responses. If other solutions are available, it can either require the Client to select one, or it automatically sends the *allocateWork* message to the selected Manager.

In choosing the optimal dispatching solution, the Dispatcher must account for geographical location of the machine and the destination of the service, machining time, and efficiency of the machining, using a specified manufacturing solution.

If the best solution cannot be evaluated automatically (e.g., different solutions with different costs), the choice is left to the Client.

The Dispatcher can serve different Client's requests simultaneously because of its parallel computing design. This situation is represented on the diagram with a stack of *Processing of request.*

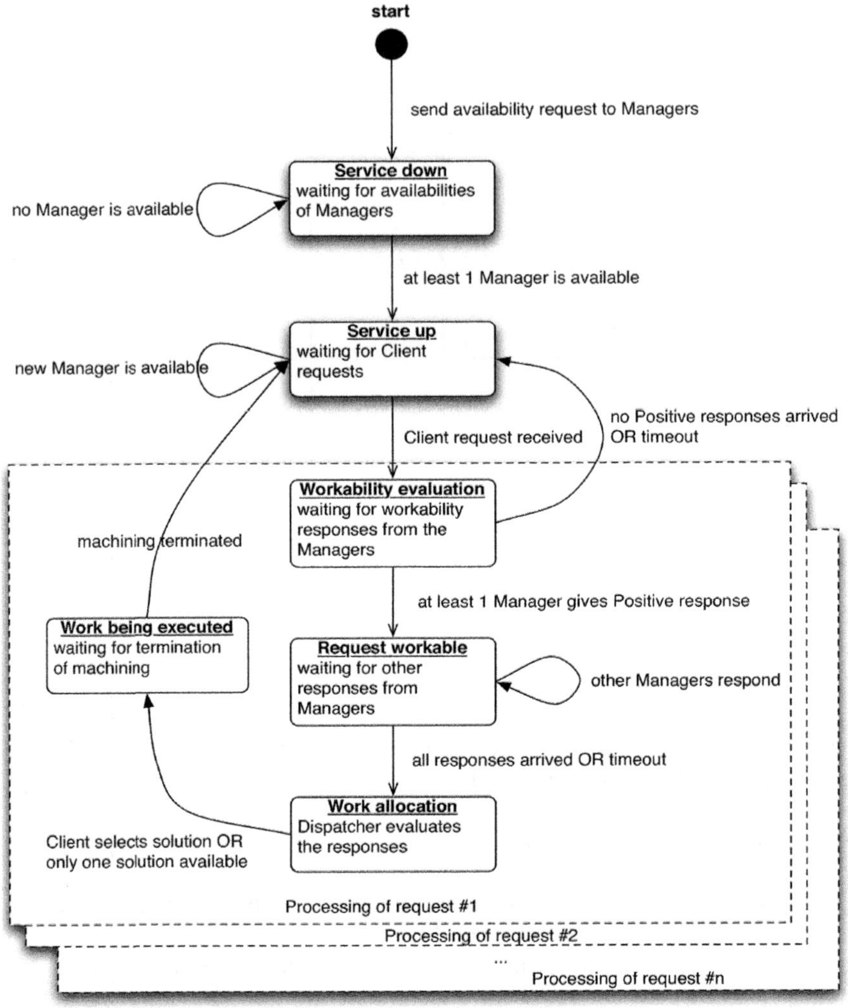

Figure 14.4. Dispatcher's state diagram

Manager

When a Manager starts, it waits for the availability of Agents (Figure 14.5). If at least one Agent is available (service up), it can positively reply to the *checkAvailabilty* message sent by the Dispatcher. When the Dispatcher sends the *checkWorkability* message, the Manager evaluates the workability of the task. If the request is positively evaluated (the work is feasible with one of the Agents) it temporary locks the resource (adding the job to the Agent's queue), replies with a positive response, and waits for the *workAllocate* message. If the Dispatcher allocates the work, then the Manager adds the job to the selected Agent's queue and waits for its availability for commanding the execution of the machining.

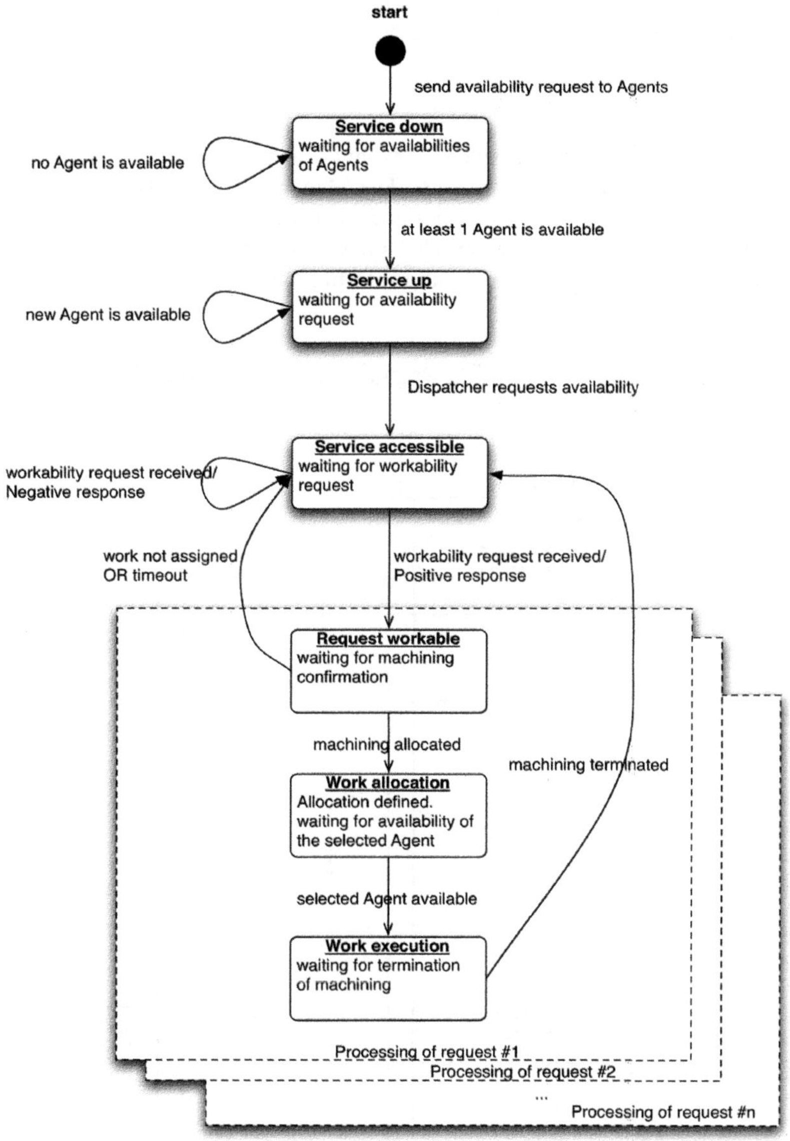

Figure 14.5. Manager's state diagram

A Manager can manage various Dispatcher requests and every request is managed in a parallel way (note the stack of Processing of requests in Figure 14.5).

In order to keep the diagram legible, the transitions happening when errors occur to the Agents in charge of executing a machining have not been drawn. In this scenario, the Manager is alerted and reallocates the work to another Agent. If no Agent is found, an error message is forwarded to the Dispatcher to restart a new dispatching process (see for instance Figure 14.15).

Agent

An example of interaction between Manager and Agents is described in the sequence diagram shown in Figure 14.6. The Manager first sends a request to return the characteristics of each machine connected to the Agents. Then it allocates the work to Agent 1, who processes the work once it becomes available. The Agent can send feedback to the Managers regarding information about tool changes, errors, and status.

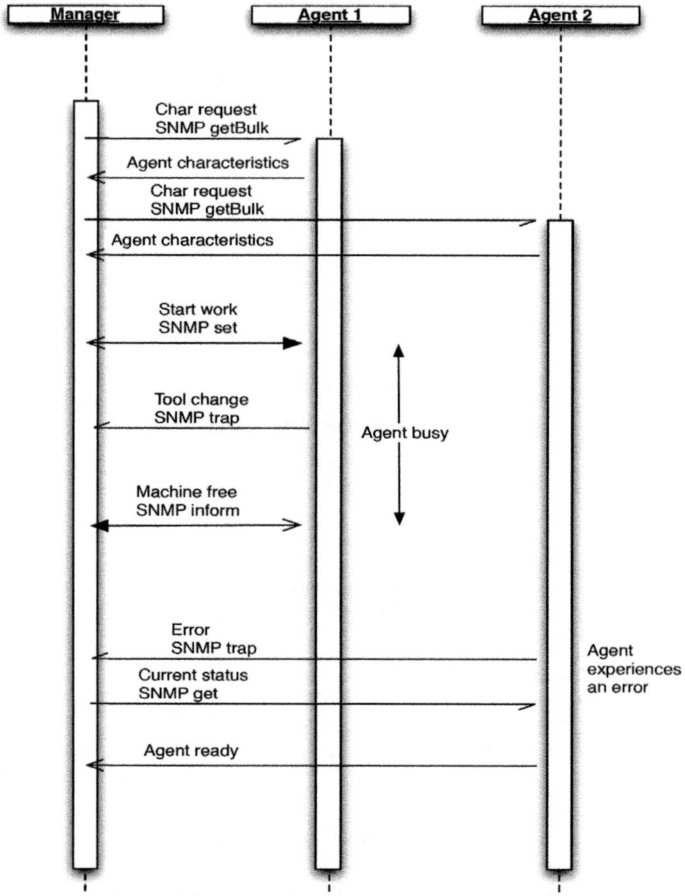

Figure 14.6. Sequence diagram of the interaction between Manager and Agents. See Figure 14.3 for arrowhead notations

14.3.3 Simplified and Hybrid Architectures

The proposed architecture is scalable, in the sense that one physical entity can implement more than one tier of the architecture, from which two simplified architectures can be derived:

- *Dispatcher – Manager/Agent:* this solution can be adopted when there are different machines and they are dislocated in different units (Figure 14.7a)
- *Dispatcher/Manager – Agent:* this solution can be adopted when there is only one manufacturing unit, and a manageable number of machines (Figure 14.7b)

Hybrid solutions can be adopted in case both aforementioned conditions hold true. One example of this solution has been adopted in the test case described in Section 14.4.

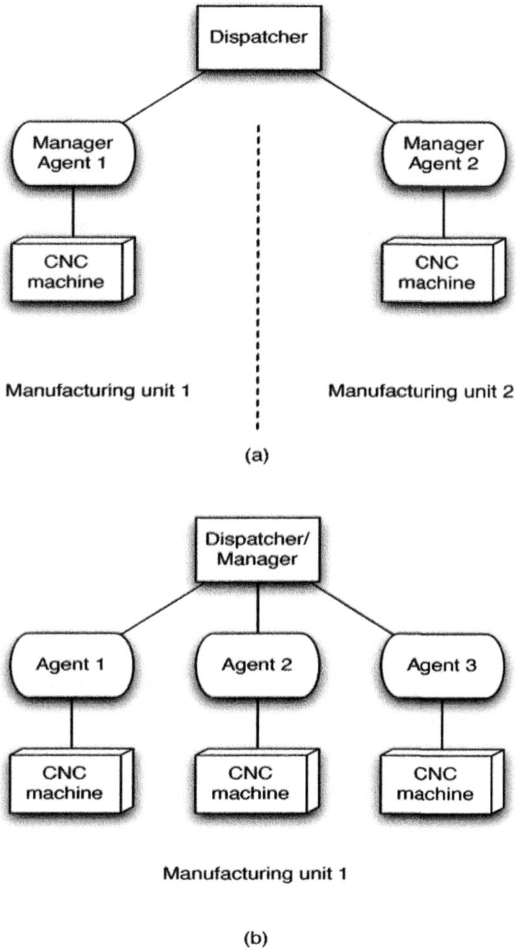

Figure 14.7. Simplified architectures

14.3.4 SNMP Compliant Controller

Here the design process of the embedded system able to control a CNC machine using the STEP-NC approach is presented. The scheme provided in [14.16] has been extended to introduce the STEP-NC Network Management Protocol (SNMP) architecture. The activities performed by a STEP-NC compliant CNC controller are divided into five main tasks (Figure 14.8):

- Interpreter
- High-level Controller
- Tool-path Generator
- Low-level Controller
- Machining Inspector

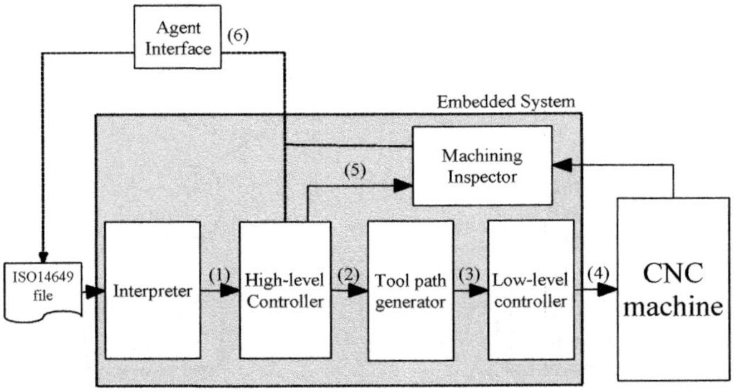

(1) workingstep description
(2) workingstep parameters
(3) path
(4) motor/tool command signals
(5) workingstep control information
(6) machining status

Figure 14.8. Scheme of the designed STEP-NC compliant CNC machine

14.3.5 Interpreter

This module interprets the STEP-NC file and memorizes the process information (features to be machined, tools to be used, and sequence of operations to be followed) in apposite data structures. Then, for each Workingstep, the module calls the High-level Controller, which will evaluate its feasibility and order the execution of the specified feature.

14.3.6 High-level Controller

This module analyzes each Workingstep description and evaluates its feasibility considering the CNC machine characteristics (Workplan dimension, available tool, maximum velocity of each motor). Based on the part material, a cutting speed is determined and compared with the speed described in the Technology description. If differences are found, the module notifies the higher-level component about the mismatch and suggests a more suitable cutting speed. If the Workingstep is feasible, the controller will decide which machining operation to execute and will call the related tool-path subroutine.

14.3.7 Tool-path Generator

This module is composed of several subroutines that implement tool-path generators for different kinds of features: pocket, slot, round hole, toolpath feature, etc. [14.15]. Observe that each machining Workingstep has a well-defined start and end point of the tool motion. If two subsequent Workingsteps require different tools, a tool change is required and the generator calculates the respective tool-paths to and from the tool change position.

14.3.8 Low-level Controller

The Low-level Controller sends suitable signals to the motor drivers and to the tool in order to realize the tracking of the trajectory specified by the Tool-path Generator.

14.3.9 Machining Inspector

This module inspects the machining execution. In particular, the following conditions are considered [14.2].

The machine has to reach the operating conditions specified in the MACHINING_WORKINGSTEP entity before the operation of the Workingstep commences. If the machine is unable to reach these conditions during the preceding Workingstep, a halt must occur before the execution of the Workingstep until all parameters are stable.

The machine predicts the part's geometry change and checks if it matches the geometry change described in the MACHINING_WORKINGSTEP entity.

This module implements the Agent part of SNMP; in particular it replies to the *SNMP getBulk* messages (requests the machine's characteristics file) and *SNMP get* (retrieves current status). The module sends event-driven messages like *SNMP inform* to communicate its availability to the Manager and *SNMP trap* messages to update its status on the bases of events (e.g., tool changes or errors in the machining).

14.4 Application of the SNMP Architecture in a Real Scenario

The SNMP architecture was implemented and tested in a real scenario composed of:

- One dispatcher – computer station, running a web server and hosting a website through which machining requests can be submitted.
- One manager – computer station, where the SNMP software is running. The Manager is connected to microcontrollers working as SNMP Agents.

- Two agents – two microcontrollers implementing the STEP-NC compliant control presented in Section 14.3.4.
- One manager/agent – a microcontroller implementing both the agent and manager components of the protocol.

The implemented solution is presented in Figure 14.9.

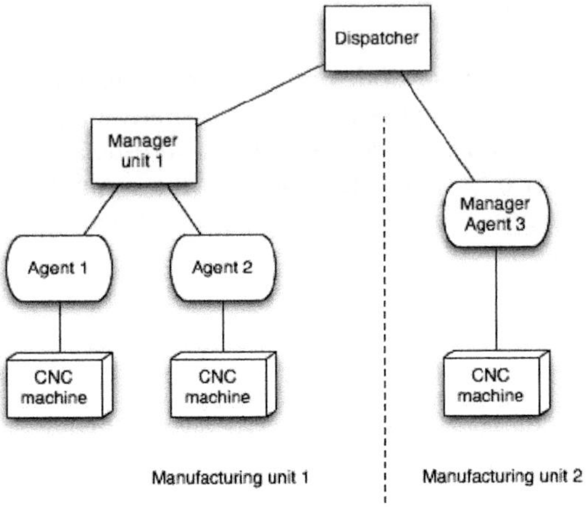

Figure 14.9. Implemented distributed manufacturing solution

Dispatcher

The Dispatcher interfaces with the Clients through a website (Figure 14.10), where the client can register, submit machining requests, and check the status of the process.

The site allows the user to upload the STEP-NC file, and specify an address and desired time of delivery.

Moreover, the Dispatcher implements an interface through which the Managers can report on the status of the machining.

To store information about the dispatching system, a MySQL database has been implemented (Figure 14.11), containing the following tables:

- Manager: information about status and availability of managers
- Client: credentials of registered clients
- Order: order's information and machining status
- Workability: log of workability responses from the Managers, useful for statistics and planning

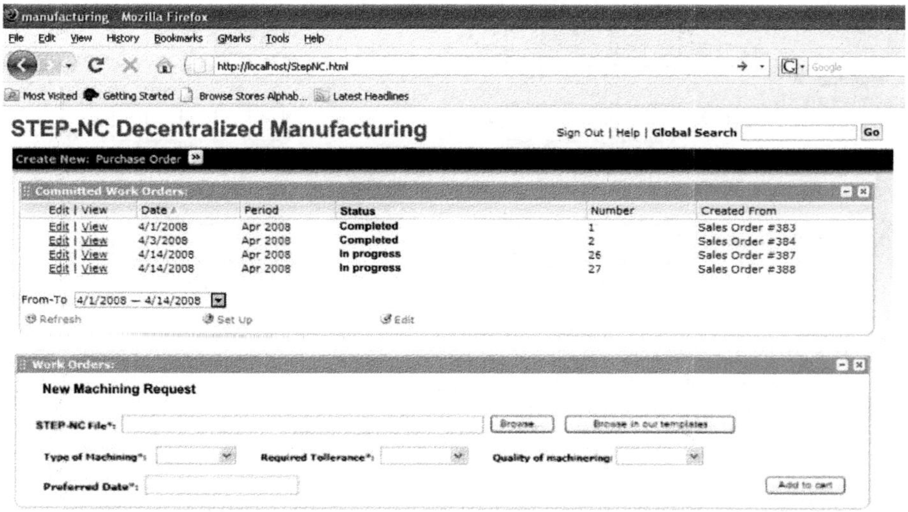

Figure 14.10. User interface for machining requests

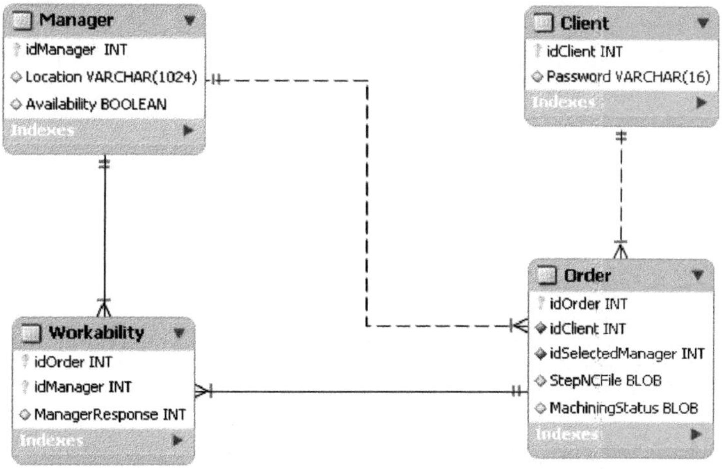

Figure 14.11. Dispatcher's database schema

The current implementation of the Dispatcher does not perform any dispatching choice, but gives Client the opportunity to select from similar machining possibilities, as provided by the two Managers (Figure 14.12). However, if only one possibility is available (only one positive response received from the Mangers), the Dispatcher automatically allocates the work, sidestepping client input (Figure 14.13).

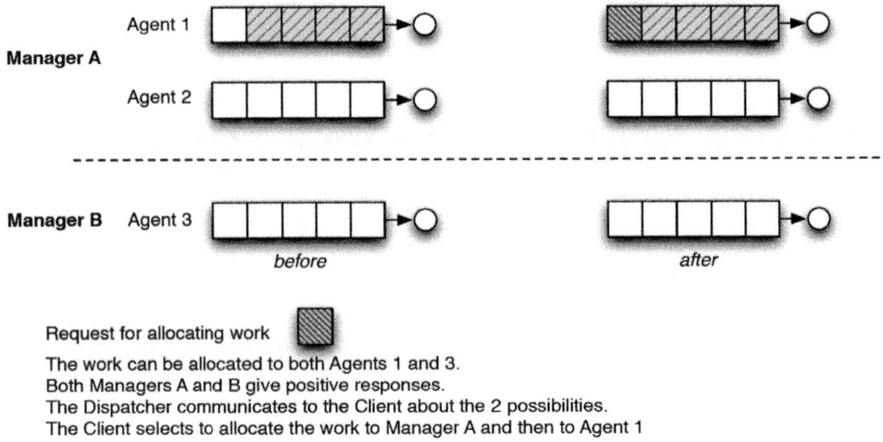

Figure 14.12. Scenario 1: the Client selects among different machining solutions

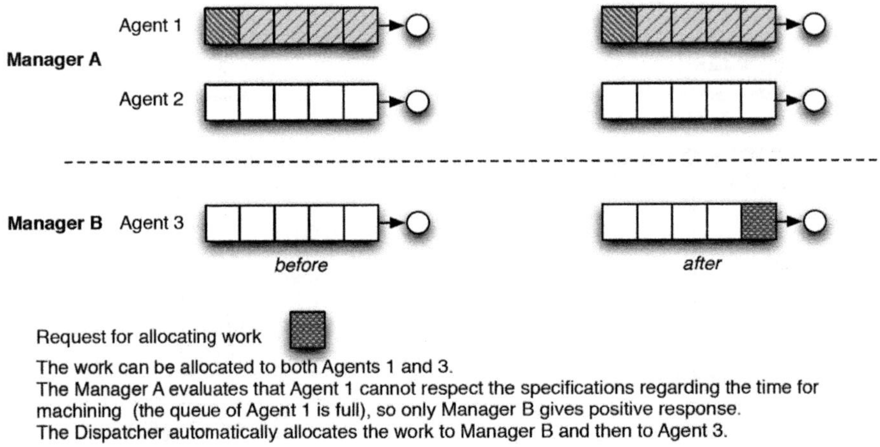

Figure 14.13. Scenario 2: the Dispatcher automatically allocates the work to a Manager

Manager

The Manager interfaces with the Dispatchers through a Web service, in order to receive:

- Manager availability requests
- Workability requests
- Machining order

The Manager also implements an interface through which the Agents can report the machining status (e.g., machining terminated, errors, tool changes).

To store information about the management system, a database has been implemented using MySQL (Figure 14.14), containing the following tables:

- Agent: information about availability and characteristics of Agents
- Dispatcher: credentials of registered Dispatchers
- Order: order's information, selected Agent, allocation
- Machining: log of terminated, in progress and planned machining, useful for workability's evaluation, statistics and planning

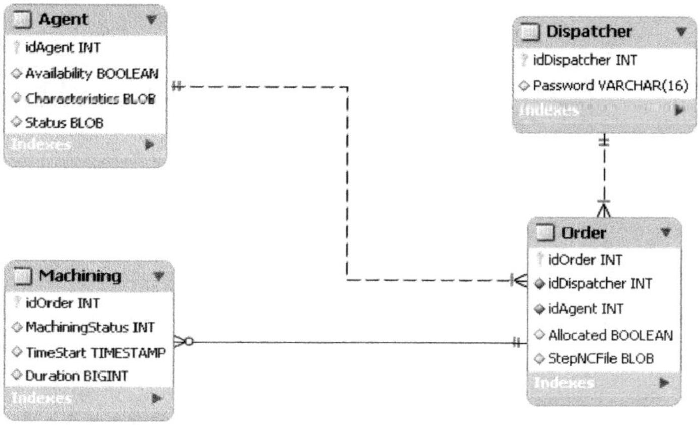

Figure 14.14. Manager's database schema

Critical aspects in the implementation of the Manager are:

- Evaluating the feasibility of the request based on the characteristics of the available Agents
- Evaluating an approximate time for the machining using a specified Agent
- Error management in the machining

The first two aspects have been implemented by simulating the machining process and analyzing the results of the High-Level Controller (see Section 14.3.4). The total machining time is then calculated by summing the predicted times of all remaining operations of the CNC machine connected to that Agent, and this information is then stored in the Machining table.

The third aspect has been tackled by forwarding the error message to the Dispatcher, which will then re-start a dispatching process that finds a new Manager/Agent to handle the request (see scenario in Figure 14.15).

Observe that for the Manager of unit 2, implemented in the microcontroller running the Agent component, the database can be simplified, since the Manager is connected to only one Agent through an ad hoc data management system.

Agent

The Agent implements a Web interface through which the Manager can submit:

- Agent availability requests
- Machine's characteristics request
- Machining allocation

Moreover, the Machining Inspector (see Section 14.3.4) pushes to the Manager information about machine's status, errors and tool changes.

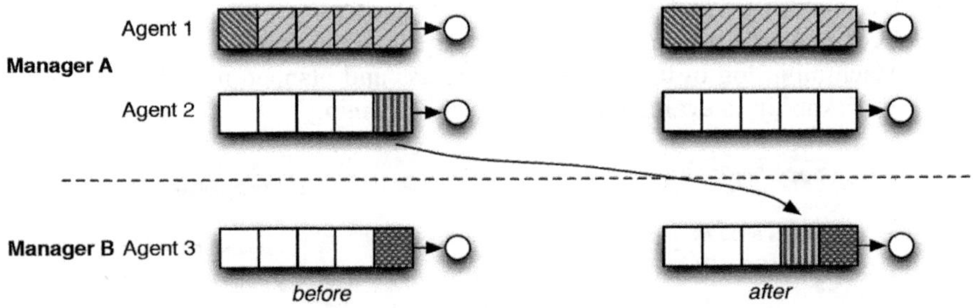

Figure 14.15. Scenario 3: due to an error in the machining using an Agent, the Dispatcher is contacted to reallocate the work to a different Manager and Agent

14.4.1 Evaluation of the Performance of the System

In this section a performance evaluation of the designed decentralized manufacturing solution is presented. Queuing theory [14.24] is used to understand the overall availability of the system, the use of the different components, and possible bottlenecks that can result in redesigning the distribution of agents/machines or in adding new agents/machines.

Let us hypothesize that the Client's requests can be grouped in categories, and for each category (called Job in the sequel) let us group the requests that can be machined with the same agents/machines. Consider the matrix $Work_m$ associated to each Manager, where

$$Work_m(j,a) = \begin{cases} 1 & \text{the Job } j \text{ can be machined by the Agent } a \\ 0 & \text{the Job } j \text{ cannot be machined by the Agent } a \end{cases}$$

Let us indicate with

$\bar{t}_s^{DISP}$ – the average service time of the Dispatcher

$\bar{t}_s^{MAN\,WORK}$ – the average service time for a Manager, in order to evaluate the workability of a Job (for simplicity it has been considered that all Managers have the same average service time)

$\bar{t}_s^{MAN\,ALLOC}$ – the average service time for a Manager, to allocate a Job to an Agent

$\bar{t}_s(j)$ – the average service time for machining the Job j (for simplicity it has been hypothesized that all Job can be machined with the same time regardless of the used Agent)

λ – the Client requests arrival rate (it has been hypothesized that the Job requests are uniformly distributed, i.e., $\lambda(j) = \lambda / n\_jobs$

Let us hypothesize that the requests, arrival can be modeled with a Poisson process while the service times of the servers can be modeled as exponential processes.

The decentralized system in Figure 14.9 can be modeled using the queuing theory as depicted in Figure 14.16. In this simplified model the feedback between tiers is not considered, but it is hypothesized that this interaction can be taken into account in the average service times of the servers. Moreover, in the model, the Dispatcher service and the Managers' workability evaluation have been combined in a single server, as each service has the same number of requests.

Using this simplified model, it is possible to evaluate some performance indicators of the single servers and of the overall system.

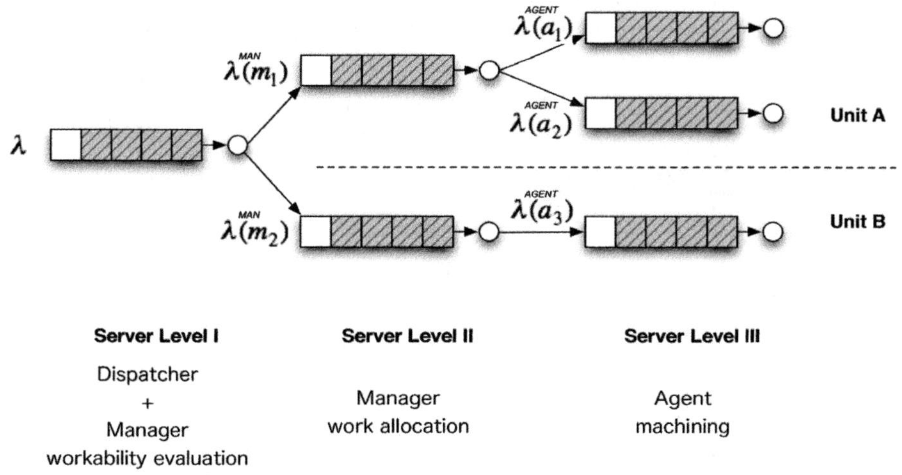

Figure 14.16. Simplified queuing model of the decentralized manufacturing solution shown in Figure 14.9

Server level I – Dispatcher-Manager workability evaluation

The request arrival ratio to the Dispatcher is the total request arrival ratio:

$$\lambda^{DISP-MAN\,WORK}(d) = \lambda, \tag{14.1}$$

and its utilization is

$$\rho^{DISP-MAN\,WORK}(d) = \lambda^{DISP-MAN\,WORK}(d) \cdot (\bar{t}_s^{DISP} + \bar{t}_s^{MAN\,WORK}). \tag{14.2}$$

Server level II – Manager m work allocation
The request arrival ratio to the Manager is

$$\lambda^{MAN}(m) = \sum_{j=1}^{n\_jobs} \left(\lambda(j) \cdot \frac{\sum_{a=1}^{n\_agents} Work_m(j,a)}{\sum_{m'=1}^{n\_managers} \left(\sum_{b=1}^{n\_agents'} Work_{m'}(j,b) \right)} \right), \tag{14.3}$$

considering that the *n_agents* represents the number of Agents managed by the Manager m and *n_agents'* is the number of Agents managed by Manager m'.

Then the utilization of the Manager m is

$$\rho^{MAN}(m) = \lambda^{MAN}(m) \cdot \bar{t}_s^{MAN\,ALLOC}. \tag{14.4}$$

Server level III – Agent a machining
Let m be the Manager connected to Agent a. The request arrival ratio is

$$\lambda^{AGENT}(a) = \sum_{j=1}^{n\_jobs} \left(\lambda(j) \cdot \frac{\sum_{b=1}^{n\_agents} Work_m(j,b)}{\sum_{m'=1}^{n\_managers} \left(\sum_{c=1}^{n\_agents'} Work_{m'}(j,c) \right)} \cdot Work_m(j,a) \right). \tag{14.5}$$

In this case, the utility of the Agent a is

$$\rho^{AGENT}(a) = \sum_{j=1}^{n\_jobs} \left(\lambda(j) \cdot \frac{\sum_{b=1}^{n\_agents} Work_m(j,b)}{\sum_{m'=1}^{n\_managers} \left(\sum_{c=1}^{n\_agents'} Work_{m'}(j,c) \right)} \cdot Work_m(j,a) \cdot \bar{t}_s(j) \right). \tag{14.6}$$

The utility rates calculated above can help us understand the individual component occupancy in the decentralized manufacturing solution, and detect overloaded components and bottlenecks. Another important quantifiable parameter to evaluate the performance of the decentralized system is the *response time*. It is easy to recognize that the average response time for each server can be calculated as (see [14.24])

$$\bar{t}_r = \frac{\bar{t}_s}{1-\rho}, \tag{14.7}$$

while the average response of the whole system is:

$$\bar{t}_r^{TOT} = \bar{t}_r^{DISP-MAN\,WORK} + \sum_{m=1}^{n\_managers} \left(\frac{\lambda^{MAN}(m)}{\lambda} \cdot \bar{t}_r^{MAN}(m) \right) + \sum_{a=1}^{n\_agents} \left(\frac{\lambda^{AGENT}(a)}{\lambda} \cdot \bar{t}_r^{AGENT}(a) \right). \tag{14.8}$$

The performance analysis theory tells us that, if the system is overloaded, the result is $\bar{t}_r^{TOT} >> 2 \cdot \bar{t}_s^{TOT}$. In this case, it is necessary to add other components to balance the requests.

The described model and indicators can be recalculated over time in order to detect changes in the component loads. If the results obtained indicate that the performance objectives are met, then the decentralized manufacturing solution can continue. Otherwise, some hardware and/or software alternatives should be considered and the performance validation task repeated for the new hardware/software system.

14.5 Conclusion

In this chapter, an architecture and protocol for allowing decentralized manufacturing solutions have been presented. An architecture composed of three logical tiers has been proposed, and can be adapted based on the manufacturing unit scale.

The proposed solution also highlights the advantages of using microcontrollers for implementing the CNC control [14.16]. In fact, there is no need for a specific computer for each machine. Instead, a single computer running the Manager module can control and monitor different machines simultaneously. This solution is easily to scale, since the addition of new CNC machines implies only the updating of the database in the Manager.

A STEP-NC Network Management Protocol has been defined to describe the communications among the tiers for the purposes of supervision, dispatchment, feasibility, and availability evaluation.

To test the architecture's effectiveness, a simplified version of the dispatching/management system has been implemented to shown its applicability.

Several improvements in the test-bed architecture will be part of our future works, where the goal will be to simplify the process of Client's request and to optimize the allocation choices, based on previously collected information.

The Web interface provided by the Dispatcher to the Client can be enhanced with a list of frequently used STEP-NC files, with the possibility to customize some parameters. The Dispatcher can plan for the best manufacturing solutions if requests are already known. Also, the Managers can publish their availability to a service listing which provides the Dispatchers with the possibility to find the most appropriate Manager based on the Client's request. The use of a log of machining in the Manager's database in order to evaluate the most frequent requests can allow for computing in advance the feasibility check and Agent's selection. The implementation of this knowledge base allows for a quicker Manager response time and better task allocation (by taking accounting for real processing times and eventual errors). In general, by mining the logs of the different components, it is possible to plan reconfiguration or installation of new machines. Finally, more complex dispatching strategies can be implemented, such as the dispatching algorithm used for Mixed VLSI products [14.20].

References

[14.1] ISO 6983/1. 1982. Numerical control of machines—program format and definition of address words—part 1: data format for positioning, line and contouring control systems, 1st ed.

[14.2] ISO 14649-10. 2003. Data model for computerized numerical controllers: part 10—general process data.

[14.3] ISO 14649-1. 2003. Data model for computerized numerical controllers: part 1—overview and fundamental principles.

[14.4] Xu X.W. and Newman S.T. 2006. "Making CNC machine tools more open, interoperable and intelligent—a review of the technologies," *Computers in Industry*, 57(2): 141-152.

[14.5] http://www.steptools.com, accessed on: 30/07/2008.

[14.6] Manufacturing Data Systems Inc., *OpenCNC Brochure* www.mdsi2.com/Solutions/CNC_Controls/Brochure/OpenCNCbrochure.pdf, available on: 30/07/2008.

[14.7] Weck M. 2003. "STEP-NC – A new interface closing the gap between planning and shop-floor", WZL RWTH Aachen, http://www.step-nc.org/, STEP-NC Workshop, Aachen, Germany.

[14.8] Xu X.W. 2006. "Realization of STEP-NC enabled machining," *Robotics and Computer Integrated Manufacturing*, 22(2): 144-153.

[14.9] Lee W. and Bang Y.B. 2003. "Design and implementation of an ISO14649-compliant CNC milling machine," *International Journal of Production Research*, 41,(13): 3007–3017.

[14.10] Suh S.H., Cho J.H. and Hong H.D. 2003. "On the architecture of intelligent STEP compliant CNC," *Computer Integrated Manufacturing*, 15(2): 350-362.

[14.11] ISO 10303-1. 1994. Industrial automation systems and integration—product data representation and exchange—part 1: overview and fundamental principles.

[14.12] ISO 10303-203. 1994. Industrial automation systems and integration—product data and exchange—part 203: application protocol: configuration controlled 3D designs of mechanical parts and assemblies.

[14.13] ISO 10303-21. 2002. Industrial automation systems and integration—product data representation and exchange—part 21: implementation methods: clear text encoding of the exchange structure.

[14.14] STEP-NC Newsletters. Issue 3, Nov. 2000.

[14.15] Lin A.C., Lin S.-Y. and Cheng S.-B. 1997. "Extraction of manufacturing feature from a feature-based design model," *International Journal of Production Research*, 35: 3249–3288.

[14.16] Calabrese F. and Celentano G. 2007. "Design and Realization of a STEP-NC Compliant CNC Embedded Controller," IEEE Conference on Emerging Technologies and Factory Automation, Patras, Greece. 1010-1017.

[14.17] Suh S.H., Chung D.H., Lee B.E., Cho J.H., Cheon S.U., Hong H.D. and Lee H.S. 2002. "Developing an integrated STEP-Compliant CNC prototype," *Journal of Manufacturing Systems*, 31(5).

[14.18] RFC 3411 – An Architecture for Describing Simple Network Management Protocol (SNMP) Management Frameworks, http://tools.ietf.org/html/rfc3411

[14.19] Thompson T., Weil R. and Wood M.D. 2003. "CPXe: Web Services for Internet Imaging," IEEE Computer, 36(10): 54-62.

[14.20] Saito K. 2007. "A Robust Dispatching Algorithm for Autonomous Distributed Manufacturing of Mixed VLSI Products," *IEEE Transactions on Semiconductor Manufacturing*, 20(3): 299-305.

[14.21] Lan, H., Liu R. and Zhang C. 2008. "A multi-agent based intelligent STEP-NC controller for CNC machine tools," International Journal of Production Research, 46(14): 3887-3907.

[14.22] Xu X. 2007. "STEP into Distributed Manufacturing with STEP-NC," Process Planning and Scheduling for Distributed Manufacturing, edited by L. Wang and W. Shen, Springer Verlag.

[14.23] Li L., Fuh J.Y.H., Zhang Y.F. and Nee A.Y.C. 2005. "Application of genetic algorithm to computer-aided process planning in distributed manufacturing environments," *Robotics and Computer-Integrated Manufacturing*, 21(6)" 568–578.

[14.24] Jain R. 1991. "The Art of Computer Systems Performance Analysis: Techniques for Experimental Design, Measurement, Simulation, and Modeling", Wiley-Interscience.

15

A Generic Product Modelling Framework for Rapid Development of Customised Products

Shane Q. Xie[1] and Wan-Lin Chen[2]

Department of Mechanical Engineering, University of Auckland, 20 Symonds Street, Auckland, New Zealand
Email: [1] s.xie@auckland.ac.nz, [2] steven.chen@auckland.ac.nz

Abstract
In response to the current rapidly changing manufacturing environment, product modelling technology has been widely researched to provide information for supporting the development of customised products. Traditional product modelling technologies are unable to support the information exchange and share in various stages of product development processes that could be taking place in a distributed manufacturing environment. This has caused many problems such as information loss, data format incompatibility and reduced efficiency and effectiveness of product data applications. This has consequently created bottlenecks for the integration of product development processes. In this chapter, a Generic Product Modelling Framework (GPMF) is proposed to overcome the problems of information exchange and sharing. This framework uses the Standard for the Exchange Product Model Data (STEP) as a foundation, and consists of four functional components: an EXPRESS Data Model namely EDM, a STEP–based modelling environment, a "five-phase" modelling method, and three EDM data exchange and sharing methods. The case study shows that the product models built based on the GPMF are capable of integrating information in product design, manufacturing and assembly. The GPMF is compatible, comprehensive, and flexible, and it is able to support information exchange and sharing.

15.1 Introduction

Nowadays more and more manufacturing companies have realised that the ability to quickly develop a customised product quickly through an efficient and cost-effective way is critical for them to survive in the increasingly competitive international market, particularly for those one-of-a-kind production companies [15.1–15.3]. In these one-of-a-kind production companies, due to a high customisation and involving a large amount of uncertainties, their product development cost is normally higher and development lead-time is much longer than those product-focused manufacturing companies. To respond rapidly to market pressure, these manufacturing companies require new systems, tools and technologies to manage their product development processes through a whole product development cycle.

These processes include, for example, customer requirements interpretation to decide how to address customers' needs, product design to define products that satisfy these needs, conceptual and functional prototyping to prove the effectiveness of the designs, manufacturing planning to determine how the product is to be made, design of tooling used to manufacture the components of the product, manufacture of these tools, pre-production testing, manufacturing, and finally delivering the product to the customer or launching the product into the market.

The one-of-a-kind production companies normally wish to meet their customers' needs in one go and minimise the costs spent on the intermediate tests and mock-up manufacture. A key for their success lies in the methods of managing and utilising product information and knowledge throughout their production processes. Product modelling technology is one of the key and indispensable technologies in the product development processes. One of the major functions of product modelling technology is to provide a well-organized information structure to model product data and information. Product development team members can then utilize an information model to store, exchange and manipulate product data. Well-structured product data can be used to support the integration of product development systems and can be reused as knowledge for development of similar products. A data model of a product can support similar product development by making reusable data and information available. Product modelling has been regarded as the key for supporting rapid development of customized products [15.2]. It determines engineering productivity and eventual industrial competitiveness [15.4] and has become an important research issue in supporting the development of any product that requires the efficient integration of its lifecycle activities.

Conventional product modelling technologies have significantly enhanced the performance of product development processes in the last few decades. For example, geometric models are utilised to model product geometric information for supporting CAD systems to exchange and share product data. However, the conventional modelling technologies cannot meet the requirements of current product development processes due to the nature of customized products and new challenges from the ever-changing manufacturing environment. The limitations of conventional product modelling technologies have become the main hurdles for the development of modern manufacturing systems.

Problem of Integration

Conventional product modelling technologies are normally used to model products for supporting the integration of one or two systems such as a CAD or Computer Aided Process Planning (CAPP) system. They cannot be directly used in the integration of the systems that are employed in other stages of product development processes such as Computer Aided Manufacture (CAM) or Computer Aided Engineering (CAE). Hence, more work should be carried out to establish seamless product data exchange and sharing throughout the entire product development processes. On the other hand, for downstream systems, there are normally information conversions between a CAD based product model and models of CAPP or CAM systems. Information loss may result from incompatible formats of product models in different systems. This has become a major problem when attempting to integrate these systems. The representation format of product data impedes the fluent information flow between systems, and leads to high development costs as a result of unnecessary costly rework and redesign. Consequently, the inefficient product development process will definitely influence the competitiveness of the manufacturing company.

Problem of Cooperation

Nowadays complex customised products are usually developed by combining the strength of several manufacturing companies. Data exchange and sharing between these companies needs to be very efficient and effective. Most companies structure their products using different modelling methods. Hence, it has become an issue for them to cooperate with each other in support of the development of a particular product. Normally, an extra data conversion process should be carried out. This is very inefficient and sometimes useful data might be lost due to conflicts between model structures. Therefore, a non-compatible product model has become a barrier to cooperative product development. New modelling methods are required to develop for establishing highly compatible product models.

15.2 Product Modelling: A Review

STEP is currently considered as a promising product modelling resource since it provides a standardised mechanism for product data representation and exchange. Considerable effort has been placed on STEP-based product modelling in recent years. Earlier focus was placed on the definition of classes and the design of the user interfaces with CAD software tools. Li et al. developed a feature based parametric product modelling system, which employed a product model based on the STEP and managed by an object-oriented database [15.5]. This system was suitable for being applied in a computer integrated manufacturing (CIM) environment. Gu et al. developed a STEP-based generic product modelling (GPM) system, which was designed and implemented according to the generic resources of STEP [15.6]. The system can therefore be used to integrate manufacturing activities, such as process planning and inspection planning in the concurrent engineering environment. They presented an object-oriented approach for building product models for supporting product design. Usher et al. presented a STEP-based object-oriented product model based on STEP AP 224 [15.7]. This model was proposed for supporting CAPP analysis. Chin et al. presented a STEP-based part information model for process planning purpose [15.13]. Their models included a process planning information model and a production resource information model. Song et al. presented a STEP-based die and product integrated information model (DPIIM), in which integrated resources of STEP were utilised to model six EXPRESS schemas [15.14]. These models could support the concurrently developing stamp and die products. Zha et al. presented a product data exchange using STEP (PDES)/STEP-based assembly model for the concurrent integrated design and assembly planning [15.29]. It can be concluded that STEP-based modelling method has become the core of product modelling processes to organize product data in the standardised representation, which greatly enhances the capability of data exchanging and sharing in the integrated manufacturing environment. To utilise the modelling resources defined in the STEP, various modelling methods are integrated with STEP to form an integrated product modelling environment.

Geometry Based Modelling Methods

Application Protocols (APs) are used to develop information models for the integration of STEP with different geometric modelling methods, such as AP204 [15.30] and AP203 [15.19]. Shaharoun et al. utilised STEP to describe geometric data of a particular plastic product [15.31]. The geometrical descriptions of the product were transferred into a CAD system to assist the design and machining of a suitable mould for the plastic product. Cai et al. proposed a method to build self-defined APs for all kind of machine parts based on the STEP [15.8]. They implemented this method to develop two APs for presenting the geometric data model in the cone gear product for final driver of automobile driving axle system.

Feature-based Modelling Methods

STEP provides a suitable representation method for different features for supporting feature-based product modelling. For instance, AP224 [15.9] illustrates the mechanical product definition of process plans using Machining Features. Other APs, such as AP214 [15.32] and AP218 [15.10] also contain the STEP expressions for the specific features in the particular application areas. Meng et al. presented a STEP-based feature modelling system, which was based on a user-defined AP development on the basis of AP214 [15.11]. Zhao et al. delineated an object-oriented feature-based aero-engine blade product modelling system [15.12]. In this modelling system, the design platform utilized STEP to standardise data modelling and to support the information transmission from design platform to analysing system.

Integrated Modelling Methods

There have been many research projects combining STEP-based product modelling methods with integrated modelling methods. For example, Chin et al. proposed a multiple view methodology for integrated product modelling based on STEP [15.13]. Song et al. utilised a STEP-based integrated product model to support the proposed design for manufacturing (DFM) system [15.14]. The aim was to extract the design information of parts from CAD system for automatically evaluating the manufacturability of those parts. Jasnoch et al. developed a collaborative working virtual prototyping environment to integrate existing CAD systems [15.25]. The underlying product model of this environment was a STEP-based integrated product model.

The purpose of our research is to develop a generic product modelling framework (GPMF) for supporting the integration of various product development activities. The focus is placed on the modelling methodologies and the definition of the structure of the schemas for manufacturing, inspection, etc., and the integration of the schemas with other resources defined within STEP. There are 25 schemas defined to enable that the proposed GPMF is compatible and can be used in modelling various types of products. These aspects, to our best knowledge, are still at their very early stages of research.

15.3 Generic Product Information Framework

This chapter introduces a generic product modelling framework (GPMF) that attempts to provide an infrastructure for modelling various types of customised products. The output of the GPMF is a set of data models defined to model a product at different stages of its development processes. Figure 15.1 shows the structure of the GPMF developed based on STEP. It consists of four functional components including an EXPRESS data model EDM, a STEP based modelling environment, a 'five-phase' modelling method, and three EDM data exchange and sharing methods.

The EXPRESS data model (EDM) is the core of the proposed GPMF. It defines a complete product data structure and uses the standardized data format. There are 11 defined EXPRESS schemas defined and STEP AP 203 included in the EDM. Each EXPRESS schema utilises either STEP resources or STEP-based compatible resources defined to model a particular type of product information. The STEP-based modelling environment is developed for the GPMF. Within the environment, a modelling language-EXPRESS and its graphical representation method EXPRESS-G are used to model product structure. STEP generic resources are utilised to model product information defined by STEP. STEP AP 203 is used to model product geometric information, and new modelling resources are defined for modelling product information that is not covered in STEP. The 'five-phase' modelling method is proposed to develop the EDM. The method defines a formal approach to organise logically all the tasks for building up the EDM in the modelling processes.

Three EDM data exchange and sharing methods are used in the GPMF. As shown in Figure 15.1, product data is exchanged and shared through either exchange files, or working forms, or database management systems. The product models defined within the GPMF are exchanged or shared using one of the methods. These three methods are easily integrated into any application software environment, which makes it easy to implement the product models defined by the GPMF in applications.

15.3.1 STEP-based Modelling Environment

The STEP-based modelling environment in the GPMF is to facilitate data sharing and exchange through well defined STEP-compliant product data models. The EXPRESS data modules and the modelling methods are used to develop STEP-compliant data models for the integration of product development systems. A number of methods and resources are developed for establishing such an environment.

Modelling Language: EXPRESS and EXPRESS-G

EXPRESS modelling language consists of language elements that can allow unambiguous data definition and specification of constraints on the data defined, and is one part of STEP defined in Part 11 [15.15]. The resources in STEP, including generic resources and APs, are normally represented as EXPRESS schema. In the GPMF, EXPRESS modelling language is utilised to develop the EDM, which represents the structure of product data. Figure 15.2 shows the main elements of an EDM. The data model in the EDM is represented by one or more schemas, which "group together the modelling objects with related meaning and purpose" [15.16]. The most important EXPRESS language element is the ENTITY data type, which defines the objects of interest in the domain being modelled. The ENTITY is characterised by its attributes and constraints. EXPRESS language also supports various kinds of data types, including simple types, aggregations types, and constructed types as shown in Figure 15.2.

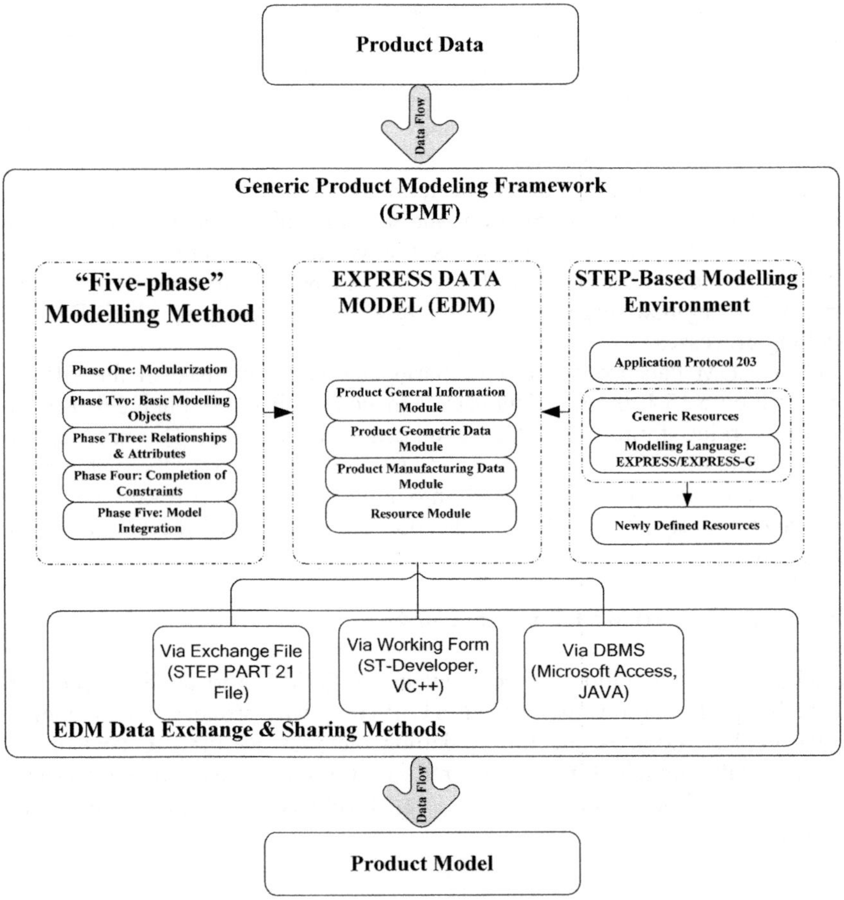

Figure 15.1. Structure of STEP-based GPMF

Generic Resources and STEP AP 203

The generic resources are directly utilised to define the basic data objects in an EDM. For example, entities defined in product_definition_schema in STEP Part 41 [15.17] are utilised to present the product general information in the EDM; STEP Part 45 [15.18] is used to define the EXPRESS schema that represents the data structure for material information. The generic resources are defined in STEP. The use of the generic resources enhances the compatibility of the GPMF. The STEP AP 203 [15.19] is utilised to model product geometric information in the EDM. It is employed by different CAD systems as the data model to structure product geometric data. Thus, the product geometric data module of EDM can be integrated with CAD systems using STEP AP 203.

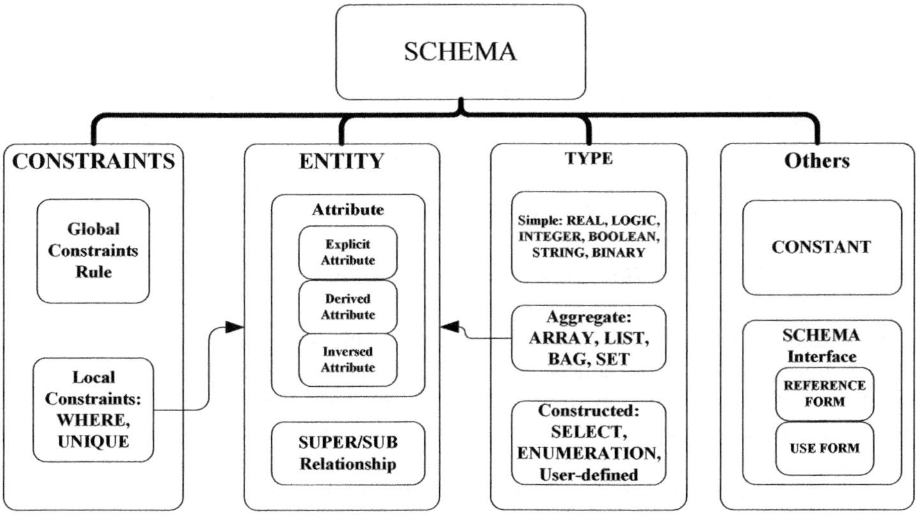

Figure 15.2. Main elements of EXPRESS data model

New Modelling Resources

STEP has already defined many modelling resources to support the product modelling. However, it is still under development. There are still not sufficient resources defined to present information of different types in product development processes. Hence, two types of new modelling resources are defined and utilised in the GPMF as the supplementation of STEP-defined modelling resources.

One type of new resources is called hybrid modelling resources. They are modified from the STEP generic resources. For instance, the *document_schema* in Part 41 [15.18] is modified to generate a data model for the document information. The method_definition_schema and process_property_schema in Part 49 [15.34] are modified to define the data model in the EDM for the process information. New STEP-compliant modelling resources are defined by the authors to model the product data which are not covered in STEP [15.3].

15.3.2 'Five-phase' Modelling Methodology

The 'five-phase' modelling method is proposed for the development of the EDM. The method is to standardise the modelling of different types of product information. It is composed of five phases: (1) modularisation, (2) basic modelling objects, (3) relationship and attributes, (4) completion of constraints and (5) module integration. The five steps are discussed in the following sub-sections.

Phase One: Modularization

This phase is to define and modularize an EDM. The main tasks in this phase include: (1) product modelling objective analysis, (2) classifying the product data, (3) modularizing the EDM. Product modelling objective analysis is very complex and involves a large quantity of data of different types. Modelling these data without analysis, classifying and modularising could result in data loss or repetition. Consequently, the EDM will not be able to represent correctly a product. In the modularisation phase, the EDM is divided into four modules including a product general information module, a product geometric data module, a product manufacturing data module, and a resources module that is developed to present the sharable basic modelling objects extracted from the other modules.

Phase Two: Basic Modelling Objects

This phase defines the basic modelling objects and the general structure of each EDM module. The phase starts by analysing the structure of the defined EDM modules. The fundamental elements of the modules are identified and defined as the basic modelling objects. After defining the basic modelling objects, the way to structure them is analysed and applied in the EDM. For example, in the product manufacturing data module, product assembling information is normally defined by four basic modelling objects including an assembly product object, a product component object, a subassembly component object, and an object called connector is defined to represent the connections between these objects. These objects are named *assembly_product*, *part*, *subassembly*, and *connector* respectively.

Phase Three: Relationships and Attributes

This phase refines and enriches the basic modelling objects defined in the second phase by adding relationships to the defined objects. The following tasks are included in this phase: (1) definition of attributes and relationships between entities, (2) enhancement of the defined entities, (3) definition of new entities and (4) correctness checking. This phase must be continued until the EDM has reached the desired level of details for representing the content of objects and their relationships with other objects. For instance, the basic modelling object *part* normally has seven attributes: *id*, *name*, *description*, *of_assembly*, *super_item*, *level_in_assembly_hierarchy*, and *connecting_information*. Their value types are defined as identifier, label, text, *assembly_product*, super_item_type, integer, and *connector* respectively.

Phase Four: Constraints Modelling

This phase models the constraints of an EDM. It has the following three tasks: (1) definition of constraints of objects and their relationships, (2) addition of global constraints to the model and (3) model error- checking. Constraints-modelling defines the objects and their relationships based on requirements. A complete EDM module is developed after this phase. For example, one local constraint named WR1 is defined in the *part*. This constraint specifies the range of the *level_in_assembly_hierarchy* attribute, which defines that the part must be in the second or under the second level of the Assembly Model Tree (AMT). The error-checking task detects possible grammar errors, missing information, and conflicts of constraints.

Phase Five: Model Integration

This phase integrates the modules defined after the four steps to form an EDM. The main tasks involved in this phase are to check how data is represented in each module, define the inputs and outputs the four modules, and evaluate whether EDM is complete, minimally redundant, unambiguous and error free.

15.3.3 EDM Data Exchange and Sharing Methods

The EDM discussed in Section 3.2 provides the basic framework for establishing product models. To support data exchanging and sharing, it needs to be mapped to a model that is accessible through commercial software tools. Lee et al. demonstrated three implementation methods for the information model that have been currently used in practice [15.21]. These three methods were indicated by [15.22] and [15.26] as the three implementation methods for EXPRESS data model, which are: (1) implementation via an exchange file, (2) implementation via a working form and (3) implementation via a Database Management System (DBMS). In the GPMF, these three methods are utilised as the EDM data exchange and sharing methods to support to meet the requirements for implementing the GPMF in the product modelling processes.

Figure 15.3 shows that EDM-defined data structures can be used to support various stages of a product development process, e.g. computer aided customer user interface (CACUI), CAD, CAPP, CAM, etc. These systems can be integrated via the three implementation methods for EXPRESS data model. An integrated product development platform can be then developed. Product data can be made available by using API and data is presented as EDM-defined structures. A DBMS based on the proposed models and available tools has been developed for manipulating the product database, including data query, data update, data retrieval. Different application systems can access the product database to read and write product data. These product data are exchanged and shared within these systems as shown in Figure 15.5. The data structure defined in the EDM should be mapped into the product database. This structure information can support the possible DBMS applications to implement this level method.

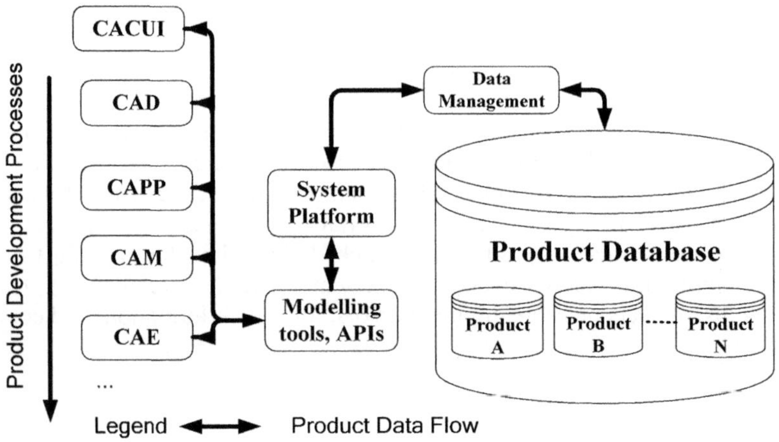

Figure 15.3. Product development platform

15.4 EXPRESS Data Model

Figure 15.4 shows the structure of the EDM. It consists of four modules. The product general information module, product geometric data module and product manufacturing data module are defined and based on the results of classifying product data in the first modelling phase. The resources module is developed by grouping the sharable basic modelling objects to support the development of other modules.

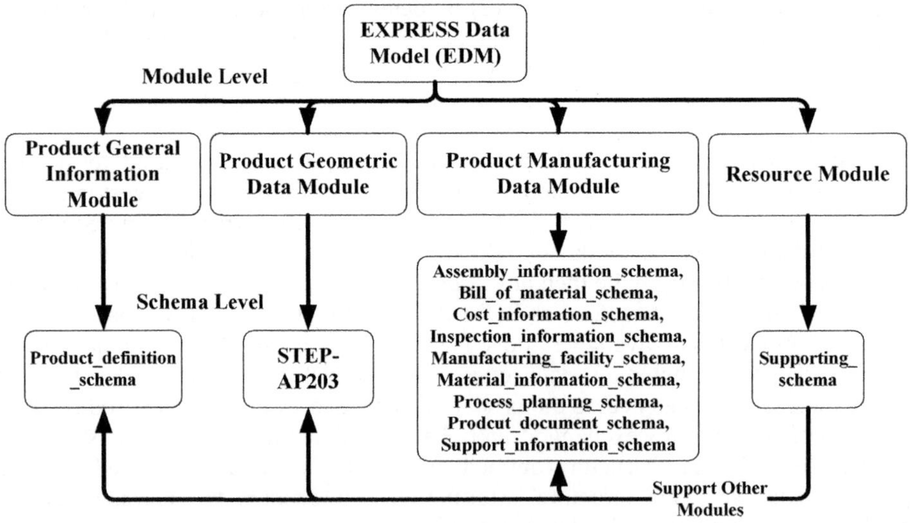

Figure 15.4. Structure of the EDM

The product general information module represents the product data, which are not directly related to the product manufacturing, such as the product identity, product property, and the relationship between products. In this module, a *product_definition_schema* is defined to support modelling of this aspect of the product data. As shown in Figure 15.4, a product geometric data module supports modelling product geometric data, such as shape information and dimension information. This module is a key to integrate different computer aided systems. In the product geometric data module, STEP AP 203 is directly utilised for representing and exchanging product 3D geometric information. The product manufacturing data module is the core of the proposed EDM, and consists of nine EXPRESS schemas to support modelling different manufacturing data.

A *supplier_informaton_schema* is defined to model the supplier information. Manufacturing companies need information about suppliers and their products to arrange the manufacturing processes. A *manufacturing_facility_schema* is defined to model the facility information. The manufacturing facilities, such as machines, tools and fixtures are directly involved in the product manufacturing processes. They are the key resources determining the product manufacturing processes and influencing the product quality. A *product_document_schema* is defined for documents aspect. The documents are defined as the industry standards and documented criteria that regulate and guide the manufacture of a product. A *bill_of_material_information_schema* is defined for representing Bill of Material (BOM) information. The BOM, which can be called part list as well, is a list of product components or resources for manufacturing or assembling this product. The BOM is utilised to determine the product cost and used as an effective tracking source during the manufacturing and assembling processes.

A *material_information_schema* is defined to represent the material information. The material is the basic element for a product and it directly influences the selection of proper manufacturing facility and manufacturing processes. A *process_planning_information_schema* is defined for modelling process data. The manufacturing process information is the detailed description about the actual manufacturing activities. It is essential to consider this aspect information in design stage, which leads to optimise the manufacturing processes. A *assembly_information_schema* is defined for these two aspects of product data. Assembly product is one of the most important types of products. To develop an assembly product, the components information and the assembling method are required.

An *inspection_information_schema* is defined to model inspection information. The inspection processes are essential parts of product development processes to control product quality. The inspection results can assist to indicate the problems of a product and its production processes. The *cost_information_schema* is defined for the cost information occurred in product development processes. The inclusion of cost information is critical for any engineering or manufacturing organisation. Managing cost information can help the firm to increase its own competitive ability.

The resources module in Figure 15.4 defines basic modelling objects that are shared by the other modules. All these basic modelling objects are grouped into the *supporting_schema*. The resources in the *supporting_schema* are represented as EXPRESS entity, the constructed TYPE, and FUNCTION. Through the EXPRESS schema interface, these resources in this schema are utilised by other schemas to structure an effective and efficient data model representation. The four modules of EDM are developed by applying the STEP generic resources including Part 41, Part 45, and Part 49, STEP AP 203 as well as the new defined STEP compatible modelling resources. The relationships among EDM schemas and these resources are developed in our research group with the schemas defined [15.3].

15.5 Case Study

Injection moulding products are highly customer involved and regarded as typical one-of-a-kind products [15.23]. In this case study, they are used to validate the proposed GPMF and its related models and methods. A STEP-based injection moulding product data model defined in EXPRESS language is developed. It covers the life-cycle information for the development of injection moulding products. As a result, a STEP compliant Product Data Management System (SCPDMS) is developed by an international injection moulding company for the efficient management of its product development processes [15.27]. The company has its departments distributed in four different countries. The product model is used for the company to management its product development activities. The model developed for the company is made up of the following basic data models: (1) product and its assembling information, (2) injection tooling information, (3) machine tool information, (4) manufacturing and workshop information, (5) material information and (6) supplier information. These data models are defined based on the proposed GPMF.

15.5.1 Product and its Assembling Information

A product data model is created for representing the information of an injection mould product. Figure 15.5 shows the data structure of the model. The root of the model is entity *product*. This entity is defined to represent both general product information and manufacturing information of an injection moulding product. The first type of information included in this entity are: (1) identifier of a product (attribute *product*.cat_number), (2) product family information (*product*.product_family), (3) product performance and its description, and (4) drawing file information. The second part of information included are: (1) assembling methods (defined by entity *product_fastener_and_clip*), (2) design information (by entities *design_standard* and *product_designer*), (3) material information (by entity *product_materials*), (4) project appraisal information (defined by entity *project_appraisal*) and (5) the relationship with injection tool (defined by entity *product_tool_relationship*).

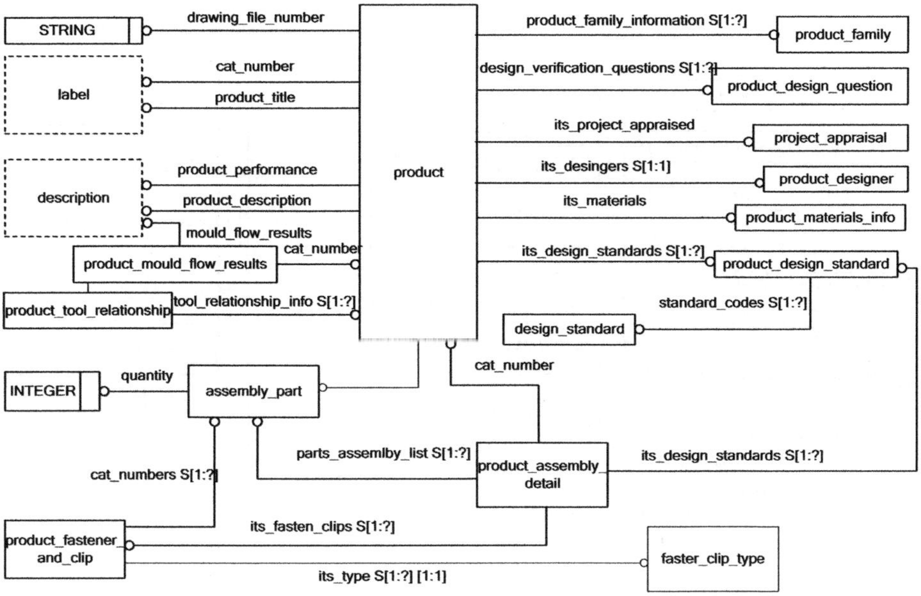

Figure 15.5. Product and assembling information

If an injection moulding *product* has a number of assembled products, its detailed assembling information is represented by entity *product_assembly_detail*. The individual components of an assembly are defined by entity *assembly_part*, which is a subtype of entity *product*. All the interrelationships between *assembly_parts* are defined in entity *product_fastener_and_clip* and the details of the method for assembling them are represented in entity *fasten_clip_type*, as shown in Figure 15.5. Entity *fasten_clip_type* has following attributes defined: (1) *fasten_clip_type*.id_name is defined to identify this assemble type, (2) *fasten_clip_type*.product_designing_feature determines whether this assembling feature is an existing design feature or not. (3) *fasten_clip_type*.its_description presents demonstration of assembling information, (4) *fasten_clip_type*.subassembly_menbers is defined for listing all the subassemblies in this assembling type and (5) *fasten_clip_type*.suppliers for the supplier information.

15.5.2 Tooling Information

Another data model is defined to represent information of injection tool (injection mould) and its tooling information. The structure of the model is shown in Figure 15.6. The root of this data model is entity *tool*, and it includes attributes such as tool code, complete data information, tooling descriptions, tooling cost, designer, etc. Attribute *tool*.its_material lists all the material utilized for a *tool*. Attribute *tool*.its_components collects all the individual component of a tool and defines their information by using entity *tool_component*.

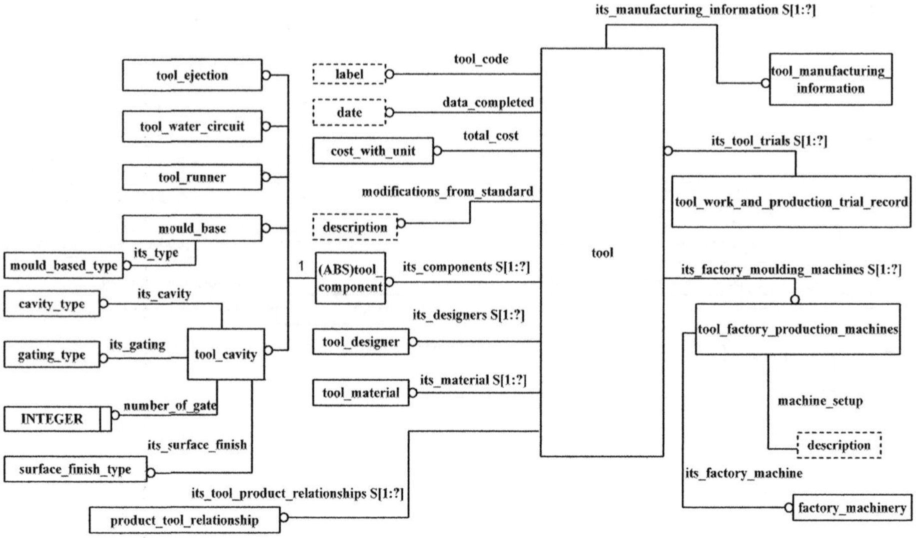

Figure 15.6. Data model of tooling information

Entity *tool* defines information about its manufacturing processes. Attribute *tool*.its_tool_trials represents all of the tool trial results (by using entity *tool_work_and_production_trial_record*). The information of manufacturing a *tool*, such as machine tool utilised, machining and programming time, operator, etc. is defined by an entity *tool_manufacturing_information* in the attribute *tool*.its_manufacturing_information. The Entity *tool_factory_production_machine* utilised in the attribute *tool*.its_factory_moulding_machines is to define the injection machine tool, which utilises this *tool*.

A *tool* is further divided into the following components: *tool_ejection*, *tool_water_circuit*, *tool_runner*, *tool_cavity* and *mould_base*. Entities *runner_type*, *water_circuit_type* and *ejection_type* are used to further detail *tool_runner*, *tool_water_circuit* and *tool_ejection* respectively. Entities *cavity_type*, *gating_type* and *surface_finish_type* are used to specify *tool_cavity*. Entity *mould_based_type* is used to define *mould_based*. This data model also represents information of ejection pin component. Entity *ejector_pin_product_family* summarises the family-oriented information of an ejection pin, and Entity *ejector_pin_parameters* lists all its technical parameters, i.e. a full set of product dimensions, material properties including strength, hardness, and heat resistance.

15.5.3 Machine Tool Information

A data model is defined to represent two types of machine tool information utilised in the manufacturing process of an injection moulding product, which includes *injection_moulding_machine_tool* and *machining_machine_tool*. As shown in Figure 15.7, an injection moulding machine tool is modelled by a defined entity *injection_moulding_machine_tool*, which includes three types of information: injection unit information, clamping unit information and other specified information. Injection unit information is defined by entity *injection_unit_information*, which contains information about screw diameter, torque, stroke and speed range, and also includes injection pressure and injection capacity. Entity *clamping_unit_information* defines the properties of an injection mould clamping device. The main parameters of the entity include mould information (height and thickness), clamping force, opening stroke, space between tie bars, ejection quantity and ratio of ejection force and stroke. Other information of injection moulding machine tool is defined by entity *other_unit_information*, which describes other parameters of an injection moulding machine, e.g. size and weight.

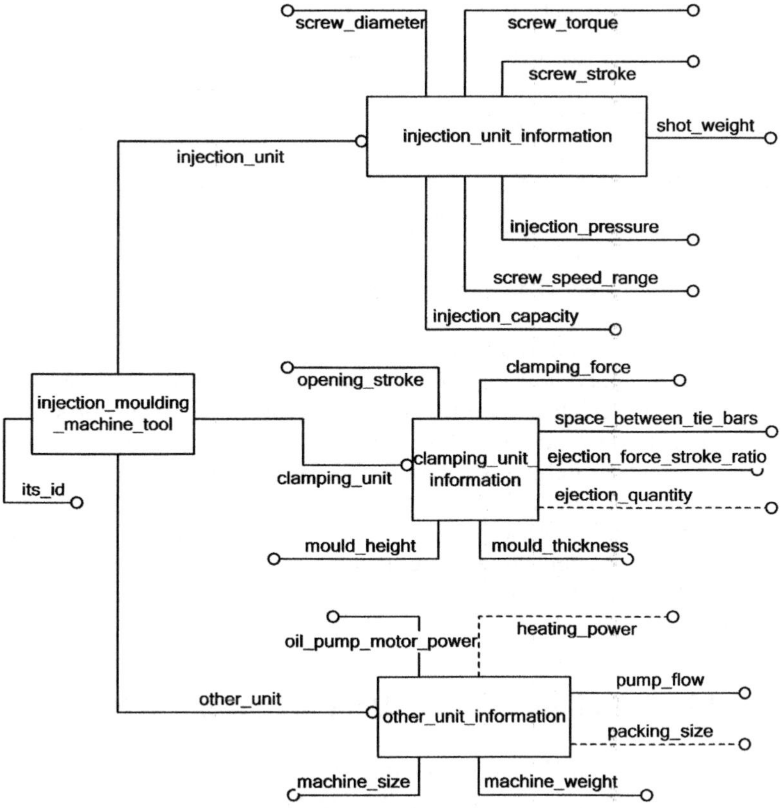

Figure 15.7. Machine tool information data model

15.5.4 Manufacturing Information

Figure 15.8 shows the structure of the defined manufacturing information data model. This data model is defined to represent workshop related machinery capability, which is important to define knowledge for machine selection. The root entity of this data model is *manufacturing_information*, which contains four types of information: (1) tool trail information, (2) mould manufacturing information, (3) machine tool information and (4) manufacturing cost information.

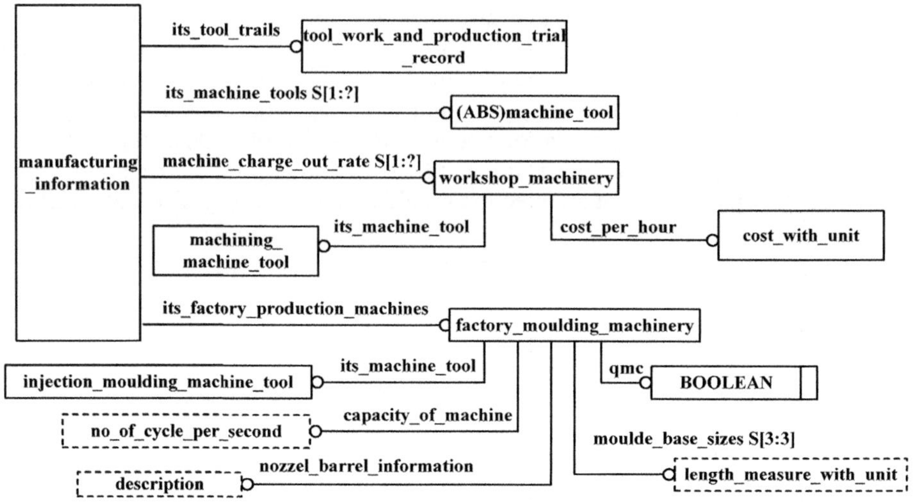

Figure 15.8. Manufacturing information data model

Tool trail information is defined by the entity *tool_work_and_production_trial_record*. This entity records both administration information and technical information for its tool trail process. Entity *factory_moulding_machinery* is defined to model the injection machine tools on the shop-floor. The model defines injection moulding machine information, cycle time of injection, size of mould base, quality control information, nozzle and barrel information, etc. Entity *workshop_machinery* models the cost information for a machine tool. Data models are also defined for modelling other information of the company so that the developed system covers the whole life cycle of its product development processes.

15.6 STEP Compliant Product Data Management System

An SCPDMS has been developed in this research as an on-line data library to support the whole life cycle of product development. It was developed according to the STEP-compliant data models defined above. Figure 15.9 shows the basic structure of the data structure of the database for supporting product development in a cooperative environment [15.28]. The data structure is made up of data models defined for modelling, product, tools, resources, partners and suppliers. "product" means all the design and manufacturing data about a product. The "supplier" includes all information about the company's suppliers, which are contact information, product specifications, reference prices, delivering lead times, reputation assessment and quality assessment. These data are stored in distributed computers that can be accessed via Internet or Intranet.

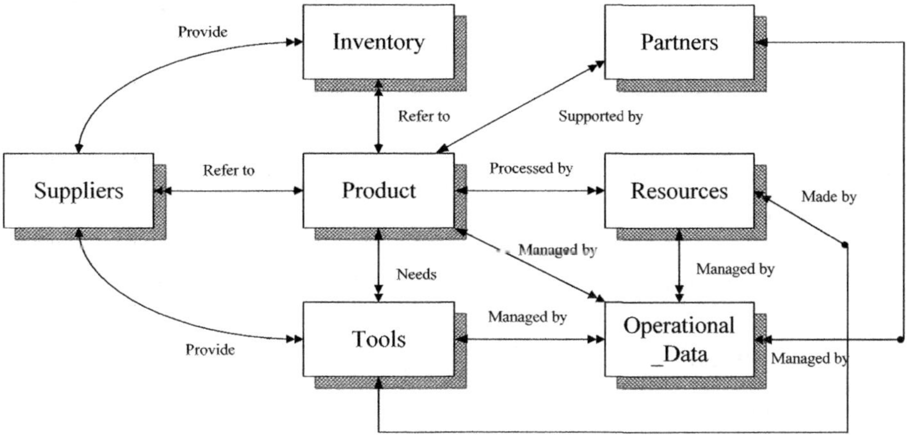

Figure 15.9. Data structure of the database

The SCPDMS is developed by (1) mapping the EXPRESS entity to the table in the database and (2) mapping the relationships between different EXPRESS entities to the corresponding tables in the database. Figure 15.10 shows the user interfaces developed for testing the SCPDMS and evaluate the proposed data models. The product database not only contains product information defined but also objects to help manage these data, such as tables, forms, reports and so on. The SCPDMS developed in this research stores all the product data following the structure defined by the defined product data models. There are 48 tables in the database which could be categorized into 6 types: (1) general information, (2) product information, (3) tool information, (4) manufacturing information, (5) material information and (6) supplier information. From the user interface of the database, users could easily and conveniently look up the tool information, product information, product tool relationship information, manufacturing information and supplier information of a product. Meanwhile, users can enter information on to the database or edit the existing information of the database directly from these interfaces or enter/edit detail information from the Data Enter/Edit Board as well.

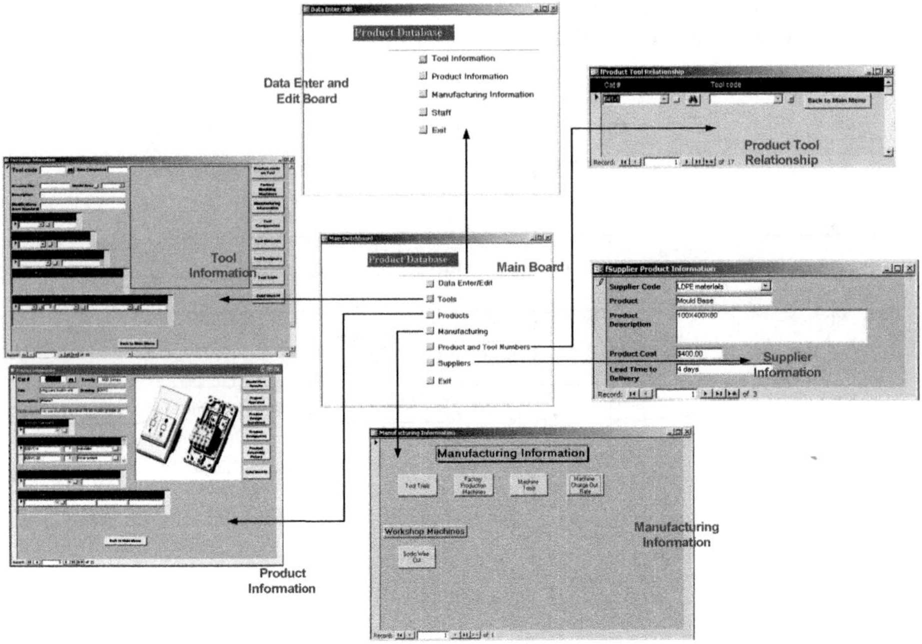

Figure 15.10. User interface of the SCPDMS

15.7 Conclusion and Future Work

This chapter presented a STEP-enabled GPMF for efficient information sharing and exchange in support of rapid development of customised products. The focus of the study is to develop a generic modelling methodology for modelling products of different types. This is achieved through the definition of the schemas and the proposed modelling methodologies. A case study is carried out to validate the proposed GPMF. A prototype SCPDMS system has been developed to demonstrate how the GPRF works. From this research work, the following conclusions are drawn:

- The proposed STEP-enabled GPMF is compatible for modelling products of different types. This has been validated by applying the proposed framework in modelling a typical type of one-of-a-kind production products, injection moulding products.
- The GPMF is able to support the modelling of a wide range of product data. In the case study, five aspects of product data are modelled through the proposed GPMF. The corresponding product models provide a comprehensive view of the product.
- The case study shows that all the parts of the GPMF are associated with a core: the EDM. The entire product modelling processes is tightly dependent on the data structure defined in this data model.

- The EDM is flexible to be implemented. Each module of EDM can be considered as an individual EXPRESS data model. The model can be applied with the EDM data exchange and sharing methods to model the corresponding product data. In the case study, the product general information module, the product geometric data module, the inspection_ information_schema, the proess_planning_shcema, and the assembly_infomration_shcema are utilized individually to model product data.
- A prototype SCPDMS is developed to demonstrate how the proposed STEP-enabled GPMF works with general database systems for modelling and managing product data. This is based on the EDM data exchange and sharing methods.

The GPMF provides a well-established mechanism to support the integration of systems through the proposed product modelling methodologies. However, more work needs to be carried out before it is applied in an actual integrated manufacturing environment for supporting the integration of product development systems in a modern manufacturing environment. Future work in the area is enormous and not limited to the following three areas;

- The first area is to develop the prototype system further to integrate with customers and downstream product development systems. This includes the input/output interfaces for the integration of various computer aided systems. The system needs to provide a standard interface to transfer product data with proper format to the end user.
- The second area is to validate the proposed GPMF further by applying it in modelling products of other types. Our research work in modelling sheet metal and injection moulding products has shown that this is a complicated process [15.23]. Future work in this area requires great effort to define new schemas as STEP itself is still at its development stage.
- The third is to explore how to extend the proposed GPMF for supporting Web/Internet-based manufacturing. Integrating web technologies into the modelling framework can intensively improve the information exchanging effectiveness through internet/intranet. This enables the proposed GPMF for the distributed product development environment for supporting the web-based product development activities. An implementation method also needs to be developed to map between extensible mark-up language (XML) and EXPRESS, which is defined in STEP Part 28 [15.20].

Acknowledgment

The author would like to acknowledge the support of the International Investment Opportunities Fund (IIOF) from the Foundation for Research, Science and Technology (FRST) of New Zealand.

Reference

[15.1] Tu, Y L, Fung, R Y K, Tang, J F and Kam, J J , 2003, Computer aided customer interface for rapid product development, *The International Journal of Advanced Manufacturing Technology*, Vol. 21, pp.743-753.

[15.2] Xie, S. Q. and Tu, P. L., " Rapid One-of-a-Kind Product Development", *International Journal of Advanced Manufacturing Technology*, 2005, Vol. 27, No.5-6, pp. 421 - 430.

[15.3] Xie S. Q., Yang, W. Z., Tu, Y L, "Towards a generic product modeling framework", *International Journal of Production Research*, Vol. 46, Issue 8, April 2008 , pages 2229 – 2254.

[15.4] Krause, F. L., Kimura, F., Kjellberg, T., Lu, S. C. Y., Van der Wolf, A. C. H., Ating, L., ElMaraghy, H. A., Eversheim, W., Iwata, K., Suh, N. P., Tipnis, V.A. and Weck, M., 1993, "Product modelling", *CIRP Annals: Manufacturing Technology*, 42, 695-706.

[15.5] Li, H.L., Han, J.H., Dong, J.X. and Wang, Y., Feature-based, parametric modelling system for CAD/CAPP/CAM integrated system. J.Engng Appl. Sci., 1996, 1, 329–333

[15.6] Gu, P. H. and Chan, K., 1995, Product modelling using STEP, *Computer-Aided design*, 27, 163-179.

[15.7] Usher, J. M., 1996, STEP-based object-oriented product model for process planning, *Computers and Industrial Engineering*, 31, 185-188.

[15.8] Cai, C.T., Li, Y.Y., Dai, Y.H. and Liu, X.Y., Design method of application protocol of the machine parts based on STEP. Jisuanji Jicheng Zhizao Xitong/Comput. Integr. Mfg Systems, CIMS, 2002, 8, 892–895.

[15.9] International Organization for Standardization, Industrial automation systems and integration: Product data representation and exchange: Part 224: Application protocol: Mechanical product definition for process planning using machining features, Reference number: ISO/DIS 10303-224, Third Edition, 2005 (ISO: Geneva, Switzerland).

[15.10] International Organization for Standardization, Industrial automation systems and integration: Product data representation and exchange: Part 218: Application protocol: Ship structures, Reference number: ISO 10303-218, First Edition, 2004 (ISO: Geneva, Switzerland).

[15.11] Meng, M.C., Yang, L. and Bai, L.K., Feature modeling system based on STEP. Jishanji Jicheng Zhizao Xitong/Comput. Integr. Mfg System, CIMS, 1997, 3, 34–38.

[15.12] Zhao, W. and Ma, W., Feature modeling for aeroengine blades according to STEP. Beijing Hangkong Hangtian Daxue Xuebao/J. Beijing University of Aeronaut. Astronaut., 1999, 25, 535–538

[15.13] Chin, K.S., Zhao, Y. and Mok, C.K., STEP-based multiview integrated product modelling for concurrent engineering. Int. J. Adv Mfg Technol., 2002, 20, 896–906.

[15.14] Song, Y. Y., Chu, X. P. and Cai, F. Z., 1999, Real-time concurrent product and process design system for mechanical parts, *High Technology Letter*, **5**, 74-80.

[15.15] ISO, 1994, *Industrial automation systems and integration: Product data representation and exchange: Part 11: Description methods: The EXPRESS language reference manual, Reference number: ISO 10303-11:1994(E)*, First edition, Switzerland.

[15.16] Kahn, H., Filer, N., Williams, A. and Whitaker, N., 2001, A generic framework for transforming EXPRESS information models, *Computer-Aided Design*, **33**, 501-510.

[15.17] ISO, 2000, *Industrial automation systems and integration: Product data representation and exchange: Part 41: Integrated generic resource: Fundamentals of product description and support, Reference number: ISO 10303-41:2000(E)*, Second edition, Switzerland.

[15.18] ISO, 1998, *Industrial automation systems and integration: Product data representation and exchange: Part 45: Integrated generic resource: Materials, Reference number: ISO 10303-45:1998(E)*, Second edition, Switzerland.

[15.19] ISO, 1994, *Industrial automation systems and integration: Product data representation and exchange: Part 203: Application protocol: Configuration controlled 3D designs of mechanical parts and assemblies, Reference number: ISO 10303-203:1994(E)*, First edition, Switzerland.

[15.20] ISO, 2003, *Industrial automation systems and integration: Product data representation and exchange: Part 28: Implementation methods: XML representations of EXPRESS schemas and data, Reference number: ISO 10303-281:2003(E)*, First edition, Switzerland.

[15.21] Lee, Y. T., 1999, *An overview of information modelling for manufacturing system integration, NISTIR 6382*, National Institute of Standard and Technology, Gaithersburg, MD, USA.

[15.22] Wilson, R. R., 1990, *Information modeling and PDES/STEP, Technical Report 90017*, Rensselaer design Research Center, Pensselaer Polytechnic Institute, Troy, New York.

[15.23] Xie, S. Q., Tu, Y. L., Aitchison, D., Dunlop, R. and Zhou, Z. D. , "A WWW based Product Development Platform for Intelligent and Concurrent Sheet Metal Products Design and Manufacturing", *International Journal of Production Research*, Vol. 39, Number 6, 3829 – 3852, 2001.

[15.24] Xu, X, Wang, H., Mao, J., Newman, S. T., Kramer, T. R., Proctor, F. M. and Michaloski, J. L., "STEP-compliant NC research: the search for intelligent CAD/CAPP/CAM/CNC integration", *International Journal of Production Research*, Vol. 43, No. 17, 2005, 3703–3743.

[15.25] Jasnoch, U. and Haas, S., Collaborative environment based on distributed object oriented databases. Comput. In Industry, 1996, 29, 51–61.

[15.26] Loffredo, D., 1998, *Efficient Database Implementation of EXPRESS information models*, PHD Thesis, Rensselaer Polytechnic Institute, Tory, New York.

[15.27] Xie, S. Q. and HUANHG, H., TU, Y. L. , "a WWW-based information Management System for Rapid and Integrated Mould Product Development", *International Journal of Advanced Manufacturing Technology*, Vol.20 No 1, 50-57, 2002.

[15.28] Tu, Y L, Xie, S Q and Kam, J J, 2006, "Rapid one-of-a-kind production", *International Journal of Advanced Manufacturing Technology*, Vol. 29, Number 5 / June, 2006, pp. 499-510.

[15.29] Zha, X. F. and Du, H., 2002, "A PDES/STEP-based Model and System for Concurrent Integrated Design and Assembly Planning", *Computer-Aided Design*, 34, 1087-1110.

[15.30] ISO, 2002, *Industrial automation systems and integration: Product data representation and exchange: Part 204: Application protocol: Mechanical design using boundary representation, Reference number:* ISO 10303-204:2002(E), First Edition, Switzerland.

[15.31] Shaharoun, A.M., Razak, J.A. and Alam, M.R., *A STEP-based geometrical representation as part of product data model of a plastics part.* J. Mater. Proc. Technol., 1998, 76,115–119.

[15.32] ISO, 2003, *Industrial automation systems and integration: Product data representation and exchange: Part 214: Application protocol: Core data for automotive mechanical design processes, Reference number:* ISO 10303-214:2003(E), Second edition, Switzerland.

[15.33] ISO, 1998, *Industrial automation systems and integration: Product data representation and exchange: Part 49: Integrated generic resources: Process structure and properties*, Reference number: ISO 10303-11:1998(E), First edition, Switzerland

16

STEP in the Context of Product Data Management

Vijay Srinivasan

IBM and Columbia University, New York, N.Y., U.S.A.
Email: vasan@us.ibm.com

Abstract
The ISO STEP suite of standards is quite vast and covers many domains. One such domain of interest is Product Data Management (PDM), which deals with product metadata and related business objects along with several engineering and business processes. The schemas that deal with PDM are spread over several Application Protocols of STEP, and have been collected together informally as STEP PDM Schema. In recent times, other standards development organizations, such as the Object Management Group (OMG), have taken up the task of developing standardized services that involve PDM information models and related processes. These services exploit the fast growing web services over the Internet. These services are also built upon the STEP standards, among other things. This chapter describes the PDM domain covered by STEP and how it is influencing other standards such as the OMG PLM Services.

16.1 Introduction

Historically, Product Data Management (PDM) had a humble beginning in the 1970s and the early 1980s as a computer file management system for application software that dealt with computer-aided drafting and design. It soon grew in scope and definition to an extent that the current market size of PDM software systems is almost as large as that of mechanical computer-aided design (MCAD) software systems. In fact, it is predicted that the worldwide PDM market share will exceed that of the worldwide MCAD market share by the year 2010.

So what does a PDM software system do? Leading PDM software products such as TeamCenter, MatrixOne, Enovia, Windchill, and Agile claim a dizzying array of functionalities from deep integration with their CAD systems all the way to enterprise-wide integration of business processes. While the scope of these offerings keeps growing, it is useful to pause and abstract a minimal set of capabilities that can be expected of a modern PDM system today. These include:

- *Vaulting* – secure storage and retrieval of product data
- *Cataloguing and Searching* – indexing and finding relevant product information

- *Design collaboration* – multiple party edits and synchronization, version control
- *Configuration control* – maintaining and tracking different configurations and variants of the product
- *Change management* – requesting, authorizing, and tracking engineering change in products
- *Workflow management* – routing, branching, iterating, exception handling, and executing engineering and business processes
- *Project management* – task allocation and progress monitoring

Suffice it to say that PDM systems of today have come a long way from merely managing CAD files. And they are capturing greater market share.

This trend is reflected in the STEP suite of standards, at least in spirit if not in scale. Initial outpouring from the STEP standards developers catered predominantly to the needs of exchanging two-dimensional drawings (e.g., AP 201 and 202) and three-dimensional geometric models (e.g., AP 203 and AP 214). However, these APs (Application Protocols) also included some functionality for configuration control of the documents and mechanical designs of parts and assemblies. Thus, of necessity over time, standardized schemas for document and product metadata, and product structure were developed and collected under 'STEP PDM Schema' so they can be used by other modules and APs of STEP.

While the ISO STEP development community was busy churning out several APs, other standards development organizations (SDOs) sprang up to meet the demands of industry that was getting networked by the World Wide Web. Some of these SDOs started taking an active interest in Product Lifecycle Management (PLM), thereby encroaching or enlarging the PDM and other domains that had previously been the sovereign territory of STEP. Notable among these SDOs are the Object Management Group (OMG), the Open Applications Group Inc (OAGi), and the Organization for the Advancement of Structured Information Standards (OASIS).

It is useful to reflect upon the roles of the new SDOs in so far as their work is related to that of STEP. In the case of OMG, a new genre of standards specification called PLM Services has been issued; in this specification, the hard work of STEP has been augmented and rendered in a form that is more readily consumable by the Internet and the World Wide Web. OASIS is fostering the evolution of STEP AP 239 into a richer PLCS (Product Life Cycle Support) standard by embracing modern information technology. OAGi has followed a different path; it has deep roots in the Enterprise Resource Planning (ERP) domain and its BODs (Business Object Documents) contain business objects that overlap some of the data models of STEP PDM Schema. These SDOs publish freely downloadable XML (eXtensible Markup Language) schemas of their standardized data models. In addition – and this is quite important – they define standardized queries and freely publish them as WSDL (Web Services Description Language) documents. These outputs make their work very enticing for implementing engineering and business processes across the entire, worldwide enterprise as workflows that employ service-oriented architecture (SOA). In short, these SDOs are meeting the market demand for standardized data models in the form of XML schemas and standardized communication queries in the form of WSDL documents.

In this chapter the author will focus on STEP's PDM-related standards and their relationship to OMG's PLM Services. Section 16.2 explains the contents of product data and product metadata. It sets the stage for a discussion of STEP PDM Schema in Section 16.3. This is followed in Section 16.4 by a description of OMG's PLM Services. Section 16.5 draws attention to a few impending and important developments. Finally, some concluding remarks are made in Section 16.6.

A simple clarification of terminology is in order before we proceed. When PLM is used in this chapter and in external literature, it refers to a superset of PDM. Roughly speaking, a PLM system consists of CAD, CAM, CAE, and PDM as subsystems.

16.2 Product Data and Metadata

To position STEP in the right context of PDM, it is useful to examine the types of data associated with products. Broadly speaking, it is possible to classify them into product data and product metadata as explained in the following two subsections.

16.2.1 Product Data

Seasoned engineers in industry will recognize geometric specifications in the form of two-dimensional projections in a drawing, as in Figure 16.1. It is a prime example of the type of product data that visually conveys detailed geometric design of a part and provides necessary information for manufacturing it. In addition to the nominal geometry, such specifications include critical geometric dimensions and tolerances (GD&T). In fact, it is impossible to obtain even a cost quotation from manufacturing shops in the absence of such GD&T data. Instead of relying on two-dimensional projections, which may be viewed as somewhat archaic in some advanced applications, it is now possible to present similar geometric data in a three-dimensional view as shown in Figure 16.2.

The syntax and semantics of various symbols and annotations found in these figures are defined using prose, graphics, and mathematics in the standards of ASME (American Society of Mechanical Engineers) [16.1, 16.2] and ISO [16.3, 16.4]. Generating and interpreting product data such as those presented in Figures 16.1 and 16.2 need considerable engineering education and training in these standards and in MCAD systems that support them. Thus they remain firmly in the engineering domain. Hence such product data also fall under the category of *engineering objects*.

It is possible to exchange product data found in Figures 16.1 and 16.2 as standardized graphics and image files; in fact, it is the most common practice today. An open standard for PDF (Portable Document Format) has been issued by ISO for exchanging and archiving two-dimensional graphics such as Figure 16.1 [16.5].

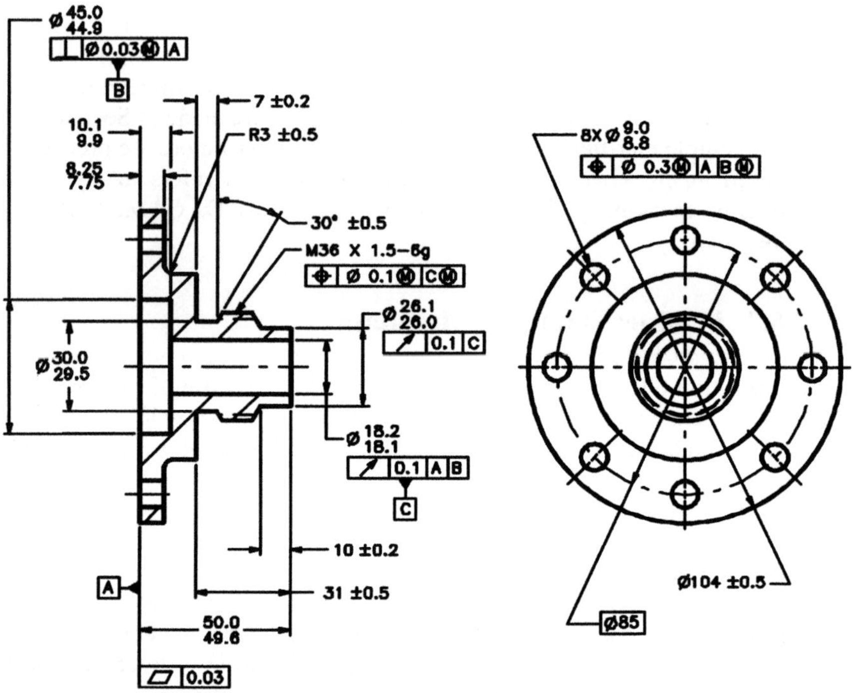

Figure 16.1. Example of an industrial drawing with dimensioning and tolerancing annotations on projected views of a part. (Courtesy Archie Anderson.) All dimensions are in mm

ISO has also issued an open standard based on PDF/E for three-dimensional graphics such as Figure 16.2 so that they can be zoomed, panned, and rotated to provide three-dimensional interactivity; the graphic elements can also be structured to selectively turn some of them on or off [16.6]. The underlying three-dimensional part geometry can be sectioned and selected distances queried (measured). While these files convey the data syntax faithfully in the form of symbols and stylized indications, they require a human being (well versed in [16.1–16.4]) at the receiving end to interpret the semantics of these technical specifications.

To alleviate this problem, ISO STEP has developed semantic standards for three-dimensional nominal geometry and GD&T. For more information on these product data technologies and related standardization, the reader is referred to articles [16.7, 16.8] in recent special issues of the Computer-Aided Design journal that are devoted to them.

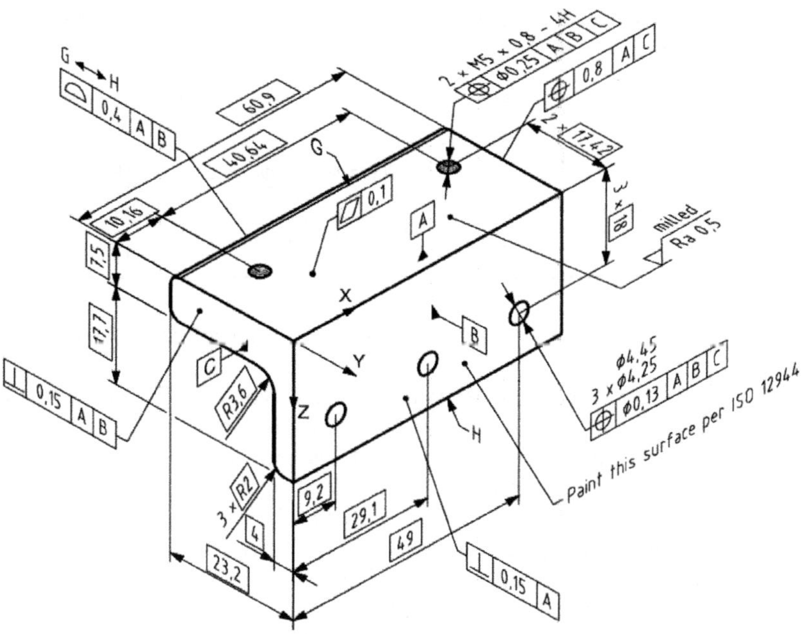

Figure 16.2. Example of standardized three-dimensional presentation of GD&T on a part

STEP and its predecessor IGES (Initial Graphics Exchange Specification) can be used as open standard formats to exchange three-dimensional geometric product data among different PLM systems. In actual practice, people use several combinations of open and proprietary data formats in product data exchange across a partner network. What we can surmise with confidence is that people will continue to exchange files that contain detailed product data and other technical information in a combination of open standard and proprietary data formats, leaving the task of interpreting their content to the receiving systems or human beings as appropriate. These files and 'technical data packages' can then be referred to by the product metadata, to which we now turn.

16.2.2 Product Metadata

Metadata is data about data. In our context, we are interested in product metadata. Historically, such metadata evolved from engineering drafting practices. The venerable 'title block' in an engineering drawing initially held much of the metadata about that drawing document and the part/product represented by that drawing.

Just as we saw in the case of product data in Section 16.2.1, it is instructive to go back to our engineering roots and examine what types of metadata are captured in a 'title block' of a drawing and related 'document headers.' A title block is located at the right-hand bottom corner of an engineering drawing. If the graphics in the drawing are two-dimensional projected views – such as those shown in Figure 16.1 – the title block will contain information about the scale, the style of projection (first-angle or third-angle projection), and default tolerances associated with the graphics. In addition, it will contain some information such as shown in Figure 16.3.

<table>
<tr><td>Responsible dept.
ABC 2</td><td>Technical reference
Patricia Johnson</td><td>Created by
Jane Smith</td><td colspan="3">Approved by
David Brown</td><td></td></tr>
<tr><td colspan="2" rowspan="3">Legal owner</td><td>Document type
Sub-assembly drawing</td><td colspan="4">Document status
Released</td></tr>
<tr><td rowspan="2">Title, Supplementary title
Apparatus plate
Complete with brackets</td><td colspan="4">AB123 456-7</td></tr>
<tr><td>Rev.
A</td><td>Date of issue
2002-05-14</td><td>Lang.
en</td><td>Sheet
1/5</td></tr>
</table>

Figure 16.3. Example of a title block per ISO 7200 [16.9]

ISO 7200 provides a classification of the type of information that one may find in the title blocks and headers of technical product documents. These are some of the basic, minimal metadata of interest to us. Table 16.1 summarizes the standardized classification of this metadata per the latest ISO 7200. This standard was produced by ISO TC 10/SC 1 to facilitate the exchange of documents and ensure uniform interpretation by defining field names, their contents, and their (character) lengths. Some of the fields are mandatory and others are optional. Detailed definitions of the filed names can be found in ISO 7200 [16.9].

A few general observations about ISO 7200 are in order. First, it takes a document-centric approach; incidentally, many industries still follow this approach to manage their product data, either partially or fully. However, several of the fields defined by this standard can also be used in a product-centric approach in a PDM system, and the rest can be used for documents associated with the product. As we will see in Section 16.3, STEP PDM Schema treats a product either as a *part* or as a *document*. Second, the metadata about the documents and products can be used by the entire business enterprise and not be restricted to engineering design departments. In other words, such metadata do not require formal engineering education and training. Thus in this standard we see the emergence of elements of *business objects*. Finally, there are other standards, such as those from the ASME, that overlap and extend the product and document metadata. Some of these standards are:

- ASME Y14.24 Types and Applications of Engineering Drawings [16.10]
- ASME Y14.34 Associated Lists [16.11]
- ASME Y14.35 Revision of Engineering Drawings and Associated Documents [16.12]
- ASME Y14.42 Digital Approval Systems [16.13]
- ASME Y14.100 Engineering Drawing Practices [16.14]

These standards represent decades of experience and best practices from industry. Taken together, these ISO and ASME standards form a very rich source of standardized metadata. In addition, they provide elements of standardized engineering processes (such as revision control, change management, approval processes) related to product data and metadata. In an interesting development over the past decade, most of these engineering processes are also treated as business processes because of their impact on overall business performance.

Table 16.1. Classification of basic technical document metadata per ISO 7200 [16.9]

Class	Field name	Obligation
Identifying data	Legal owner	Mandatory
	Identification number	Mandatory
	Revision index	Optional
	Date of issue	Mandatory
	Segment/sheet number	Mandatory
	Number of segments/sheets	Optional
	Language code	Optional
Descriptive data	Title	Mandatory
	Supplementary title	Optional
Administrative data	Responsible department	Optional
	Technical reference	Optional
	Approval person	Mandatory
	Creator	Mandatory
	Document type	Mandatory
	Classification/key words	Optional
	Document status	Optional
	Page number	Optional
	Number of pages	Optional
	Paper size	Optional

Thus far, the author has focused on various open standards other than STEP for product and related document metadata. So what does STEP say about metadata? The answer can be gleaned from how some of the conformance classes (CC) are defined in STEP AP 214, as shown in Table 16.2. While the ISO 7200 and the ASME Y14 series of standards mentioned earlier describe standardized product metadata in prose and figures, STEP rigorously defines the standardized product data and metadata models using EXPRESS language. In Table 16.2 we see that CC 1 and CC 2 cover product geometric data models, which we can also relate to engineering objects. CC 6 and CC 8 cover product metadata models that have evolved into business objects. It is useful to note here that product metadata are only a subset of business objects, as business objects in industry cover many more types of information.

Separation of product metadata from product data is quite convenient when it comes to industrial implementation. For example, it is possible to handle product metadata using the STEP standards while keeping the data files that metadata 'point to' either in standardized format or in proprietary format. Such flexibility allows asynchronous development and adoption of new technologies as appropriate for business enterprises. In fact, several new PDM software vendors (who did not have their own CAD systems) sprang up to support multiple CAD systems exploiting the separation between product data and metadata, and proved the efficacy of their business approach by capturing a good portion of PDM market share.

Table 16.2. Standardized product data and metadata models in ISO STEP AP 214

	Class	Description	Remarks
Data (Engineering objects)	CC 1	Component design with 3D shape representation.	Covers 3D geometry of single parts, including wire-frame, surface, and solid models.
	CC 2	Assembly design with 3D shape representation.	Covers 3D geometry of assemblies of parts, including the assembly and model structure.
Metadata (Business objects)	CC 6	Product data management (PDM) without shape representation.	Covers product data management systems that manage geometric models as files. It also covers administrative data of parts, assemblies, documents, and models.
	CC 8	Configuration controlled design without shape representation.	Covers CC 6, with additional requirements for product configuration control.

STEP AP 214 does not define designations such as data and metadata, and their associations with engineering and business objects, respectively. These designations are presented here to provide an entrée to PDM Schema covered in the next section.

16.3 STEP PDM Schema

STEP AP 214 is not the only application protocol of STEP that deals with product metadata models. Figure 16.4 illustrates an intersection of four different STEP APs that contain a common, core 'PDM Schema.' Here AP 203 deals with 'configuration controlled 3D design of mechanical parts and assemblies,' AP 212 with 'electrotechnical design and installation,' AP 214 with 'core data for automotive mechanical design processes,' and AP 232 with 'technical data packaging core information and exchange' [16.15]. The STEP PDM Schema as a common subset is maintained by two prominent standards consortia, PDES Inc [16.16] and ProSTEP iViP [16.17].

The STEP PDM Schema [16.18] contains some of the most important standardized product metadata models that came out of the ISO STEP efforts, and so it deserves a closer look. In the STEP PDM Schema, a general product can be conceptually interpreted either as a *part* or as a *document*. In this way, parts and documents are managed in a consistent and parallel fashion. As we saw in Section

16.2.2, metadata about documents are widely used and they play an important role in the overall product data management.

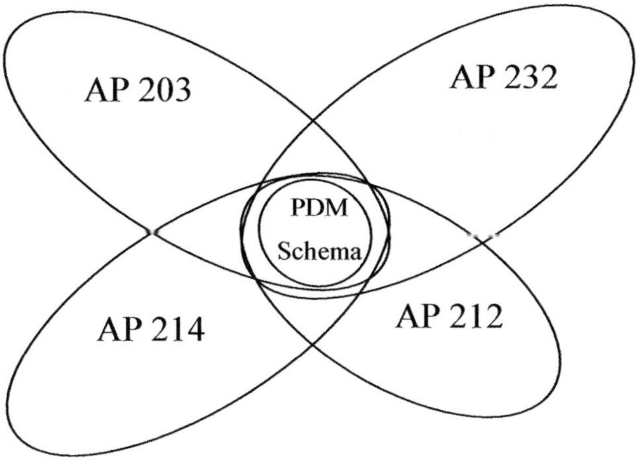

Figure 16.4. PDM Schema culled from the intersection of four STEP APs

The units of functionality for parts and documents covered by the STEP PDM Schema include identification, versioning, product structures (including transformations), approvals, authorization, project, work order, work request, affectivities', classification, and properties. These are further elaborated in Table 16.3.

Two hundred and ten entities are defined to compose these units of functionality in the STEP PDM Schema. Just the names of these entities are listed in Table 16.4. Each of these 210 entities is defined in greater detail using the EXPRESS schema language. A quick perusal of Tables 16.3 and 16.4 will give the reader an appreciation of the richness of the product metadata defined by the STEP PDM Schema.

It is also instructive to compare these entities with those of ISO 7200 and ASME standards referred to in Section 162.2. It can be confidently said that the STEP PLM Schema cover previously standardized 'title block' and 'document header' information, and more. This is the result of a major investment by industry, and has been exploited and reused in other related efforts such as the OMG PLM Services that will be described in Section 16.4.

The use of STEP PDM Schema can be gleaned from a small example, illustrated in Example 16.1, of a part master file with context and type classification [16.18]. The purpose behind the Example 16.1 is not to describe the details of the standardized product metadata, but to illustrate the style and content of a typical product metadata as defined by the STEP PDM Schema. This example is a schema instance, written in EXPRESS Part 21 language. Only about a dozen of the 210 entities (such as PRODUCT, PERSON, and ORGANIZATION) are used in this illustrative example.

Table 16.3. Units of functionality covered by the STEP PDM Schema

Units of functionality	Content
Part identification	Part as Product Product master identification Context information Type classification
Specific part type classification	Classification of parts and managed documents
Part properties	General part properties External part shape External geometric model structure Relative orientation and location of related geometric models
Part structure and relationships	Explicit assembly Bill of Material Multi-level assembly digital mock-up Different views on assembly structure Relating part shape properties to product structure Other relationships between parts
Document identification	Document as Product Document master identification Context information Type classification
Specific document type classification	Product related product category and product category relationship
External files	External file identification
Relationship between documents and constituent files	Product definition with associated documents
Document and file properties	Product definition or document representation Document content property Document creation property Document size property Document source property Additional document properties Document type classification
Document and file association with product data	Document reference External file reference Constrained document or file reference
Document and file relationships	Sequence relationships between document versions Relationships between document representations Relationships between external files

Alias identification	Associating with additional identifier
Authorization	Organization and persons Approval Dates, times, and event references Security classification Certification
Configuration and effectivity information	Configuration identification Configuration composition management General validity period
Engineering change and work management	Request for work Work order and work definition Project identification Contract identification
Measure and units	Measures with unit specification Unit definition

Table 16.4. Entities defined in the STEP PDM Schema

2	Action	3	cartesian_transformation_operator_3d
	action_assignment		certification
	action_directive		certification_assignment
	action_method		certification_type
	action_property		characterized_object
	action_property_representation		configuration_design
	action_relationship		configuration_effectivity
	action_request_assignment		configuration_item
	action_request_solution		context_dependent_shape_representation
	action_request_status		context_dependent_unit
	action_status		contract
	address		contract_assignment
	alternate_product_relationship		contract_type
	amount_of_substance_measure_with_unit		conversion_based_unit
	amount_of_substance_unit		coordinated_universal_time_offset
	application_context		date
	application_context_element		date_and_time
	application_context_relationship		date_and_time_assignment
	application_protocol_definition		date_assignment
	applied_action_assignment		date_role
	applied_action_request_assignment		date_time_role
	applied_approval_assignment		dated_effectivity
	applied_certification_assignment		derived_unit
	applied_contract_assignment		derived_unit_element
	applied_date_and_time_assignment		description_attribute
	applied_date_assignment		descriptive_representation_item

applied_document_reference	dimensional_exponents
applied_document_usage_constraint_assignment	directed_action
applied_effectivity_assignment	direction
applied_event_occurrence_assignment	document
applied_external_identification_assignment	document_file
applied_identification_assignment	document_product_association
applied_organization_assignment	document_product_equivalence
applied_organizational_project_assignment	document_reference
applied_person_and_organization_assignment	document_relationship
applied_security_classification_assignment	document_representation_type
approval	document_type
approval_assignment	document_usage_constraint
approval_date_time	document_usage_constraint_assignment
approval_person_organization	document_usage_role
approval_relationship	effectivity
approval_role	effectivity_assignment
approval_status	effectivity_relationship
area_measure_with_unit	electric_current_measure_with_unit
area_unit	electric_current_unit
assembly_component_usage	event_occurrence
assembly_component_usage_substitute	event_occurrence_assignment
axis2_placement_2d	event_occurrence_role
axis2_placement_3d	executed_action
calendar_date	external_identification_assignment
cartesian_point	external_source
cartesian_transformation_operator	founded_item
cartesian_transformation_operator_2d	functionally_defined_transformation

Table 16.4. Entities defined in the STEP PDM Schema (continued)

4	general_property	**5**	product_definition_context_association
	general_property_association		product_definition_context_role
	general_property_relationship		product_definition_effectivity
	geometric_representation_context		product_definition_formation
	geometric_representation_item		product_definition_formation_relationship
	global_uncertainty_assigned_context		product_definition_formation_with_specified_source
	global_unit_assigned_context		product_definition_relationship
	id_attribute		product_definition_shape
	identification_assignment		product_definition_usage
	identification_role		product_definition_with_associated_documents
	item_defined_transformation		product_related_product_category
	length_measure_with_unit		promissory_usage_occurrence

length_unit
local_time
lot_effectivity
luminous_intensity_measure_with_unit
luminous_intensity_unit
make_from_usage_option
mapped_item
mass_measure_with_unit
mass_unit
measure_representation_item
measure_with_unit
name_attribute
named_unit
next_assembly_usage_occurrence
object_role
organization
organization_assignment
organization_relationship
organization_role
organizational_address
organizational_project
organizational_project_assignment
organizational_project_relationship
organizational_project_role
person
person_and_organization
person_and_organization_assignment
person_and_organization_role
personal_address
placement
plane_angle_measure_with_unit
plane_angle_unit
point
product
product_category
product_category_relationship
product_concept
product_concept_context
product_context
product_definition
product_definition_context
property_definition
property_definition_representation
quantified_assembly_component_usage
ratio_measure_with_unit
ratio_unit
relative_event_occurrence
representation
representation_context
representation_item
representation_map
representation_relationship
representation_relationship_with_transformation
role_association
security_classification
security_classification_assignment
security_classification_level
serial_numbered_effectivity
shape_aspect
shape_aspect_relationship
shape_definition_representation
shape_representation
shape_representation_relationship
SI_unit
solid_angle_measure_with_unit
solid_angle_unit
specified_higher_usage_occurrence
thermodynamic_temperature_measure_with_unit
thermodynamic_temperature_unit
time_interval
time_interval_based_effectivity
time_interval_with_bounds
time_measure_with_unit
time_unit
uncertainty_measure_with_unit
value_representation_item
vector
versioned_action_request
volume_measure_with_unit
volume_unit

```
ISO-10303-21;HEADER;
FILE_DESCRIPTION(('test', 'file'), '2;1');
FILE_NAME('pid2_p21a.stp', '1999-05-03T21:03:29+00:00', ('N.N.'), (''), '', '', '');
FILE_SCHEMA(('PDM_SCHEMA {1.2}'));
ENDSEC;
DATA;

/* part #1 */
#10 = PRODUCT('K01-42051', 'Bicycle Bell RX 3', $, (#20));

/* part context */
#20 = PRODUCT_CONTEXT('', #30, '');
#30 = APPLICATION_CONTEXT('');
#40 = APPLICATION_PROTOCOL_DEFINITION('version 1.2', 'pdm_schema', 2000, #30);

/* part versions for part #1 */
#50 = PRODUCT_DEFINITION_FORMATION('02', 'lever modified', #10);
#60 = PRODUCT_DEFINITION_FORMATION('03', 'upper housing modified', #10);
#70 = PRODUCT_DEFINITION_FORMATION_RELATIONSHIP('', 'sequence', $, #50, #60);

/* definition of view on version 03 of part #1 */
/* primary life_cycle_stage = design, primary application_domain = mechanical design */
#80 = PRODUCT_DEFINITION('/NULL', $, #60, #90);
#90 = PRODUCT_DEFINITION_CONTEXT('part definition', #100, 'design');
#100 = APPLICATION_CONTEXT('mechanical design');

/* association of the id owner for part #1 */
#130 = APPLIED_ORGANIZATION_ASSIGNMENT(#140, #150, (#10, #160, #170));
#150 = ORGANIZATION_ROLE('id owner');

/* information on person and organization */
#140 = ORGANIZATION('ABC27166', 'Onyx AG', 'location');
#540 = PERSON('sarah.rijker@onyx.com', 'Rijker', 'Sarah', $, $, $);
#550 = PERSON_AND_ORGANIZATION(#540, #140);

/* part #2 and part #3 */
#160 = PRODUCT('H24-1123.1', 'Fixture RX25B', '', (#20));
#170 = PRODUCT('DIN 932', 'Screw M3x15', '', (#20));

/* part versions for part #2 and part #3 */
#180 = PRODUCT_DEFINITION_FORMATION('B', 'larger screw holes', #160);
#190 = PRODUCT_DEFINITION_FORMATION('15', '', #170);

/* view definition for version of part #2 */
#200 = PRODUCT_DEFINITION('/NULL', $, #180, #210);
#210 = PRODUCT_DEFINITION_CONTEXT('part definition', #215, 'design');
#215 = APPLICATION_CONTEXT('mechanical design');

/* view definition for version of part #3 */
#220 = PRODUCT_DEFINITION('/NULL', $, #190, #230);
#230 = PRODUCT_DEFINITION_CONTEXT('part definition', #240, 'design');
#240 = APPLICATION_CONTEXT('mechanical design');

/* part discriminator for parts #1 - #3 */
#250 = PRODUCT_RELATED_PRODUCT_CATEGORY('part', $, (#10, #160, #170));
#260 = PRODUCT_CATEGORY_RELATIONSHIP('', $, #250, #270);
#270 = PRODUCT_RELATED_PRODUCT_CATEGORY('detail', $, (#160));
#280 = PRODUCT_CATEGORY_RELATIONSHIP('', $, #250, #290);
#290 = PRODUCT_RELATED_PRODUCT_CATEGORY('assembly', $, (#10));
#300 = PRODUCT_CATEGORY_RELATIONSHIP('', $, #250, #310);
#310 = PRODUCT_RELATED_PRODUCT_CATEGORY('standard', $, (#170));

ENDSEC;
END-ISO-10303-21;
```

Example 16.1. A product metadata instance [16.18] in EXPRESS Part 21 language

Table 16.5 shows the attributes of just one such entity called PRODUCT. In practice, an actual schema instance can contain a lot more entities and cover much more metadata. Details of the attributes for each of the 210 entities in Table 16.4 can be found in [16.18].

Table 16.5. Attributes of the PRODUCT entity in the STEP PDM Schema

ENTITY product	Attribute population	Remarks
id	type: identifier = string product number identification	The *id* attribute stores the unique base identifier for a product, the product number
name	type: label = string	The *name* attribute stores the nomenclature, or common name, of the product
description	type: text = string	The *description* attribute should contain an expanded name or description of the product
frame_of_reference	type: entity = product_context	Context information is stored in the *frame_of_reference* attribute. All products must be found in some product_context

The EXPRESS Part 21 syntax, although effective to the task at hand, lacks extensibility, can be hard for humans to read, and – perhaps most limiting – is computer interpreted by a relatively small number of software development toolkits that support STEP. It won't also surprise us if a reader familiar with recent Internet technologies finds the code snippet in Example 16.1 somewhat archaic. A better approach, the reader might argue, would be to capture this type of data in the more ubiquitous XML.

In order to capitalize on XML's popularity and flexibility and to accelerate STEP's adoption and deployment, ISO has developed the standard ISO 10303-28 for representing EXPRESS schemas and instance populations in XML [16.19]. This will not only enable developers to use low-cost, ubiquitous XML software tools to implement STEP definitions, but will also facilitate the use of STEP standards in XML-based web services described later in this chapter.

The ISO STEP community has developed another standard, ISO 10303-25, for transforming EXPRESS schemas into UML models [16.20]. This will enable developers to use familiar UML tools to see the contents of STEP EXPRESS schemas, and to specify relationships between STEP data models and other UML models they use. This is especially useful for the so-called 'model-based' approaches to developing standards and architecting service-oriented applications.

Although the XML and UML representation of EXPRESS schemas are theoretically applicable to both product data and product metadata, it is in the context of product metadata they have shown the greatest promise. To exploit recent advances in the Internet-based standards, protocols, and middleware, companies prefer to encode the product metadata in XML schema instances and implement the workflow using web services. These, they believe, allow them to achieve the desired level of granularity in their integration of engineering and business information systems, without ripping and replacing part or all of the application software they use. A response to this market need is the OMG PLM Services is described in the next section.

16.4 OMG PLM Services

OMG PLM Services is a specification built on two levels of architecture [16.21]. The first is a general *model-driven architecture* applicable to all distributed software system development. The second is an architecture for a *PLM Services* standard – it itself is built on other standards. The author will describe each of these in the following subsections.

16.4.1 OMG's Model Driven Architecture

One of the earliest efforts to provide standardized services to share information over a computer network was spearheaded by the OMG even before the Internet became a commercial success. Starting in the early 1990s, OMG standardized object request broker (ORB) and a suite of object services. This work was guided by its Object Management Architecture (OMA), which provided a framework for distributed software systems. It also delivered the Common Object Request Broker Architecture (CORBA) as a part of that framework.

The OMA framework identifies types of components that are combined to make up a distributed system and, together with CORBA, specifies types of connectors and the rules for their use. CORBA was an architecture designed to enable software components written in multiple computer languages and running on multiple computers on different hardware platforms to interoperate. In fact, hard lessons learned in the design and implementation of CORBA provided valuable input to the evolution of successors to this architecture.

In 2001, OMG adopted a second framework called the Model Driven Architecture (MDA) [16.22]. It is an approach to using models in various levels of abstraction in software development. MDA has three important viewpoints and associated models:

A *computation independent* viewpoint focuses on the environment of a software system, and the requirements for the system. The details of the structure and processing of the system are hidden or as yet undetermined. A *computation independent model* (CIM) is a view of a system from this computation independent viewpoint. A CIM does not show details of the structure of software systems. A CIM is sometimes called a domain model. A vocabulary that is familiar to the practitioners of the domain in question is used in its specification.

A *platform independent* viewpoint focuses on the operation of a software system while hiding the details necessary for a particular implementation platform. A platform independent view shows that part of the complete specification that does not change from one implementation platform to another. A platform independent view may use a general purpose modelling language, or a language specific to the area in which the system will be used. A *platform independent model* (PIM) is a view of a system from the platform independent viewpoint. A PIM exhibits a specified degree of platform independence so as to be suitable for use with a number of different platforms of similar type.

A *platform specific* viewpoint combines the platform independent viewpoint with an additional focus on the details of the use of a specific platform by a software system. A *platform specific model* (PSM) is a view of a system from the implementation platform specific viewpoint. A PSM combines the specifications in the PIM with the details that specify how that system uses a particular type of platform.

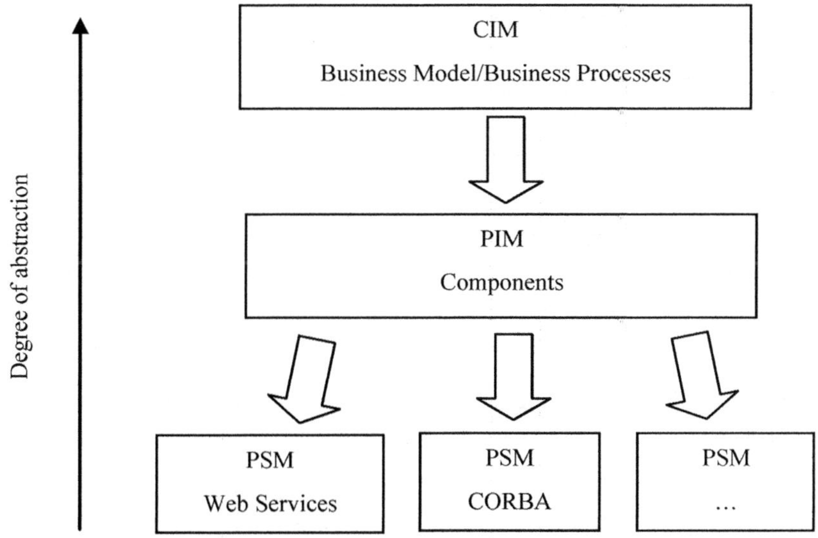

Figure 16.5. OMG's Model Driven Architecture

In the model hierarchy described above, one model is transformed into another by a model transformation. For example, a platform independent model and other relevant information can be combined by the transformation to produce a platform specific model. Figure 16.5 illustrates the three models of OMG's MDA and their transformation from more abstract models to more concrete models. Note that MDA is applicable to software development in several application domains, such as manufacturing, finance, telecom, and healthcare. In the manufacturing industry domain, this standardized, hierarchical model driven architecture guided the development of a standardized PLM Services described in the next section.

16.4.2 OMG PLM Services Architecture

PLM Services is an OMG standard specification, developed by an industrial consortium under the umbrella of the ProSTEP iViP Association [16.21, 16.23].

OMG has just issued PLM Services 2.0 specification, which is a superset of its earlier PLM Services 1.0 specification; PLM Services 2.0 also supersedes PLM Services 1.0. In the rest of the chapter, whenever the author refers to PLM Services it means PLM Services 2.0.

OMG PLM Services specification is architected so that it is based on other standards, as shown in Figure 16.6. Several of the base standards at its bottom layer, such as STEP AP 214 and STEP PDM Schema, were already described in Section 16.3. These are ISO standardized metadata models of industrial products. We also see PDM Enablers and PDTnet as other standards and specifications in this bottom layer; the reader might be less familiar with these. So the author will now describe them briefly.

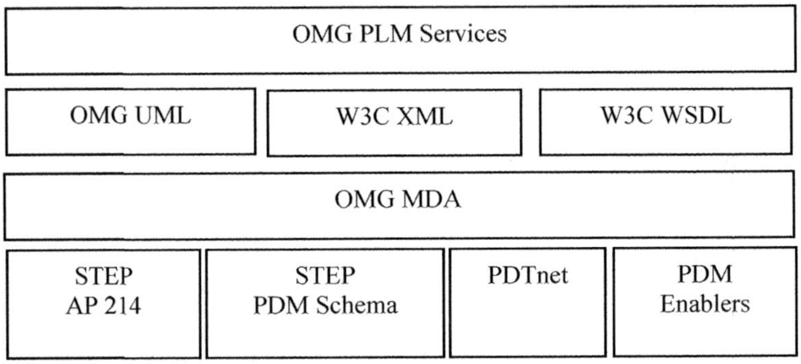

Figure 16.6. OMG PLM Services is architected as a standard specification based on other standards

PDM Enablers [16.24] is an OMG standard specification intended to provide standardized interfaces to PDM systems, or other software systems providing similar services, from other manufacturing software systems. It is quite extensive in its scope and contents. The PDM Enablers specification is organized as 12 interface description language (IDL) modules, as shown in Table 16.6. These modules also provide various UML diagrams and CORBA interfaces. In addition, the PDM Enablers specification provides mapping of product development processes (broken down into 11 sub-processes) to these modules. The 11 sub-processes are outlined in Table 16.7. A quick perusal of Tables 16.6 and 16.7 will reveal the amount of work that has gone into the OMG standardization of product metadata and product development process elements. The ISO STEP PDM Schema standards and the OMG PDM Enablers are different in purpose, scope, abstraction level, and operational characteristics. However, they have common and complementary components that could be used for a PLM Services standard.

Table 16.6. Modules in OMG PDM Enablers

	Module	Remarks
1	PdmResponsibility	Defines minimum interfaces for references to people and organizations as owners, contacts or some other responsibility
2	PdmFoundation	Defines basic foundation classes for units of behaviour such as manageable, lockable, revisionable, iteratable, statable, and security classifiable
3	PdmFramework	Provides PDM-specific object and relationship classes. It collects basic units of behaviour into higher level PDM-specific abstract objects such as ItemMaster, ItemRevision, and ItemIteration
4	PdmBaseline	Describes a collection of items and their relationships to enable their configuration to be reconstructed and audited
5	PdmViews	Provides framework and explicit classes for indicating what items and relationships apply or qualify in particular use cases
6	PdmDocumentManagement	Describes a conceptual model for document management such as check-in and check-out
7	PdmProductStructureDefinition	Defines parts and the bill of material relationships between items for discrete manufacturing
8	PdmEffectivity	Provides a range in which an item revision, component, or other qualified item should be used in production
9	PdmChangeManagement	Provides classes for engineering change issue, engineering change request, engineering change order, engineering change notice
10	PdmManufacturingImplementation	Describes objects for manufacturing process specification and relationships between process specification, items produced, manufacturing location, and engineering change orders
11	PdmConfigurationManagement	Extends the PdmProductStructureDefinition. Provides the ability to define the features or specifications that distinguish the different configurations and the ability to define rules to select specific components based on the feature values
12	PdmSTEP	Provides interface to STEP import and export capabilities to read and write data conformant to specific APs such as AP 203 and AP 214

Table 16.7. OMG PDM Enablers' eleven sub-processes to support product development.

	Sub-processes	Description
1	Develop strategic product plan	Determine product areas for business, market decisions, design and manufacturing responsibilities; project cost and volumes
2	Develop product business plan	Determine approach to product design to meet the cost and volume estimates; create development plan with schedules and estimates
3	Develop product definition	Develop product general specifications, conceptual layout, mock-ups and visuals; identify relationship to existing parts and manufacturing capabilities; establish development plans and objectives; estimate costs of major components; manage existing parts/modules library
4	Define product marketing configuration and rules	Define features and options; project sales volumes and cost by options
5	Develop product design	Develop major component general specifications, major component layout; retrieve, incorporate and establish reusable parts and major components; establish design plan and record status; define module/sub-assembly content with optional components; develop part design; select preferred features
6	Develop process design and procurement agreements	Develop assembly process design and manufacturing facility plan; assess manufacturing capabilities; determine part sourcing; procure machine tools, firm tools and services; develop production tooling design, part fabrication process design, and procurement contracts
7	Coordinate design change	Identify design and process changes; process external design changes; identify implantation dependencies; notify required team members and design changes across all products
8	Evaluate product design	Develop model/prototype of proposed design, and test plans and prototype configurations; run tests and report results; analyze test results
9	Implement production changes	Plan for future manufacturing and product change implementation; generate material requirements for prototype; alert change and restrict inventory/investment; build prototype using production facilities; distribute prototype build documentation; schedule tool tryout; establish change implementation schedule; generate material requirements for pilots; notify and distribute updated manufacturing documentation
10	Develop product service methods	Develop service assembly procedures and diagnostic procedures
11	Develop service distribution plan	Create part catalogs and service inventory plans

Table 16.8. Predefined queries of PDTnet

	Query	Description
1	AliasIdentificationQuery	Returns alias information for a selected item
2	AlternativeSolutionQuery	Returns information about the alternative solutions of a product component
3	ApprovalQuery	Returns approval information for a selected element
4	ApprovalRelationshipQuery	Returns specified relationships of a selected element approval
5	AssemblyStructureQuery	Returns the assembly structure for a selected item
6	AssociatedDocumentQuery	Returns all documents associated with a selected item
7	AssociatedOrganisationQuery	Returns all organizations associated with a selected item
8	ConnfigurationQuery	Returns configuration information for a selected item
9	CreateProductStructure	Creates a product structure including its metadata under a selected target element
10	DateAndTimeQuery	Returns date and time information for a selected item or document
11	DocumentClassificationQuery	Returns the specific document classification for a selected document
12	DocumentPropertyQuery	Returns specified properties of a selected document
13	DocumentStructureQuery	Returns the subdocuments for a selected document
14	DownloadDigitalFile	Returns an URL. Sending a HTTP request using this URL will download the file
15	DownloadSetOfDigitalFiles	Gets a set of files as a package
16	EffectivityAssignmentQuery	Returns effectivity information for a selected element
17	ExternalFileQuery	Returns the external files referred by a selected document representation
18	GetProductStructure	Returns the assembly structure of a selected item and all its metadata
19	DocumentQuery	Returns document information for a selected item
20	IntiateOfflineUpload	Sends selected files to a target system by an offline data transfer
21	ItemClassificationQuery	Return the classification for a selected item
22	ItemPropertyQuery	Returns the property values for a selected element
23	ItemQuery	Returns a selected item with specified content
24	ItemRelationshipQuery	Returns specified relationships of a selected item
25	ItemWhereUsed	Returns all parent nodes of a selected item
26	OrganizationQuery	Returns organization information for a selected item
27	ProductStructureQuery	Returns product structure
28	SearchInDesignSpace	Returns a list of times that are positioned within a specified design space of a selected item
29	StartNodeQuery	Returns a list of items that can be defined by their part number, part name, and version number
30	SubscriptionListAcknowledgeContent	Acknowledges changes on elements of the subscription list
31	SubscriptionListAddContent	Adds new elements to the subscription list
32	SubscriptionListGetContent	Returns the current content and state of the subscription list
33	SubscriptionListRemovalContent	Removes elements from the subscription list
34	UploadDigitalFile	Uploads a file to a server
35	UploadSetOfDigitalFiles	Upload a set of files to a server

We will now turn to a brief description of PDTnet [16.25], also found at the bottom layer of Figure 16.6. It was a project undertaken by an industrial consortium under the umbrella of the ProSTEP iViP Association during the year 2000 to enable neutral, system-independent product data communication between automobile manufactures and suppliers based on the ISO STEP AP 214 standards and the emerging XML technologies. Three years later, the lessons learned from this project in using STEP product information standards and Internet technologies were documented and used as an input for the OMG PLM Services development. Of particular importance are the PDTnet queries. Table 16.8 lists some of these predefined queries.

By now, the reader has been exposed to various standards and specifications that contributed to the OMG PLM Services. It will not be surprising if they seem somewhat confusing. A visual timeline shown in Figure 16.7 is very useful to clarify the picture and understand their evolution. These developments contributed to the overall architecture of OMG PLM Services as shown in Figure 16.8. Recall that the OMG's MDA (Model Driven Architecture) demands a strict separation between PIM (Platform Independent Model) and PSM (Platform Specific Model). In some sense, the PIM is more important because it can be used to derive a particular PSM if the underlying platform for implementation changes. The informational PIM (shown in Figure 16.8) in OMG PLM Services currently supports the following 27 use cases, modelled using UML:

- Export assembly data
- Import assembly data
- Authentication/Start-Up of session
- Authorization
- Start node identification
- Browsing down product structure data
- Browsing up product structure data
- Download product data
- Download metadata including structures
- Download a single digital file
- Generic object query
- Search in design space
- Upload product data
- Upload a single digital file (simple user interaction)
- Upload metadata including structures
- Change notification
- Display content of subscription list and confirm changes
- Change content subscription list
- Product class identification
- Browsing abstract product structures
- Browsing alternative solutions within an abstract product structure
- Retrieve configuration data within an abstract product structure
- Viewing change management information
- ECM (Engineering Change Management) participant proposal for a change
- ECM participant comments and approval
- ECM participant detailing and comments

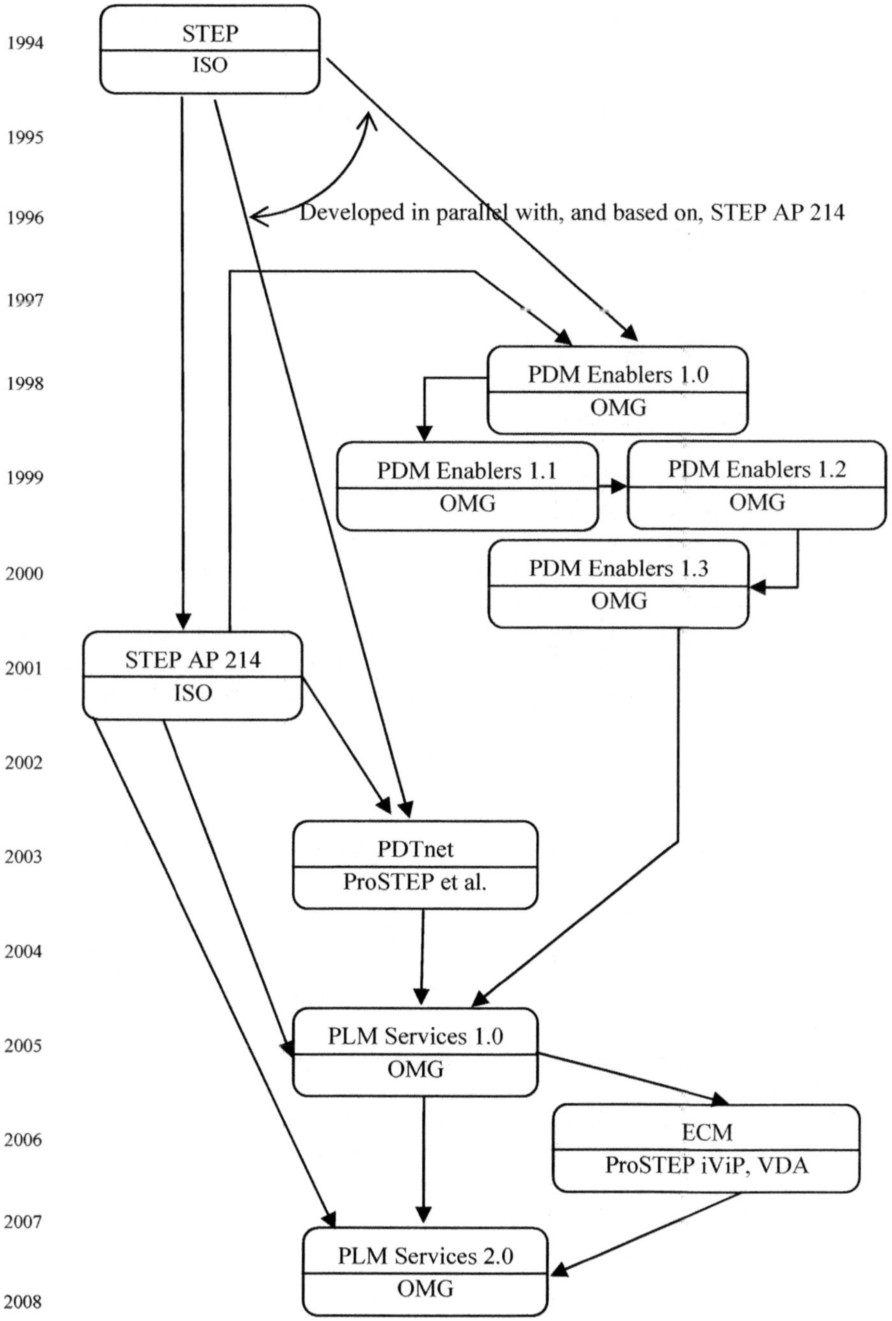

Figure 16.7. A timeline of development of OMG PLM Services and related standards

These use cases are classified as 'informative' in the specification, because they don't dictate the precise use of the 'formative' components of the standard. Rather they provide the empirical evidence and support for the necessity of the 'formative' information models. Thus complementing these use cases, the 'formative' informational PIM supports the following 13 product information models as packages:

- PLM base
- Part identification and structure
- Document and file management
- Shape definition and transformation
- Classification
- Properties
- Alias identification
- Authorization
- Configuration management
- Change and work management
- Process planning and multi-language support

While the 13 packages of information models and the 27 use cases focus on the product metadata and how they can be used, the 'formative' computational PIM (also pictured in Figure 16.8) takes a computational viewpoint. It provides the necessary computational functionality to create, read, update, and delete (usually referred to as CRUD functions in software engineering) instances of the data schemas defined in the information models. It defines mechanisms to query and traverse instances of these data and, therefore, it depends on the information viewpoint.

To start with, the computational PIM defines PLM connectors and PLM connection interfaces. Once the connection has been made, several queries defined in the following query conformance points can be invoked:

- *Utilities queries conformance point*: Provides specialization of an abstract base type called PLM_query with the possibility of concatenated, recursive, and batch queries.
- *Generic queries conformance point*: Defines a toolset of classes to query data from a PLM system.*XPath queries conformance point.* Provides the possibility to use arbitrary XPath expressions conforming to the W3C XPath specification as queries.
- *Proxy queries conformance point*: Provides the possibility to avoid redundant data transfers in multiple query results.
- *Specific queries conformance point*: Defines a rich set of low level specialized queries. It starts with defining some common queries classes (Class Specific_query, Class Query_with_relating_type_predicate, and Class Relationship_query) as base types for the rest of the specific queries. Table 16.9 lists the rest of the 70 specific queries that support PLM-related atomic tasks.
- *PDTnet queries conformance point*: Defines a set of specialized queries that fulfil the requirements of the uses cases described above.
- *Message queries conformance point*: Defines queries derived from an abstract class called PLM_message_query.

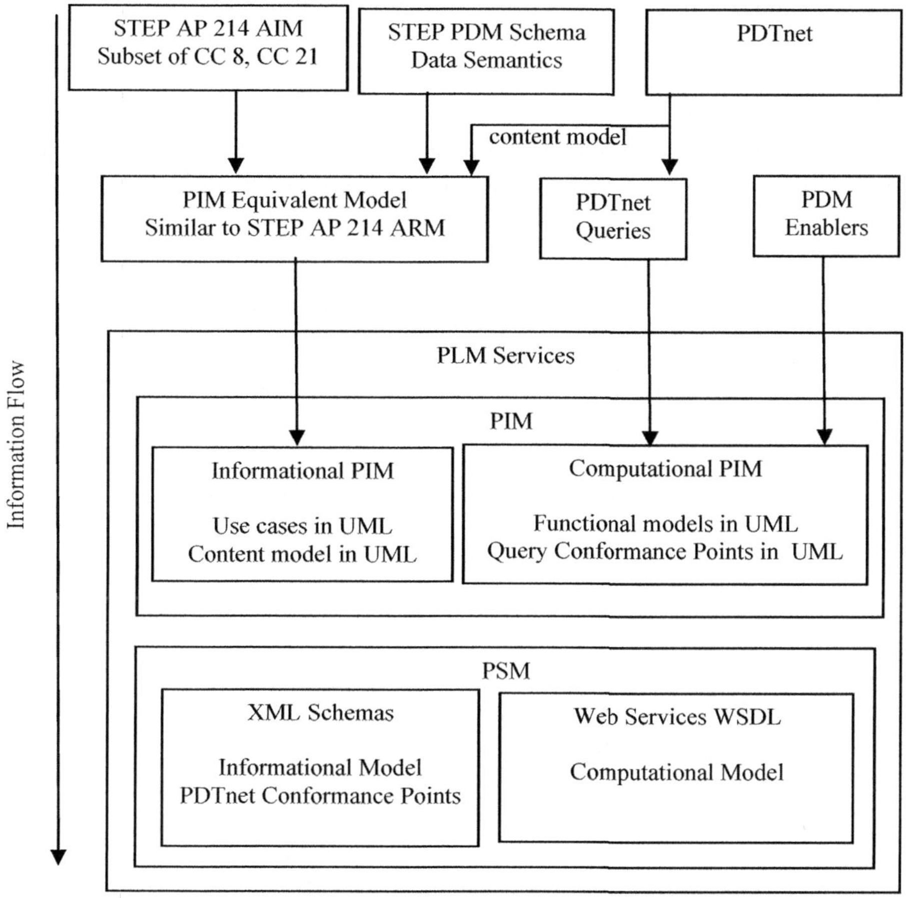

Figure 16.8. Derivation and architecture of OMG PLM Services

It should by now be clear to the reader just how strong influences of ISO STEP metadata models, OMG PDM Enablers, and the PDTnet project on the PIM of OMG PLM Services specifications, as captured in Figure 16.8. In particular, the role of ISO STEP PDM Schema as part of the CIM for the OMG PLM Services is apparent.

The PSM for PLM Services is defined in XML Schema and in WSDL (Web Services Description Language). The informational viewpoint (in XML Schema) and the computational viewpoint (in WSDL) are derived from the corresponding 'formative' PIM specifications described above. After mapping the PIM into PSM, these standardized product metadata models can be used as domain-specific messages in the messaging layer of SOA, and the standardized use cases realized by careful selection of queries can be used in the management and composition layer of SOA to compose higher-level business processes.

One of the major additions in PLM Services 2.0 on the top of PLM Services 1.0 is the support for the engineering change management (ECM) process. Several 'informative' use cases address this process; in addition, in the PSM layer, several 'formative' informational models as XML Schemas and computational models in WSDL support the ECM process.

As we close this section, it is worth noting that the ECM process defined in the OMG PLM Services also appears as part of the latest edition 3 of STEP AP 214. In fact, it forms the new CC 21 in AP 214. It is a testimonial to the close cooperation between the STEP community and the OMG.

Table 16.9. Specific queries of OMG PLM Services that deal with PLM-related atomic tasks

2. Activity_query	3. Document_version_query
Activity_authorization_query	Effectivity_assignment_query
Activity_element_query	Effectivity_query
Activity_relationship_query	Event_reference_query
Activity_resolved_request_query	External_model_query
Alias_identification_query	File_property_query
Alternative_solution_query	Geometric_model_relationship_query
Application_context_query	Item_classification_query
Approval_relationship_query	Item_classification_hierarchy_query
Assembly_component_placement_que	Item_definition_instance_relationship_qu
ry Associated_activity_query	ery
Associated_approval_query	Item_instance_query
Associated_classification_query	Item_query
Associated_date_organization_query	Item_use_query
Associated_date_time_query	Item_version_query
Associated_document_query	Item_version_relationship_query
Associated_effectivity_query	Object_by_uid_query
Associated_file_query	Objects_by_uids_query
Associated_general_classification_que	Organization_query
ry Associated_item_property_query	Organization_relationship_query
Associated_person_organization_quer	Person_in_organization_query
y Associated_process_property_query	Person_in_organization_relationship_que
Associated_project_query	ry
Associated_property_query	Person_organization_assignment_query
Class_structure_query	Person_query
Complex_product_query	Product_class_query
Configuration_query	Product_identification_query
Date_organization_assignment_query	Product_structure_query
Design_discipline_item_definition_qu	Project_assignment_query
ery Document_classification_query	Project_query
Document_classification_hierarchy_qu	Simple_property_value_query
ery	Work_order_query
Document_property_query	Work_order_is_controlling_query
Document_query	Work_request_activity_query
Document_representation_query	Work_request_query
Document_structure_query	Work_request_relationship_query
	Work_request_scope_query

16.5 Others to Watch

In the Introduction section of this chapter we encountered several SDOs. Of these, only the works of ISO STEP and OMG received top billing in the rest of the chapter due to their relative maturity. There are other noteworthy developments spearheaded by other SDOs, each deserving a separate full treatment.

OASIS PLCS PLM web services: after the initial STEP AP 239 standard was issued by ISO, a technical committee was formed in the OASIS organization to develop this further. A set of PLCS web services has been developed by a private company (Eurostep) as part of the European Union funded VIVACE project [16.26]. Eurostep has put this forward on behalf of VIVACE to the OASIS PLCS committee for consideration as the basis for an OASIS PLCS PLM web services standard.

OAGIS BODs: OAGi is working with ISA-95 [16.27] and the STEP community to harmonize the OAGIS BODs with various design and manufacturing standards.

SDM: simulation data management is gaining momentum, and standards are emerging to share and exchange SDM metadata with PDM systems.

16.6 Concluding Remarks

This chapter focused on the role of STEP in the context of PDM. It described the distinction between product data and product metadata, both in industrial practice and in the standardization activities of STEP. This distinction allows the asynchronous development and deployment of PDM functionalities separate from those of CAD. In addition, product metadata are viewed as business objects and they are used in several business processes (which subsume a considerable amount of what have been traditionally considered engineering processes) implemented over the entire, worldwide enterprise.

Product metadata as standardized by STEP PDM Schema and other APs are influencing emerging standards such as OMG PLM Services. This has enabled the deployment of standardized product metadata and standardized engineering/business process queries using modern information technologies based on service-oriented architecture. Such a trend is fortunate, because otherwise the enormous amount of work that has gone into the development of PDM-related STEP standards would have stagnated or even been consigned to the scrapyard of technology.

While we rejoice at the fortunate convergence of standardized product metadata and Internet-based technologies, we should also be aware of the challenges we face. We need to keep track of fast developments in information technology and update our STEP standards, work with other SDOs mentioned in this chapter that are as important as ISO, and spend more resources on industrial deployment of these standards.

Acknowledgment

The author is grateful for the help and support of numerous colleagues in national and international standards organizations. However, the opinions expressed here are his own and do not reflect the official position of these standards bodies.

References

[16.1] ASME Y14.5M, 1995, *Dimensioning and Tolerancing*, American Society of Mechanical Engineers, New York.

[16.2] ASME Y14.41, 2003, *Digital Product Definition Data Practices*, American Society of Mechanical Engineers, New York.

[16.3] ISO 1101, 2004, *Geometrical Product Specifications (GPS) – Geometrical Tolerancing – tolerances of form, orientation, location and run-out*, 2nd ed. International Organization for Standardization, Geneva.

[16.4] ISO 16792, 2006, *Technical Product Documentation – Digital Product Definition Data Practices*, International Organization for Standardization, Geneva.

[16.5] ISO 19005-1, 2006, *Document Management – Electronic document file format for long-term preservation – Part 1: Use of PDF 1.4 (PDF/A-1)*, International Organization for Standardization, Geneva.

[16.6] ISO 24517-1, 2008, *Document Management – Engineering document format using PDF – Part 1: Use of PDF 1.6 (PDF/E-1)*, International Organization for Standardization, Geneva.

[16.7] Srinivasan, V., 2008, "Standardizing the specification, verification, and exchange of product geometry: Research, status and trends", *Computer-Aided Design*, **40** (7): 738-749.

[16.8] Srinivasan, V., 2009, "An integration framework for product lifecycle management", to appear in *Computer-Aided Design*.

[16.9] ISO 7200, 2004, *Technical Product Documentation – Data fields in title blocks and document headers*, 2nd ed. International Organization for Standardization, Geneva.

[16.10] ASME Y14.24, 1999, *Types and Applications of Engineering Drawings*, American Society of Mechanical Engineers, New York.

[16.11] ASME Y14.34M, 1996, *Associated Lists*, American Society of Mechanical Engineers, New York.

[16.12] ASME Y14.35M, 1997, *Revision of Engineering Drawings and Associated Documents. The American Society of Mechanical Engineers*, New York.

[16.13] ASME Y14.42, 2002, *Digital Approval Systems*, American Society of Mechanical Engineers, New York,

[16.14] ASME Y14.100, 2000, *Engineering Drawing Practices*, American Society of Mechanical Engineers, New York.

[16.15] STEP application handbook ISO 10303. Version 3. SCRA (South Carolina Research Authority). June 2006. Free download from http://isg-scra.org/STEP/STEPHandbook .html.

[16.16] http://pdesinc.aticorp.org

[16.17] http://www.prostep.org

[16.18] Usage guide for the STEP PDM Schema V1.2, Release 4.3, PDM Implementor Forum, January 2002.

[16.19] ISO 10303-28, 2003, Industrial Automation Systems and Integration – Product data representation and exchange – Part 28: Implementation methods: XML representations of EXPRESS schemas and data, International Organization for Standardization, Geneva.

[16.20] ISO 10303-25, 2005, Industrial Automation Systems and Integration – Product data representation and exchange – Part 25: Implementation methods: EXPRESS to XMI binding, International Organization for Standardization, Geneva.

[16.21] Srinivasan V, Lämmer, L, and Vettermann S., 2008, "On architecting and implementing a product information sharing service". ASME Journal of Computing and Information Science in Engineering, **8** (1), pp. 011006-1 – 011006-11.

[16.22] Model Driven Architecture Guide Version 1.0.1, 2003, Object Management Group.

[16.23] http://www.prostep.org/en/standard-info/omg-plm-services.html

[16.24] PDM Enablers Version 1.3, 2000, Object Management Group.

[16.25] PDTnet Implementation Guide Version 1.6, 2003, PROSTEP AG.

[16.26] http://www.vivaceproject.com

[16.27] http://www.isa-95.com

17

STEP in the Context of PLM

Chandresh Mehta[1], Lalit Patil[2] and Debasish Dutta[3]

[1,2] Department of Mechanical Engineering, University of Michigan, 2250 G.G. Brown, 2350 Hayward St, Ann Arbor, MI 48105, USA
Email: [1]mehtacr@umich.edu, [2]lpatil@umich.edu

[3] Department of Mechanical Science and Engineering, University of Illinois at Urbana-Champaign, 204 Coble Hall, MC-322, 801 S. Wright St, Champaign, IL 61820, USA
Email: ddutta@illinois.edu

Abstract
This chapter explores the suitability of STEP to enable Product Lifecycle Management (PLM). It summarizes the relevant standards and discusses their potential to support PLM. After evaluating that STEP is best suited to support PLM, the chapter focuses on engineering change management an activity that is typical of PLM. In order to capture the change evaluation data, STEP-compliant extensions to the existing engineering change data model are proposed. A physical STEP file derived from the proposed information model to capture and represent an example engineering change is demonstrated. The chapter closes by identifying issues internal to ISO 10303 that potentially hinder lifecycle-wide exchange and reuse of product information.

17.1 Introduction

Globalization is leading to new drivers in product development. Co-located and monolithic product development teams and software can no longer efficiently manage the growing complexity and variety of knowledge required to develop products for global markets. Several distributed stakeholders, including designers, manufacturers, source vendors, and sales partners with varying specialties create and use an enormous amount of product data throughout the lifecycle – right from the specification of its requirements to its disposal. As a result, the boundaries of the enterprise extend beyond traditional design and manufacturing. Organizations – both private and federal – are increasingly embracing the view of Product Lifecycle Management (PLM) as a philosophy to view the extended enterprise as one product-centric enterprise, and not a set of siloed activities.

While rooted in Computer Aided Design (CAD) and Product Data Management (PDM) systems, PLM is being continually differentiated in terms of its technological expectations. It is expected to provide a shared platform to capture, represent, and exchange information among different stakeholders across all stages of the product lifecycle. At its core, PLM envisions to create and utilize a computational environment to manage the product lifecycle by ensuring that the *right information* is available at the *right time* in the *right context* throughout the lifecycle.

In order to generate the *right information*, the stakeholders use specialized information resources and software to develop the physical form, attributes, specifications, key interrelationships, and other information that defines a product well beyond its geometry. Individual activities within the enterprise function more effectively as each of these tools improves. Figure 17.1 highlights the variety of platforms used in the United States automotive supply chain in 1997 [17.1]. As seen from the figure, the number of software increases significantly from the Original Equipment Manufacturer (OEM) to the sub-tier suppliers. These software tools have evolved not only in number, but also in their individual capabilities. As a result, product data is created and stored in multiple, frequently incompatible formats.

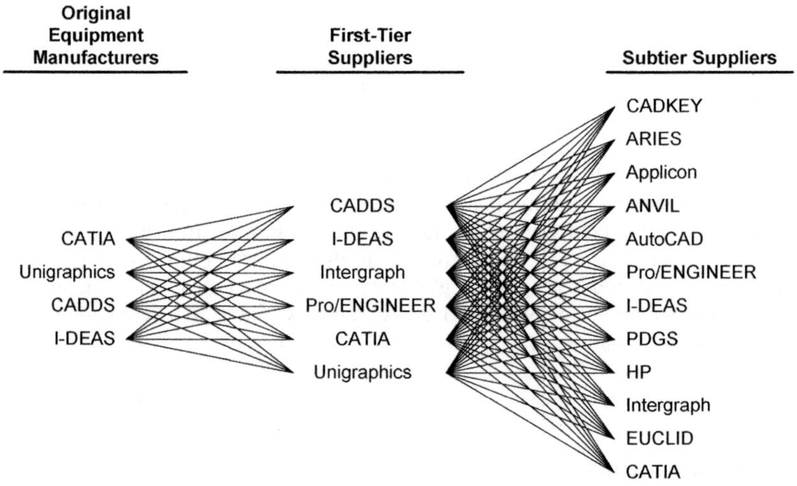

Figure 17.1. Multiple CAD/CAM tools in the automotive supply chain (adapted from [17.1])

The disparity of information, discussed above, is one of the two primary challenges in PLM. Another key component is the enormous complexity of products and associated processes, often tailored to a specific manufacturer. A typical automotive vehicle is known to contain more than 300,000 lines in its Bill of Material (BOM). Adding the systems information and attributes to the systems and sub-systems exponentially increases the size and complexity of the information.

This multiplicity of the stakeholders and their goals, the heterogeneity and complexity of the products, and the disparity across information resources – all make PLM, a very knowledge intensive process and emphasize the need to deduce the right information from the entire knowledge base.

The right product information could be easily accessible to all stakeholders, if there was a single tool to support the entire activity. However, geographic, functional, and cultural boundaries prevent the development of such a tool.

Therefore, in order to enable effective management of the product lifecycle, the complex product knowledge must flow seamlessly from one phase to another, and from one application to another throughout the lifecycle. Recent advances in the Internet make it the right medium to ensure connectivity throughout the value chain, i.e., right information could be available at the right time.

Yet conservative estimates in a study for the National Institute of Standards and Technology (NIST) suggest that imperfect interoperability costs at least 1 billion dollars per year to the United States Automotive Supply Chain [17.1]. The majority of these costs are due to resources spent in correcting and recreating data that is not usable by the receiving applications. A follow-up study documents a similar impact of the lack of interoperability for the Aerospace and Shipbuilding industries [17.2].

It is necessary for the success of OEMs, federal departments such as the U.S. Army, their suppliers, and software vendors that interoperability is correct and efficient, i.e., it is not merely syntactic data exchange. This becomes even more important in the context of increasing and changing capabilities of existing systems in distributed product development. Figure 17.1 shows that the use of multiple software systems by the different OEMs and the lack of interoperability forces sub-tier suppliers to purchase, learn, and maintain multiple systems. Therefore, it is a costly problem, not only for the OEMs, but also for the small shops that produce the parts and small assemblies.

Thus, a fundamental problem in PLM is the need to ensure that the exchange and reuse of product knowledge across the extended enterprise takes place in the right context. In addition to the issues (product complexity, multiple information sources, etc.) discussed above, the right context is difficult to deduce because product knowledge, unlike text data, is not self-descriptive. The lack of standard metadata severely restricts the reuse and repurposing of product knowledge. In view of this, there is a need for a standard that will capture, represent, and enable exchange of the lifecycle-wide components of product data.

This chapter explores the potential of ISO 10303 (STEP) to support PLM. Section 17.2 summarizes widely-accepted standards that might be relevant to the exchange and reuse of product lifecycle data. In particular, discussions in this chapter are presented in the context of Engineering Change Management (ECM), since it represents a typical PLM activity. Section 17.3 first explores the suitability of STEP and then utilizes its information modelling methodology to represent and exchange Engineering Change (EC) evaluation data. Section 17.4 presents issues and future directions in using STEP for effective information exchange and interoperability in PLM.

17.2 Overview of Standards for PLM

As mentioned earlier, standards play a pivotal role in enabling effective Product Lifecycle Management (PLM) system by facilitating representation and exchange of lifecycle-wide data. Standards that focus on, and are successful in, enabling interoperability within certain domains of PLM are identified in [17.3]. However, due to the complexity and variety of inter-linked data and domains within PLM, such stovepipe efforts are not enough and present additional problems of in communication between the standards. PLM needs either a single standard that can account for the lifecycle-wide activities and information resources, or a set of complementary standards linked through a common framework that enables interoperability among them.

This section summarizes some of the widely-accepted standards considered relevant to data exchange in PLM.

17.2.1 EIA-649 National Consensus Standard for Configuration Management

Configuration Management (CM) is a PLM activity for establishing and maintaining consistency of a product's performance, functional, and physical attributes with respect to its requirements, design, and operational information throughout its lifecycle. Electronics Industries Alliance (EIA)-649, a standard by the American National Standards Institute (ANSI), defines and describes a CM process that comprises five functions and their underlying principles [17.4]. The five functions are: CM planning and management, configuration identification, configuration change management, configuration state accounting, and configuration verification and audit. Although EIA-649 is meant to be used as a standard to guide the process and does not contain any information models for capturing and representing product data, the principles can be useful in planning information models for CM.

17.2.2 ANSI/GEIA GEIA-859-2004 Data Management

The Government Electronics and Information Technology Association (GEIA)-859 is a voluntary standard for planning and implementing a data management program for a product, project, or enterprise. [17.5] defines data management as sharing, integration and management of product lifecycle data to ensure customer requirements are met or exceeded – consistently and uniformly. GEIA-859 comprises nine high-level principles to enable high-quality data management over the product and data lifecycles. This standard does not incorporate any information models for capturing and representing the data; its focus is on providing guiding principles.

17.2.3 ISO/IEC 12207 Software Life Cycle Processes

ISO/IEC 12207 is a standard created to establish a common international framework to acquire, supply, develop, operate, and maintain software products [17.6]. The common framework is established by defining 23 processes and 95 activities across the software lifecycle. The focus of this standard is primarily on software and extensions to products in the mechanical design space are not yet available.

17.2.4 PLM-XML

PLM XML is an emerging XML-based format created and supported by Siemens PLM Software with a goal of offering a standardized protocol for data interoperability across various phases of product lifecycle [17.7]. It extends Siemens Parasolid XT data format to non-geometric product model information, including product structure, shape, and process information. It is published, open, and compliant with the World Wide Web Consortium (W3C) XML schema. The aim of the PLM XML format is not to develop a single persistent data model; rather, wherever appropriate, it aims to integrate/reuse other representations.

17.2.5 ISO 10303-239 (STEP AP 239)

STEP, the Standard for the Exchange of Product model data, is an international standard focusing on the representation of product information throughout its lifecycle (please see Chapter 1 for several details). In the context of PLM, the most suitable Application Protocol is AP 239, informally known as Product Life Cycle Support (PLCS). It addresses the requirements of exchanging product information over the entire lifecycle, i.e., from planning and conception to retirement [17.8]. In this context, AP 239 is an enabler of PLM and addresses following key areas:

- Work management: requesting, defining, justifying, approving, scheduling, and capturing feedback on product lifecycle activities and associated resources
- Product definition: defining product requirements and configurations, including relationships between parts and assemblies in multiple product structure (as-designed, as-built, and as-maintained)
- Operational feedback: describing and capturing feedback on product properties, operating states, behaviour, and usage
- Support solution and environment: defining facilities, personnel, organizations, and necessary support in a specified environment.

PLCS incorporates a generic and large information model that is also flexible and extensible. The large information model makes PLCS larger in scope than most business processes require or most information technology application can manage; this problem is overcome by defining Data EXchange specifications (DEXs). DEX is a way of partitioning the PLCS information model into sections suited for a specific business process. For example, DEX: D0007 is a subset of information models just for activities related to operational feedback. The components, e.g., date and time, of the PLCS information model common to many DEXs are packaged together into modules called "Capabilities" that are reused across different DEXs.

Through the use of Reference Data, the generic PLCS information model can be extended or tailored to fulfil the specific needs of particular industries or organizations. A reference data is the collection of pre-defined class definitions, e.g., types of task, grades of people, types of products, and types of documents, representing a concept, e.g., task, people, product, and document. In this context, by adding a classification scheme to the basic constructs, reference data refine or add semantics to the PLCS model. A Reference Data Library (RDL) is a managed collection of reference data classes that is specified separately to any data exchange file. Applications that have different reference data will not interoperate until the two sets of RDLs have been harmonized or mapped. Many existing standards could be used as potential sources of reference data by PLCS. DEXs and reference data may be standardized at any level, e.g., project, company, cross-company, or cross-domain, and at any time during or after the development of PLCS. The PLCS Technical Committee at Organization for the Advancement of Structured

Information Standards (OASIS), which is coordinating the development of PLCS, has created an initial set of RDL, called OASIS PLCS RDL, using Web Ontology Language (OWL)-XML syntax [17.8]. However, there are several aspects (see section 17.3.3) of PLM that still need to be captured and represented in PLCS.

17.2.6 STEP-based Standards

PLM Services

STEP Product Data Management (PDM) schema, which is based on a core set of PDM-related entities in STEP design APs, lacks a functional model, i.e., methods to access data. PLM Services (PLMS) [17.9], a STEP PDM schema-based standard by Object Management Group, enhances STEP PDM schema by a functional model that permits web services access (synchronous and asynchronous) to lifecycle-wide data. In addition, PLMS links data modelling capabilities of STEP with industry need of cross-company collaboration by means of pre-defined use cases and transactions. PLMS is intended to be used for exchanging product lifecycle-wide data between stakeholders by using XML and Web services technologies.

Other STEP-based Data Models

While there are several standards that support one or more aspects of PLM, none can match STEP APs in the depth and breadth of coverage for product lifecycle data. At the same time, due to complexity and variety of data inherent in PLM, it is unfeasible for STEP APs to predefine a data model for each activity of PLM. As a solution to this problem, STEP is made extensible, so that new information models can be added as need. STEP's information modelling methodology has been successfully used for enhancing or adding new STEP-compliant data models. For example, [17.10] uses STEP's information modelling methodology to present a STEP-based data model for engineering data management.

In spite of several distributed efforts to enhance STEP's information models, there is no comprehensive model for effective exchange and reuse of data expected from a PLM system. The following section explores the suitability of STEP and then utilizes its information modelling methodology to capture and represent data for typical activity in PLM.

17.3 Applying STEP to Data Exchange and Reuse in PLM

While there are several aspects of PLM that still need to be captured and represented in STEP, this chapter focuses on Engineering Change Management (ECM), because it is an activity that is representative of the issues in PLM.

17.3.1 Engineering Change Management as a Typical PLM Activity

Every product undergoes several Engineering Changes (ECs) throughout its lifecycle. An EC refers to any change to the shape, dimensions, structure, material, manufacturing process, etc., of a part or assembly after the initial design has been released (and often after the part is already in production). Engineering Change Management is a typical PLM activity since it is characterized by the following:

- *Product complexity:* several components interact in several ways to form a product. Therefore, changing one component is very likely to affect several other parts and processes. This requires that any part changes must be further coordinated with the designs or changes of all the affected parts.
- *Multiple information resources:* globally distributed, cross-functional product development teams use multiple software tools to create and use knowledge of a product during all the phases of its development. These multiple information resources are disparate and capture domain-specific information. However, EC management requires them to be used and studied simultaneously.
- *Causes from across the lifecycle:* an EC depends on several phases throughout the product's lifecycle. It may be a result of growing and changing end-of-life requirements, or customer usage, or a suggestion from a supplier, etc.
- *Effect on different phases of the lifecycle:* a seemingly simple EC such as changing the shape of an object can potentially affect several phases of the product's lifecycle, e.g., end-of-life treatment planning, manufacturing processes, etc.

Traditionally, ECM systems were used for organization and control of documentation associated with the process of making ECs. With the advent of PLM, the ECM systems are now looked upon as decision support tools that enable systematic evaluation of the impacts of proposed Engineering Changes based on the knowledge about past changes.

The following section presents the requirements for exchange and reuse of Engineering Change (EC) data. The next sections propose a STEP-compliant Change Evaluation Model (CEM) for representing EC-related data and demonstrate the Part 21 syntax for an example Engineering Change.

17.3.2 Requirements for Exchange and Reuse of ECM Data

A typical EC process flow involves: identification of potential for change, creation of change request, development of alternative solutions, technical analyses of change, decision on change request, and creation and implementation of change order. Past knowledge of ECs, if captured and represented, could be useful in developing alternative solutions and analyzing the change.

For example, consider a proposed EC between the cover and housing of a butterfly valve as shown in Figure 17.2 taken from [17.11]. The proposed change will have downstream effects and impacts on several lifecycle-wide components. Before approving the change, enterprise will want to evaluate and determine all its downstream effects so that the correct decision about acceptance or rejection of change can be made. To enable reuse of past EC knowledge for evaluating a proposed change, standards for representation should incorporate information models for capturing and representing not only change-related concepts, e.g., engineering change request, but also change evaluation-related concepts, such as change evaluation process type, priority, effects/impacts on various lifecycle-wide components, and impact analysis reports/documents.

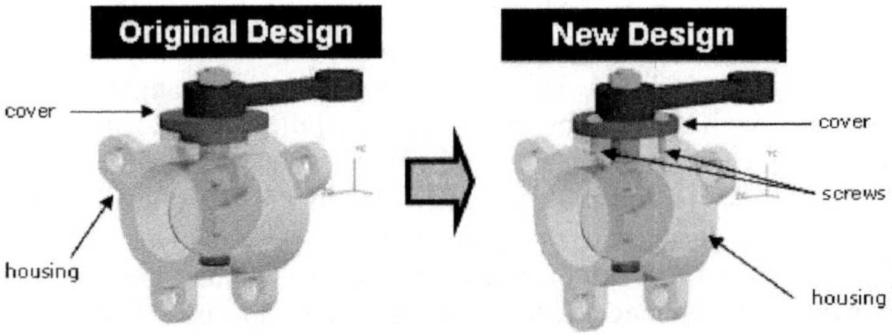

Figure 17.2. Change in joint between cover and housing of a butterfly valve from press-fit to screw (source [17.11])

In the following section we review current capabilities of ISO 10303 (STEP) for exchange and reuse of ECM data.

17.3.3 Suitability of STEP for Exchange and Reuse of ECM Data

Information models for representing data relevant to engineering changes appear in multiple APs, e.g., manufacturing APs: AP 224 and AP 240, systems engineering AP: AP 233, lifecycle AP: AP 239, and product data management APs: AP 203, AP 212, AP 214, and AP 232. The corresponding (EC-related) information models for are almost similar across these APs. Therefore, the remaining part of this section focuses on only AP 240. The reader is encouraged to refer to Chapter 1 to obtain information on the meaning of certain concepts, such as EXPRESS-G, Part 21 file, and AIM in the context of STEP.

Figure 17.3 shows a partial EXPRESS-G illustration of the core set of objects and associated attributes for representing EC data; the corresponding application interpreted model (AIM) entities are shown in parentheses. Application objects relevant to ECs include *Design_exception_notice*, *Engineering_change_proposal*, and *Engineering_change_order* [17.12]. The object, *Design_exception_notice* represents a notification of a design discrepancy identified while creating the process plans for a given part such that process planning cannot continue until a technical recommendation is made to correct the problem. Each *Design_exception_notice* could have *issues* defined *Engineering_change_proposal* objects. An *Engineering_change_proposal* is a document that describes potential alterations to a part and is linked to one or more *Engineering_change_order* that represents an authorization for modification of the product data that will result in a new process plan for a part.

However, STEP does not completely capture or represent knowledge related to EC evaluation process and outcomes. For example, there are no concepts/attributes to capture the data about items, such as total cost impact, change evaluation process type, change classification, priority, change effectivity, and so on. Such information is essential to exchange and reuse the knowledge about past ECs for evaluating a proposed change. In addition, absence of this information obstructs the STEP-based exchange of EC data among standards, for example PLM-XML and VDA4965 [17.13], that support, although partially, specification of change evaluation process and outcomes.

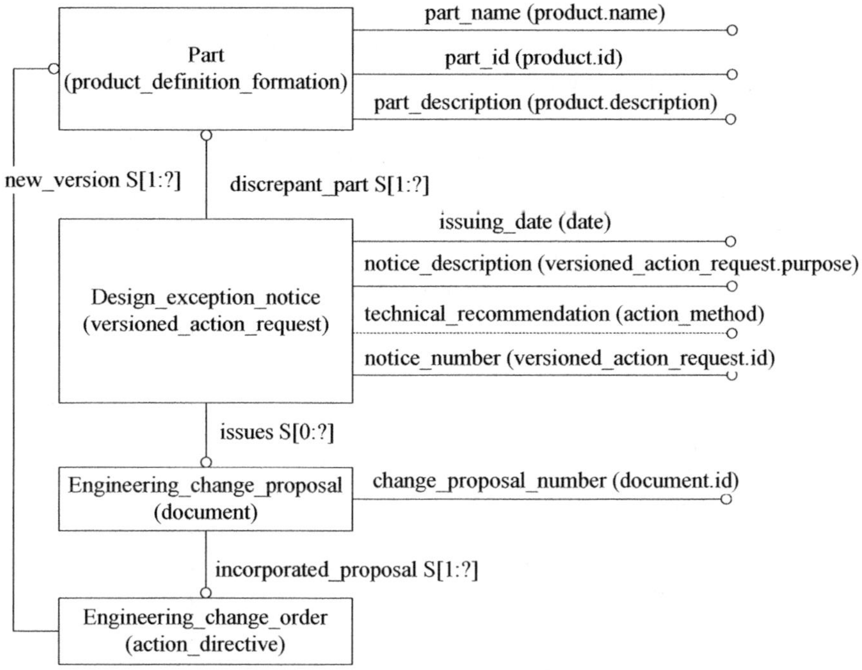

Figure 17.3. EXPRESS-G diagram of core EC objects and associated attributes in AP240. AIM elements are shown in parentheses

While STEP does not currently support representation of change evaluation data, it provides fundamental data structures that can be used and extended to capture such information. The following section discusses a proposed STEP-compliant change evaluation data model.

17.3.4 Enhancing EC Representation in STEP – Change Evaluation Model

In this section, the information modelling methodology of STEP is utilized to define additional concepts related to change evaluation process. For the sake of simplicity of explanation, only some of the relevant concepts are discussed. This proposed data model is referred to as Change Evaluation Model (CEM).

Figure 17.4 depicts an EXPRESS-G illustration of concepts related to change evaluation process. The entity *change_proposal* represents an EC request. The consequence of the change request on items, such as associated parts, assemblies, processes, etc., is captured by entity *effect*. The measurable consequence of a proposed change or its effects on an item is represented by entity *impact_evaluation*.

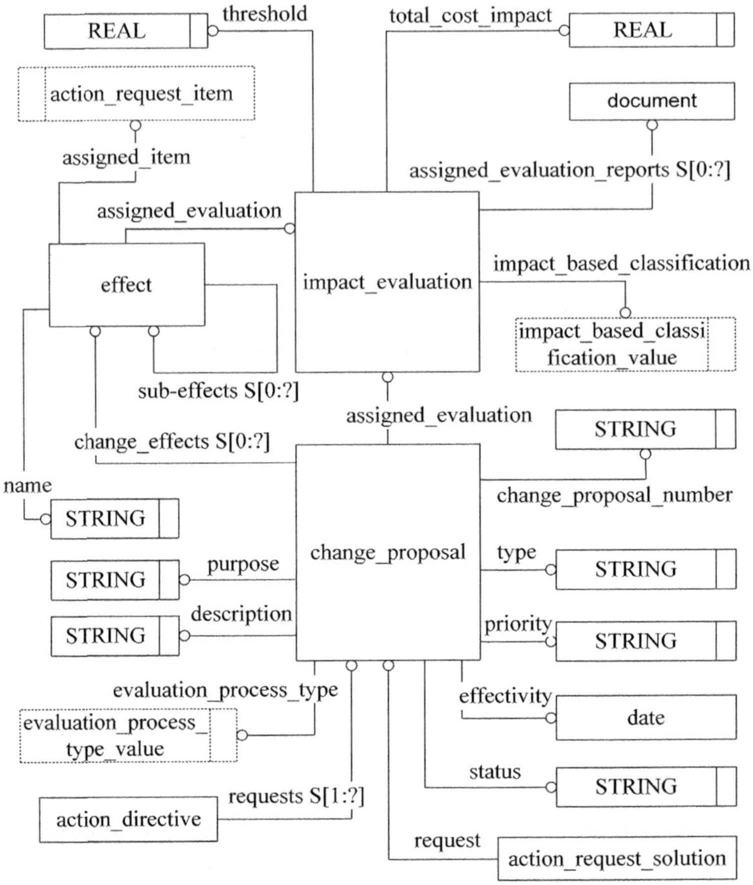

Figure 17.4. EXPRESS-G diagram of change evaluation model (CEM) to represent concepts related to change evaluation process

The entity *impact_evaluation* comprises attributes *total_cost_impact*, *threshold*, *impact_based_classification*, and *assigned_evaluation_reports*. The entity *total_cost_impact* captures the monetary consequence of proposed change on affected item/s. Depending on an enterprise set *threshold* and the value of *total_cost_impact*, the *impact_based_classification* takes a value of either HIGH or LOW. The *assigned_evaluation_reports* is an optional attribute that links *impact_evaluation* to a set of instances of entity *document* that references to the detailed evaluation reports.

An *effect* has attributes: *name*, *assigned_item*, *assigned_evaluation*, and *sub-effects*. The attribute *assigned_item* links an *effect* to an *action_request_item* that is affected by proposed change. The evaluated impact of *effect* on *action_request_item* is represented by attribute *assigned_evaluation*. Each EC effect might have its own *sub-effects* which are of type *effect*.

The change_proposal has attributes: change_proposal_number, purpose, description, type, priority, effectivity, status, evaluation_process_type, change_effects, and assigned_evaluation. The change_proposal_number specifies a unique identification for proposed change. The attributes purpose and description capture the reason and details of EC request, respectively. An enterprise may pre-define the values for purpose by specifying descriptive labels, such as corrective action, problem prevention, technical improvement, and customer request that informally describe the reason for change. The type and priority represents the change type (e.g., change in joint and change in material) and change evaluation/implementation urgency (e.g., high, medium, and low), respectively. The attribute type allows an enterprise to classify the large repository of changes in to smaller subsets. This will in turn reduce the number of past ECs that are searched in the process of determining those that are similar to a proposed engineering change. The date from which the change will be in effect is captured by effectivity. The proposed change request might be evaluated by fast or detailed track; this information is represented by evaluation_process_type. The change_effects links the change_proposal to its effects, whereas assigned_evaluation links the proposed change to an entity impact_evaluation, which represents the overall impact due to change. This is typically determined as a function of impacts due to all its effects. The entity status, whose values may be pre-defined, characterizes the level of completion of change request. The entities action_request_solution and action_directive represent the potential solutions to an EC request and engineering change order, respectively. The entities document, date, action_request_solution and action_directive are defined in Part 41 [17.14] of the ISO 10303 documentation.

17.3.5 Example Application of CEM

Figure 17.5 shows part of the Part 21 file format for the engineering change example shown in Figure 17.2 and its impact as per the CEM discussed in the previous section. It should be noted that Figure 17.2 illustrates a change in joint between the cover and housing of a butterfly valve from press-fit to screw. Assume this was a past EC; therefore all information about its evaluation is available.

The two instances of entity *product_definition_formation* in Figure 17.5 represent the two parts, i.e., housing and cover, being affected by proposed change. The *document_file* instances, i.e., #401 and #402, represent the detailed evaluation report of impacts on affected parts. The effects on housing and cover are represented by #450 and #451, respectively. One of the sub-effects of effect on housing is represented by #452. The #500, #501, and #503 represents the impact_evaluation of change in housing, cover, and housing manufacturing process, respectively. Entity *change_proposal* in #600 models the proposed engineering change. As indicated by #600 EC1, Change in Joint, Normal, technical improvement, change in joint between the cover and housing from press-fit to screw, Fast Track and In-review are the instance values for attributes *change_proposal_number*, *type*, *priority*, *purpose*, *description*, *evaluation_process_type,* and *status*, respectively. The total impact due to proposed change is represented by #502.

```
#10=APPLICATION_CONTEXT('mechanical design');
.
#220=PRODUCT_CONTEXT('', #10, '');
#230=PRODUCT_DEFINITION_CONTEXT('part definition', #10, 'design');
.
#360=PRODUCT('H01', 'housing', 'housing of butterfly-valve' , (#220));;
#361=PRODUCT_DEFINITION_FORMATION('H01_V01', 'version 1 housing ', #360);
.
#370=PRODUCT('C01', 'cover', 'cover of butterfly valve', (#220));
#371=PRODUCT_DEFINITION_FORMATION('C01_V01', 'version 1 of cover', #370);
.
#399= DATE('2007');
#400=DOCUMENT_TYPE(' ');
#401=DOCUMENT('R-H01',' housing evaluation report', $, #400);
#402=DOCUMENT('R-C01',' cover evaluation report', $, #400);
#403=DOCUMENT('R-MP01',' housing manufacturing process evaluation report', $,
#400);
.
#450=EFFECT('change in housing ', #500, #361,(#452, #453));
#451=EFFECT('change in cover ', #501, #370, (#454,#455));
#452=EFFECT('change in manufacturing process', #503, #361, '');
.
#500=IMPACT_EVALUATION('50000','80000', 'high', #401);
#501=IMPACT_EVALUATION('50000','10000', 'low', #402);
#502= IMPACT_EVALUATION('50000','90000', 'high', (#401, #402) );
#503=IMPACT_EVALUATION('20000','25000', 'high', #403);
.
#600=CHANGE_PROPOSAL('EC1', 'Change In Joint', 'Normal', 'technical improvement', '
change in joint between the cover and housing from press-fit to screw ', (#450, #451),
#502, 'Fast Track', 'In-review', #399);
```

Figure 17.5. Part 21 syntax of EC from Figure 17.2 as per Change Evaluation Model (CEM)

Once stored in PLM knowledge base, this change instance can be reused for evaluating similar future change requests of type Change in Joint.

17.4 Further Issues and Directions

The previous sections demonstrate how STEP's information modelling methodology can be extended to capture and represent lifecycle-wide data. However, there are several other issues and challenges for using STEP for effective information exchange and interoperability in PLM. This section highlights some of these issues.

17.4.1 Conflicts Within the Standard

Different Working Groups in ISO 10303 seem to work on developing APs for different domains. There is no well-defined approach for finding or reconciling the similarities and differences across them. However, in the context of PLM, it is important to have the ability to communicate across the different domains.

Undetected overlaps, particularly those that are supposed to mean similar concepts, but use different descriptions, can be a significant hindrance.

At a conceptual level, the real world concept of engineering change request has different representations in AP 214 and AP 240. This data representation schematic conflict is summarized in Table 17.1. In AP 214, the object *Work_request* represents an engineering change request while *Design_exception_notice* and *Engineering_change_proposal* together represent this concept in AP 240. The object *Work_request* is defined as solicitation for some work to be done; *Design_exception_notice* is a notification of a design discrepancy discovered during the creation of the process plan for a given part; and *Engineering_change_proposal* is a document that describes potential modifications to a part. In addition, the different number of attributes for the concept of engineering change request in these APs gives rise to what is known as schema isomorphism conflict.

Table 17.1. Schematic conflict across AP 214 and AP240 to represent EC data

	AP 214	**AP 240**
Application object	*Work_request*	a. *Design_exception_notice* and b. *Engieering_change_proposal*, together
Represents the real world concept of	Engineering change request	Engineering change request
which is/are defined as	Solicitation for some work to de done	a. Notification of a design discrepancy discovered during the creation of the process plan for a given b. Document that describes potential modifications to a part.
And has corresponding AIM entity	*versioned_action_request*	a. *versioned_action_request* and b. *document*

There are some discrepancies at a definitional level as well. For example, both AP 203 and AP 214 define *ENTITY coordinated_universal_time_offset*, with a sense represented by the entity *ahead_or_behind.* AP 203 defines *ahead_or_behind* as *ENUMERATION OF (ahead, behind).* However, AP 214 defines the same entity, *ahead_or_behind* as *ENUMERATION OF (ahead, exact, behind).* AP 239, which is specifically designed for PLM uses yet another approach. It uses an entity called *time_offset* with a sense *offset_orientation*, which is defined as *ENUMERATION OF (ahead, exact, behind).* These conflicts are represented in Table 17.2.

Table 17.2. Semantic conflicts across APs 203, 214, and 239 to represent local time zone

	AP 203	**AP 214**	**AP 239**
Entity for local time zone	coordinated_universal_time_offset	coordinated_universal_time_offset	*time_offset*
Has sense	ahead_or_behind	ahead_or_behind	*offset_orientation*
Sense is enumeration of	*ahead, behind*	ahead, exact, behind	ahead, exact, behind

It is possible to resolve some of these conflicts after a careful analysis of the APs; however, currently, this is a manual task and is practically infeasible because each of the APs is very well-developed and makes complicated uses of the basic entities. In particular, in the above example, the semantically differing definitions of the term *sense* can affect PLM negatively. In addition, in the absence of reasoning support, it is unclear how changing the definition of such a concept propagates.

17.4.2 Abstract/Ambiguous Definitions

Some of the concepts in ISO 10303 are abstract, which contradicts the view of it being useful and representative of real world.

In its definition of Part 41/AP239 [17.14], the usage of a *product* is described through the following examples:

- *EXAMPLE 1* The SS Titanic is a product that could be represented by the entity data type Product.
- *EXAMPLE 2* Lifeboat is a class of products that could be represented by the entity data type Product. Each lifeboat on the SS Titanic is a member of this class.

From the above examples, the term *product* can be used to represent a specific product, as well as a class of products. Therefore, it seems that Product as defined in ISO 10303 should be an abstract concept, since it cannot be used without further specialization. Yet it is not defined as an abstract concept. Furthermore, it is not clear what impact such modifications would have, considering that the term Product is used in every APf of ISO 10303.

17.5 Concluding Remarks

This chapter explores the potentials of ISO 10303 (STEP) standard in enabling Product Lifecycle Management (PLM). The focus is on Engineering Change Management as a typical PLM activity. In order to capture the change evaluation data STEP-compliant extensions to the existing engineering change data model are proposed. A physical STEP file derived from the proposed information model to capture and represent an example engineering change is demonstrated.

It is noted that, as a result of variety and complexity of inter-linked data and activities inherent in PLM, there lacks a single monolithic standard that can support all its components. While there are several standards that support one or more aspects of PLM, none can match STEP APs in the depth and breadth of coverage for product lifecycle data. However, STEP still lacks several information models that are essential for effective PLM. The chapter highlights some additional issues that can provide directions for future research in enhancing STEP's capabilities to support PLM.

Acknowledgment

The authors would like to thank the anonymous reviewers. This material is based in part upon work supported by the University of Michigan's PLM Alliance, the National Institute of Standards and Technology (NIST) under Grant No. 60NANB6D66204, and the National Science Foundation under Grant Numbers CMMI-0653838 and CMMI-0758150.

References

[17.1] Brunnermeier, S. and Martin, S. 1999. Interoperability cost analysis of the U.S. automotive supply chain. *Research Triangle Institute: Report prepared for the National Institute of Standards and Technology.* 99-1.

[17.2] Gallaher, M. P., O'Connor, A. C. and Phelps, T. 2002. Economic Impact Assessment of the International STandard for the Exchange of Product model data (STEP) in Transportation Equipment Industries. *Research Triangle Institute: Report prepared for the National Institute of Standards and Technology.*

[17.3] Sudarsan R., Eswaran S., Abdelaziz B., Steven J. F., Sebti F. and S., R. D. 2006. The Role of Standards in Product Lifecycle Management Support, In *Proceedings of International Conference on Product Lifecycle Management PLM'06.* Bangalore, India, 122-136.

[17.4] ANSI/EIA-649-A:2004 National Consensus Standard for Configuration Management. American National Standards Institute (ANSI).

[17.5] ANSI/GEIA GEIA-859:2004 Data Management. American National Standards Institute (ANSI).

[17.6] ISO/IEC 12207:1995 Software Life Cycle Processes. International Organization for Standardization (ISO), Geneva, Switzerland.

[17.7] Siemens PLM Software. 2007. Open product lifecycle data sharing using XML. [online] available: http://www.plm.automation.siemens.com/en_us/Images/plm xml wp W 3_tcm53-11521.pdf [20 January 2009]

[17.8] Eurostep Group. 7 July 2008. *Product Life Cycle Support (PLCS).* [online] available: http://plcs-resources.org [20 January 2009].

[17.9] Object Management Group (OMG). 2007. Product Lifecycle Management Services, Version 2.0 - Draft Adopted Specification. [online] available: http://www.prostep.org/en/downloads/recommendations-standards.html

[17.10] Peng, T.-K. and Trappey, A. J. C. 1998. A step toward STEP-compatible engineering data management: the data models of product structure and engineering changes. *Robotics and Computer-Integrated Manufacturing.* 14(2): 89-109.

[17.11] Joshi, N. 2006. Methodologies for improving product development phases through PLM. Ph.D. Dissertation. University of Michigan, Ann Arbor. MI.

[17.12] ISO10303-240:2005 Product data representation and exchange: Application protocol: Process plans for machined products, Geneva, Switzerland: International Organization for Standardization (ISO).

[17.13] Verband der Deutschen Automobilindustrie (VDA). 2006. Recommendation 4965, Version 2.0 [online] available:

[17.14] http://www.vda.de/de/downloads/382 [20 January 2009]

[17.15] ISO 10303-41:2005 Industrial automation systems and integration - Product data representation and exchange - Part 41: Integrated generic resources: Fundamentals of product description and support, Geneva, Switzerland: International Organization for Standardization (ISO)

18

Usage of Agent Technology to Coordinate Data Exchange in the Extended Enterprise

Omar López-Ortega[1] and Karla López de la Cruz[2]

Centro de Investigación en Tecnologías de Información y Sistemas, Instituto de Ciencias Básicas e Ingeniería, Universidad Autónoma del Estado de Hidalgo, Carretera Pachuca – Tulancingo Km. 4.5, Pachuca, Hidalgo, México.
Email:- [1] lopezo@uaeh.reduaeh.mx, [2] karla@hotmail.com

Abstract
In the current competitive and interconnected market place, manufacturing organizations are exploring varied forms of collaboration in order to sustain the creation of wealth. The *Extended Enterprise* is one of the proposed paradigms to facilitate collaboration among individual business units. Such a paradigm mainly claims that core capabilities are to be shared by participating firms, and exploited by the new, extended enterprise. In this setting, it is pivotal to store, advertise, and share capabilities that are offered and required by firms in the joint organization. The authors claim that merging EXPRESS data models, Business Processes and agent technology, innovative systems for coordinating data sharing along the networked enterprise are constructed. EXPRESS data models permit the unambiguous representation of core capabilities. Therefore, STEP-based repositories are built to store and retrieve information about them. Also, agent communication protocols, obtained after defining suitable Business Processes, are modelled and implemented to coordinate the flow of data. The validation is performed by a MAS in charge of sharing and exchanging data in a distributed setting.

18.1 Introduction

J. Browne has already claimed that the global economy is leading the entire manufacturing system to be considered in the context of the value chain [18.1]. Hence, the focus for improving design and manufacturing is being directed to facilitate inter-enterprise networking and collaboration models in distributed manufacturing networks [18.2]. Several approaches have been developed in order to materialize inter-company collaboration, namely the Virtual Enterprise, the Extended Enterprise, and Supply Chains. Although the three of the approaches to create networks of enterprises differ in their goals and means, all of them reflect the urgency to cooperate among distributed units. However, the authors' research is focused on the Extended Enterprise.

An *Extended Enterprise* is defined as a long-term agreement among individual and complementary industrial units. An Extended Enterprise is formed to provide products and services in a well-defined yet evolving market segment. Also, in an Extended Enterprise, companies remain separate legal entities that keep control over their own systems [18.3] and resources. Thus, each legal entity provides its own *core capabilities*, the sum of which come together for the purpose of producing a customer-defined product. Within an Extended Enterprise, organizations share and exchange information of the product, processes and resources for which they show consistent and outstanding performance. To do so, Extended Enterprises are grounded on confidence and long-term contracts [18.4].

Nevertheless, sharing technical expertise among organizations is a complex task because industrial information is highly heterogeneous. Heterogeneity reflects the fact that information models are developed and implemented under different formalisms and platforms, which in turn are closely bound to particular practices within each company. To overcome this incompatibility problem, the use of standardized data models has been proposed as a solution. On this matter, ISO 10303, ISO 15531, and ISO 13584 are fundamental standards to facilitate sharing and exchanging of product design and manufacturing data. However, models within them grew separately from a semantic modelling perspective and, consequently, an integrated framework is needed not only to facilitate data sharing but also to enhance cooperation in the Extended Enterprise. Therefore, to promote cooperation among firms in a distributed manufacturing setting, the authors propose to address the following three challenges.

One major problem concerns the unambiguous electronic representation of firms' core capabilities. Standardization is vital to enhance interoperability of the information systems that are used to store and exchange design and manufacturing data along the product life-cycle [18.5]. Section 18.2 presents the main ideas for building an integrated EXPRESS model used to represent and further store firms' core capabilities

The second issue is about the supporting software for sharing and exchanging core capabilities. Stand alone applications no longer possess the required adaptability to deal with the distributed nature of the Extended Enterprise. Therefore, it is necessary to use state of the art technology in order to integrate systems that are operations along the network. This means that information systems remain locally owned by the constituent companies, but cooperation among them is needed. Hence, coordination must be carried out by a central unit with the capacity to acquire local data and propagate it along the network. Multi-Agent Systems (MAS) are the appropriate paradigm to achieve distributed information processing [18.6] and maintain the stability of the extended enterprise. The proposed solution for this matter is explained in Section 18.3.

Coordination of data flow is the third major issue that the authors envision for facilitating inter-enterprise cooperation. In the Extended Enterprise, coordination is needed to avoid conflicts on resource usage. Business Processes (BP) have been proposed extensively in the scientific literature in order to model the dynamics of cooperation between firms. We propose to use a well-defined set of BPs to establish communication mechanisms in a MAS environment, by employing an agent communication language. Thus, communication among agents must mirror the underlying Business Processes so that proper coordination is reached.

To summarize, this chapter presents a solution to coordinate data exchange in the Extended Enterprise by merging three major topics: (1) the unambiguous representation of core capabilities; (2) a MAS architecture that contains a central coordinator while respecting, at the same time, the locally owned information systems of the participating firms; and (3) the definition of communication models after Business Processes.

The organization of the chapter is as follows. Section 18.2 presents the rationale behind the integrated EXPRESS model. Description and validation of the MAS dynamics is given in Section 18.3. Discussion on the suitability and limitations of this proposal is done in Section 18.4. Concluding remarks and future work are outlined in Section 18.5.

18.2 Integrated EXPRESS Model

This section presents the main ideas supporting the mathematical framework that is employed to integrate entities of ISO 10303, ISO 15531, and ISO 13584. The rationale of this framework has been originally depicted in [18.7]. The scope of these standards is briefly mentioned to understand their importance for this research. Yet, a more elaborate explanation regarding STEP is already presented in Chapter 1 of the present book.

18.2.1 STEP-related Standards

MANDATE (MANufacturing management DATa Exchange) contains EXPRESS models that represent industrial manufacturing management data [18.8–18.10]. ISO 15531 primarily addresses the modelling of (1) resources management data and (2) processes related features. The standard is intended at facilitating information exchange between software applications such as E.R.P., manufacturing management software, maintenance management software, and quotation software. Similarly, any given product consists of one or numerous *parts*, and information about them is provided by *suppliers*. Data regarding *parts* is exchanged and shared among different organizations which generate and use this information. ISO 13584 Parts Library (PLIB) provides EXPRESS data models for exchanging, operating, and revising *parts* information and their *suppliers*. Non-geometric information about parts, which is the purpose of ISO 13584, is transferred through a STEP physical file [18.11, 18.12]. Finally, EXPRESS models have been developed and proposed for representing flexible manufacturing resources [18.13]. Flexible manufacturing resources are paramount due to the high degree of automation found in the manufacturing domain.

Consequently, a subset of these models is useful to share and exchange manufacturing data along networked enterprises. The resultant integrated model is presented next.

18.2.2 Semantic Integration to Represent Core Capabilities

Core capabilities are defined as any means, physical or logical, that are offered to the net of enterprises in order to enhance the overall production potential. Four major classes of core capabilities are proposed: Product, Resources, Organizations, and Processes. These four major classes were defined after analyzing the scope of ISO 10303, ISO 15531, ISO 13584, and the EXPRESS models for flexible manufacturing resources.

Hence:

- ISO 10303 and ISO 13584 provide data regarding Products and Product Components, respectively
- ISO 15531 provides data regarding Resources and Processes
- ISO 13584 provides data regarding Organizations in charge of delivering components

The concept of Resource has been extended to specify Flexible Manufacturing Resources [18.13].

The scope of the research is limited to analyzing the mentioned EXPRESS models, even though modularity and extensibility are possible should more components be appended.

An Extended Enterprise must be formed by the appropriate combination of the following: (1) product configuration, (2) resources, (3) organizations, and (4) manufacturing processes. The mentioned categories of data are defined in *Extended Enterprise Elements Set*, or E^3S. Based on this concept, an Extended Enterprise can be seen as a family of sets, as described by the following equations:

$$E = \{E^3S_r, E^3S_p, E^3S_o, E^3S_m\} \tag{18.1}$$

$$E^3S_i = \{e_i \mid P(e_i)\} \in E \tag{18.2}$$

In Equation (18.2) sub-index i can be substituted by r (resources), p (product), o (organization), m (manufacturing processes). Formula $P(e_i)$ states the defining property of set $\boldsymbol{E^3S_i}$. Equation (18.1) is a construct to define the elements that must be established when individual firms decide to join. It is necessary to explore what relationships can be formed between the E^3S_i sets. The *power set* of $\boldsymbol{E}$ is computed to generate the entire number of combinations that can be formed with $E^3S_i \in \boldsymbol{E}$. The computation of the power set is employed to see what relations or *liaisons* can be generated. Thus, the set of liaisons, $\boldsymbol{L}$, of an Extended Enterprise is given by

$$L = 2^E \tag{18.3}$$

$\boldsymbol{L}$ is a set that contains 16 subsets representing the possible combinations of E^3S_i. More precisely, it contains the possible groups of core capabilities. However, only one element of the set $\boldsymbol{L}$ defines the actual set of data of a particular Extended Enterprise. This represents the *Shape* of the Extended Enterprise. Therefore, once the individual enterprises reach an agreement on what core capabilities they share, the necessary data to support this unique formation is always included in the liaison set $\boldsymbol{L}$, and the Extended Enterprise acquires a certain shape. That is:

$$L = \{S \mid S \subseteq E\} \tag{18.4}$$

The set S contains the actual core capabilities to be shared by the members of the Extended Enterprise. Based on its contents the information system must be modeled and implemented.

According to Equation (18.2), each element E^3S_i that belongs to set $\boldsymbol{E}$ is also a set. To define its elements it is necessary to state what properties must be fulfilled. Thus, formula $P(e_i)$ must be made explicit. Four sets are defined according to the following equations:

$$E^3S_r = \{e_r \mid P(e_r)\} \tag{18.5}$$

where $P(e_r)$ = every entity that contains information of manufacturing resources.

$$E^3S_p = \{e_p \mid P(e_p)\} \tag{18.6}$$

where $P(e_p)$ = every entity that describes product design data.

$$E^3S_o = \{e_o \mid P(e_o)\} \tag{18.7}$$

where $P(e_o)$ = every entity representing an individual firm.

$$E^3S_m = \{e_m \mid P(e_m)\} \tag{18.8}$$

where $P(e_m)$ = every entity that possesses information about manufacturing processes.

Equations (18.5–18.8) are the formal representation of core capabilities regarding Resources, Product, Organizations, and Processes.

The notion E^3S_i is an abstract representation which describes core capabilities. However, it must be established what entities need be taken into account in order to enhance data sharing. A concept called Information Set (IS) is defined in the following equation:

$$E^3S_i = \{xIS_1, xIS_2, \ldots, xIS_j\} \tag{18.9}$$

$$\bigcap_j xIS_j = \varnothing$$

$$\bigcup_j xIS_j = E^3S_i$$

where i = resources, product, organizations, manufacturing processes; j is a positive integer. Table 18.1 summarizes the main Information Sets on which core capabilities are represented. These ISs play the role of host entities where EXPRESS components are adhered to construct the integrated model.

Table 18.1. List of Extended Enterprise Sets

Set	Equation
E	$E = \{E^3S_r, E^3S_p, E^3S_o, E^3S_m\}$
E^3S_r	$E^3S_r = \{RIS, RMIS\}$
E^3S_p	$E^3Sp = \{DRIS, CIS\}$
E^3S_o	$E^3S_o = \{SIS, FMIS\}$
E^3S_m	$E^3S_m = \{MPIS\}$
RIS	$E^3S_{r1} \subset E^3S_r := RIS$
RMIS	$E^3S_{r2} \subset E^3S_r := RMIS$
DRIS	$E^3S_{p1} \subset E^3S_p := DRIS$
CIS	$E^3S_{p2} \subset E^3S_p := CIS$
SIS	$E^3S_{o1} \subset E^3S_o := SIS$
FMIS	$E^3S_{o2} \subset E^3S_o := FMIS$
MPIS	$E^3S_{m1} \subset E^3S_m := MPIS$

18.2.3 The Integrated EXPRESS Model

According to the formalization, Equation (18.1) clearly establishes that an Extended Enterprise is composed of data regarding resources, product, organizations, and manufacturing processes. In the model, the entity E is the core concept, and it is related to entities E3SR, E3SP, E3SO, E3SM. Also, the EXPRESS representation contains the entities RIS and RMIS, which are subtypes of E3SR. In the same way, entities DRIS and CIS depict the hierarchy of E3SP. The EXPRESS equivalent for *organizations* is represented by entities SIS and FMIS, which belong to E3SO. Finally, the information set for the manufacturing processes constructs is represented in the model by the entity E3SM, inherited by MPIS.

The data model of Figure 18.1 possesses four dimensions, each containing the corresponding Information Set where standardized EXPRESS models are attached. For instance, to store information of resources, elements of standard ISO 15531 part 32 [18.14] are attached to entity RMIS. In a similar manner, the entity called RIS is used to integrate information of flexible manufacturing resources, as stated in [18.13]. The integrated EXPRESS model also contains entities found in parts 41 and 44 of STEP to store product data [18.15, 18.16]. They are attached to the hierarchy of entity DRIS. Accordingly, the elements in part 10 of PLIB are glued to entity CIS [18.17]. This approach provides a meta-model for sharing data of products and components.

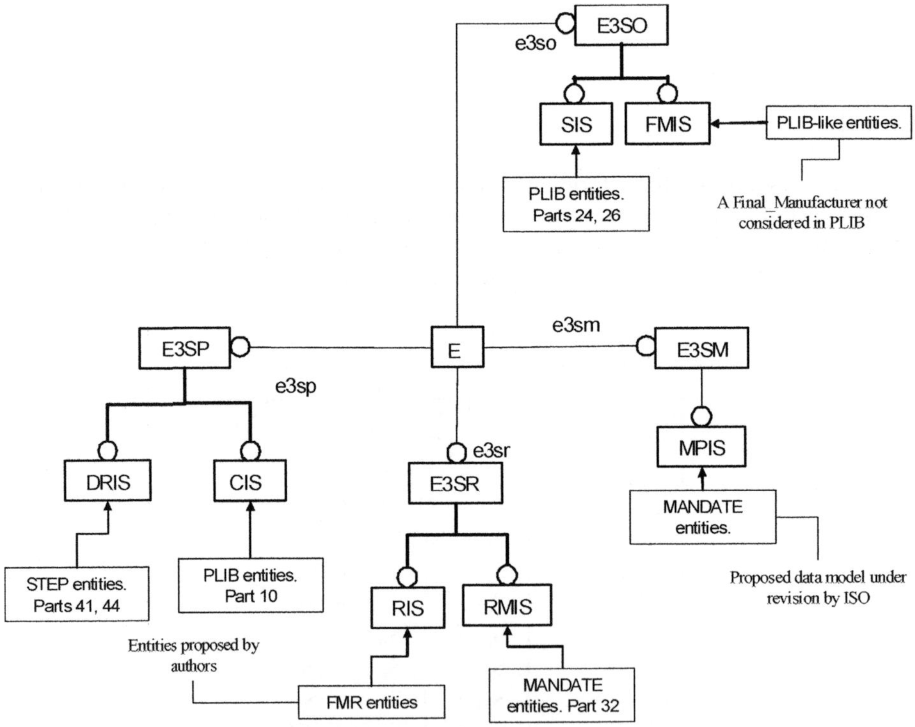

Figure 18.1. Base model for core competences

Data regarding suppliers is included on the PLIB project. These models are contained in the parts 24 and 26 of ISO 13584, which altogether constitute the logical model of suppliers [18.18, 18.19]. This standard defines a `Basic Semantic Unit (BSU)` as an unambiguous and universal identifier to build a dictionary of suppliers. PLIB defines the following: (1) `Supplier`, (2) `Supplier_BSU`, (3) `Supplier_Code`. Using these three entities, the name, address, and identification of the Supplier can be stored. We extend this notion to incorporate more organizations, such as a Final Manufacturer, by means of the `Final_Manufacturer_BSU` entity. Employing this approach, the entity E3SO connects two entities: SIS and FMIS. SIS serves to define the hierarchy of suppliers, whereas FMIS does the same for organizations whose role is that of a final manufacturer.

Extended Enterprises must possess core methods to transform product design data into a series of operations. Consequently, the information system has to be designed to store information about the state of the process or the elements that are required to complete them. Despite the importance of this type of information, a complete EXPRESS model has not been constructed so far. However, part 41 of the MANDATE standard aims to develop an EXPRESS model to store and exchange information about manufacturing processes [18.20].

The integrated framework is appropriate to store information about core capabilities by employing standardized model components. Nevertheless, the ultimate purpose of this research is not only to store core capabilities but also to establish appropriate mechanisms for coordinating data sharing and exchanging. While core capabilities are to be represented by the framework described, the coordination mechanisms are defined after Business Processes, and implemented on a Multi-Agent System. Such design and implementation is presented in the following section.

18.3 Model and Implementation of the Multi-agent System

To obtain a robust Multi-agent System for coordinating the exchange of data in the Extended Enterprise, we depart from a series of Business Processes, some of which have been already proposed [18.21] and some that we define. Business Processes contain sequences of interactions to achieve a design and manufacturing goal, along the product life-cycle. Their main feature is, therefore, the specification of coordinated activities. We take advantage of this feature to determine communication protocols between agents. The execution of those protocols yields the output predicted by the corresponding BP.

18.3.1 Business Processes as Inspiration for Communication Protocols

Although any list of Business Processes might not ever be considered definitive, the one proposed here is broad enough to clarify what interactions take place in distributed manufacturing environments. According to [18.22] a Business Process Reference Model defines three conceptual levels. The first regards Business Process Types (BPT). This level presents business processes at a general level. The second level considers Business Process Variations (BPV), which refers to specializations of Business Process Types. The third and last of the levels deals with Business Sub-processes (BSP). They define sequences of actions to implement a Business Process Type.

In this chapter, the following Business Process Types are proposed (Figure 18.2):

1. Setting design parameters
2. Formalizing core manufacturing resources
3. Generating an extended enterprise *Shape*
4. Configuring an extended enterprise executive system
5. Coordinating production planning
6. Scheduling manufacturing processes
7. Tracking manufacturing processes along the extended enterprise
8. Adapting the system to unforeseen circumstances
9. Tracking customer demands
10. Back-feeding clients' new requirements

Moreover, the interaction between any pair of Business Process Types is bi-directional. To achieve proper coordination it is compulsory to confirm what data is being exchanged vs what data has been solicited. Therefore, Business Processes interactions are given in pairs, and each of them models a server – client communication protocol. A handshake is necessary to initiate communication and appropriate messages are further exchanged. Validation of the data source, the content, and the receiver are required to automate this dynamics.

As can be seen in Figure 18.2 the BPTs form a cycle. The departing point is the setting of design parameters finally to back-feed newer requirements that the market might demand for the product. To operate the Extended Enterprise, core capabilities must be set up. These include definition of the appropriate manufacturing resources on which the product will be manufactured. Then it is necessary to set up plans and a proper scheduling. To automate these activities, BPVs must be further specified from their source BPTs. For instance, Figure 18.3 and Table 18.2 contain the BPT "formalizing core manufacturing resources", while Table 18.3 presents the BPV "Receive Product Specification."

Communication models and actual agent-based programming are the proposed solution to coordinating data flow.

18.3.2 Communication Protocols among Agents to Support Data Exchange

Let A = {Organization Agent, Product Agent, Resources Agent, Manufacturing Management Agent, Coordinator Agent} be an ensemble of agents for the extended enterprise. This ensemble complies with the following features:

- Distribution of data about core capabilities
- Collective decision making
- Central coordination
- Local information systems
- Local ownership of data and knowledge

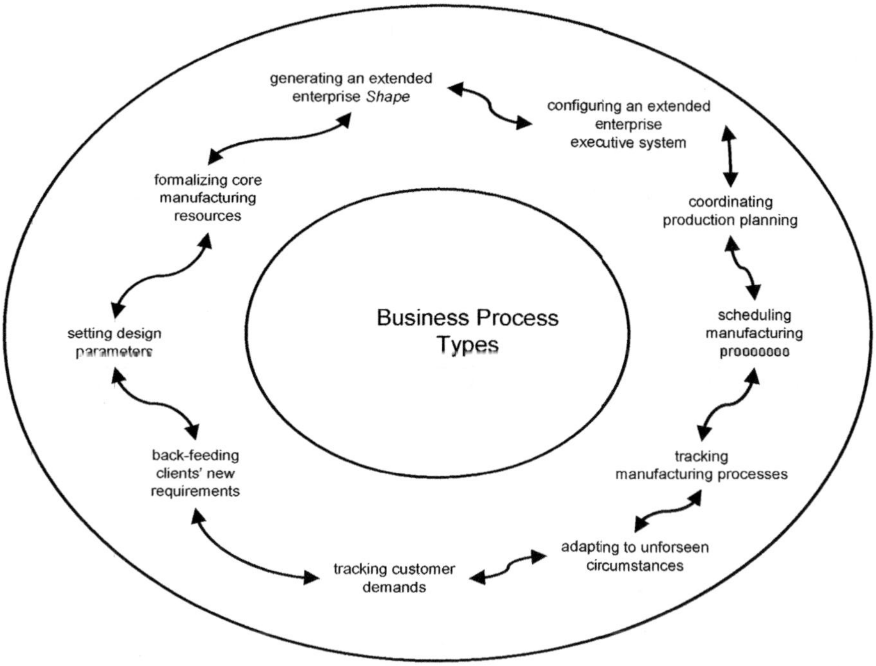

Figure 18.2. Business Process Types for the Extended Enterprise

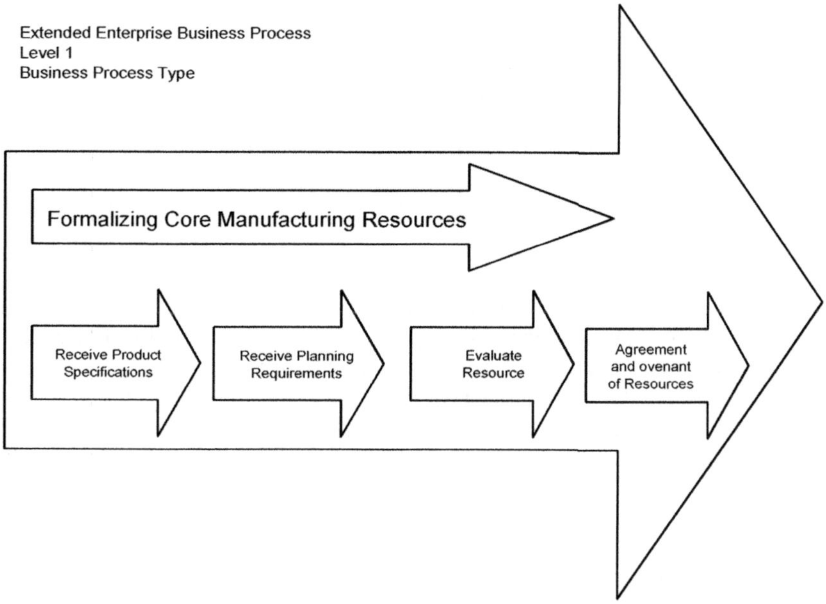

Figure 18.3. Business Process Type: Formalizing core manufacturing resources

Table 18.2. Definition of a Business Process Type

Business process ID	ExEntbp1	
Business Process Type Name	Formalize core manufacturing resources	
Process objective	Publicize, share and agree on the core manufacturing resources	
Inputs	Product specification Dimensional constraints Lot size Due date	
Outputs	Core resources that comply to specification of products and production planning	
Performance attributes	Resource Capability Resource Capacity Resource Tolerance Resource Repeatability Resource Availability Resource Status Resource Configuration Resource Constitution	These attributes are defined in MANDATE.

Table 18.3. Receive Product Specifications

Business Process Variation	Receive Product Specifications
Actors	Product Agent, Resource Agent, Coordinator Agent
Purpose	To receive product data and to propose manufacturing resources that comply with the received specifications, informing the acceptance to the coordinator agent
Summary:	The product agent informs the resource agent about the product specifications. The resource agent proposes candidate resources Upon acceptance, the coordinator agent is informed

Figure 18.4 illustrates the structure of the Multi-Agent System. The coordinator agent is the interface to the real world. It controls the Shape of the Extended Enterprise (Equation 18.4), which is stored in a central repository. Each node is in charge of acquiring and offering information for Resources, Products, Organizations, and Manufacturing Processes. This is done by the corresponding software agent, who obtains data from the local STEP-based information systems.

Agents access and modify STEP-based repositories by employing the SDAI functions. The dynamics of the entire system is dictated by the execution of Business Process Variations. The implementation of these BPVs is actually carried out by the exchange of communicative acts. A communicative act consists of three components: the illocutory component, the locutory component, and the perlocutory component, as explained in [18.23], following the formal language theory formulated by Austin. The locutory component refers to how a message is created on the sender's side. The illocutory component is further composed by an illocutory force (the actual performative) and the illocutory content (the actual value transmitted). The perlocutory component refers to the performed response of the receiver. This response could be either another communicative act initiated by the receiver, in which case it now becomes a sender, or an action, such as modifying a database or performing a calculation. The Agent Communication Language standardized by the Foundation for Physical Agents (FIPA) is closely supported on this notion of communication and, consequently, the JADE platform complies to this theory of communication.

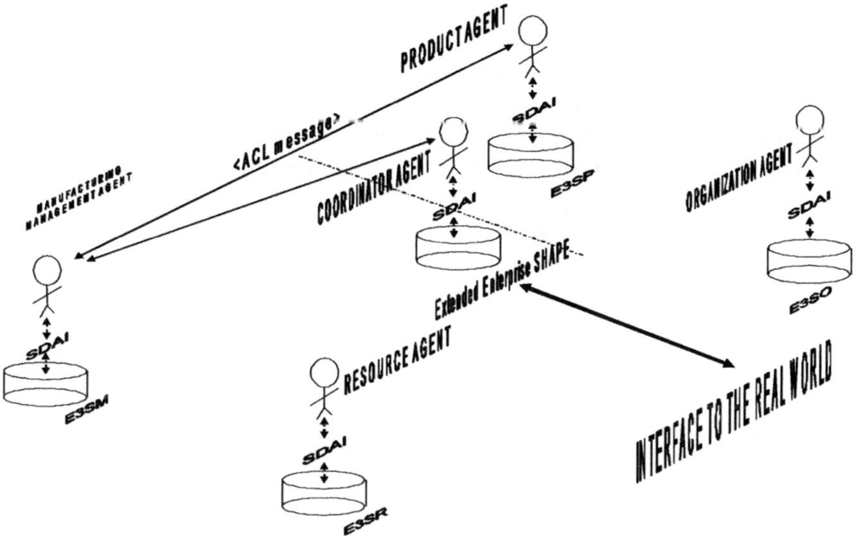

Figure 18.4. Structure of the MAS for the Extended Enterprise

Therefore, let a communicative act between two agents be a tuple consisting of:

<Performative *Sender Agent* *Receiver Agent* Message_Content>

To exemplify, the communication protocol between the Product Agent and the Resource Agent, after Business Process Variation `Receive Product Specifications` is:

<Inform *Product_Agent Resource_Agent* product_specifications>
<Request *Product_Agent Resource_Agent* resource_capabilities>
<Propose Resource_Agent Product_Agent capabilities set>
if (accepted)
{<Accept *Product_Agent Resource_Agent* acceptance_message>}
else
{<Refuse Product_Agent Resource_Agent rejection message>}
<Inform Resource_Agent Coordinator_Agent resource_id>

18.3.3 Agent-oriented Programming

To actually implement the proposed MAS, two issues have to be considered. One refers to the realization of information systems that adhere to the Step Physical File specification. The second is the realization of Multi-Agent Systems according to FIPA. Both of them come separately. Development tools exist that facilitate modelling and programming STEP-based applications. Similarly, FIPA-compliant platforms such as JADE [18.24] are robust enough to support the intense data flow that is expected in this setting. A complete implementation of the proposed architecture might require a strategic decision on which tools are best fitted to construct distributed MAS to exploit STEP-based repositories. We chose JADE as a development tool for MAS. The decision on what STEP tool is left to the readers.

The following lines of code illustrate how to receive and send messages on the JADE platform. The action to be executed depends on whether the corresponding message template matches the message received or not. A call to SDAI functions, implemented on the ST-Developer software, is performed either to store or retrieve data from a STEP physical file. Its contents are then transmitted via ACL messages.

A request for resources capabilities is performed by the product agent, via an ACL message to the resource agent:

```
   class requestResourceCapabilities extends CyclicBehaviour
   {
     public void action()
     {
        ACLMessage msg_rqt=new ACLMessage(ACLMessage.REQUEST);
        msg_rqt.addReceiver(new
AID("resource_agent",AID.ISLOCALNAME));
        msg_rqt.setContent(getProductDimension());//by using
Java binding of SDAI
       ((product_agent)myAgent).send(msg_rqt);
       block();}}
```

In turn, the resource agent informs the list of resources that match what it's been requested by the coordinator agent:

```
   class sendResources extends CyclicBehaviour
   {
   public void action()
   {
   MessageTemplate
tem1=MessageTemplate.MatchPerformative(ACLMessage.REQUEST);
   MessageTemplate tem2=MessageTemplate.MatchSender(new
AID("product_agent",AID.ISLOCALNAME));
   MessageTemplate tem3= MessageTemplate.and(tem1,tem2);
   ACLMessage
msg_requested=((resource_agent)myAgent).blockingReceive(tem3);
   if(msg_requested!=null)
   {
   ACLMessage msg_propose=new ACLMessage(ACLMessage.PROPOSE);
   msg_propose.addReceiver(new
            AID("product_agent",AID.ISLOCALNAME));
   ((resource_agent)myAgent).send(msg_propose);
   msg_propose.setContent(getResources()));//by using Java
binding                                //of SDAI
   }
   Else {block();}}}
```

In the previous two pieces of code, methods `getProductDimensions()` and `getResources()` contain specific instructions to access and modify the STEP-based information system.

JADE is advantageous due to its capability to wrap an *agent container* in Java code. The following excerpt is illustrative for this purpose:

```
public class agentCreator extends Agent
{
    Runtime rt = Runtime.instance();
    AgentController ac;
    ContainerController cc;
    /** Creates a new instance of agentCreator*/
    public agentCreator(String host, String port)
    {
        Profile p = new ProfileImpl();
        p.setParameter(Profile.MAIN_HOST, host);
        p.setParameter(Profile.MAIN_PORT, port);
        cc = rt.createMainContainer(p);
    }
    public AgentController launchAgent(String name, String
         nameClass)
    {
        if (cc != null)
        {
            try
            {
              ac = cc.createNewAgent(name, nameClass, null);
                ac.start();
                return ac;
            }
            catch (Exception e)
            {e.printStackTrace();}}
        return null;
    }
    public AgentController launchAgent(Agent MyAgent, String
         name, String nameClass)
    {
        cc = MyAgent.getContainerController();
        if (cc != null)
        {
            try
            {
              ac = cc.createNewAgent(name, nameClass, null);
                ac.start();
                return ac;
            }
            catch (Exception e)
            { e.printStackTrace();}
       }
        else {System.out.println("-----> EMPTY CONTAINER");}
        return null;}}
```

The previous coding is thought to be used co-ordinately with a *main* class in order to execute the MAS. This is illustrated next:

```
public class Main {
     public Main() {
     }
     public static void main(String[] args) {
         String nameC="Coordinator";
         String hostName="localhost";
         String portName=null;
         agentCreator agcre = new
agentCreator(hostName,portName);
          AgentController coord =
     agcre.launchAgent(nameC,"extended_enterprise.Coordinador"
);
     }
}
```

The corresponding Java packages for creating and manipulating STEP-based repositories through the SDAI functions must be imported so that the JADE agents operate as desired.

18.3.4 Exemplification of a Business Process Type

This subsection presents the communication between Coordinator Agent and Product Agent. This process is carried out by establishing the appropriate ACL messages, as presented in previous coding. This, however, is fully transparent to the final user, who only interacts via GUIs.

The BPT "Setting Design Parameters" is chosen to exemplify. Even though the ultimate goal is to actually exchange data along the product life-cycle after defining the shape of the extended enterprise, we concentrate on data that is exchanged along this BPT. "Setting Design Parameters" is one of the most complex BPTs, because it is intended to merge qualitative product data and technical product data in order to provide further input to manufacturing, i.e., a list of suitable materials, processes, and resources. We intend to focus on data exchange, rather than explaining the inner algorithms used as part of the intelligence given to Product Agent. Thus, by setting a list of qualitative aspects of the final product, and taking into account the specific manufacturing domain, Product Agent determines the prioritization of design objectives to fulfil during a further manufacturing stage.

There must be an extended enterprise manager (a human) who is in charge of starting the system. Input to this BPT relates to data that has been gathered from the target market segment, which is filled to the MAS. This is done through a GUI attached to the Coordinator Agent. The MAS is first launched (Figure 18.5) by the Extended Enterprise Manager. Then a new product's identification is established (Figure 18.6). The manager uses menus of Figure 18.7 to provide the set of qualitative functions, which are received by Coordinator Agent.

The qualitative functions are arranged in four categories: (1) usability, (2) performance, (3) tactile sensation, and (4) hygiene. Each category, in turn, possesses the actual functions along with a measure of how the market desires such functions in the final product. In Figure 18.7, under the Performance category (desempeño), a product not too heavy is given an evaluation of 2 on a scale of 1 to 5.

Shortly after Coordinator Agent sends Product Agent the entire set of qualitative functions, Product Agent runs an algorithm to obtain the priorities of the design objectives. Product Agent's task is to transform these qualitative functions into a series of design objectives to fulfil during the manufacturing stage.

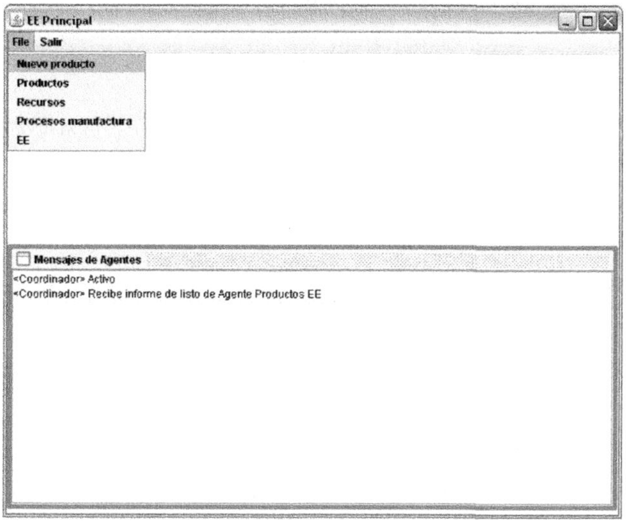

Figure 18.5. Extended Enterprise manager launches the MAS

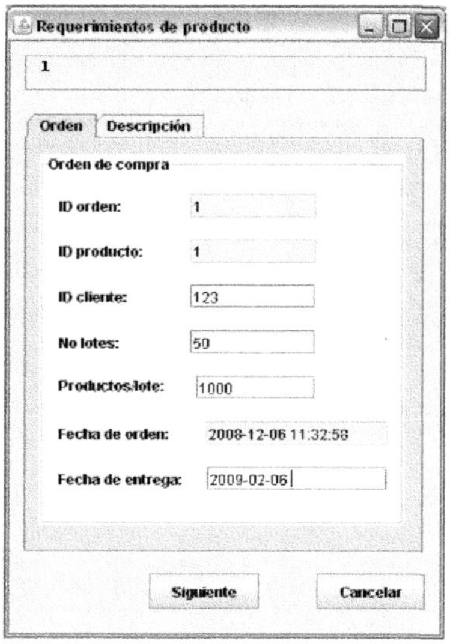

Figure 18.6. Coordinator Agent receives data for product identification

The final results for this example are displayed in Figure 18.8, where the ranking assigned to design objectives in order to comply with all the qualitative functions can be seen. According to the results given by Product Agent, the material to use (10 points), is followed by the shape of the product (Forma), with an evaluation of 7.704 points. Dimensions (Dimensiones) comes as the third most important design objective with 7.049 points. Finishing Type (Acabado de Superficie) gets a score of 6.721 points. Weight (peso) gets 3.606 points.

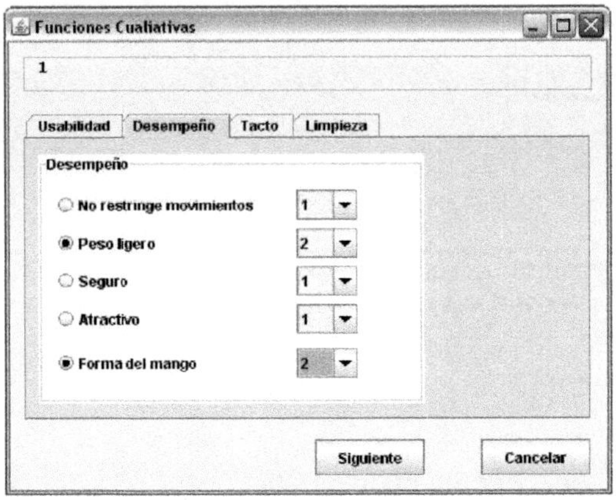

Figure 18.7. Coordinator Agent receives product's qualitative functions

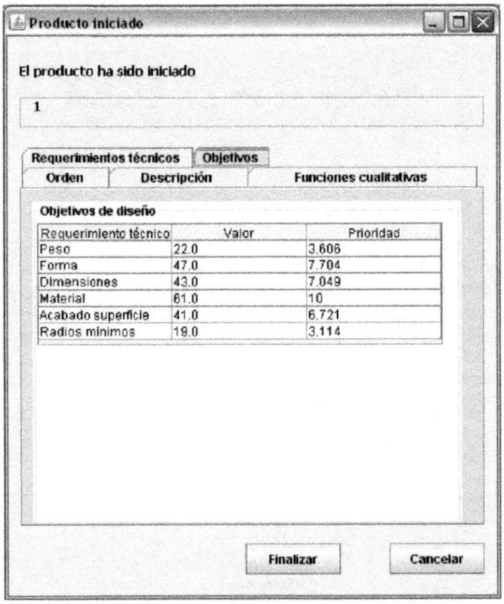

Figure 18.8. Product Agent establishes priorities for design objectives

These priorities determine important pieces of data. They establish that the final product must be made of a material that can be shaped properly as the client determines, it must fit a wide range of dimensions, and finishing type must be attractive. Weight, however, must be taken into consideration but it is not as important as the previous objectives. On these data, the right kind of material, machining processes, and flexible manufacturing resources must be set. Once the appropriate types are elucidated, Coordinator Agent searches what individual enterprises are assigned the contract to supply material and manufacture the product. This dynamics, though, will be presented in a research article dedicated to the matter.

18.4 On the Networking of Enterprises

18.4.1 Multi-agent Systems on Distributed Design and Manufacturing

Several proposals to include agent technology for virtual enterprises have been reported. [18.25] reported an agent-based supply chain management system. Symeonidis et al. [18.26] employ a multi-agent system in order to plan enterprise resources through data mining techniques. An MAS aimed at planning enterprise-wide resources is depicted in [18.27]. Park [18.28] also introduces the notion of a coordinator agent to smooth the activities being performed in a distributed manufacturing setting. Agent systems have also been developed to optimise utilization of resources within existing manufacturing systems, and deal with changing demand and products [18.29]. An agent-based architecture is assigned to support the scheduling of distributed manufacturing resources in a Virtual Enterprise [18.30]. Recent work focuses on how to mix Web services and agents to integrate inter-organizational business processes [18.31, 18.32]. In [18.33] they consider two types of agents: the leader agent, who receives a customer request and accepts achieving it, and the member agent, who agrees to cooperate after being asked by a leader agent indicating the capability in which it is interested. The Multi-Agent approach has been explored to acquire data from heterogeneous information systems [18.34]. A hybrid architecture consisting of four different types of workspaces to integrate different sources of product and process information from a network of enterprises is proposed. All those perspectives reinforce the idea of core competences that we present here.

18.4.2 Covenants in the Extended Enterprise

The relatively high degree of absence of face-to-face interactions and cultural differences of distributed organizations stress the need for creating a common identity [18.35]. Identity building is the process of developing meaningful patterns of interpretation related to one or several cultural settings. The aim of identity building in modern organizations is to combine collective and individual interests in such a way that the organization improves its capacity to act and react to changes of external economic, technological and political settings. An Extended Enterprise is viewed as a chain of enterprises on a pair-wise relationship between customers and suppliers [18.36]. In [18.37] an Extended Enterprise is perceived as a community obliged to follow the terms of contracts, which are specified on XML, stating specific roles and responsibilities to fulfil by each individual enterprise within the community. Boardman [18.38] attempts to develop a framework based on the term *structured engagement*, which specifies how individual firms acquire provisional understandings between suppliers, customers, and partners before engaging in the development of new products. With the concept of structured engagement, expertise of individual firms is established by a series of steps: Capability assessment, Concept definition, Product realization and Operation. To realize effectively the quality control of extended enterprises, it is necessary to investigate on suitable quality control models [18.39].

18.5 Conclusions

Bringing together STEP-based information systems, business processes and Multi-Agent Systems is challenging yet achievable. The appropriate technological tools exist to construct STEP-based information systems and to build software under the Multi-Agent Systems paradigm. On this, we successfully defined the Shape of an Extended Enterprise, which is reflected in the Extended Enterprise Executive System. This Shape is formed by sharing core capabilities. Also, theoretical contributions have defined interactions occurring when a group of enterprises share abilities and produce value-added goods and services. There are still some gaps in knowledge to be addressed. It is necessary to discuss whether classic planning techniques (ERP, for instance) are adequate to plan and control production in the entire network. Another research opportunity regards optimizing the network performance prior and during the existence of the extended enterprise. It is also of great interest to couple existing CAD, CAM, and CAPP systems to STEP-based repositories, and integrate the MAS so that data coming from design is transferred and transformed along the product life-cycle. Yet the proposal here presented is valuable since it illustrates some of the possible models and dynamics that are to be used if MASs and STEP is the favored approach to networking enterprises.

References

[18.1] Browne, J., 1995, "Future manufacturing systems – Towards the extended enterprise", *Computers in Industry,* **25**, 235 - 254.

[18.2] Kovács, George and Paganelli, Paolo, 2003, "A planning management infrastructure for large, complex, distributed projects: beyond ERP and SCM" *Computers in Industry*, **51**, 165 – 183.

[18.3] Goethals, F., Vandenbulcke, J., Lemalhieu, W., 2004 "Developing the extended enterprise with FADEE", In *Proceedings of the 2004 ACM Symposium on Applied Computing*, 1372 – 1379.

[18.4] Kuczynski, A., Stokic, D., Kirchoff, U., 2006, "Set-up and maintenance of ontologies for innovation support in extended enterprises", *International Journal of Advanced Manufacturing Technology*, **29**, 398 – 407.

[18.5] Chen, D., and F. Vernadat, 2004, "Standards on enterprise integration and engineering – state of the art", *International Journal of Computer Integrated Manufacturing,* 17, 235-253.

[18.6] Jennings, N. R., 1999, "Agent oriented software engineering", *Lecture Notes in Artificial Intelligence*, **1661**, 4–10.

[18.7] López-Ortega, Omar and Ramírez, Moramay, 2007, "A formal framework to integrate EXPRESS data models in an extended enterprise context", *Journal of Intelligent Manufacturing*, **18**, 371 – 381.

[18.8] Cutting-Decelle, A. F., J.J. Michel, 2003, "ISO 15531 MANDATE: a standardised data model for manufacturing management" *International Journal of Computer Applications in Technology*, **18**, 43 – 61.

[18.9] Cutting-Decelle, A. F., R.I.M. Young, J.J. Michel, R. Grangel, J. Le Cardinal, J.P. Bourey, 2007, "ISO 15531 MANDATE: A Product-process-resource based approach for managing modularity in production management" *Concurrent Engineering*, **15**, 217-235. DOI: 10.1177/1063293X07079329.

[18.10] International Organization for Standardization, 2004, "ISO 15531-1 Industrial automation systems and integration – Industrial manufacturing management data -- Part 1: General overview", Geneva, Switzerland.

[18.11] Pierra, Guy, 2000, "Représentation et échange de données techniques", *Mec. Ind.*, **1**, 397 – 414 (In French).

[18.12] Pierra, Guy, 2003, "Context-explication in conceptual ontologies: The PLIB approach", In *Proceedings of Concurrent Engineering 2003, Special track Data Integration in Engineering*, Madeira, Portugal, July 2003.

[18.13] López-Ortega, Omar and Ramírez, Moramay, 2005, "A STEP-based manufacturing information system to share flexible manufacturing resources data", *Journal of Intelligent Manufacturing*, **16**, 287 – 301.

[18.14] International Organization for Standardization, 2005, "ISO 15531-32 Industrial automation systems and integration – Industrial manufacturing management data -- Part 32: Conceptual model for resources management data", Geneva, Switzerland.

[18.15] International Organization for Standardization, 1997, "ISO 10303 Industrial Automation Systems and Integration – Product data representation and exchange – Part 41: Generic Description of Products", Geneva, Switzerland.

[18.16] International Organization for Standardization, 1997, "ISO 10303 Industrial Automation Systems and Integration – Product data representation and exchange – Part 44: Product Configuration", Geneva, Switzerland.

[18.17] International Organization for Standardization, 2001, "ISO 13584 Industrial automation systems and integration – Parts library – Part 10: Conceptual models of Parts Library", Geneva, Switzerland.

[18.18] International Organization for Standardization, 2003, "ISO 13584-24 Industrial automation systems and integration – Parts library – Part 24: Logical resource: Logical model of supplier library", Geneva, Switzerland.

[18.19] International Organization for Standardization, 2000, "ISO 13584-26 Industrial automation systems and integration – Parts library – Part 26: Logical resource: Information supplier identification", Geneva, Switzerland.

[18.20] International Organization for Standardization, 1998, "ISO 15531 Industrial Automation Systems and Integration – Industrial Manufacturing Management Data – Part 41: Manufacturing flow management data – Overview and Fundamental Principles", Geneva, Switzerland.

[18.21] Jiang, P.Y., Zhou, G., Zhao, G., Zhang, Y.F., Sun, H.B., 2007, "e2-MES: an e-service-driven networked manufacturing platform for extended enterprises", *International Journal of Computer Integrated Manufacturing*, **20**(2-3), 127-42

[18.22] Choi, Y., Kang, D., Chae, H., Kim, K., 2008, "An enterprise architecture framework for collaboration of virtual enterprises", *International Journal of Advanced Manufacturing Technology*, **35**, 1065 – 1078.

[18.23] Jacques Ferber, 1999, "Multi-Agent Systems: An Introduction to Distributed Artificial Intelligence" 1st English version, London, UK. Addison – Wesley.

[18.24] Bellifemine, F. L., Caire, G., and Greenwood, D., 2007, "*Developing Multi-Agent Systems with JADE",* John Wiley and Sons, USA.

[18.25] Julka, N., Srinivasan, R., and Karimi, I., 2002, "Agent-base supply chain management -1: framework", *Computers and Chemical Engineering*, **26**, 1755 – 1769.

[18.26] Symeonidis, A. L., Kehagias, D., and Mitkas, P., 2003," Intelligent policy recommendations on enterprise resource planning by the use of agent technology and data mining techniques", *Expert Systems with Applications*, **25**, 589 – 602.

[18.27] Lea, B. R., Gupta, M. C., and Yu, W. B., 2005, "A prototype multi-agent ERP system: an integrated architecture and a conceptual framework", *Technovation*, **25**, 433 - 441.

[18.28] Park S, Sugumaran V., 2005, "Designing multi-agent systems: a framework and application", *Expert Systems with Applications*, **28,** 259–271

[18.29] Anosike, A I., Zhang, D. Z., 2007, "An agent-based approach for integrating manufacturing operations", *International Journal of Production Economics*.

[18.30] Wang, D, Nagalingam, S-V, Lin, G-C-I., 2007, "Development of an agent-based Virtual CIM architecture for small to medium manufacturers", *Robotics and Computer Integrated Manufacturing*, **23:** 1-16.

[18.31] Shen, W, Qi, H, Shuying, W, Yinsheng, L, Hamada, G., 2006, "An agent-based service-oriented integration architecture for collaborative intelligent manufacturing", *Robotics and Computer Integrated Manufacturing*, **23:** 315-325.

[18.32] Fung, S. C., 2005, "An agent-based infrastructure for virtual enterprises using web-services standards", *Journal of Advanced Manufacturing Technology*, DOI: 10.1007/s00170-007-1243-1.

[18.33] Zarour, N., Boufaida, M., Seinturier, L., Estraillier, P., 2005, "Supporting virtual enterprises systems using agent coordination", *Knowledge and Information Systems*, **8**, 330 – 349.

[18.34] Nahm, Y.-E and Ishikawa, H., 2005, "A hybrid multi-agent system architecture for enterprise integration using computer networks", *Robotics and Computer-Integrated Manufacturing*, **21**, 217 – 234.

[18.35] Rasmussen, L. B., Wangel, A., 2007, "Work in the virtual enterprise – creating identities, building trust, and sharing knowledge", *Artificial Intelligence and Society*, **21**, 184 – 199.

[18.36] Bititci, Umit, S., Meddibil, Kepa, Martínez, Verónica and Albores, Pavel, 2005, "Measuring and managing performance in extended enterprises", *International Journal of Operations and Production Management*, **25** (4), 333 – 353.

[18.37] Linington, P. F. et al, 2004, "A unified behavioural model and a contract language for extended enterprise", *Data and Knowledge Engineering*, **21**, 5 – 29.

[18.38] Boardman, John T. and Clegg, Ben, 2001, "Structured engagement in the extended enterprise" *International Journal of Operations and Production Management*, **21** (5-6), 795 – 811.

19

An XML Implementation for Data Exchange of Heterogeneous Object Models

X.Y. Kou[1] and S.T. Tan[2]

Department of Mechanical Engineering, The University of Hong Kong, Pokfulam Road, Hong Kong, China
Email: [1]kouxy@hku.hk, [2] sttan@hku.hk

Abstract
Heterogeneous objects refer to objects with spatially different material compositions or structures. Tremendous research efforts have been devoted to modelling heterogeneous objects and many heterogeneous object representations have been proposed. Regardless of the diversity of these CAD models, there are needs to transport and exchange the included *geometry*, *topology* as well as *material distribution* between CAD modellers, CAE tools and CAM facilities. In literature and practical applications there have been lots of STEP (STandard for the Exchange of Product model data) based tools and implementations for the exchange of the geometric/topological data. However, there has been only limited research on the data exchange of material distributions. This chapter focuses on an XML implementation for data exchange of heterogeneous CAD models. The proposed heterogeneous CAD model is described by Extensible Markup Language and detailed approaches to represent the *voxel based*, *explicit function based* and *heterogeneous feature tree based* models are described. The idea is to introduce self-descriptive, customised tags/vocabularies to fit the specific needs of material modelling. The structure of the heterogeneous CAD model is specified with XML schemas and related data validations can accordingly be checked to ensure the model correctness. A prototype CAD module is developed to construct XML-based heterogeneous material model, and the XML model is then exported to SolidWorks to test the validity of the proposed approach. Results show the proposed XML based model can facilitate the data exchanges of heterogeneous material distributions.

19.1 Introduction

Heterogeneous object modelling [19.1,19.2] is a relatively new research direction in the CAD community. Different from traditional homogeneous solid modelling, in which the material distribution of an object is assumed uniform in geometric domain, heterogeneous object modelling incorporates and utilises spatially varying material distributions as additional design freedoms.

The advantages of using heterogeneous material distributions in CAD design have been getting increased recognition in recent years. One primary reason is that the users' design requirements are usually manifold and can seldom be fulfilled with a single homogeneous material. For instance, in artificial finger joint replacement, the implanted finger joint should be both strong and biocompatible. Homogeneous metals (stainless steel, for instance) can provide sufficient strengths; however, the bonding of the artificial joint with the human bones are not ideal. Using biocompatible materials (e.g. titanium, etc.) can enhance the bonding of the artificial joints with the fingers, but the wear resistance of the joint is relatively poor, which may deteriorate the finger movement accuracy and long life use. To alleviate the problems, a heterogeneous finger joint can be developed, as shown in Figure 19.1. In this heterogeneous finger joint, two primary materials are used: material A (cylindrical in shape) has good biocompatibility and can gradually grow into bones (thus offering enhanced integrity with the fingers); and material B (kidney shape) has good strength and wear resistance properties and can contribute to the long life use of the joint. The gradient material region in Figure 19.1 is designed to eliminate possible cracks due to abrupt material changes in the material interfaces.

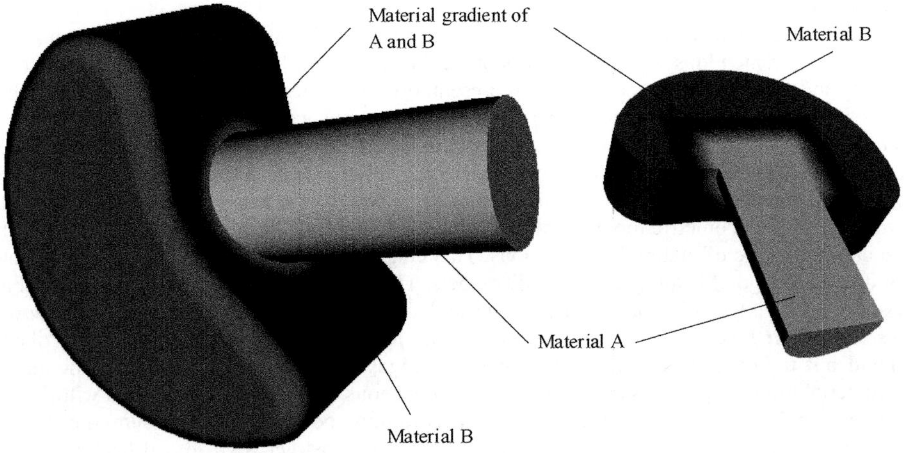

Figure 19.1. An example heterogeneous object

The example shown in Figure 19.1 demonstrates a typical heterogeneous object application, in which the material heterogeneities contribute significantly to the object's functionality and performance. In recent years, similar applications have also been found in mechanical, electrical, thermal and other interdisciplinary areas [19.1].

The wide applications of heterogeneous objects have aroused very active research in modelling, analysis and fabrication of heterogeneous objects in the past few decades. In the literature, many heterogeneous object models have been proposed [19.1,19.2]. These heterogeneous object models differ in their *representational capacities, design intuitiveness, model exactness, compactness, etc.* and many of them target to one or two specific application fields, for instance, *visualisation* or *finite element analysis* of heterogeneous objects.

In practical engineering applications, a heterogeneous object model usually needs to be converted to another format to share/exchange the geometry, topology as well as the associated material heterogeneities. This naturally calls for a standard data format to accomplish such goals. The exchange of geometric/topological data has been extensively studied by many researchers and engineers, mostly under the framework or STEP (STandard for the Exchange of Product model data), and numerous such papers have been published, as reviewed in other chapters of this book. In the literature, however, there is only limited research which tackles the issues of data exchange of material distributions/heterogeneities. This chapter discusses XML based heterogeneous CAD models to cater for such needs.

A brief discussion on XML (Extensible Markup Language) and ISO 10303 is first provided in Section 19.2, where the scopes of current XML-based data exchange are analysed. Before delving into the proposed XML-based heterogeneous model, a concise review on existing heterogeneous object representations is presented to provide sufficient backgrounds in this field. Our Heterogeneous Feature Tree (HFT) based model is particularly elaborated as it is taken as an example to demonstrate the proposed approach. Section 19.4 focuses on the implementation issues and a case study is provided. Finally this chapter is concluded in Section 19.5.

19.2 XML Technologies and ISO 10303

The Extensible Markup Language is a general-purpose specification for creating custom markup languages. It allows users to define semantic constraints for specific applications and enables versatile data to be described in plain text. The plain-text nature facilitates data sharing and exchange across different languages and platforms. By defining meaningful markups, the data can be self-descriptive and much easier to understand.

Using XML to describe/exchange product data is relatively new [19.3, 19.4] and EXPRESS language (ISO 10303-11) [19.5] has always been the formal language to describe product data and their relationships. Recent studies, however, show that XML based data exchange is much easier, more flexible and systematic [19.4–19.7]. In the past few years, there have been many investigations that target on mapping EXPRESS schema/data to XML schema/data for information exchanges. Stephen Chan et al. [19.4] discussed product data exchanges using XML-based mediators, and they analysed the pros and cons of "early and late binding mappings from STEP to XML". Barkmeyer and Lubell [19.7] discussed the issues of reformulating EXPRESS model as XML and tackled some mismatch problems during the mapping, for instance the issues encountered during name mapping, attribute mapping and element mapping. ISO 10303-28:2007 standard [19.8] documents detailed specifications on the use of XML to represent EXPRESS schema as well as the data governed by EXPRESS schemas. The latest release of ST-Developer SDK v12 [19.9], developed by STEP Tools Inc., provides programmatic interfaces to read and write STEP Part-28 XML files to and from many commonly used data formats (inclusive of STEP Part-21 file modelled in EXPRESS). These efforts take advantage of both the rich semantics of EXPRESS language and the widespread infrastructure of XML; however these existing approaches are mostly targeted to objects/products with homogeneous material definitions. Although ISO 10303 has included many specific Application Protocols (AP), for instance, the AP 202 (Associative draughting), AP 203 (Configuration –controlled design) , AP207 (Sheet metal die planning and design), AP 227 (Plant spatial configuration), etc. [19.10], so far there are no application protocols which can be used for data exchanges of heterogeneous object models. To the best of our knowledge, the series of work done by Patil et al. [19.11, 19.12] seem to be the only available investigations along this research direction. Patil et al. [19.11, 19.12], however, use EXPRESS-G language (a subset of EXPRESS family language) to construct STEP-compliant heterogeneous object models and therefore cannot take full advantage of XML's merits, for instance self-descriptive property and explicit hierarchies [19.4, 19.7]. The approach presented in this chapter reflects our attempt towards this goal and a prototype CAD module is proposed to construct and parse XML-based heterogeneous CAD model, as is detailed below.

19.3 An XML Implementation for Data Exchange of Heterogeneous Object Models

19.3.1 Existing Heterogeneous Object Models

Most conventional CAD models assume the material of the product under design is homogeneous on and inside the object's boundaries. The modelling space is usually three-dimensional Euclidean space E^3 or its subspace [19.1], and the major focus in geometric modelling is the shape and spatial relations [19.13,19.14]. In addition to geometric information, heterogeneous object modelling also deals with material heterogeneity defined over the geometric domain. Under such assumptions, for any two points inside a heterogeneous object, their material compositions might be, in general, distinct.

To model the material compositions at a given location, the most common way is to user a k-dimension vector $(r_1, r_2, ..., r_k)$, where r_i represents the volume fraction of the i-th primary material of interest. Within the framework of heterogeneous object modelling, a heterogeneous object can be regarded as a point set $\{P\}$ [19.1]:

$$P = (P_g, P_m), P_g = (x, y, z) \in \Omega_g \subset E^3,\ \ P_m = (r_1, r_2, \cdots r_k) \in \Omega_m \subset E^k$$
$$0 \le r_i \le 1,\ \ 1 \le i \le k, \sum_{i=1}^{k} r_k = 1 \qquad (19.1)$$

where P_g is the location of the point P in the geometric domain Ω_g, P_m is the material composition defined at P_g, and Ω_m is the material domain (subspace of E^k) [19.1]. Note that all the scalars r_i are constrained to sum up to unity such that the material composition P_m is physically meaningful.

There are many ways to represent heterogeneous material distributions in E^3 or its subspace. The voxel based model is perhaps the most intuitive one. In a voxel model, a heterogeneous object is represented as a collection of heterogeneous voxels, each of which represents a small cube in space with a homogeneous or interpolated material distribution [19.15–19.18]:

$$O = \{V_i\} = \{(x_i, y_i, z_i, m_i)\},\ 1 \le i \le n \tag{19.2}$$

where V_i is a representative voxel in object O, (X_i, Y_i, Z_i) is the voxel's location in space, m_i denotes its material compositions and n is the number of voxels. Figure 19.2 shows an example heterogeneous object modelled in a voxel array.

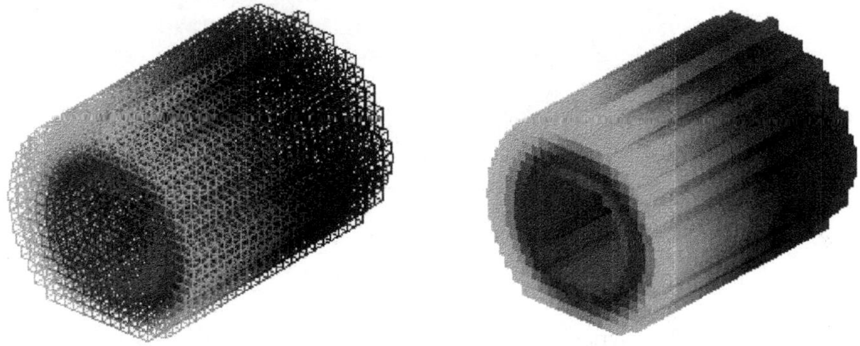

(a) (b)

Figure 19.2. A voxel representation of a heterogeneous hollow cylinder (a) The wireframe view (b) The shaded view

In the above voxel model, it is seen that the material heterogeneity is represented as enumeration of material compositions, as rendered in different colours in Figure 19.2. Apart from such direct enumerations, it is also natural to use analytic functions $V = f(x, y, z)$ to denote the material composition at the position (x, y, z), for instance power-law functions [19.19] and exponential and parabolic functions [19.20]. Figure 19.3 shows a component whose heterogeneity follows a linear material gradation along the Y-axis.

Despite many merits such as the intuitiveness and easy implementations, it is not always possible, or sometimes non-trivial, to describe the material distributions with explicit mathematical functions, for instance, the hollow cylinder shown in Figure 19.4. The cylinder is "extruded" from a heterogeneous disk (Dsk, see Figure 19.4), which has a graded material transition between two circles (Cir1 and Cir2). Along the extrusion direction, the material distributions of these two circles are governed by two heterogeneous lines (L1 and L2). The lines' material are in turn dependent on the location and material definition of the point pairs (P1, P2) and (P3, P4) respectively. The final material distributions of the hollow cylinder are shown in Figure 19.4a from two different viewing directions.

Figure 19.3. Material heterogeneity represented with analytic functions

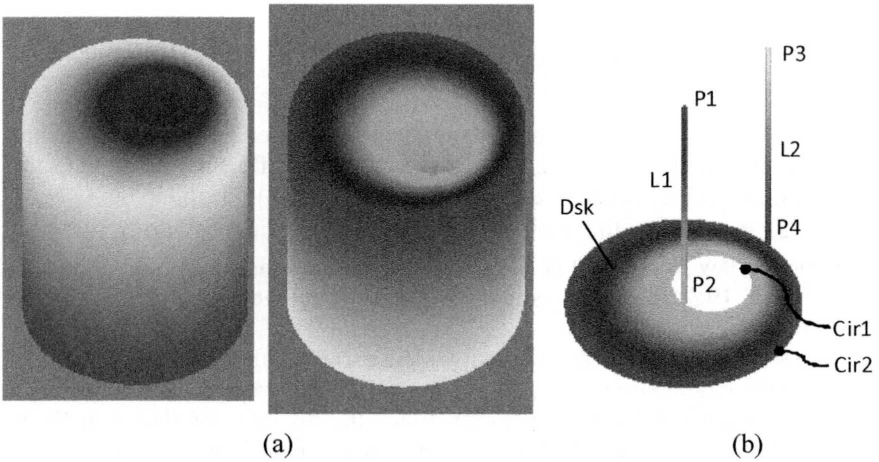

Figure 19.4. A heterogeneous cylinder with trivariate material gradations (a) 3D shaded view from two viewing directions (b) Dependent heterogeneous features

To represent the material heterogeneity shown in Figure 19.4, an explicit function based model is usually inadequate or less flexible. In our previous papers [19.21,19.22], we proposed a generic data structure called Heterogeneous Feature Tree (HFT) to model such material variations. The key idea is to encode the material variation dependency into a tree structure, where the parent feature's material distribution is defined to be dependent on its child features' spatial locations as well as their material definitions. For instance, the material distribution of the cylinder is dependent on the heterogeneous disk (from which it is extruded) and therefore, the disk is modelled as the child feature of the cylinder, as shown in Figure 19.5. The disk's material is defined as a linear material gradation between the two circles, so Cir 1 and Cir 2 are saved as the children of the "Dsk" feature. Such hierarchies are continuously constructed until the leaf nodes (the four points, whose material definitions do not depend on any other features, i.e. without any child features) are populated into the HFT structure, as illustrated in Figure 19.5.

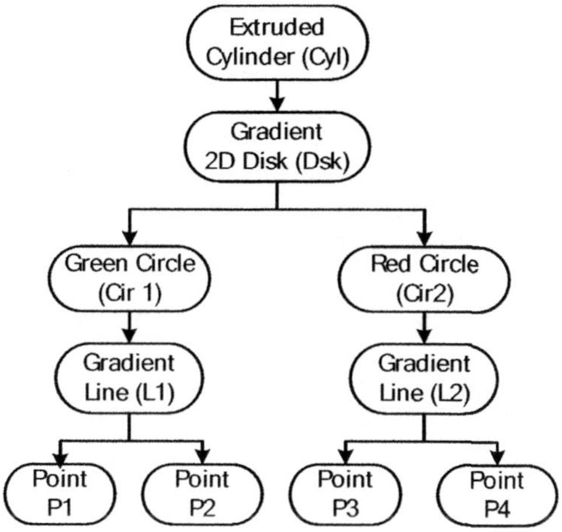

Figure 19.5. A simplified heterogeneous feature tree structures for representing the material heterogeneity shown in Figure 19.4

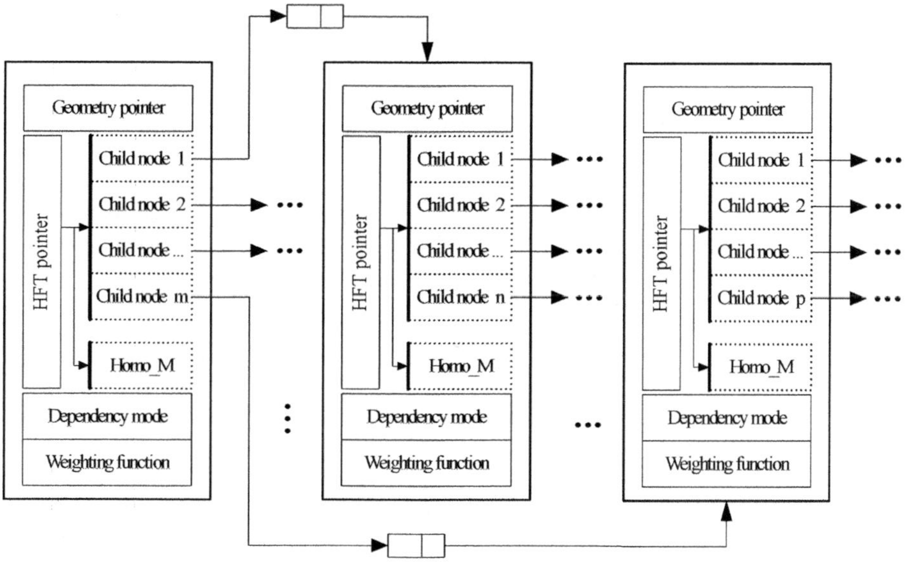

Figure 19.6. The Heterogeneous Feature Tree structure (HFT)

The heterogeneous feature tree structure shown in Figure 19.5 only qualitatively describes the material variation dependency relationships. To represent the material heterogeneity of interest precisely, we propose to associate a list of helper directives with each HFT node, as is graphically illustrated in Figure 19.6. For instance, each HFT node has an associated material gradation function which controls how child features' material compositions are blended to define the parent's material heterogeneities. More detailed descriptions of these parameters and their usages are beyond the scope of this chapter and readers can refer to [19.21–19.23] for more details.

Apart from the aforementioned data models, there are many other different heterogeneous object representations and it is beyond the scope of this chapter to offer an exhaustive analysis on each of the models. For vigorous computer models in the field of heterogeneous object modelling, refer to [19.1] for details.

19.3.2 Representing Material Heterogeneity with XML

Because of the many merits of XML, we propose to use XML based models to describe heterogeneous material distributions defined over the geometric domain. The voxel model, the explicit function model and the HFT based model are taken as examples to illustrate the proposed approach.

Representing Homogeneous Materials with XML

To represent the material vector (r_1, r_2, ..., r_k), a custom defined XML element type called "HomoMaterial" is proposed. In what follows, we assume k=3, i.e., only three primary materials will be considered. An example XML element for the description of a homogeneous material is shown in Figure 19.7a. The "HomoMaterial" element has three sub-elements, each of which represents the volume fractions of three primary material. The underlying data type of Material 1, 2 and 3 is the built-in simple data type specified in the standard XML schema [19.24], as shown in Figure 19.7b.

```
<HomoMaterial>
  <Material1>1</Material1>
  <Material2>0</Material2>
  <Material3>0</Material3>
</HomoMaterial>
```

(a)

```
<simpleType name="double">
  <restriction base="double" />
</simpleType>
```

(b)

Figure 19.7. An example XML element for the description of a homogeneous material (a) XML elements (b) Related XML schema snippet.

Note that the markup "*HomoMaterial*", "*Material1*" etc. in Figure 19.7a are custom tags that are specifically defined for representing material definitions. The vocabulary and structure of these custom data are defined in a proposed XML schema, as shown in Figure 19.9.

Representing Voxel and Explicit Function Based Models with XML

Given the definition of the "HomoMaterial" as described above, a voxel model can be easily represented using XML aggregate or collection data types. Figure 19.8a illustrates an XML element containing two voxels. As the element names and attributes are self-descriptive, we will not elaborate on these details. By sequentially populating all the voxels in similar formats, the heterogeneous object shown in Figure 19.2 can be easily modelled.

```
- <VoxelModel VoxelSize="0.1">
  - <Voxel>
      <XCoordinate>0</XCoordinate>
      <YCoordinate>0</YCoordinate>
      <ZCoordinate>0</ZCoordinate>
    - <HomoMaterial>
        <Material1>0.6</Material1>
        <Material2>0.4</Material2>
        <Material3>0.0</Material3>
      </HomoMaterial>
    </Voxel>
  - <Voxel>
      <XCoordinate>0.1</XCoordinate>
      <YCoordinate>0</YCoordinate>
      <ZCoordinate>0</ZCoordinate>
    - <HomoMaterial>
        <Material1>0.5</Material1>
        <Material2>0.5</Material2>
        <Material3>0.0</Material3>
      </HomoMaterial>
    </Voxel>
  </VoxelModel>
```

(a)

```
- <math>
  - <mrow>
    - <msup>
        <mi>x</mi>
        <mn>3</mn>
      </msup>
      <mo>+</mo>
    - <mrow>
        <mn>0.5</mn>
        <mo />
        <mi>x</mi>
      </mrow>
    </mrow>
  </math>
```

(b)

Figure 19.8. XML based heterogeneous model (a) XML element of an illustrative voxel model; (b)) XML element for the representation of an explicit function

For the case of explicit function based models, the key issue is to represent the mathematical functions with XML markups. By using the MathML [19.25], the recommended XML specification for describing mathematics by the World Wide Web Consortium (W3C), there is no need to reinvent the wheel from scratch. For instance, the function $f(x) = x^3 + 0.5x$ can be represented with the XML elements shown in Figure 19.8b. Here the "*msup*" tag indicates superscript, "*mi*" indicates "identifier", "*mn*" means "number" and "*mo*" refers to "operators". Following the specification of MathML, it is straightforward to transform more complex mathematical formulae into XML elements and then use the functions to model precisely the desired material distributions.

Representing HFT-Based Model with XML

Transforming the hierarchical HFT structures into XML requires the introduction of additional custom data types, since there is no off-the-shelf constructs in standard XML schema.

```
<?xml version="1.0" encoding="UTF-8" ?>
- <xs:schema xmlns:xs="http://www.w3.org/2001/XMLSchema">
  - <xs:element name="Children">
    - <xs:complexType>
      - <xs:sequence>
          <xs:element ref="HeterogeneousFeature" maxOccurs="unbounded" />
        </xs:sequence>
      </xs:complexType>
    </xs:element>
  + <xs:element name="GeometryID">
  - <xs:element name="HeterogeneousFeature">
    - <xs:complexType>
      - <xs:sequence>
          <xs:element ref="Name" />
          <xs:element ref="GeometryID" />
          <xs:element ref="MaterialDescriptor" />
        </xs:sequence>
      </xs:complexType>
    </xs:element>
  + <xs:element name="HomoMaterial">
  + <xs:element name="Material1">
  + <xs:element name="Material2">
  + <xs:element name="Material3">
  - <xs:element name="MaterialDescriptor">
    - <xs:complexType>
      - <xs:choice>
          <xs:element ref="Children" />
          <xs:element ref="HomoMaterial" />
        </xs:choice>
      </xs:complexType>
    </xs:element>
  + <xs:element name="Name">
  </xs:schema>
```

Figure 19.9. A snippet of the proposed XML schema for the representation of HFT based material distributions

Figure 19.9 shows a snippet (parts of the schema are suppressed for brevity) of the proposed XML schema, where the required kernel data types are defined. In this schema, the "*Children*" type is defined as XML aggregate type with zero or more element items (see the "maxOccurs" attribute value in Figure 19.9). The item of the "Children" sequence is of type "*HeterogeneousFeature*", which contains a "*Name*" element for the textual description of the feature, a "*GeometryID*" element which contains the identification of the geometry referred to (i.e. the XML equivalent of the "Geometry pointer" in Figure 19.) and a sub-element of "*MaterialDescriptor*". The "MaterialDescriptor" type element, as indicated by the keyword "xs:choice", is either a "*HomoMaterial*" element or contains a "*Children*" element for modelling the material distributions, as described earlier. Note that the type "xs:choice" is prefixed by the *namespace prefix* associated with the XML Schema, indicating that the data type is of a predefined one.

This XML schema defines an XML equivalent of the data structure and constraints depicted in Figure 19.6. Based on such definitions, the object in Figure 19.4a can be then modelled with an XML file shown in Figure 19.10.

```
- <HeterogeneousFeature xmlns="http://www.CAX4D.com/HOM_XML">
    <Name>Extruded Feature 1</Name>
    <GeometryID>8</GeometryID>
  - <MaterialDescriptor>
    - <Children>
      - <HeterogeneousFeature>
          <Name>Region 1</Name>
          <GeometryID>3</GeometryID>
        - <MaterialDescriptor>
          - <Children>
            - <HeterogeneousFeature>
                <Name>Arc 1</Name>
                <GeometryID>1</GeometryID>
              - <MaterialDescriptor>
                - <Children>
                  - <HeterogeneousFeature>
                      <Name>Line 3</Name>
                      <GeometryID>6</GeometryID>
                    + <MaterialDescriptor>
                    </HeterogeneousFeature>
                  </Children>
                </MaterialDescriptor>
              </HeterogeneousFeature>
            + <HeterogeneousFeature>
            </Children>
          </MaterialDescriptor>
        </HeterogeneousFeature>
      </Children>
    </MaterialDescriptor>
  </HeterogeneousFeature>
```

Figure 19.10. A snippet of XML file for the model shown in Figure 19.4.

The hollow cylinder is assigned a descriptive name "Extruded Feature 1", and the cylinder's geometry points to an entity whose ID is "8". By tracing the entity with the provided ID, the geometric details of the hollow cylinder can be interrogated. In our implementation, the "ID" is modelled with ACIS custom attribute which is associated with internal geometric entities, such as BODY, FACE, and EDGE, etc. [19.26]. The sub-element "MaterialDescriptor" corresponds to the HFT representation shown in Figure 19.5. For the hollow cylinder, the sole child feature on which its material distribution depends is the "Region 1". Similarly, in terms of material heterogeneity, "Region 1" relies on "Arc 1", which corresponds to the "Green Circle (Cir 1)" in Figure 19.5, and "Arc 1" relies on "Line 3" and so forth.

Note that the XML file in Figure 19.10 is not illustrated in full details and some less important elements are suppressed in the XML tree for brevity reasons. In addition, some low level details such as the "Weighting function" etc. (see Figure 19.6) are not elaborated here to avoid unnecessary complexities. Technically, there should be no difficulties in integrating such information into the proposed XML model using MathML languages, as described in previous sub-section. Throughout this chapter we assume that linear material blending function is used for all the HFT nodes by default.

Using XML File for Exchanges of Material Heterogeneities

The merits of representing the material distributions with XML include enhanced interoperability, flexibility and ease of use. The interoperability comes from XML's plain text nature which enables computer models constructed with different languages to be shared and exchanged. The flexibility is benefited from XML's widespread infrastructure and extensibility. The ease of use is attributed to its self-descriptive property, many ready-for-use markup tags and the available rich development tools offered by commercial vendors as well as a huge user community.

In our previous work [19.21–19.23], we have proposed a prototype heterogeneous CAD modeller based on the hierarchical HFT representation. The proposed modeller, CAD4D [19.27], uses ACIS [19.26] as the geometric modelling kernel and the HFT structure is implemented using C++ class. The native CAD4D model is of a binary format, which is efficient in data serialization, but not convenient for information exchanges with other CAD/CAE packages. For instance, we have encountered quite a few challenges when attempting to export the material distributions to COMSOL Multiphysics [19.28] to perform numerical analysis on some designed heterogeneous objects. In the work reported in [19.29, 19.30], material compositions of points at regular grids (imported from COMSOL Multiphysics) must be evaluated by CAD4D and then exported to COMSOL Multiphysics via text files, as illustrated in Figure 19.11.

The approach illustrated in Figure 19.11 has certain limitations. First, the heterogeneous CAD model is not self-contained and the material composition evaluation relies on CAD4D's functionalities. In other words, such a heterogeneous model is meaningless outside CAD4D and its semantics cannot be interpreted by applications other than CAD4D. Second, as CAD4D runs on Windows platform only, CAD/CAE packages running on other platforms such as Linux etc. are therefore unable to utilise such models in downstream applications. The reason for such problems is that our heterogeneous CAD model is private and not interchangeable. By definition, however, the proposed hierarchical HFT representation could be implementation and platform independent, i.e. the model can be constructed by C, C++, C# or other well known programming languages without any constraints imposed, and such CAD models should also be accessed and maintained by software packages running on Windows, Linux and other operating systems.

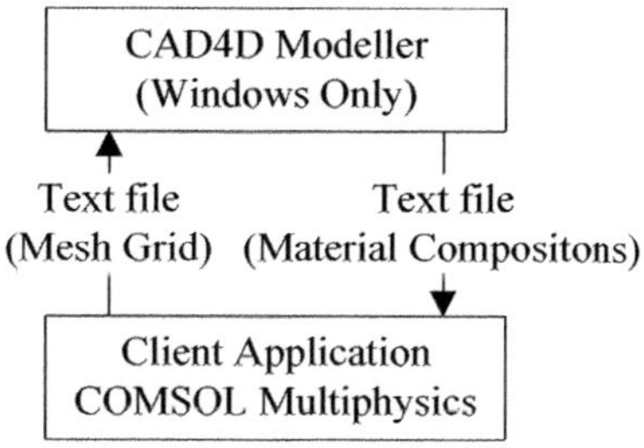

Figure 19.11. Data exchanges between CAD4D and COMSOL Multiphysics

One way to accomplish this goal is to exchange such information using XML files. To do this, heterogeneous CAD modellers need to convert their internal data representations using the data types defined in the XML schemas. Other applications such as COMSOL Multiphysics or SolidWorks, etc. can then retrieve the data from the neutral file, as shown in Figure 19.12. In this approach, XML is taken as an intermediate layer to exchange the material information between different applications.

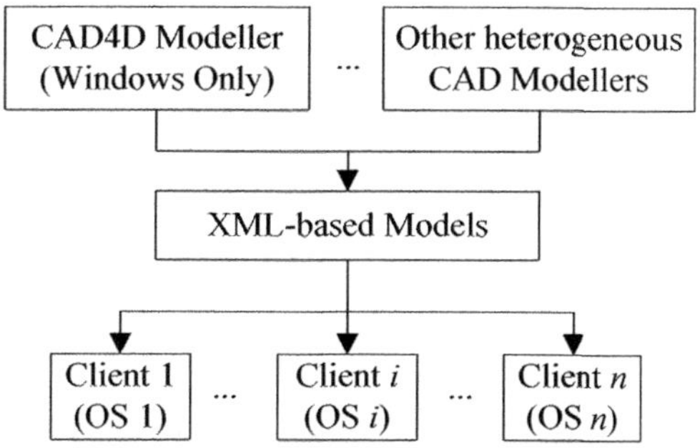

Figure 19.12. Data exchanges using XML-based heterogeneous object model

Note that when using the XML-based models, it is no longer the role of CAD4D to offer the material interrogation services; instead, the role is passed to the clients which retrieve, parse and utilise the data from the XML models.

Material Composition Interrogation from XML Files

Given an XML-based heterogeneous model, for an arbitrary point of interest, its material composition can be interrogated.

For the voxel-based model, all the XML elements are first read into a dynamic list and the voxel which contains the point under interrogation (i.e. whose distance to the input point is smaller than the voxel size) is identified, if it exists. The point under evaluation is said to be contained by the voxel and the "HomoMaterial" sub-element of the voxel (see Figure 19.8) is then returned as the interrogated material composition. If the material composition inside a voxel is also represented by interpolation functions [19.15–19.18], the point can be input as the function parameter, and the function value is returned as the output material. This is essentially the same as the material evaluation process for the explicit function based models.

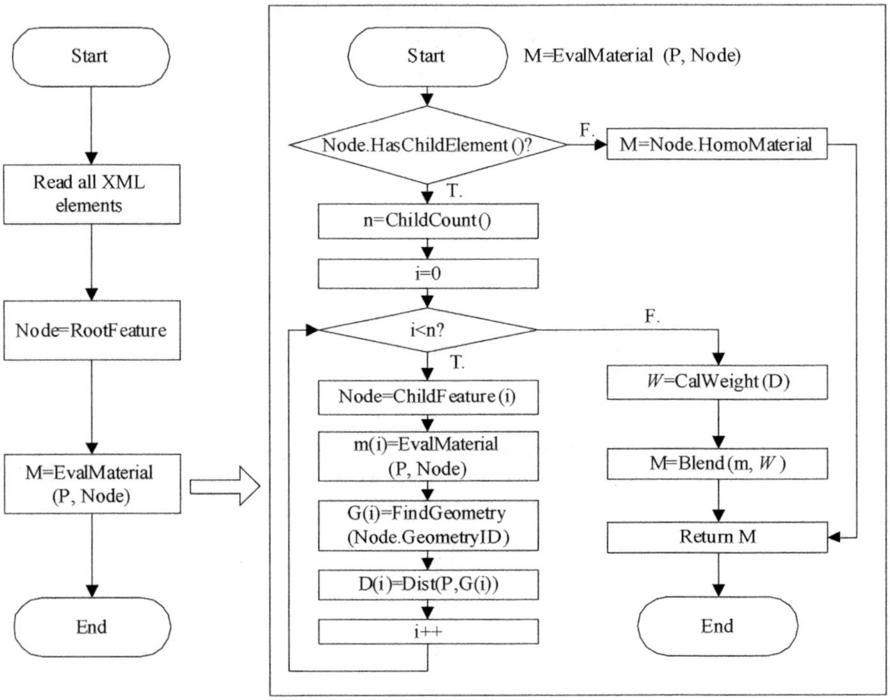

Figure 19.13. Material evaluation from XML-based HFT model

For the heterogeneous feature tree based model, the material evaluation can be obtained by calling a recursive function [19.21], as explained in Figure 19.13. The material composition evaluated on the root XML node is returned as the output material. According to the definition of the heterogeneous feature tree representation, the parent-child relationship is used to encode the material variation dependencies. So, to get the material evaluated on a specific node, the materials evaluated from all its sub-nodes (child nodes) must be known. This recursion continues until a leaf node which has no child element is reached. For a leaf node, its material can be directly retrieved from its HomoMaterial sub-node.

The materials evaluated from all the XML sub-nodes contribute to the material evaluated on their parent XML node, and the weights are related to the distances from each feature's geometry to the input point. The feature's geometry can be traced via the node's GeometryID element, which is associated with the geometric entity by custom attributes. Finally, all the materials evaluated from each sub-element are blended with calculated weights applied, and such blending is repeated until the material evaluated from the root XML node is returned. Figure 19.13 shows a flow chart depicting the above algorithm.

19.4 Implementations and a Case Study

To test the validity of the proposed XML-based heterogeneous object model, we have developed a prototype module to convert our private model files to XML files. The geometry and material information in the CAD4D model are respectively represented with ACIS B-Rep and the HFT structures. CAD4D is developed with Microsoft Visual C++ 6.0 on Windows XP platform, as shown in Figure 19.14.

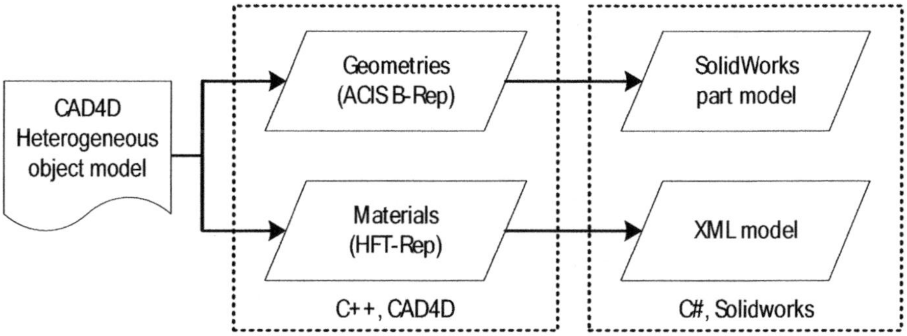

Figure 19.14. Implementation of the proposed software module

In CAD4D, the heterogeneous feature tree structure is populated to a XML file using MSXML API 4.0 SP1 [19.31]. The XML file is then imported by SolidWorks using SolidWorks API. Several C# classes are developed to de-serialise the material information from the XML file[19.32], and the internal C# representations of these classes are shown in Figure 19.15.

In Figure 19.15, the classes "*HeterogeneousFeature*", "*MaterialComposition*" and "*MaterialRep*" correspond to the XML constructs "*HeterogeneousFeature*", "*HomoMaterial*" and "*MaterialDescriptor*" in the XML schema shown in Figure 19.9.

The ACIS based geometry is converted to SolidWorks [19.33] part model using SolidWorks API and C# language.

As discussed in Section 19.3, the "GeometryID" element in the XML file refers to a feature's geometry ID and such geometric information is necessary in the material composition evaluation process. In our implementation, we attached such ID information to relevant ACIS entities, and to trace the associated geometric entities in SolidWorks we developed a custom ACIS attribute parser to enable entity tracking via these ID attributes.

An example heterogeneous model is first constructed in CAD4D, as shown in Figure 19.16. The geometry and material data are then exported to an ACIS SAT file and an XML file. Custom C# Addin is then developed to read, parse and manipulate the related data in SolidWorks. Figure 19.17 shows the translated heterogeneous model in SolidWorks graphics area. The right side of the figure shows the source XML file which is used to represent the material heterogeneities. This example shows that the proposed model is valid and effective.

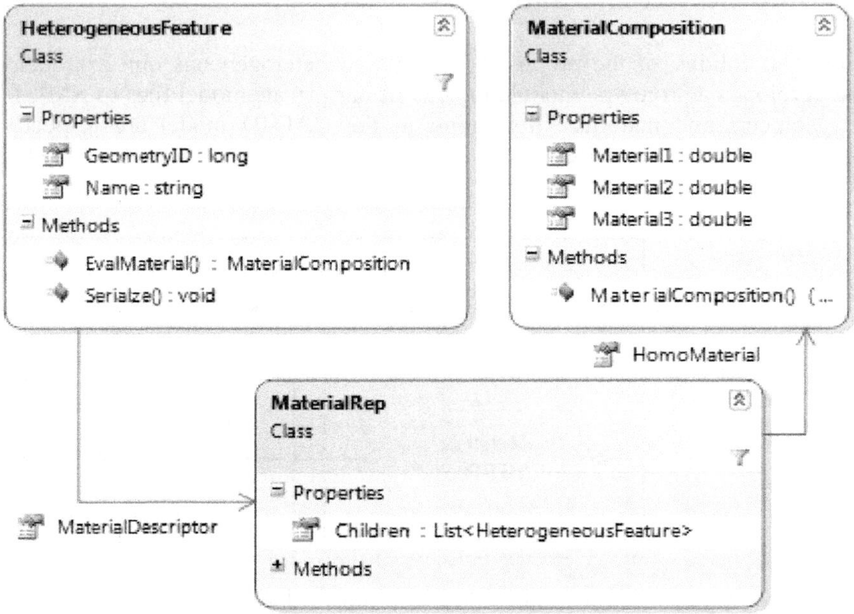

Figure 19.15. A diagram of related C# classes

Note that the geometry of the heterogeneous object is not represented in XML format because, even if it is, it will ultimately be translated to SolidWorks part file in this case study. Technically, there is no difficulty in representing the geometries in XML and there have been mature data conversion tools for such purposes. For instance, SolidWorks has related APIs to convert the part model to Dassault Systèmes 3D XML format [19.34]. In this regard, there is no difficulty in representing the entire heterogeneous object (inclusive of the geometries and material distributions) in a single, unified XML file.

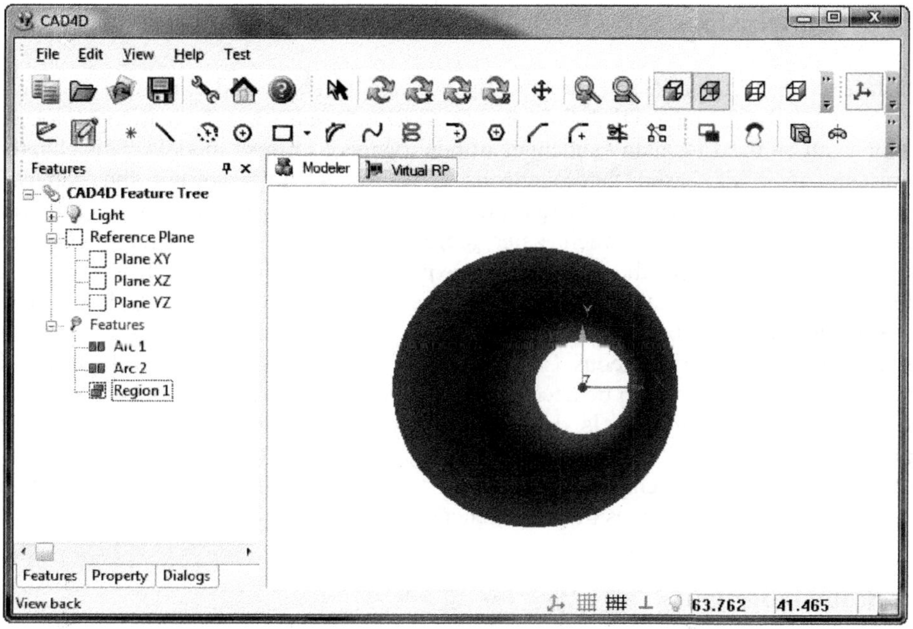

Figure 19.16. A heterogeneous model constructed by CAD4D

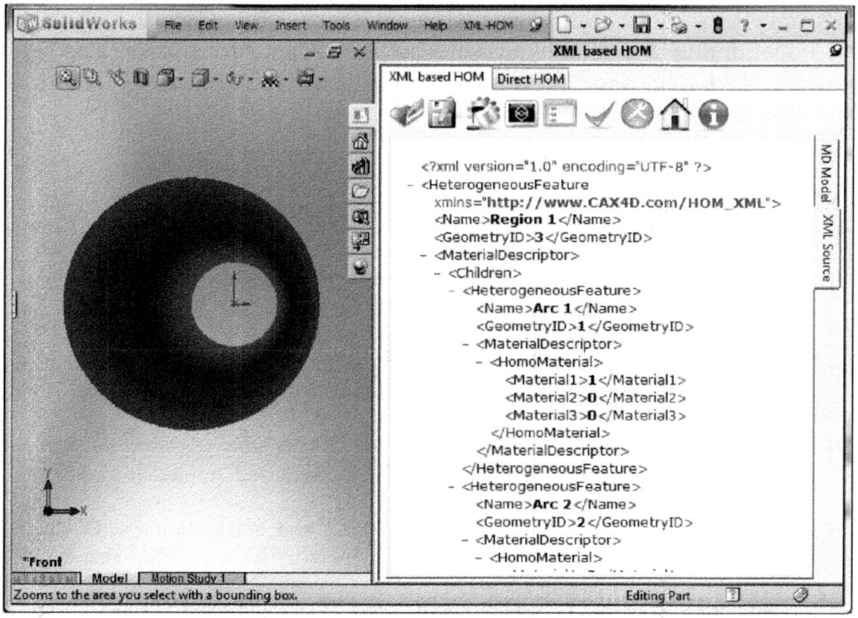

Figure 19.17. A heterogeneous model imported by SolidWorks and parsed by the XML-HOM Addin

19.5 Conclusions

Data exchanges under STEP are mostly targeted to share and exchange products with homogeneous material definitions. In STEP, there are no application protocols which can be used for data exchanges of heterogeneous object models. This chapter presents an XML-based heterogeneous CAD model to represent the material heterogeneities defined over 3D Euclidean space. Using XML language to represent material heterogeneities is much easier, more flexible and more systematic. Detailed approaches to represent the voxel based, explicit function based and heterogeneous feature tree based models are described. The idea is to introduce self-descriptive, customized tags/vocabularies to fit the specific needs of material descriptions. The structure of the heterogeneous CAD model is specified with XML schemas and related data validations can be accordingly checked to ensure the model correctness. A prototype CAD module is developed to construct XML-based heterogeneous object model, and the XML model is then exported to SolidWorks to test the validity of the proposed approach. Results show the proposed XML based implementation can facilitate data exchanges of heterogeneous material distributions.

Acknowledgment

The authors would like to thank the Department of Mechanical Engineering, The University of Hong Kong and the Research Grant Council for supporting this project (Project No: HKU 7200/04E). We would also like to thank Mr. Olivier Lejardinier, from TraceParts S.A. (http://www.traceparts.com) for his discussions and suggestions during the prototype software development. Thanks also to the anonymous reviewers and the editors of this book for their many constructive comments and suggestions.

References

[19.1] Kou, X. Y. and Tan, S. T., Heterogeneous object modeling: A review, Computer-Aided Design, vol. 39, pp. 284-301, 2007.

[19.2] Pasko, A., Adzhiev, V., and Comninos, P., Heterogeneous Objects Modelling and Applications, in Collection of Papers on Foundations and Practice Series, Lecture Notes in Computer Science. vol. 4889 2008.

[19.3] Burkett, W. C., Product data markup language: A new paradigm for product data exchange and integration, CAD Computer Aided Design, vol. 33, pp. 489-500, 2001.

[19.4] Chan, S. C. F., Dillon, T., and Ng, V. T. Y., Exchanging step data through XML-based mediators, Concurrent Engineering Research and Applications, vol. 11, pp. 55-64, 2003.

[19.5] ISO10303-11, Industrial automation systems and integration — Product data representation and exchange— Part 11: Description methods: The EXPRESS language reference manual, 2004.

[19.6] Peak, R. S., Lubell, J., Srinivasan, V., and Waterbury, S. C., STEP, XML, and UML: Complementary technologies, Journal of Computing and Information Science in Engineering, vol. 4, pp. 379-390, 2004.

[19.7] Barkmeyer, E. and Lubell, J., XML Representation of EXPRESS Models and Data, in ICSE Workshop on XML Technologies and Software Engineering Toronto, Ontario, Canada, 2001.

[19.8] ISO10303-28, Industrial automation systems and integration--product data representation and exchange, 1st ed. Geneva: ISO, 2007.

[19.9] http://www.steptools.com/.

[19.10] Michael, J. P., Introduction to ISO 10303 - the STEP standard for product data exchange, J. Comput. Info. Sci. Eng., vol. 1, pp. 102-103, 2001.

[19.11] Patil, L., Dutta, D., Bhatt, A. D., Jurrens, K., Lyons, K., Pratt, M. J., and Sriram, R. D., Representation of heterogeneous objects in ISO 10303(STEP), American Society of Mechanical Engineers, Manufacturing Engineering Division, MED, vol. 11, pp. 355-363, 2000.

[19.12] Patil, L., Dutta, D., Bhatt, A. D., Jurrens, K., Lyons, K., Pratt, M. J., and Sriram, R. D., A proposed standards-based approach for representing heterogeneous objects for layered manufacturing, Rapid Prototyping Journal, vol. 8, pp. 134-146, 2002.

[19.13] Requicha, A. A. G. and Voelcker, H. B., Solid modeling - A historical summary and contemporary assessment, IEEE Computer Graphics and Applications, vol. 2, pp. 9-16, 18-20, 22-24, 1982.

[19.14] Requicha, A. G., Representations for rigid solids: theory, methods, and systems ACM Computing Surveys, vol. 12, pp. 437-464, 1980

[19.15] Hu, Y., Blouin, V. Y., and Fadel, G. M., Design for manufacturing of 3D heterogeneous objects with processing time considerations, Proceedings of ASME 2005 Design Engineering Technical Conferences, September 24-28, 2005, Long Beach, California USA, 2005.

[19.16] Cho, J. R. and Ha, D. Y., Optimal tailoring of 2D volume-fraction distributions for heat-resisting functionally graded materials using FDM, Computer Methods in Applied Mechanics and Engineering, vol. 191, pp. 3195-3211, 2002.

[19.17] Shin, K.-H., Representation and process planning for layered manufacturing of heterogeneous objects: PhD Thesis, University of Michigan, 2002.

[19.18] Chen, M. and Tucker, J. V., Constructive volume geometry, Computer Graphics Forum, vol. 19, pp. 281-293, 2000.

[19.19] Elishakoff, I., Gentilini, C., and Viola, E., Three-dimensional analysis of an all-round clamped plate made of functionally graded materials, Acta Mechanica, vol. 180, pp. 21-36, 2005.

[19.20] Eraslan, A. and Akis, T., On the plane strain and plane stress solutions of functionally graded rotating solid shaft and solid disk problems, Acta Mechanica, vol. 181, pp. 43-63, 2006.

[19.21] Kou, X. Y. and Tan, S. T., A hierarchical representation for heterogeneous object modeling, Computer-Aided Design, vol. 37, p. 307, 2005.

[19.22] Kou, X. Y., Tan, S. T., and Sze, W. S., Modeling complex heterogeneous objects with non-manifold heterogeneous cells, Computer-Aided Design, vol. 38, pp. 457-474, 2006.

[19.23] Kou, X. Y., Computer-Aided Design of Heterogeneous Objects: PhD Thesis, The University of Hong Kong, 2006.

[19.24] http://www.w3.org/2001/XMLSchema-datatypes.

[19.25] http://www.w3.org/Math/.

[19.26] Corney, J., 3D modeling with the ACIS kernel and toolkit. Chichester ; New York: J. Wiley & Sons, 1997.

[19.27] Kou, X. Y. and Tan, S. T., An interactive CAD environment for heterogeneous object design, in Proceedings of ASME 2004 Design Engineering Technical Conferences, September 28-October 2, 2004, Salt Lake City, Utah USA.

[19.28] http://www.comsol.com/.

[19.29] Kou, X. Y. and Tan, S. T., Heterogeneous Object Design: An Integrated CAX Perspective, in Heterogeneous Objects Modeling and Applications, Lecture Notes on Computer Science, V. A. Alexander Pasko, Peter Comninos, Ed., 2007, pp. 42-59.
[19.30] Kou, X. Y. and Tan, S. T., A systematic approach for Integrated Computer-Aided Design and Finite Element Analysis of Functionally-Graded-Material objects, Materials & Design, vol. 28, pp. 2549-2565, 2007.
[19.31] http://www.microsoft.com/downloads/details.aspx?familyid=3144B72B-B4F2-46DA-B4B6-C5D7485F2B42&displaylang=en.
[19.32] Albahari, J., Albahari, B., and Drayton, P., C# 3.0 in a nutshell, 3rd ed. Beijing ; Cambridge: O'Reilly, 2007.
[19.33] http://www.solidworks.com/.
[19.34] http://www.3ds.com/3dxml.

20

Module-based Platform for Seamless Interoperable CAD-CAM-CNC Planning

Christian Brecher[1], Wolfram Lohse[2] and Mirco Vitr[3]

Werkzeugmaschinenlabor (WZL) of RWTH Aachen University, Steinbachstraße 19, 52074 Aachen, Germany
Email: [1] c.brecher@wzl.rwth-aachen.de
[2] w.lohse@wzl.rwth-aachen.de
[3] m.vitr@wzl.rwth-aachen.de

Abstract
Software tools enabling enterprises to plan manufacturing processes efficiently can be turnkey solutions for gaining advantages in global competition. Nevertheless, existing CAM systems complementing each other in functionality are often not interoperable without supplementary technical and organisational effort. Open computer-based manufacturing, which is introduced in this chapter, is an approach to solving the problems of software inhomogeneity along the CAD-CAM-NC chain via a common platform. With an extension of the STEP standards for CAM related data and functional interfaces, this framework is able to embed software modules which encapsulate specific functionalities. Possible realisation can rely on known techniques from service-oriented architecture. The developed platform can be used, for example, in coupled simulation systems or process data acquisition, which can improve NC planning processes.

20.1 Challenges of Production Industries in High-wage Countries

From year 2005 to 2008, the worldwide demand on machine tools and plants has increased constantly. This requires manufacturers in countries with high wages (hereafter referred to as "high-wage countries"), in particular Japan and Germany, to maintain leadership in the medium and premium sector market for production facilities. In many cases, manufacturers had to reject orders because of insufficient capacities. However, this situation has changed. In order to maintain leadership, manufacturers of production facilities in high-wage countries need to find ways to both survive and gain profit in the current global competition.

There are four types of economies that an enterprise needs to strike a balance with: economies of scale, economies of scope, planning-orientation and value-orientation [20.1]. Production enterprises need to find an optimal position between these four strategies considering the global market. For example, the position strongly depends on workers' wages. A company running its production in countries with a low-wage level (in the following referred to as "low-wage countries") will in general focus on economies of scale, whereas a company in a high-wage country must trade-off scale effects and customised products, thus emphasising economies of scope.

Moreover, economic efficiency of planning processes constantly gains importance. Also of relevance is the location of the company. In high-wage countries, production plants are typically complex and require expensive planning tools in order to achieve continuous improvement. In other countries, robust and value-oriented processes are common, making changes easier with less capital tied up. The combination of the two dimensions – scale against scope, planning against value-orientation – results in the so-called polylemma of production.

The Cluster of Excellence "Integrated Production Technology for high-wage countries" at RWTH Aachen University aims at providing solutions to resolving the presented polylemma. Figure 20.1 shows the concept. It is assumed that an optimal position on the base of the pyramid, i.e. between the four strategies, does not suit production enterprises in high-wage countries in today's global competitive market. Instead, it is necessary to reduce the distance between the two opposing poles so that methods and tools of two strategies can be applied at the same time [20.1].

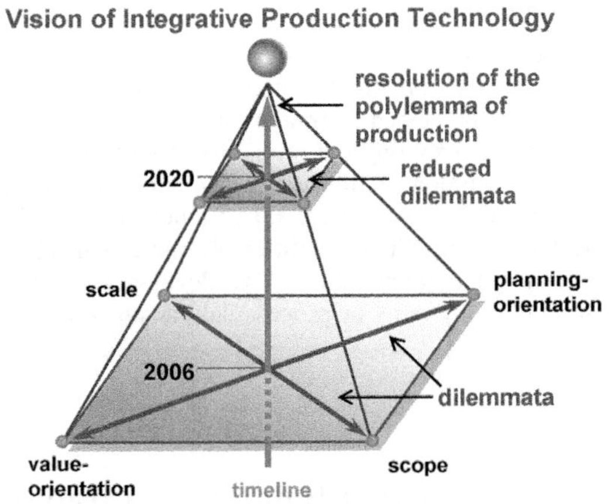

Figure 20.1. Method of resolution for the polylemma of production [20.1]

Economies of scale may refer to the quantity of the produced units by reducing part-specific fixed costs; it may also refer to improved mastering of processes by gathering know-how at shop-floor and about planning processes. For example, the knowledge about production changeovers may lower costs for small batch sizes. The second strategy, economies of scope, aims at customising products in order to address individual requirements of customers. Two key features for combining these two strategies are standardisation and integration. Enhancement of planning and simulation systems contributing to the polylemma is discussed later in the chapter. A solution for the second dimension of the polylemma – planning-orientation against value-orientation – must include the advantages of planning with specialised tools and concepts of self-organising operations of production facilities. Simulation-based methods and standardisation activities are also apt to solve this dilemma.

Since neither the four single poles nor the two dilemmata are independent of each other, the methods developed in the Cluster of Excellence must consider the entire polylemma. Four research fields have been identified that may lead to effective solutions: “Individualised Production”, “Virtual Production Systems”, “Hybrid Production Systems” and “Self-optimising Production Systems”.

Computer-Aided Manufacturing (CAM) systems may contribute to both dilemmata presented above. Existing tools support process engineers to improve the efficiency of the planning processes by offering numerous functionalities. However, these tools, often specialised for limited purposes, are not combinable, making it hard to integrate them. Planners are forced to choose over others, thus losing the capabilities of other IT tools (see Section 20.2). The platform concept presented in Sections 20.3 and 20.4 constitutes a method to overcome this disadvantage and enables engineers to compose systems of specialised modules. Sample applications are given in Section 20.5.

20.2 Deficits in the Interoperability of Existing CAM Tools

In today’s industry, managing business processes requires the support of various IT systems. Product development and work preparation particularly depend on IT tools, often referred to as “CAD-CAM-NC chain”. This family of tools contributes to an increasing efficiency that is necessary for solving the dilemma between planning- and value-orientation.

The CAD-CAM-NC chain starts with CAD that provides the necessary functionality for creating parts and assemblies. CAM enables users to plan manufacturing processes by designing tool-paths and/or selecting tools. The last part of the chain consists of NC simulation systems and NC controls. The NC simulation systems allow for analysis of manufacturing processes by detecting collisions and visualising tool-paths. The resulting NC programs are finally transferred to NC controls, which transform the NC commands into lead values to drive a machine tool. Other software products such as product data management (PDM) systems or computer-aided quality (CAQ) control systems also take part in the chain and may have significant influence on the efficiency of the whole planning and manufacturing process. Since the software solutions are developed by numerous companies, the IT infrastructure is characterised by different architectures, data formats, user interfaces, databases, etc.

A survey among 251 executive officers of German-speaking enterprises concluded that the deficiencies of the IT-supported business process are significantly determined by the software systems employed [20.2]. The main difficulties are presented in Figure 20.2.

The main problem identified by the executive officers is related to the data stream along the engineering process. About 90% confirmed that their enterprises face difficulties when data has to be exchanged. The causes of the problem include differing data formats, converting failures, different software versions or systems. The problem becomes more complex with cross-company processes [20.2].

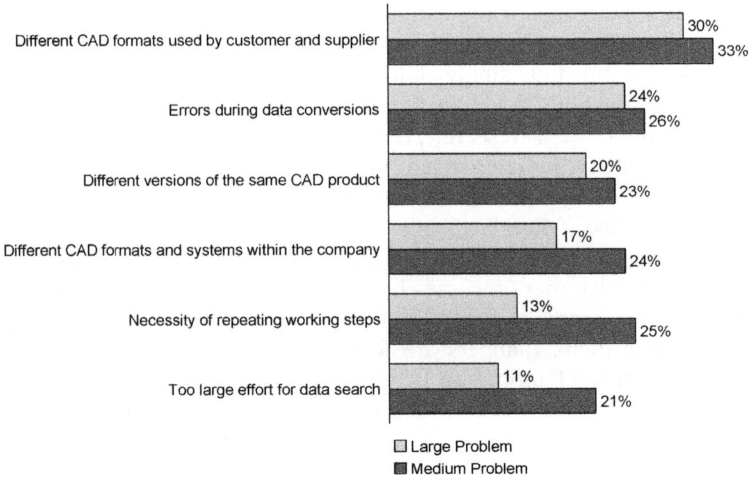

Figure 20.2. Interoperation difficulties of systems in the CAD-CAM-NC chain [20.2]

20.2.1 CAM Tools in Today's Business Processes

The issue is not just centred around CAD systems, on which the survey focussed. CAM and NC applications are also problematic. Regarding the CAD-CAM-NC chain, the data formats can be grouped into proprietary and standard formats. Proprietary formats are designed for specific tools, allowing fast and efficient processing of information in the proprietary system. In many cases, information cannot be ported to other software tools. On the contrary, standard data formats allow users to exchange data. STandard for Exchange of Product (STEP) data [20.10] is an example. Because of its versatile application protocols, it covers a large number of geometric features, settings, technology information, etc. and simplifies data exchange between software systems. STEP is still an evolving standard. For example, free-form surfaces are often approximated by using substituting elements if their proprietary representation is unknown to the standard [20.3]. Difficulties may also occur in case of constraints used in construction assemblies that are not yet entirely supported by STEP.

Information losses along the CAM-NC-chain with a focus on manufacturing data may concern raw part data and process plans, in which a step-by-step creation of a final part is documented. Here, it is necessary to include information about applied technologies, machine tools, tolerances, etc. STEP-NC [20.4, 20.5] is an approach to overcome these losses. Since its first publication in 2003, development of STEP-NC is still ongoing. It is not yet supported by the majority of CAM systems. Therefore, the potential benefits of STEP-NC in the CAM-NC-chain have not materialised.

To recapitulate, data loss is expected when transferring data between systems in the CAD-CAM-NC-chain. Additional work is always needed to remedy the problems. Standards such as STEP can significantly reduce the cost of data transfer and exchange. Yet, compelling commercial applications are still few and far between.

20.2.2 Limits of Current CAM Systems

Current software systems for production industries are characterised by an unequal range of functionality [20.6–20.8]. Different software systems have different strengths and weaknesses. A process using a large number of interpolated NC axes through several control channels may require a different CAM system than does a simple process. Furthermore, modern tools, fixtures and manufacturing processes such as laser hardening and laser coating are progressively integrated into machine tools. This presents a need to use different CAM systems, keeping in mind that the entire operation planning also needs to be optimised. Data exchanges and compatibility among the concerning CAM systems are no doubt a key factor in this type of manufacturing process. Most of the CAM systems cater for either rotational parts (turning lathes) and prismatic parts (drill/mill centres) process planning; some can do both. For complex geometry, e.g. impellers or "blisks" (bladed disks), simultaneous 5-axis milling is to be used and specific CAM tools are needed.

The above discussions lead to an inevitable use of multiple CAM systems. This presents many problems. There may be redundant 3D visualisation data, redundant simulation functionality or incompatible data management systems. Once a company chooses a subset of IT tools, it will be limited to the corresponding functionality. Moreover, budgetary restrictions of a company may lead to choices of less expensive systems that compromise the functionality.

In summary, the planning tasks along the CAD-CAM-NC chain are often supported by multiple, stand-alone software tools, presenting severe interoperability deficiency. This in turn often leads to elongated throughput time and loss of valuable data [20.9]. The inherited functionality deficiency with an IT tool gets imposed onto the company that decided to use the tool. It also diminishes scale effects regarding processes because different software systems lead to a rising number of process variants. Clearly, lack of interoperability is part of the two dilemmata of production.

20.3 IT Platform for Open Computer-based Manufacturing

Planning departments dealing with complex processes and resources need a way to combine different CAM systems effectively. An approach for achieving this aim is "Open Computer-based Manufacturing" (openCBM).

20.3.1 The Open Computer-based Manufacturing Approach

The core of openCBM is a configurable platform which enables modularised software with different functionality in support of co-operative process planning. Once combined, these modules add up to an NC planning system independent of the requirements of specific users.

openCBM has three basic requirements. First, existing CAM functionalities have to be separated into hierarchical groups according to their output results and the context/domain (turning, milling, laser-supported processes, etc.) in which they are employed. These groups form the basis for defining encapsulated software modules, which possess comprehensive functionalities required by a specific task. CAD utilities, operation planning (for example milling processes with multiple axes and turning), feature recognition, time prediction, simulation and tool management are examples of the encapsulated functionalities. Since tasks may depend on each other, the modules are enabled to co-operate, i.e. modules can use the functionality offered by others. For instance, a module calculating manufacturing times may request the duration of an NC program run from an NC simulation. This separation step must take the views of users and developers into account. On one hand, functionality belonging to specific manufacturing methods must be merged in order to guarantee that users can effectively and efficiently work with the software. On the other hand, the concerned functionality must yield a sufficient degree of standardisation – the main aspect on which modularisation is based on.

Second, a consistent data exchange between modules is a major requirement for this approach. Therefore, openCBM needs a data format that covers a broad spectrum of applications in the area of machining. An adequate solution is the STEP standard, which meets these requirements due to its flexible architecture using Application Protocols, and its range of already considered domains. Another advantage of STEP is its extensibility. It enables the platform to include a growing number and variety of modules.

Third, module interfaces for software programming must be defined to follow a specific syntax. These interfaces are implemented as part of a common platform whose topology must enable multiple modules to cooperate. Possible tasks and features within this scope are:

- Module identification: according to its configuration, the platform identifies available modules and receives a description of their functionality.
- Establishing references: in the initialisation stage, modules need the platform as broker if they depend on external functionality. They submit service requests to the platform which returns references to suitable modules.
- Common database: the platform is provided with a common database on which the modules operate. For example, the data could be stored with the syntax and semantic of STEP.
- Avoiding functional redundancies: the platform and its base modules may obtain a range of basic functionality, for example a 3D visualisation context with mouse handling, which is used by extended modules.

Apart from STEP being used as data exchange format, this approach should also adopt the concept of service-oriented architecture. Functionalities offered by the platform or by modules can be seen as services that are accessible for clients. The introduction of services results in an open process-oriented system architecture which simplifies maintaining, updating and adding modules.

Current CAM systems present differences in their software architecture, programming languages, data formats, etc. Since a successful realisation of openCBM depends on the integration of existing software solutions, the approach has to be independent of the implementation technology. In addition, the platform must support the integration of new developments from vendors willing to contribute to openCBM. Therefore, it is necessary to define a meta-language to describe such modules. The elements of the meta-language can be implemented compatible to STEP (Figure 20.5); they provide the basis for defining module interfaces of applications along the CAD-CAM-NC chain.

Though all modules can be combined syntactically, some combinations do not make sense from a semantic point of view. The platform is hence responsible for leading a user through the configuration process. It has to interpret existing rules and report errors in case of violated restrictions. Hence, the variety of modules offered can evolve rapidly, requiring an adaptable description of the rules.

In a visionary context, openCBM includes a digital market place where modules are offered with a list of their services. According to existing connection points and requirement specifications, the platform can use agents to choose suiting modules.

20.3.2 Application of the Platform for Open Computer-based Manufacturing

The application of the openCBM approach can exemplarily be explained by introducing the following scenario (see Figure 20.3).

Module providers publish characteristics and functionalities of their offered modules in a digital market place (module database). These modules are described in the meta-language mentioned before. The user may choose whether she or he wants to select modules manually or whether the platform shall use its agents to look for the best match concerning offered and needed services.

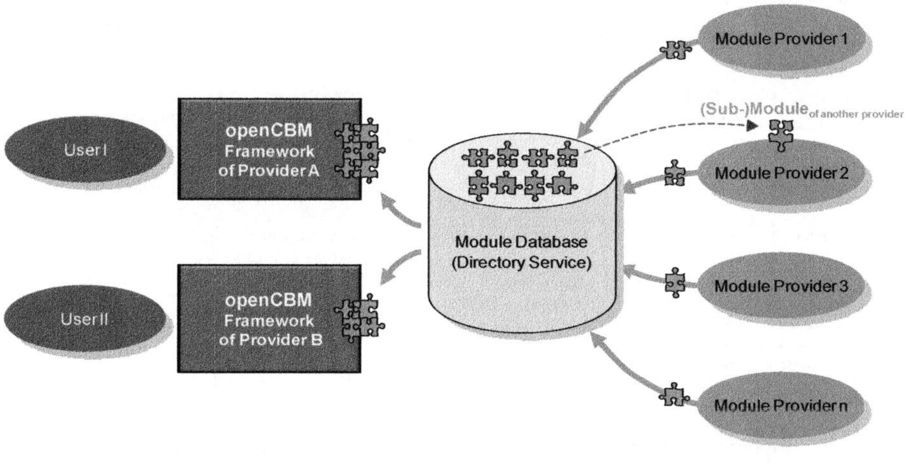

Figure 20 3. Module users and providers of the openCBM approach

As long as the openCBM interface conforms to the specification for core platforms, it is possible to have several platform implementations from different vendors. In the presented scenario, User I buys Provider A's solution while User II prefers the openCBM platform offered by Provider B. Both users are free to integrate every openCBM module available on the market.

It may also be possible to introduce a "software-on-demand" concept. A user "rents" planning and simulation software functionality from the provider by using his openCBM platform to integrate modules which run remotely on a computer of the provider. Nevertheless, this vision not only yields technical obstacles, but also affects general industrial interests. For instance, it has to be considered how transferred data can be protected from undesired access. In this case, encryption and authorisation technologies are important.

The openCBM approach can cover multiple hierarchy levels of modules and sub-modules. This leads to a structure where customers can use services in order to enhance their own modules. This is illustrated in Figure 20.3 by module provider 2.

A potential configuration of modules is shown in Figure 20.4. The framework application can integrate various modules. Their functionality concerns, e.g. visualisation, operation and path planning, simulation or tool management. The corresponding software can be developed by the framework provider (as shown in Figure 20.4, "Visualisation Geometry"), by alternative vendors or by the user himself. The hierarchical concept is, for example, applied in the case of the planning of milling processes ("OP Milling 2.5D"). The operation planning module depends on supplementary functionality of modules "TP Milling Pocket 1" and "TP Drilling".

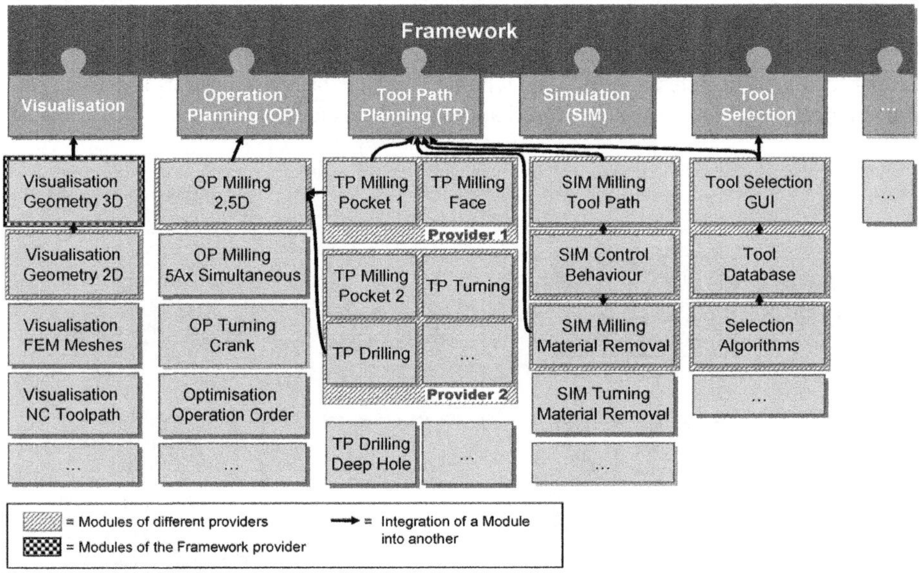

Figure 20.4. Example for a configuration of the openCBM framework

20.4 Design Concept for the Module-based Platform

This section discusses the system architecture of openCBM with a particular focus on its service-oriented nature. Interoperable data structures based on STEP standards are also presented. A critical analysis of the current functional parts of STEP and STEP-NC in the context of the openCBM platform is also given.

20.4.1 System Architecture of openCBM

As mentioned before, openCBM requires interface definitions for connecting the platform with the available modules. Therefore, it is necessary to apply an interface definition language (IDL) for openCBM. Since the common data of the platform may be represented by STEP, this IDL could be an extension of EXPRESS, the information modelling language developed with STEP [20.10]. Figure 20.5 shows a possible draft design for the basic interfaces of openCBM.

The platform ("opencbm_platform") is the centre of the architecture. It inherits properties from an abstract object ("open_cbm_object") which is also the basis for the entities "service", "module" and "provider". The platform is able to integrate an arbitrary number of modules; each "module" belongs to a specific "provider" and can have references ("module_ref") to other modules providing external services. The external services are a modularised representation for the functionality of modules and are specified by a child entity of "service_type". Services may require a list of input parameters for certain actions ("service_arguments") that are, among others, containers for input data from the CAD-CAM-NC chain. For example, a service creating a tool-path has to receive geometry information before it can be executed. Moreover, the result – in this case a tool-path – has to be returned after a service has finished its calculations. "Service_arguments" are typified by "service_arg_type", an object providing additional information concerning the task and data that are related to a service call.

The meta-model presented in Figure 20.5 constitutes the basis for openCBM. In the next step, specific modules and services (schematically depicted by entities referring to other pages, for instance "module_spec_1" and "service_spec_1") have to inherit from the corresponding base interfaces in order to fit into the architecture. Examples for definitions extending the meta-model are given later.

20.4.2 Service-oriented Architecture for the openCBM Platform

The realisation of the openCBM platform requires a flexible and extensible concept which is open to software vendors willing to integrate their applications as modules (openCBM relies on the software vendors to co-operate). The previously discussed IDL leads to standardisation which is a main prerequisite for the common platform consisting of multiple modules by multiple vendors. It nevertheless does not make any suggestions for implementing the defined interfaces.

Systems with architectures similar to openCBM have already been developed for software solutions outside production and manufacturing industries, e.g. Web applications. One such approach is "Service-Oriented Architecture" (SOA). SOA aims at adapting software tools to business processes, but not vice versa in practice. This method of resolution enables enterprises to react faster to changing demands of customers and the market. The structures and advantages of SOA can be implemented in openCBM and thus to the CAD-CAM-NC chain. This is described further in the following.

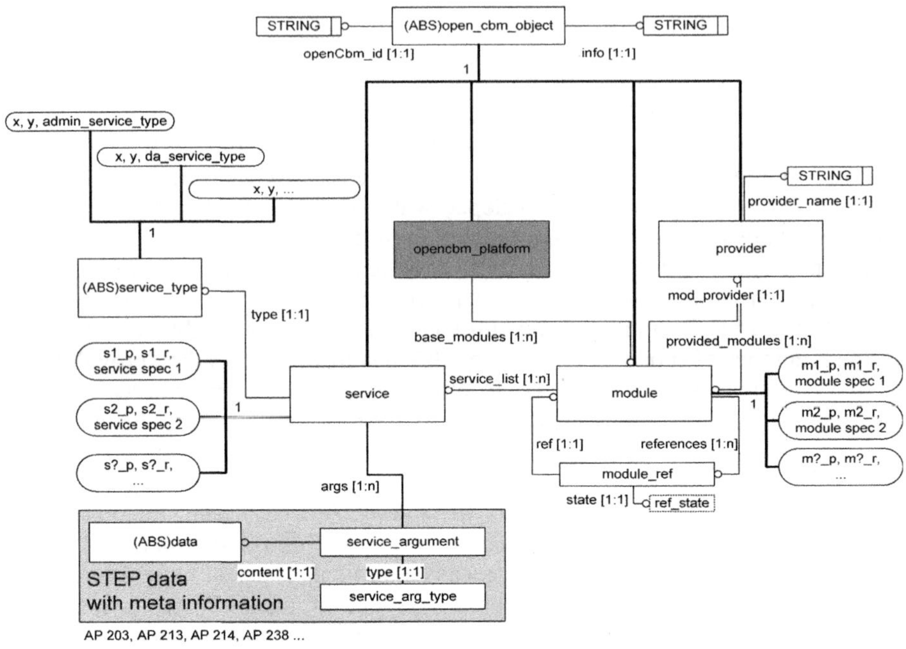

Figure 20.5. Meta-model of the openCBM architecture in EXPRESS-G

The motivation for applying SOA on openCBM lies in its modularity and reusability. Applications are not realised as monolithic programs, but as a set of services which are loosely linked to each other. These services are combined in an architecture connecting service providers and consumers (Figure 20.6). Providers are responsible for implementation and publication of the services. It is insignificant as to what programming language is applied. Hence, SOA relies on the definition of interfaces.

A unique identification of a service is necessary for external access and must be supplied by the provider. In Figure 20.5, this identification is realised by an attribute (*opencbm_id*) that is a member of the common base interface (*opencbm_object*).

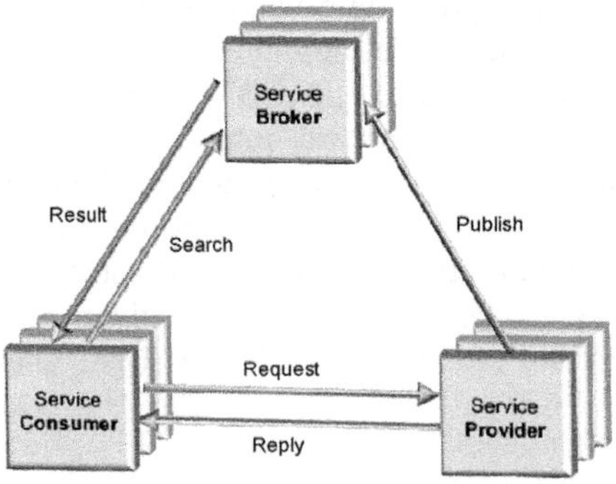

Figure 20.6. Cooperation in a service-oriented architecture

Consumers can directly subscribe to services if they know where the corresponding providers are hosted. Otherwise, consumers can use so-called brokers, which are part of the architecture (Figure 20.6). These brokers gather subscriptions of providers and store them in repositories. If consumers require specific services, they send a search request to their broker, which looks for matching entries in known repositories. These entries are returned to the consumer who establishes connections to the providers identified by the broker. Concerning Web applications, the described method is called "find, bind and execute" [20.11].

As presented below, many of the abilities provided by SOA fulfil the requirements of openCBM:

- The platform consists of modules contained in a framework. SOA is apt to realise this structure, supporting fast implementation and simplified integration of functionality developed by third-party vendors. Each openCBM module contains services that can be published and distributed as SOA services.
- SOA leads to specific functional representations of business processes and offers an adequate range of services. This is contrary to conventional software solutions whose range of functionality is often far beyond the required scope. Hence, users can learn to work with openCBM faster. Furthermore, SOA yields advantages of transparent processes.
- Current solutions for SOA systems imply a service bus which can be adapted for the purposes of openCBM.

As mentioned before, the application of SOA requires an interface definition which is independent of vendors and specific software solutions. Moreover, it is necessary to model business processes with rules for representing interdependencies. In order to use existing technologies, it is advantageous to integrate standards in this domain, e.g. SOAP[1], WSDL[2] or BPEL[3] [20.12]. These concepts have Extensible Mark-up Language (XML) as the common syntax for data exchange.

SOA and openCBM can be combined if interfaces and data of openCBM can be represented in a format supported by the SOA implementation. Since EXPRESS schema which openCBM can be modelled with (Figure 20.5) and may be converted to XML schema [20.13], service definitions and data from modules and services along the CAD-CAM-NC chain can be used with the SOA approach.

As the implementation technology must not be relevant, interoperability is a main prerequisite for applications built upon openCBM services. Apart from syntax and the meta-model, openCBM needs specifications regarding interfaces for specific tasks (one can think of standardised interfaces for tool management, for operation planning, etc.). Strategies for implementing openCBM interfaces on an SOA system have to be developed in detail after these interfaces have been defined.

20.4.3 Interoperable Data Structures Based on STEP Standards

Standardised interfaces lead to functional interoperability of the modules, but do not consider the data passed between modules. In order to use service-based

[1] Simple Object Access Protocol
[2] Web Service Description Language
[3] Business Process Execution Language

applications, the participating modules must be able to interpret the output data of preceding functions.

Take process planning for NC-based manufacturing as an example. Software systems have to deal with inputting and outputting geometric and technologic information. Seamless interoperability thus requires a standardised data format. Since it is necessary to include global production information (e.g. batch sizes and costs), geometric features and material data, etc., a collection of syntactical entries in a file seems inappropriate for solving the related difficulties. Instead, STEP and STEP-NC in the area of CAM/NC, are probably able to provide an exchange basis for openCBM because it covers a broad spectrum of CAD-CAM-NC data.

Thus, to realise the openCBM concept, the first step is to analyze the STEP and STEP-NC ISO Standards (ISO 10303, ISO 14649) regarding their suitability and their limitations for realising the desired functionalities. In particular, the requirements for implementing a meta-information database that stores the information and the results of sub-processes needed for the following planning steps have to be considered.

The STEP standards allow the description of geometric information about a part, including tolerances and feature information (e.g. AP 203, AP 214, AP 224, AP 238, AP 240). STEP-NC (ISO 10303 AP238 and ISO 14649) is capable of describing process information, in particular machining feature information, and allows the definition of tool-paths that may be transferred to and performed on the NC controller.

Development of STEP-NC is still an ongoing process. There are a number of domains that are being, or to be, considered for inclusion in the suite of the STEP-NC standards, such as coverage of: (1) some of the manufacturing technologies (e.g. laser cutting and programming of robots), (2) a comprehensive data model for machine tools, (3) inspection tasks [20.14, 20.15], (4) mechanisms of integrating the storage and feed-back of process data from the shop-floor [20.18] and (5) a model for precise influence of controls on tool-paths.

For STEP and STEP-NC to support effectively the concept and development of openCBM, the above-mentioned standardisation must be complete and ready to use. For example, the machine tool data models are important for process planning as well as detailed simulation of manufacturing process. Currently, ISO TC184/SC1/WG7 is developing a new machine tool model [20.16]. Some approaches for describing machine tool behaviours are also reported [20.17]. The development work is ongoing within the "Virtual Production Systems" project of the previously described Cluster of Excellence.

The importance of modelling the precise effect of controls on the tool-path cannot be overstated. For example, the pre-calculated data about the velocity control is to be considered if the output data of tool-path optimisation modules is to be passed on. Recently, a corresponding model is integrated with suitable machine tool models within the Cluster of Excellence. In contrast to commercial simulation systems, such as Vericut or RealNC, this model not only considers the technologically optimal feed values over the tool-path; it also takes the detailed behaviour of the controller in combination with the control loops and the machine tool structure into account (see details in a later section).

As much as the maturity-level of STEP-NC is low, so is the public acceptance. STEP-NC suggests that the controller calculates the tool-paths at runtime. This is frowned upon by many users who used to simulate the precise control behaviour off line. NC manufacturers often do not have the knowledge that CAM vendors have in the area of path planning. This makes runtime calculations impossible. The concept of openCBM is able to close these gaps because the modular structure can be extended to NC controls as well. Given a successful realisation (including real-time capability), the same modules responsible for path planning operations in CAM systems can also be applied to controls. This results in an adaptive feature-informed control.

To recapitulate, STEP and STEP-NC provide important, though still incomplete, elements for a seamless distribution of data along the CAD-CAM-NC chain. For STEP to be used in openCBM, it has to be extended so that functions of IT systems in the area of CAM can be standardised, giving vendors the ability to provide their software applications as modules in accordance with a STEP standard.

20.5 Use Cases for the Module-based Platform

This section intends to present use cases for the openCBM platform, which is effectively a CAx framework for process planning. The modular architecture of the framework is the key to the platform, which includes, for example, modules for virtual controls and modules for simulating the structural behaviour of machine tools. Discussions are also given on possible applications of openCBM. Of particular significance is process data acquisition and process information as the foundation for tool-path optimisation and knowledge-based process planning.

20.5.1 CAx Framework for Process Planning

Modular Architecture of a CAx Framework for Process Planning

As a potential application of the openCBM platform, a CAx framework is introduced that can support process planning and – as a vision – construction of machine tools. It is based on several planning and simulation tools whose reliability has been enhanced. Consequently, these tools have reached sufficient precision for what they specialise in.

Current developments enable users to analyse existing interactions by coupling these simulation systems (examples can be found in [20.19, 20.20]). Since there are numerous IT tools available offering respective functionalities, the platform approach presented in Sections 20.3 and 20.4 can provide the necessary flexibility for customising the CAx framework dependent on boundary conditions. For example, the analysis of different machining processes requires models and simulation functionality that represent physical effects adequately ("as good as necessary" – not "as good as possible"). The modular assembly of simulation tools also has advantages pertaining to scaled levels of details. The modules can be chosen dependent on the demanded accuracy, precision and calculation time.

The presented CAx framework considers a machine tool from four points of view (Figure 20.7). These are,

- CAM for tool-path planning and optimisation
- Controls, including logic and time patterns as well as control loops
- Structural mechanics with the physical representation of static and dynamic effects and
- Machining processes, emphasising cutting engagements and forces

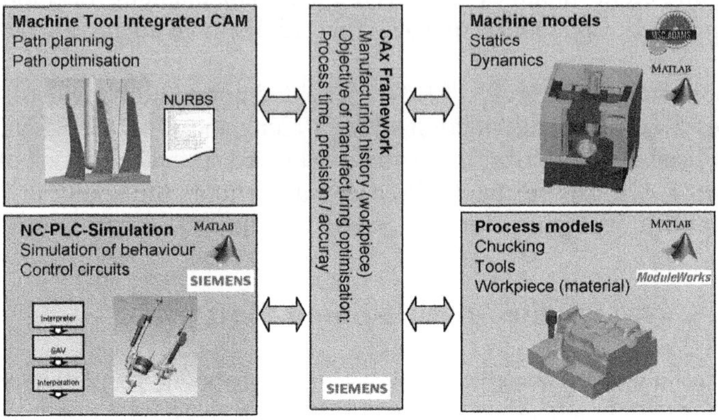

Figure 20.7. Structure of the CAx framework for planning of manufacturing processes

The modules are linked by a platform that follows the openCBM design as discussed in Section 20.3. In the current scope, the platform is developed in the Cluster of Excellence at the Laboratory for Machine Tools and Production Engineering (WZL) of RWTH Aachen University. The project focuses on the coupling of simulation and planning systems as well as on the optimisation of the NC path with predictive compensation of influences from controls and mechanics [20.21]. It does not implement the openCBM concept because the prerequisites are not yet fulfilled. Nevertheless, difficulties identified during the development could be overcome with standardised interfaces for data and functionality. The research results of this project are hence apt to narrow the requirements on an openCBM platform that is presented below for the different views considered in the CAx framework.

For integrating various simulation systems in the CAx framework, it is necessary to provide standard interfaces for the corresponding modules. A draft requirements specification for interfaces that meet the demands of the CAx framework is presented below.

In the first place, an initial NC program for a specific machining task is needed as the entry point into an iterative optimisation loop. The CAx framework hence integrates a module for planning and optimising tool-paths. Since the applied CAM system forms the basis for the framework, its functionality is not considered under the aspect of interface standardisation. In the general scope of the openCBM platform, a detailed analysis must be carried out in order to assign CAM functionalities to separate modules.

Data resulting from the simulation systems must be made available to the CAM module. The standardised interface thus requires a coupling model that specifies the data streams within the framework.

Although generality is important in order to open the platform for the majority of simulations and planning tools, complexity can be reduced by accepting some restrictions about the kind of systems to couple. The design suggestions made in the following are based upon the assumption that the simulations carried out in the IT tools be time-based as opposite to discrete event simulations. Furthermore, they have to describe operations with a similar dynamic scale; for instance it is not supported to couple control simulations with a step-size of a few milliseconds and thermal simulation systems, which have to consider several hours or days.

Regarding control simulations, another restriction is conceded. In the last few years, "hardware-in-the-loop-simulations"[1] (HILS) have progressively been applied in production industries. The main requirement demanded by HILS is the real-time capability of each member component, a characteristic that is not fulfilled by a large number of simulation systems. The HILS environments are therefore neglected and only "software-in-the-loop-simulations"[2] (SILS) are considered.

Consequently, "virtual time management" can be introduced. It allows a synchronised simulation that does not depend on system time, but uses "virtual time" instead. The latter is stretchable according to the required calculations. It is controlled by a so-called time master that distributes the current simulation time (i.e. virtual time) among connected simulation modules. It moreover defines discrete time steps, after which the modules have to be synchronised. This approach enables modules to take as much real-time as they need to complete a simulation step [20.22]. Despite this advantage, users must take the unsynchronised time interval between two discrete steps into consideration, which may affect the simulation of interactions in particular.

A standardised interface consists of configuration and execution functions. For example, modules need an initialisation function which allocates resources and imports a scenario from a file or a database. Since it cannot be assumed that resources are freed automatically, releasing calls must also be part of the interface. In addition to the general initialisation of a module, each simulation run may depend on a specific preparation as well as on operations after finishing a run (for example "reset").

Concerning its execution, a simulation using virtual time management requires a triggering function that is called by the time master. As the latter does not know when the calculation time of triggered modules is finished, a call-back mechanism for synchronisation has to be included in the standardised interface. It enables the time master to receive feedback from the simulation participants when they are ready to continue.

---

[1] Combination of hardware controls and software systems in which the behaviour of a technical system, for example a machine tool, is simulated

[2] Coupled software systems that simulate the behaviour of a technical system, for example a machine tool, and the behaviour of controlling units such as NC and PLC

Modules for Virtual Controls

The coupled simulation contains simulated controls for NC and PLC (Figure 20.8) whereby several types of modules are apt to interpret the NC program generated by the CAM system. The first one is an emulated control, and the second a virtual control. According to the boundary conditions of optimisation, it may be reasonable to switch among existing realisations by using the openCBM platform.

"Emulated controls" – as used in this context – are blackbox models of an existing hardware control. The behaviour of the hardware is hence approximated by the emulations so that the difference between these two systems concerning their output, given the same input, is reduced.

"Virtual controls" use the same algorithms as hardware controls and are developed by control manufacturers who have the knowledge about the internal structure of their controls.

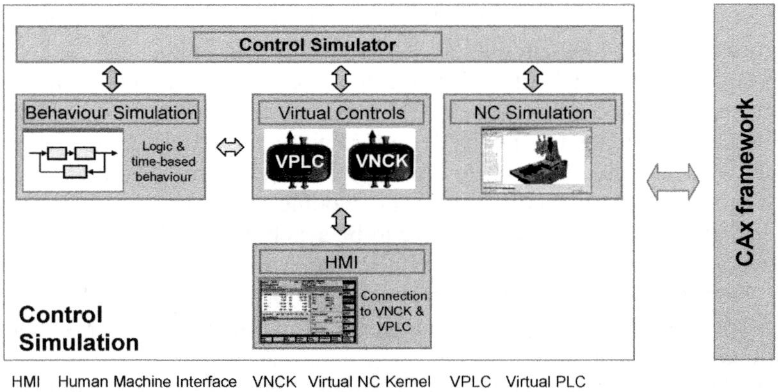

Figure 20.8. Internal structure of the control simulation module

Comparing these two types of control simulation, the deviations from real and simulated output values are reduced when virtual controls are used. Characteristics of a specific control are thus available congruent with reality. If a simulation aims at optimising complex NC programs with critical machining features, for example in the case of 5-axis milling, a virtual control will be selected as the module in the CAx framework. Nevertheless, emulations have advantages too. In general, they allow faster simulation runs than do virtual controls and offer back-step through NC programs.

Apart from selecting different types of control simulation, another motivation for standardising the interface to control simulations refers to the variety of controls. Machine tool manufacturers and production enterprises rely on controls from different vendors so that an IT tool for planning and optimising machining processes must be able to support this variety.

The interface has to support two categories of operations: data and functionality. The first category comprises a configuration for a specific machine tool that describes available NC channels, axes including their geometric and dynamic parameters, transformation routines, commands, etc. For achieving a valid model, the machine data used in NC controls and existing databases of emulation must be combined.

As required by the original system, boot and shutdown sequences must be part of the function interface. Moreover, other modules within the platform need access to functionality such as switching the current control mode (AUTO, JOG, MDA, etc.), setting override factors for spindle and feed as well as transferring, activating and running NC programs. Regarding the optimisation, control simulations must provide specific information about processing of NC programs. Therefore, access has to be given to lead positions, velocities or accelerations, current NC lines and commands, error messages, etc.

Although it is possible to request data cyclically, an event-based transfer from control simulations to subscribing modules yields advantages in performance and usability. A standardised interface for data call-back mechanisms is therefore necessary as part of the openCBM specifications. Since the development of software is always linked to debugging operations, this specification must also include the definition of formats for error tracing.

Virtual controls are offered by several manufacturers, partly providing proprietary programming interfaces. Existing functionality has to be integrated in the interface standardisation in order to increase applicability and industrial acceptance [20.23, 20.24].

Controls of a machine tool in general consist of an NC and a PLC. A manufacturing simulation has to take both types into account if its output is meant to include a precise determination of secondary process times. Furthermore, this aspect is relevant for software developers in the PLC domain who deal with peripheral devices such as tool changers or cooling systems.

A virtual PLC can be considered in different ways depending on the objective of analysis. The latter defines the abstraction which ranges from a non-technical description similar to workflows to the application of original control software [20.25]. While the abstract description is independent of the control hardware and does not require different simulation modules, the application of original software is not possible unless corresponding control simulations are available. For integrating these simulations into the presented CAx framework, they must change their state independent of the "virtual time". A standardised interface has to consider topology (definition of the hardware configuration) and functionality, for example methods for transporting I/O signals.

As part of the control simulation, a model of the behaviour of the real plant must be connected to the I/O signals of the PLC. Modules for "wrapping" the corresponding simulation software have to comprise the functionality for a time-based simulation as well as transport channels for sensor and actor signals or – in the case of kinematic simulations of peripheral elements such as tool or workpiece changers – axis positions [20.26].

The model represents a machine tool with a focus on the logic and time-based behaviour. Though there are only a few applications available, it is necessary to standardise the model data in order to use control simulations efficiently. Due to their close relationship to PLC development, this data is not considered any further in this chapter.

Modules for Simulating the Structural Behaviour of Machine Tools

In the case of complex machining processes, it is often relevant to consider the machine structure in order to apply preventive compensation for expected deviations of the tool centre point (TCP) path. The corresponding properties of a machine tool can be modelled in different ways with an increasing level of detail. If fast computation and few resources for modelling are required, frequency responses can be applied [20.19]. They are measured for several characteristic positions in the work space and are transferred to a specific matrix form so that the dynamic behaviour is obtained from the discrete points close to the considered position in the workspace of the machine tool.

Another modelling approach is moveable, flexible multi-body systems [20.27]. If parameters and physical models for damping and stiffness are chosen adequately, it is possible to achieve a close match of simulation and reality. This approach requires much more time for calculation and modelling than does the frequency response approach. As a consequence, users of the CAx framework may want to choose between these simulations, according to the progress and objective of the optimisation.

A corresponding machine tool module of the openCBM platform must be able to participate in the time-based simulation. The basic necessities for such a realisation, i.e. the support of virtual time management, are presented above. Furthermore, a dynamic simulation module needs forces and torques affecting a machine tool as input. The openCBM platform must hence offer functionality for passing these values to the dynamic simulation.

In the research project, two parameter sets have been identified as relevant output values. These are the current TCP position considering the resilience of the machine tool and the orientation of the tool, taking bending and torsion of structural parts into account.

Since the model data is specific to the simulation system and covers a wide range of elements, use of STEP and/or STEP-NC is not possible. For now, the integration of simulation systems for dynamic behaviour is hence reduced to the functional layer; in the future it might be possible to standardise these models in a new STEP part, for example "STEP-SIM".

Regarding the interface, forces and torques have to be taken into account. On the one hand, they result from the drives, which receive lead values from controls. On the other hand, a machine tool must handle forces caused by the machining process. Therefore, systems offering force simulation capabilities must also be part of the CAx framework. Simulation approaches for calculating process forces – offered by different vendors – range from simple analytical formulas to finite-element systems and thus vary in calculation time, accuracy, reliability, development state, etc.

The combination of several process simulations can be appropriate for a fast and effective optimisation. At the beginning, the cycles run through simplified models of the dynamic behaviour of machine tools and for the machining process, thereby narrowing the solution space. Once the result is closer to a satisfying result, the simulation system can switch to simulation modules with higher accuracy. The disadvantage concerning the calculation speed is acceptable due to the low number of iterations left to find the solution for the CAM problem.

The openCBM platform covers the corresponding requirements. The interface must include transmission functions for simulation progress (virtual time management), forces, torques and angular or, respectively, linear deviations of the machine structure, the tool or the workpiece.

The scenario presented above, though complex, is still incomplete. New developments can however be easily integrated later if the idea of the platform is applied.

Application of openCBM on the CAx Framework

Figure 20.9 gives an exemplary insight into a possible application of openCBM on the CAx framework discussed in this section; this application does not claim to be exhaustive for this topic.

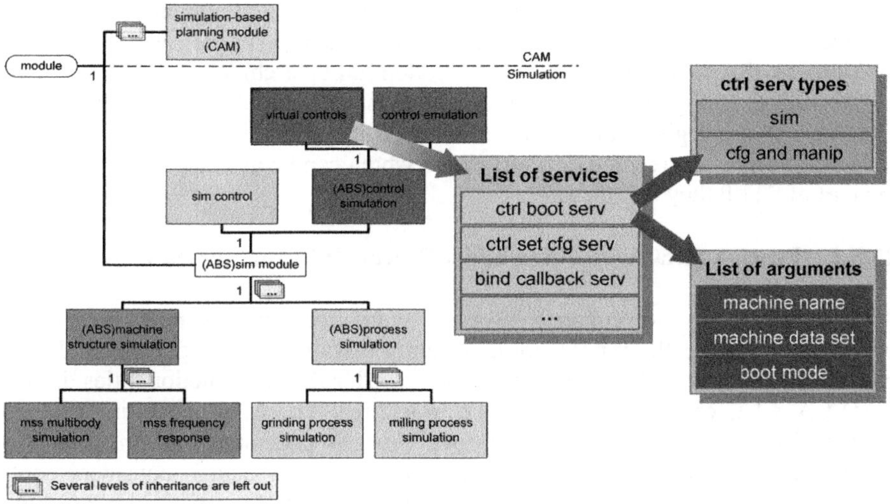

Figure 20.9. Modular architecture of the CAx framework in an openCBM platform

The modular architecture is derived from the previously discussed simulation systems and their interface descriptions. Each module contained in this architecture indirectly or directly inherits from the openCBM base module ("module" as shown in Figure 20.5). For connecting CAM and simulation layer, the architecture contains a CAM module that uses services of simulation modules in order to optimise tool-paths ("simulation-based planning module CAM") for complex machining processes.

At the simulation layer, there are four groups. The first group is needed for controlling and configuring simulation runs. In this case, it is reduced to one module called "sim control". In the CAx framework, the structural behaviour of machine tools (machine structure simulations; mss) can be represented by multibody simulations ("mss multibody simulation") or by superposition of single-degree of freedom systems ("mss frequency response"). Together, they form the second group on the simulation layer. Simulation modules for different machining processes are also part of the CAx framework (third group) and comprise milling and grinding processes. The last group consists of control simulations. As discussed above, they are divided in “emulations” and “virtual controls”.

In the following, a brief example of an openCBM service is given. Considering the module “virtual controls”, one needs different services for configuring and triggering simulation functionalities. Moreover, it is necessary to introduce call-back mechanisms for returning calculated values from the control simulation such as current TCP positions or lead values for feed and acceleration. Therefore, two types of services are introduced: “sim” for call-back services and simulation control (start, stop, etc.)[1], “cfg and manip”[2] for configuring machine data (axial limits, tolerances, etc.) and sending commands such as “NC-Start” and “Select Program”.

An implementation for the interface “virtual controls” provides services which are characterised by one of previously mentioned types. In Figure 20.9, the first service in the list, “ctrl boot serv”, starts the boot sequence of the virtual controller (similar to a hardware control). It may need parameters (“arguments”) for defining the operation, such as a machine name and machine data (in the case of the VNCK by Siemens, this data is contained in a so-called run-up archive).

According to Figure 20.5, these arguments are typified by objects (“service arg type”) which are neglected here. Nevertheless, this type identification is necessary in order to interpret aggregated data structures which can be as complex as allowed by the part of STEP they belong to.

20.5.2 Process Data Acquisition and a Process Information Database

As one possibility to enhance process simulation and, respectively, process optimisation (tool-path optimisation), currently research is carried out regarding the possible integration of new process data based software functionalities into the CAD-CAM-NC chain. The basic idea is to use real data from machining processes to optimise the process and to improve process understanding. In contrast to conventional simulation, this approach – based on real data – considers control behaviour and structural machine behaviour in interaction with the machining process without abstraction. The approach requires software modules that acquire the desired process data and provide access to this data in the CAM or simulation software. Sources for process data can be external sensors or controller-internal signals. The interfaces of the corresponding software systems have to be defined in such a way that these systems can directly be integrated into the openCBM framework, and can be replaced by versatile implementations for different NC controllers and sensor systems.

Basically, there are two ways that process data can be used for process planning and optimisation (Figure 20.10). The first is to use the data acquired while machining a part to identify any conditions and parameters leading to problems with this part directly in the CAM/simulation system. In this case, the data can be transferred from the NC controller to the simulation system and plotted in comparison with the planned tool-path in real-time. Alternatively, it can be loaded

---

[1] A simulation run differs from the execution of an NC program (for instance, a virtual control must be booted before an NC program is selectable). Consequently, “start” and “NC-Start” are not referring to the same functionality.

[2] “configuration and manipulation”

into the simulation system after finishing machining. The other way is to use process data that has been captured while machining similar parts at an earlier time. In the latter case, intelligent mechanisms are required to find suitable process data of previous manufacturing tasks and determine in what way the process data may help to optimise the current process or, respectively, to plan a completely new manufacturing task.

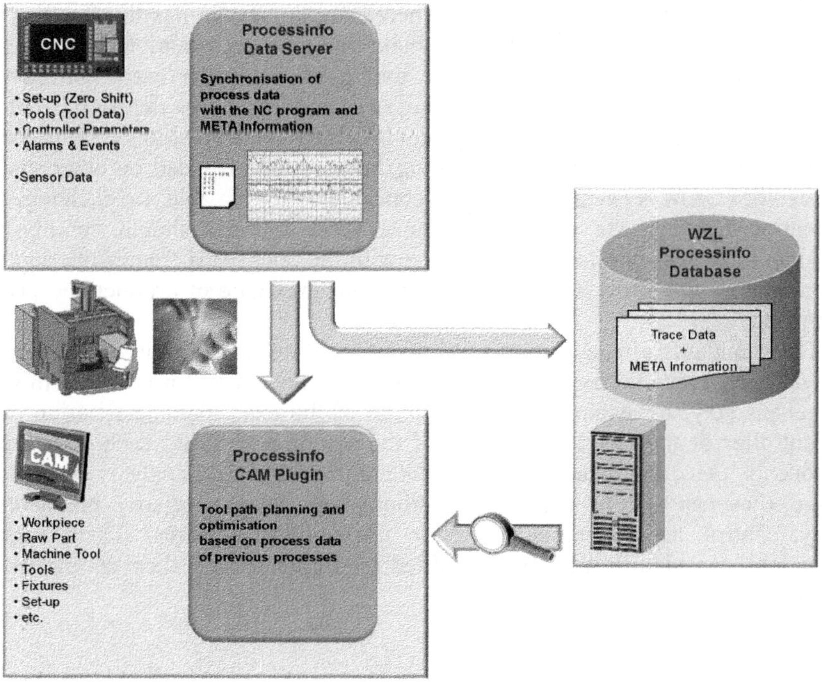

Figure 20.10. Tool-path planning and optimisation based on process data

In order to gain a better understanding of manufacturing processes, for example for enhancing simulation reliability or NC planning and optimising, the captured process data has to be matched with information about the performed process. In particular, it is necessary that for every set of time and position, the process data can be exactly assigned to the corresponding position in the NC program. This way the simulated data can be directly compared to the real data, depending on the geometric feature that has been machined and always considering the combination of controller, machine tool, cutting tools and cutting strategies that has been used.

To realise both process data based approaches, i.e. the quasi-real-time optimisation approach and the approach using a process information database, currently three software components are developed at WZL of RWTH Aachen University. The first component allows tracing process data during the manufacturing process and streaming it to a connected simulation system as well as the process information database. The second system is the process information database itself. The third system is an intelligent investigation and data analysis tool that allows finding appropriate and useful data for the current planning task easily. A corresponding data analysis tool is currently integrated into an existing CAM system (NX CAM by Siemens PLM Software) to allow data analysis and process optimisation directly in the system which is typically used for planning and where the planning data originates. The CAM system is directly coupled with the NC controller, so that a bidirectional data flow is realised (CAM-NC-coupling).

Process data can only be interpreted when all the parameters and conditions affecting the acquired data are known. Therefore, the system has to store not only the process data itself but also all the context and meta-information describing the corresponding process, e.g. the machine tool used, controller used, cutting tools used, fixture, material, geometry of the raw part, and the description of the workpiece itself. Thus, the database system and the data acquisition system need to be able to access and handle corresponding information provided by other systems such as the CAM system or the NC-controller. To be able to exchange this information between the different systems, these systems should comply with functional interfaces defined in accordance with the openCBM framework and with data interfaces that allow a complete information exchange of product and process information.

STEP standards already cover a lot of information that is required for process data feedback. Nevertheless, further extensions are necessary for covering this area comprehensively, for example the description of the exact parameterisation of the NC-controller or the exact description of the machine tool and each of its single components. Once the data acquisition, data storage and data analysis modules are realised, they can also be used for condition monitoring, predictive maintenance, quality control and documentation for traceability purposes. Their versatile applications can significantly be simplified by implementing them as part of the openCBM platform.

20.6 Conclusions

The open computer-based manufacturing approach can be used for different applications along the CAD-CAM-NC chain. Small and medium-sized enterprises will particularly profit from a large choice of software solutions whereby these enterprises have no financial resources to employ sufficient personnel for different systems or buying licenses for numerous software tools that are not individually optimal for the business processes.

Moreover, the free choice encourages competition among software vendors. Since indirect costs resulting from interoperability difficulties or media breaks significantly reduce the profit of a company, vendors of comprehensive software solutions momentarily are superior to small CAM providers whose solutions have to receive input from other software with the aid of data exchange formats. The openCBM approach would change this situation so that specific functions of CAM software must be evaluated and improved in order to remain or to become the market leader.

The platform has the potential to reduce planning costs and thus directly contribute to the second dilemma of today's production industry, i.e. value orientation-planning orientation. Since the time-span from design to manufacturing is reduced, planning activities can be accelerated at no cost of quality. This advantage decreases the distance between planning and value orientation.

Recent technical accomplishments concerning the IT infrastructure have shown that interconnectivity of enterprises is no longer out of reach so that new business models are possible. The presented platform supports this development, for example by enabling companies to offer "services-on-demand".

openCBM is still under development. The simulation platform and tools for process data acquisition as modules of the platform are not yet in place. In order to better support the openCBM concept and system development, the STEP and STEP-NC standards have to be enhanced. It is envisaged that the entry point to a customer-oriented system of different software tools is the completed interfaces and platform specifications that can guarantee seamless connectivity.

Acknowledgements

The authors would like to thank the German Research Foundation DFG for the support of the depicted research within the Cluster of Excellence "Integrative Production Technology for High-Wage Countries". They would furthermore like to thank the researchers of EXAPT, notably Mr. Karl Imbusch, and the colleagues of the Chair of Production Engineering, notably Mr. Michael Lenders, for numerous fruitful discussions regarding the openCBM idea.

References

[20.1] Brecher, C., Hirt, G., Klocke, F., Schapp, L., Schmitt, R., Schuh, G. & Schulz, W. Integrative Produktionstechnik für Hochlohnländer, in *Wettbewerbsfaktor Produktionstechnik: Aachener Perspektiven*, edited by C. Brecher, F. Klocke, R. Schmitt and G. Schuh, Eds. Aachen: Shaker. 2008: pp. 13–39.

[20.2] N. N. *Wirtschaftsfaktor Konstruktion - Ungenutztes Potential im Engineering: Status, Trends und Herausforderungen bei CAD und PDM*. 2006. München.

[20.3] N. N. ProSTEP iViP, 8th Step processor benchmark 2003, [08/2008].

[20.4] ISO 10303-238. *Industrial automation systems and integration -- Product data representation and exchange -- Part 238: Application protocol: Application interpreted model for computerized numerical controllers*. 2007.

[20.5] ISO 14649-1. *Industrial automation systems and integration -- Physical device control -- Data model for computerized numerical controllers -- Part 1: Overview and fundamental principles*. 2003.

[20.6] G. Schuh & S. Schöning, Eds. 2007. *Marktspiegel Business Software PLM/PDM 2006/2007*. Aachen: Trovarit AG.

[20.7] Friederici, G. 2006. CAM-Systeme werden intelligenter - Teil 1, *CAD-CAM-Report*, 7: pp. 58–61.

[20.8] Friederici, G. 2006. CAM-Systeme werden intelligenter - Teil 2, *CAD-CAM-Report*, 8: pp. 56–59.

[20.9] Boyson, S., Harrington, L. H. & Corsi, T. M. 2004. *In Real-time: Managing the New Supply Chain*. Westport: Praeger.

[20.10] ISO 10303-11. *Industrial automation systems and integration -- Product data representation and exchange -- Part 11: Description methods: The EXPRESS language reference manual*. 2004.

[20.11] Huhns, M. & Singh, M. P. Service-Oriented Computing: Key Concepts and Principles, [Online]. Available: http://www.cse.sc.edu/~huhns/journalpapers/V9N1soc.pdf., [08/2008].

[20.12] Liebhart, D. 2007. *SOA goes real: Service-orientierte Architekturen erfolgreich planen und einführen*. München: Hanser.

[20.13] ISO 10303-28. *Industrial automation systems and integration -- Product data representation and exchange -- Part 28: Implementation methods: XML representations of EXPRESS schemas and data, using XML schemas*. 2007.

[20.14] ISO/WD 14649-16. *Industrial automation systems and integration -- Physical device control -- Data model for computerized numerical controllers -- Part 16: Data for touch probing based inspection*. 2004.

[20.15] Brecher, C., Vitr, M. & Wolf, J. September 2006. Closed-loop CAPP/CAM/CNC process chain based on STEPand STEP-NC inspection tasks, *International Journal of Computer Integrated Manufacturing*, vol. 19: pp. 570-580(11).

[20.16] ISO/NP 14649-110. *Industrial automation systems and integration -- Physical device control -- Data model for computerized numerical controllers -- Part 110: Machine tool data model for general manufacturing processes*. 2008.

[20.17] Vichare, P., Nassehi, A., Kumar, S., Newman, S. T., Zheng, L. & Dhokia, V. September 2007. Towards a STEP-NC Compliant Model for Representation of Machine Tools, in *Proc. of the 4th International Conference of Digital Enterprise Technology*, edited by P. G. Maropoulos and S. T. Newman, Eds. Bath, UK: pp. 721–733.

[20.18] Garrido Campos, J. & Hardwick, M. 2006/5. A traceability information model for CNC manufacturing, *Computer-Aided Design*, vol. 38, 5: pp. 540–551.

[20.19] Witt, S. T. 2007. *Integrierte Simulation von Maschine, Werkstück und spanendem Fertigungsprozess*, Dissertation. Aachen: Shaker. Berichte aus der Produktionstechnik.

[20.20] Brecher, C., Fedrowitz, C., Herfs, W., Kahmen, A., Lohse, W., Rathjen, O. & Vitr, M. Durchgängiges Production Engineering - Potenziale der digitalen Fabrik, in *Wettbewerbsfaktor Produktionstechnik: Aachener Perspektiven*, edited by C. Brecher, F. Klocke, R. Schmitt and G. Schuh, Eds. Aachen: Shaker. 2008: pp. 231–261.

[20.21] Herfs, W., Vitr, M. & Lohse, W. 2007. Virtuelle Produktionssysteme, *ZWF - Zeitschrift für wirtschaftlichen Fachbetrieb*, vol. 102, 10: pp. 640–644.

[20.22] Baudisch, T. 2003. Simulationsumgebung zur Auslegung der Bewegungsdynamik des mechatronischen Systems Werkzeugmaschine, Dissertation, TU München.

[20.23] Hamm, C., Dietmair, A. & Croon, N. Gekoppelte „Offline-Simulation" mit bidirektionaler Anbindung einer NC-Steuerung (VNCK), in *Integration von CA-Techniken zur ganzheitlichen Simulation und Optimierung von Fertigungseinrichtungen vom CAD bis hin zur Hardware-in-the-Loop-Simulation (SimCAT)*, edited by N. N., Ed. Karlsruhe: Eigendruck wbk. 2006: pp. 41–45.

[20.24] Gundelach, C. DMG Virtual Machine. Aachen: WZL. 2008.

[20.25] Schneider, R. 2007. Entwicklung einer virtuellen Werkzeugmaschinensteuerung, in *Ramp-Up/2 - Anlaufoptimierung durch Einsatz virtueller Fertigungssysteme*, edited by B. Denkena and C. Brecher, Eds. Frankfurt a. M.: VDMA: pp. 91–99.

[20.26] Lohse, W., Herfs, W., Brecher, C. & Eppler, C. 2007. Anwendungsorientierte Mechatroniksimulation mit realen und virtuellen Steuerungen, in *Ramp-Up/2 - Anlaufoptimierung durch Einsatz virtueller Fertigungssysteme*, edited by B. Denkena and C. Brecher, Eds. Frankfurt a. M.: VDMA: pp. 100–128.

[20.27] Hoffmann, F. August 2008. Optimierung der dynamischen Bahngenauigkeit von Werkzeugmaschinen mit der Mehrkörpersimulation, Dissertation, RWTH Aachen

Appendix A

Software Tools for Using STEP

Table A.1. Software tools and systems for working with STEP data and EXPRESS schemas

Software name	Organization	Description of main functionality	Web site (http://)	Notes
AP203E1 to E2 converter	PDES, Inc.	Converts AP203 Edition 1 data to AP203 Edition 2 data	pdesinc.aticorp.org/vendor/203e1_e2_converter.html	It works with a long form EXPRESS schema generated by the EXPRESS Data Manager compiler
AP233 Demonstrator	Eurostep	Supports generating and validating AP233 test data.	www.eurostep.com/global/solutions/download-software.aspx#AP233%20Demonstrator	It can also be used as a way to convert and display system/requirements data. It incorporates read/write interfaces for DOOR® Requisite Pro®, Word®, and Excel®.
DEXTemplate	Eurostep	Provides DEXlib template drawing capabilities as an add-on to GraphicalEXPRESS	www.eurostep.com/global/solutions/download-software.aspx#DEXTemplate	
ECCO Toolkit	PDTec	Is an EXPRESS-based integrated development environment supporting mappings between data models such as XML, Part 21, 25 and 28	www.pdtec.de/index.php?page=7&section=7&lang=0	
EDMdeveloperSeat™	Jotne EPM Technology	Is a comprehensive package of tools for working with EXPRESS data models	www.epmtech.jotne.com/index.php?cat=78894	
EDMmodelChecker™	Jotne EPM Technology	Validates a data set against one or more EXPRESS schemas.	www.epmtech.jotne.com/index.php?id=512200	
EDMmodelConverter™	Jotne EPM Technology	Converts data from one EXPRESS schema to another by using EXPRESS-X	www.epmtech.jotne.com/index.php?id=512198	

Table A.1. Software tools and systems for working with STEP data and EXPRESS schemas (continued)

Software name	Organization	Description of main functionality	Web site (http://)	Notes
EDMServer™	Jotne EPM Technology	Is a product model server capable of storing different types of data. Includes native support for standard data models such as IFC, STEP, PLM/PLCS or Reference Data Libraries	www.epmtech.jotne.com/index.php?id=562520	It is capable of storing 3D and PLM data. It deploys ISO 10303, ISO 14721.4 (OAIS) and AECMA EN 9300 (LOTAR) in a harmonized way
EDMvisualExpress™	Jotne EPM Technology	Is a tool for creating and visualizing data models based on EXPRESS-G	www.epmtech.jotne.com/index.php?id=512195	
e-Viewer	OpenCascade	Is a plug-in for viewing STEP 3D geometries	www.opencascade.com	It supports AP203, AP214, and AP209 protocols. Open Cascade also publishes its technology as Open Source (license LGPL), which includes libraries and the API (C++) to be called from user-written applications
EXPRESS for Free	Eurostep	Converts between UML and EXPRESS	www.eurostep.com/global/solutions/download-software.aspx#EXPRESS%20for%20Free	Additional material is available at http://www.exff.org/
EXPRESS Parser (Eep)	Eurostep	Is a Windows command line EXPRESS parser	www.eurostep.com/global/solutions/download-software.aspx#EXPRESS%20Parser	It can verify EXPRESS models against ISO 10303-11:1994 or ISO DAM 10303-11:1999
Expresso	NIST/NASA	Is a development environment supporting validation of STEP data via EXPRESS schemas, and transformation of STEP data from one schema to another	sourceforge.net/projects/exp-engine/	It is Open Source software, based on EXPRESS-X The latest version is 3.2 dated April 2005

Table A.1. Software tools and systems for working with STEP data and EXPRESS schemas (continued)

Software name	Organization	Description of main functionality	Web site (http://)	Notes
gCAD3D	CADCAM-Services	Imports and views AP 203 and AP 214 data files and views the geometry	www.gcad3d.org/	It has an integrated 3D-OpenGL-viewer and an integrated NC-processor. The last updates are 2008-06-10 gCAD3D V1.146
Graphical Instance	Eurostep	Draws instance diagrams based on an EXPRESS schema	www.eurostep.com/global/solutions/download-software.aspx#Graphical%20Instance	It also supports importing and exporting (limited amounts) of data in STEP P21 format, and graphic export in SVG format
GraphicalEXPRESS	Eurostep	Is an EXPRESS-G template for use with Microsoft's Visio® drawing and diagramming software	www.eurostep.com/global/solutions/download-software.aspx#GraphicalEXPRESS	It enriches the standard template delivered with Visio® drawing and diagramming software. It enables EXPRESS models to be developed in either the published EXPRESS language (ISO 10303-11:1994), the Draft amendment (ISO DAM 10303-11:1999) or the proposed EXPRESS Edition 3
IDA-STEP v4 Viewer	LKSoftWare GmbH	Views all STEP-compliant PDM, CAD (3D) and other data	www.ida-step.net/components/viewers	It can view 3D model, parts and shapes data from CAD systems such as Allegro, Board Station and CADIF
Instance Diagram Tool	CAx-IF	Creates an EXPRESS-G representation for a Part 21 STEP file	www.cax-if.de/tools/	The latest version is dated October 11, 2004
Instance Explorer	PDTec	Provides a graphical representation, generation, and layout of STEP data (ISO 10303-21/28), and EXPRESS-G	www.pdtec.de/index.php?page=7&section=6&lang=0	The resulting instance diagrams can be pasted into a Windows application
Jen-X	Electric Boat	Generates Part 28 Edition 2 XML schemas from EXPRESS schemas	pdesinc.aticorp.org/vendor/Jen-X.html	It accepts only long form EXPRESS schemas

Table A.1. Software tools and systems for working with STEP data and EXPRESS schemas (continued)

Software name	Organization	Description of main functionality	Web site (http://)	Notes
JSDAI™	LKSoftWare GmbH	Provides Java APIs for reading, writing and runtime manipulation of data defined by an EXPRESS schema	www.jsdai.net/	JSDAI v4.1.3 is released under open source license (June 2008)
lv	Jens Kübler	Visualizes 3D CAD files in STEP or VDAFS formats, and locates individual instances in a STEP file	www.wundertools.de/	It can visualize assembly structures, check geometrical accuracy, data quality, and structural integrity of a CAD file. It can also extract portions of a STEP file for further analysis. It is only available for UNIX platforms (AIX® operating system software, HP-UX® operating system software IRIX® operating system software, Linux, and Solaris® operating system software)
Part21 to Prolog translator	KShell Analysis	Converts a STEP Part 21 file into a file of Prolog clauses	www.kshell.com/prolog	Scripts written in Prolog allow for generating XML views of the original content, searching file content, and extracting content for use by other applications. Online translation into Prolog is available
PDM Schema Exporter	Eurostep	Creates STEP-compliant files from Excel® spreadsheets - tabulated or comma separated	register.eurostep.com/pdm_schema_exporter	It can write data in ISO 10303-239 (PLCS) and PDM Schema format. It can read data from an Excel® spreadsheet. It provides an Excel® template for defining product data

Table A.1. Software tools and systems for working with STEP data and EXPRESS schemas (continued)

Software name	Organization	Description of main functionality	Web site (http://)	Notes
ST-Developer	STEP Tools, Inc.	Is a software development toolkit including: 1. Base C, C++, and Java libraries to read, write, create, search, and manipulate STEP data; 2. Custom class libraries for APs; 3. Data viewing and verification tools for STEP Part 21 and 28 files; 4. Data checkers for AP-203, AP-209, and AP-214; 5. A general EXPRESS-based data checker; 6. An EXPRESS compiler that can generate C++, Java, and HTML; 7. EXPRESS-G tools	www.steptools.com/products/stdev/	It supports AP 201, 202, 203, 203e2, 209, 214e2, 215, 216, 218, 219, 224e3, 221, 225, 227, 232, 236, 238, 239, and 240, CIMsteel CIS/2, IFC 2x2 and IFC 2x3 The current version is ST-Developer v12
STEP 2 HTML	CAx-IF	Converts STEP Part 21 files to HTML documents by adding hyperlinks to the entity references	www.cax-if.de/tools/	It supports CSS
STEP Data Browser	Eurostep	Is a Java platform that allows: browsing STEP data files in a tree view, opens a STEP model in a Part 21 or Part 28ed2 file, and exports a loaded model as a Part 21 or Part 28ed2 file	www.eurostep.com/global/solutions/download-software.aspx#STEPData	It supports ISO 10303-239 (PLCS), IFC 2x3, ISO 10303-233 (System Engineering and Design)
STEP-NC Write	STEP Tools, Inc.	Writes STEP-NC tool-path data (AP238 CC1)	www.steptools.com/products/stepncwrite/	The library writes data in XML format (ISO 10303-28) using printf(), so it does not depend on any STEP toolkit
ST-Machine	STEP Tools, Inc.	Builds STEP-NC programs; Adds tool-paths, CAD geometry, tolerances; Views and simulates cutting; and Drives a CNC	www.steptools.com/products/stmachine/	ST-Machine DLL can be added to other systems to have STEP-NC capability

Table A.2. Functionality of the software tools

Software name	Schema conversion	STEP data generation	STEP data display/ 3D viewing	STEP data conversion	STEP data validation	EXPRESS-G tools and conversions	Java-based tools, and support of HTML, XML data	STEP-based development toolkit	Software types
AP203E1 to E2 converter	+								F/W
AP233 Demonstrator		+	+		+				F/W
DEXTemplate						+			F/W
ECCO Toolkit				+			+		W
EDMdeveloperSeat™		+	+	+	+	+	+	+	W/U
EDMmodelChecker™					+				W/U
EDMmodelConverter™	+								W/U
EDMServer™								+	W/U
EDMvisualExpress™			+			+			W
e-Viewer			+						O/W/U
EXPRESS for Free	+							+	O/W
EXPRESS Parser (Eep)					+				F/W/U
Expresso	+				+				O/W/U
gCAD3D			+						F/W/U
Graphical Instance			+						F/W
GraphicalEXPRESS						+			F/W

NB: F – Free software; O – Open source (GNU); W – Windows platform; U – UNIX or Linux platform.

Table A.2. Functionality of the software tools (continued)

Software name	Schema conversion	STEP data generation	STEP data display/ 3D viewing	STEP data conversion	STEP data validation	EXPRESS-G tools and conversions	Java-based tools, and support of HTML, XML data	STEP-based development toolkit	Software types
IDA-STEP v4 Viewer			+						F/W/U
Instance Diagram Tool						+			F/W
Instance Explorer			+			+			F/W
Jen-X	+						+		F/W/U
JSDAI™		+	+		+		+		O/W/U
lv			+		+				F/U
Part21 to Prolog translator				+			+		O/U
PDM Schema Exporter		+							W
ST-Developer			+	+	+	+	+	+	W/U
STEP 2 HTML			+	+			+		F/W
STEP Data Browser			+	+			+		F/W
STEP-NC Write			+				+		O/W/U
ST-Machine			+					+	W

NB: F – Free software; O – Open source (GNU); W – Windows platform; U – UNIX or Linux platform

Index

A

B

C

G

H

I

N

T

U

V

W

X

GPSR Compliance

The European Union's (EU) General Product Safety Regulation (GPSR) is a set of rules that requires consumer products to be safe and our obligations to ensure this.

If you have any concerns about our products, you can contact us on ProductSafety@springernature.com

In case Publisher is established outside the EU, the EU authorized representative is:

Springer Nature Customer Service Center GmbH
Europaplatz 3
69115 Heidelberg, Germany

Batch number: 10755284

Printed by Printforce, the Netherlands